S0-BLS-341

Lecture Notes
in Physics

Edited by H. Araki, Kyoto, J. Ehlers, München, K. Hepp, Zürich
R. Kippenhahn, München, D. Ruelle, Bures-sur-Yvette
H. A. Weidenmüller, Heidelberg, J. Wess, Karlsruhe and J. Zittartz, Köln

Managing Editor: W. Beiglböck

357

J.L. Lumley (Ed.)

Whither Turbulence?
Turbulence at the Crossroads

Proceedings of a Workshop
Held at Cornell University,
Ithaca, NY, March 22–24, 1989

Springer-Verlag
Berlin Heidelberg New York London Paris Tokyo Hong Kong

Editor

John L. Lumley
Sibley School of Mechanical and Aerospace Engineering
Cornell University, Ithaca, NY 14853, USA

PHYS
sep lae

ISBN 3-540-52535-1 Springer-Verlag Berlin Heidelberg New York
ISBN 0-387-52535-1 Springer-Verlag New York Berlin Heidelberg

This work is subject to copyright. All rights are reserved, whether the whole or part of the material is concerned, specifically the rights of translation, reprinting, re-use of illustrations, recitation, broadcasting, reproduction on microfilms or in other ways, and storage in data banks. Duplication of this publication or parts thereof is only permitted under the provisions of the German Copyright Law of September 9, 1965, in its version of June 24, 1985, and a copyright fee must always be paid. Violations fall under the prosecution act of the German Copyright Law.

© Springer-Verlag Berlin Heidelberg 1990
Printed in Germany

Printing: Druckhaus Beltz, Hemsbach/Bergstr.
Bookbindung: J. Schäffer GmbH & Co. KG., Grünstadt
2153/3140-543210 – Printed on acid-free paper

Contents

QA913
W47
1990
PHYS

Session One

Session Two

Session Three

Opening Remarks

J.L. Lumley

Cornell University
Ithaca, New York 14853

When I sent out the first invitations to this meeting, I tried to explain the motivation in terms that I thought were clear. Particularly in the list of questions that I appended, I felt I was quite explicit. Indeed, I felt that there was probably no one that I had not offended. When I began to receive replies, however, it became apparent that I had not made myself as clear as I thought.

Several people said, in effect, "I cannot imagine why you have picked these particular topics", the implication being that there are so many others that are of greater scientific importance. What about the Renormalization Group Expansion, what about the role of experiment, and so forth? Some said "this is a strange sort of meeting; I am uncomfortable with the way you have organized it". One said "These questions have been argued in the literature, in some cases by the same people. Why do we have to go over them again now?" One said, "It is fine to invite the representatives of the funding agencies, as long as you don't let them say anything".

The questions that I have chosen for us to address during this meeting are all questions that have been the subject of controversy in the literature. I do not think that these controversies have been satisfactorily resolved. In our field, it is rare that a controversy comes to an end with one participant being carried out on his shield. In the normal course of events, our controversies conclude when the participants are exhausted, and the audience gives a verdict in the fullness of time. Something new has been added in the last several decades, however, that has changed the nature of the game in a very fundamental way. It raises questions of public policy relative to the ways in which research is funded in the United States at the present time.

The problem is that the agencies are steering the research establishment in ways that they never did before. Just after the second world war, when the notion of supporting research with Federal funds was first entertained, the agencies very politely asked the research establishment what it thought was worth investigating, and then supported those areas. They tended to be cautious and conservative, hedging their bets, looking at trendy new ideas, but not to the exclusion of more pedestrian older ones.

Now, however, the agencies are much less conservative. They no longer hedge their bets so carefully. They take strong positions on scientific issues. For example, one agency has said that it no longer supports work based on Reynolds averaging. Another agency said that since computers will be able to compute all details of turbulent combustion exactly within two decades, it will no longer support research on modeling. The agencies can exert considerable control by the simple process of making funds available for research in certain directions, and withholding funds for research in other areas. In effect they have taken sides in what were more or less harmless academic jousts; they have decided who they think should win, and have in effect stepped onto the field with a machine gun to decide the outcome.

This is why I will not muzzle the representatives of the funding agencies who are here. They are now equal partners with us, whether we like it or not, and we must have a dialog with them.

Many of us feel that the program monitors do not know enough about the field to make these decisions. There are some embarrassing Requests For Proposals which make it evident that a program monitor had heard of a trendy new area, and wanted to attract some work in that topic, but did not actually know enough about the area to write an intelligent RFP. Many of the program monitors agree that they are in over their heads in this respect. They need and want our guidance in making these decisions. The program monitors are not getting consistent, coherent guidance from the research community. At least in part, this is because the controversies in which they are involving themselves are not yet resolved. But, of course, it is no fun to support an area *after* the dust has settled. This is why it was always a good idea to hedge one's bets.

However, these decisions are being imposed on them by a new generation of managers. The program monitors are no longer allowed to adopt a conservative, hedged posture. These new managers have experience managing other things, but do not necessarily know much about fluid mechanics, turbulence, or, for that matter, research. I believe they share with the business community an unlimited respect for the good quarterly report. They rotate out to other positions in a fairly short time. They want to show that during their term in office new, different things were done.

They want to be able to say that they turned the program around. Who ever made a splash by supporting the *status quo*? Since the minimum time for a new idea to have some practical impact is of the order of ten years, and in evolved, complex areas like turbulence is often of the order of twenty years, these managers will be long gone by the time the effects of their policies are felt.

You may say, no one is forcing the research establishment to follow the lead of the program managers. However, the financial incentive is powerful. The way graduate education and research is structured in the United States today, a professor who does not have contract support has committed professional suicide. His salary is paid, but he has no students, no computer time, no laboratory equipment, no travel to meetings, no journal page charges.... He has pencil and paper only, and can do theory that does not require a computer. Unless he is a pure mathematician, that is not a realistic alternative. What happens to a colleague who sees support dry up in an area he thinks is important, and sees support available in an area he thinks is probably overblown? One colleague said to me that he was tired of swimming upstream, and was going to go with the flow. He abandoned an area that he thought was valuable, to work on something he thought was less so. Another colleague quite cynically responded to an illiterate RFP, proposing work that was just what the agency thought it wanted. He got the contract, and the work turned out splendidly. That is not the point; he is a clever guy, and an honest man, and he can do good work in many areas - he will certainly not risk his reputation by doing work that is sub-standard professionally. However, his judgment of which directions are the most productive has been ignored and discarded by the funding agencies in favor of their own. I am not advocating that the agencies take his advice about what is the right direction; after all, that is what they have already done, taken someone else's advice. The problem is that nobody's advice is completely reliable, and that bets *have* to be hedged. If Professor X thinks that a certain direction is profitable, and another not, and he has a good track record, he should be allowed to follow his nose until it is clear what the outcome will be, but the agency should at the same time be supporting work in competing directions.

Each of the areas addressed in this meeting has not only been the subject of controversy in the literature, but has resulted in a funding agency's taking a position on the proper approach to the subject. In each case some agency has said, "we will fund no more work in direction A; henceforward we will fund only work in direction

B". Less subtly, in at least one case contractors received calls from their supporting agency asking why they were not working on direction C. In effect they had to stop everything, and convince their monitor that direction C was a dead end. In the event direction C may not have been a complete dead end, but it was certainly not the best thing since sliced bread, as the monitor thought.

Not all these areas that I have selected are the freshest topics in the field. In some cases the front-runners have already moved on to more interesting things. They feel that the answers to all these controversies are obvious, and the matters no longer of interest. For the great body of us, however, that is not true. In the first place, there is no general agreement (even among the front-runners) on the right answer - there are usually multiple answers. In the second place, not everyone has thought deeply about these questions and resolved these issues in his own mind. And finally, the funding agencies, although they have taken positions, are aware that they may have made mistakes. In analyzing these areas, we will have the advantage of a slight perspective from the passage of time.

What do we hope to bring away from this meeting? At the very least it should provide the monitors with ammunition to reach more informed decisions about support. However, I hope for much more than that. I hope that the scientific community, and the monitors, will have had their consciousness raised regarding these questions of research support policy and decision making. I hope that as a result both parties will take the matter more seriously, and will think about, and talk about, these things; if they do that, who knows what might happen?

Acknowledgments. We are happy to acknowledge the generous support of the Mathematical Sciences Institute of Cornell University (which is itself supported by the Army Research Office), the Center for Turbulence Research (of Stanford University and NASA Ames Research Center), The Office of Naval Research (Mechanics Division), and the Sibley School of Mechanical and Aerospace Engineering.

Achilles and the Tortoise (and the Hare) discuss CFD and Turbulent Flow *

🐇 : Yo, Achilles... Top of the morning to you Tortie

A : Yo, 🐇 ..!

🐢 : (Softly to himself) *Whither turbulence...?*

🐇 : (Pauses; notices that 🐢 is intently reading a large poster) What are you reading Tortie?

🐢 : Oh, hi 🐇 . It's about a race.

🐇 : A race?

🐢 : Yes, a bunch of fluid dynamicists are gathered in Ithaca attempting to further the solution of the turbulence problem. The one who gets it first wins the race.

🐇 : An unlikely location...

🐢 : That's not the worst of it. I already know the answer. I have the solution. I've won the race!

🐇 : Easy does it now. This is a difficult problem. Why don't you try your solution out on me first. I promise I won't tell anyone until you've had a chance to publish.

🐢 : (Bursting with excitement) CFD is the answer...

*) Presented at the workshop *Whither Turbulence?* held at Cornell University, March 22-24, 1989, by Hassan Aref, University of California at San Diego.

🐇 : CFD??? That's a method, not a solution.

🐢 : No, no, no! CFD is the new field of *Chaotic Fluid Dynamics.* It solves many things including the turbulence problem. Just listen to my eminent colleague E. M. Lifshitz. In the preface to the second edition of *Fluid Mechanics* by Landau & Lifshitz he writes:

"There have been important changes in our understanding of the mechanism whereby turbulence occurs. Although a consistent theory of turbulence is still a thing of the future, there is reason to suppose that the right path has finally been found."

And that right path is CFD... (pauses)

🐇 : That's it? That's your solution to the turbulence problem? An appeal to authority however authoritarian... Ridiculous! Preposterous!

🐢 : I base my work on a series of theses... (pause)

🐇 : Well?

🐢 : My first thesis is:
Thesis #1
Navier-Stokes turbulence is chaotic.

🐇 : What do you mean?

🐢 : I don't really know. All I know is that Navier-Stokes turbulence is certainly not integrable like KdV or NLSE or Burgers' equation.

🐇 : That I'll grant.

🐢 : So in CFD you don't need to make *ad hoc* assumptions about stochasticity of turbulent flow at the outset. You work with deterministic equations (such as the Navier-Stokes equations) and treat the stochastic aspects of velocity and other fields as computable output rather than assumed input. The statistical hypotheses of conventional turbulence theory have always been the Achilles heal of the subject anyway.

A : You called?

🐇 : 🐢 was just expounding on some of the leg work of turbulence theory...

A : A fascinating subject. I have my own theories about it. But the subject has grown so much with the advent of CFD...

🐇 : You know the answer too?!

🐢 : (Unaffected by the other two) My second thesis is:

Thesis #2
There are flow regimes that are chaotic but not turbulent.

🐇 : (Whispers to A) There's a logical gap between his theses #1 and #2. (Loud) Give us an example if you please.

🐢 : There are several. My favorite is *chaotic advection*, the feature that flows that are very simple in Eulerian terms may, nevertheless, produce very complicated trajectories and patterns of a Lagrangian tracer. It all comes from the fact that the advection equations

$$\dot{x} = u(x,y,z,t)$$
$$\dot{y} = v(x,y,z,t)$$
$$\dot{z} = w(x,y,z,t)$$

viewed as a dynamical system are sufficiently complicated for chaotic motion to occur. To fully understand the complexity that is possible you need the notion of chaos in a low order dynamical system.

You've probably seen the spectacular images that result on the cover of *Nature*, or *Scientific American* or in the *Supercomputing Review* (shows slides). More are sure to come.

It is really quite attractive, even beyond the color pictures. There are no *ad hoc* truncations to a finite set of modes, that have always plagued applications of "chaos theory" to fluid mechanics in the usual Eulerian representation.

As a corollary, chaotic advection leads naturally to the notion of fractal structure of advectant whether it be fractality in the paths of single particles, like the float trajectories shown here (displays a figure), or in the perimeter of a cloud of tracer observed in nature or in the laboratory.

There is a sizeable literature by now. Let me show you (unrolls a large scroll; scroll is shown to audience - see next page)

References on Chaotic Advection

Aref, H. 1984 Stirring by chaotic advection. *J. Fluid Mech.* **143**, 1-21.

Aref, H. & Balachandar, S. 1986 Chaotic advection in a Stokes flow. *Phys. Fluids* **29**, 3515-3521.

Aref, H. & Jones, S.W. 1988 Enhanced separation of diffusing particles by chaotic advection. *Phys. Fluids A* **1**, 470-474.

Aref, H., Jones, S.W., Mofina, S. & Zawadzki, I. 1989 Vortices, kinematics and chaos. *Physica D* (In Press).

Arter, W. 1983 Ergodic stream-lines in steady convection. *Phys. Lett.* A **97**, 171-174.

Beloshapkin, V.V., Chernikov, A.A., Natenzon, M.Ya., Petrovichev, B.A., Sagdeev, R.Z. & Zaslavsky, G.M. 1989 Chaotic streamlines in pre-turbulent states. *Nature* **337**, 133-137.

Berry, M.V., Balazs, N.L., Tabor, M., & Voros, A. 1979 Quantum maps. *Ann. Phys.* **122**, 26-63.

Chaiken, J., Chevray, R., Tabor, M. & Tan, Q.M. 1986 Experimental study of Lagrangian turbulence in a Stokes flow. *Proc. R. Soc. Lond.* A **408**, 165-174.

Chaiken, J., Chu, C.K., Tabor, M. & Tan, Q.M. 1987 Lagrangian turbulence and spatial complexity in a Stokes flow. *Phys. Fluids* **30**, 687-694.

Chien, W-L., Rising, H. & Ottino, J.M. 1986 Laminar mixing and chaotic mixing in several cavity flows. *J. Fluid Mech.* **170**, 355-377.

Dombre, T., Frisch, U., Greene, J.M., Hénon, M., Mehr, A. & Soward, A.M. 1986 Chaotic streamlines and Lagrangian turbulence: The ABC flows. *J. Fluid Mech.* **167**, 353-391.

Falcioni, M., Paladdin, G. & Vulpiani, A. 1988 Regular and chaotic motion of fluid particles in a two-dimensional fluid. *J. Phys. A: Math. Gen.* **21**, 3451-3462.

Feingold, M., Kadanoff, L.P. & Piro, O. 1988 Passive scalars, three-dimensional volume-preserving maps, and chaos. *J. Stat. Phys.* **50**, 529-565.

Hénon, M. 1966 Sur la topologie des lignes courant dans un cas particulier. *C.R. Acad. Sci. Paris A* **262**, 312-314.

Hénon, M. 1969 Numerical study of quadratic area-preserving mappings. *Q. Appl. Math.* **27**, 291-312.

Jones, S.W. 1988 Shear dispersion and anomalous diffusion in a chaotic flow. *J. Fluid Mech.* (submitted)

Jones, S.W. & Aref, H. 1988 Chaotic advection in pulsed source-sink systems. *Phys. Fluids* **31**, 469-485.

Jones, S.W., Thomas, O.M. & Aref, H. 1987 Chaotic advection by laminar flow in a twisted pipe. *J. Fluid Mech.* (submitted) and *Bull. Amer. Phys. Soc.* **32**, 2026.

Khakhar, D.V., Franjione, J.G. & Ottino, J.M. 1987 A case study of chaotic mixing in deterministic flows: the partitioned pipe mixer. *Chem. Eng Sci.* **42**, 2909-2926.

Khakhar, D.V. & Ottino, J.M. 1986 Fluid mixing (stretching) by time periodic sequences of weak flows. *Phys. Fluids* **29**, 3503-3505.

Khakhar, D.V., Rising, H. & Ottino, J.M. 1986 Analysis of chaotic mixing in two model systems. *J. Fluid Mech.* **172**, 419-451.

Knobloch, E. & Weiss, J.B. 1987 Chaotic advection by modulated travelling waves. *Phys. Rev. A* **36**, 1522-1524.

Lichter, S., Dagan, A., Underhill, W.B. & Ayanle, H. 1987 Mixing in a closed room by the action of two fans. *J. Appl. Mech.* (submitted)

McLaughlin, J.B. 1988 Particle size effects on Lagrangian turbulence. *Phys. Fluids* **31**, 2544-2553.

Muzzio, F.J. & Ottino, J.M. 1988 Coagulation in chaotic flows. *Phys. Rev. A* **38**, 2516-2524.

Osborne, A.R., Kirwan, A.D., Provenzale, A. & Bergamasco, L. 1986 A search for chaotic behavior in large and mesoscale motions in the Pacific Ocean. *Physica D* **23**, 75-83

Ott, E. & Antonsen, T.M. 1988 Chaotic fluid convection and the fractal nature of passive scalar gradients. *Phys. Rev. Lett.* **25**, 2839-2842.

Ottino, J.M. 1989 The mixing of fluids. *Scient. Amer.* **260**, 56-67.

Ottino, J.M., Leong, C.W., Rising, H. & Swanson, P.D. 1988 Morphological structures produced by mixing in chaotic flows. *Nature* **333**, 419-425.

Pasmanter, R.A. 1988 Anomalous diffusion and anomalous stretching in vortical flows. *Fluid Dyn. Res.* **3**, 320-326.

Pierrehumbert, R.T. 1988 Large eddy energy accretion by chaotic mixing of small scale vorticity. *Phys. Rev. Lett.* (submitted)

Smith, L.A. & Spiegel, E.A.1985 Pattern formation by particles settling in viscous fluid. *Springer Lect. Notes in Phys.* **230**, 306-318.

Zimmerman, J.T.F 1986 The tidal whirlpool: A review of horizontal dispersion by tidal and residual currents. *Netherlands J. Sea Res.* **20**, 133-154.

🐢 : (continues) A lot of this was anticipated qualitatively in earlier papers, but except for the work of Arnold and Hénon the dynamical systems viewpoint, including the possibility of chaos, was generally ignored. Also many papers, in particular in geophysical fluid dynamics, would show results that cry out for an interpretation in terms of chaos. I mention by way of example:

Eckart, C. 1948 An analysis of the stirring and mixing processes in incompressible fluids. *J. Mar. Res.* 7, 265-275.

Welander, P. 1955 Studies of the general development of motion in a two-dimensional, ideal fluid. *Tellus* 7, 141-156.

Hama, F.R. 1962 Streaklines in a perturbed shear flow. *Phys. Fluids* 5, 644-650.

Perry, A.E. & Fairlie, B.D. 1974 Critical points in flow patterns. *Adv. Geophys.* **18**, 299-315.

Regier, L. & Stommel, H. 1979 Float trajectories in simple kinematical flows. *Proc. Nat. Acad. Sci. (USA)* **76**, 4760-4764.

Cantwell, B.J. 1981a Organized motion in turbulent flow. *Ann. Rev. Fluid Mech.* **13**, 497-515.

McKenzie, D.P. 1983 The Earth's mantle. *Scient. Amer.* **249**, 66-78.

Applications to problems where chaotic advection is coupled to other processes such as diffusion (as considered by Jones, Young & Aref), or particle inertia (McLaughlin), or chaotic advection of interacting particles, such as coagulating particles (as considered by Muzzio & Ottino), are also promising.

A : (Agitated) There are entire new fields of fluid mechanics here. There are immediate applications of chaotic behavior to *vortex dynamics*, and to *flow-structure*

interactions, e.g. in the coupling of shedding to forced or induced vibrations. There are manifestations of chaos in the *sound radiation* from simple vortex systems. Let me show you this calculation that Kimura and Zawadzki have done. (Pulls out a diagram = Fig.2 of formal paper) On the left you see the sound or far-field pressure signal and below it the vortex trajectories. On the right is the sound spectrum. Superimposed on the sound are the locations of poles in the complex time plane obtained by a Painlevé analysis of the point vortex equations. The location of these singularities is an excellent predictor of the sound signal. Several features of the parametric motion of the singularities in the complex time plane may be understood analytically...

🐇 : This is all very interesting gentlemen, but what does it have to do with turbulence?

A & 🐢 : (Looking at one another, puzzled) Turbulence? Who said anything about turbulence? This is *chaotic* fluid dynamics.

🐇 : (Firmly) I want to know what your "chaos theory" has done for turbulence. Please, tell me.

🐢 : I was just getting to that... My next thesis is

Thesis #3
"Chaos theory" has stimulated model-building in fluid mechanics.

🐇 : That doesn't address my question...

A & 🐢 : But it is very important! Turbulence has been considered much too special, as if nothing else in the world is like it. The focus on chaotic behavior across a wide variety of systems has given us a much greater appreciation for the kinds of stochastic behavior possible in mechanical problems. For example, it is good to know that subharmonic bifurcations can be set up in a convection experiment. It is equally good to know that you need not in general see subharmonic bifurcations every time you increase the Rayleigh number in a convection cell.

People have become much bolder in exploring deterministic models related to fluid mechanical equations. The challenges of "chaos theory" can take some credit for this healthy development.

🐇 : Granted, but what has "chaos theory" contributed to turbulence proper?

🐢 : My fourth thesis is:
Thesis #4
"Chaos theory" has produced new ways of analyzing experimental data.

🐇 : (Annoyed) At the risk of sounding repetitive that doesn't address my question either... Really, if you don't give me something tangible, I have more important races to run.

A & 🐢 : But it is again very important! Signals that just a few years ago would have been labelled "noisy, unsteady flow, but clearly not fully developed turbulence" are being re-considered with the new perspective of "chaos theory." Data is being analyzed in many new ways stimulated by the concepts of "chaos theory."

🐇 : Alright, alright, but I must insist that you address the real issue...

🐢 : My fifth thesis will please you:
Thesis #5
The turbulent cascade is a chaos phenomenon.

🐇 : (Stunned) My, oh my! I must say that *is* impressive. Inertial range cascade dynamics, K41 and on, is all a consequence of your "chaos theory." You must give me the references.

🐢 : Well, it is really only a conjecture... But a very plausible one! Chaotic systems have shown a tendency to produce hierarchical structures not dissimilar from the ones we associate with the turbulent cascades. For example, the images drawn by von Weiszäcker of interlocking eddies and the island structure of chaotic area-preserving maps have a lot in common. It seems plausible that there is a connection here.

Indeed, recently Moffatt and Gilbert have studied the statistics of "whorls" in a two-dimensional fluid and elucidated the relations between spatial structure produced by a distribution of elliptic fixed points and conventional turbulence spectra.

In a similar vein Novikov gave an argument many years ago relating the inverse cascade mechanism in two dimensions to the phase space motion of many-vortex systems...

A : And, of course, there is Lundgren's deterministic model of the $k^{-5/3}$ cascade in three dimensions, and there is all the work on vortex reconnection and realignment, "Beltramization" as they call it, where the notion of chaos plays an important role...

🐢 : One might speculate that the reason we can have an apparent limit of "inviscid dissipation" in high Reynolds number turbulence is that chaos on the very smallest scales takes up the role of molecular, dissipative mechanisms. We are seeing something of this kind for diffusivity when we add a slight amount of molecular diffusion to a chaotic advection problem...

🐇 : Gentlemen, I have enjoyed our little discourse. You have certainly convinced me that this "chaos theory" of yours can stimulate a lot of intriguing fluid mechanics. As for turbulence, I am heading for the crossroads. There I'll decide whether to take the CFD route (with the usual meaning of the acronym) and subsequently whether to opt for LES, CA or k-ε.

A & 🐢 : We're not finished! We haven't told you about Lyapunov exponents, cantori, Poincaré sections, Melnikov's method, strange attractors...

But 🐇 had already run off the fields nearby where he was nibbling on some clover. "Cascade, fractal, chaos...," he thought, "a new framework seems to be emerging... Only technical details are missing."

Meanwhile, 🐢 and A, his program manager, had some paperwork to attend to.

* * *

Session One

The Utility and Drawbacks of Traditional Approaches

Discussion Leader: G. Comte Bellot, Ecole Centrale de Lyon

The Utility and Drawbacks of Traditional Approaches.

R. Narasimha

National Aeronautical Laboratory
and Indian Institute of Science
Bangalore 560 017, India

Summary

This survey argues that traditional approaches to turbulence have led to some notable but incomplete successes, such as the establishment of a plausible connection between diffusion and correlation, and the similarity and matchability arguments of Millikan and Kolmogorov. Experience with shear flows may be summarised in the form of some basic working rules, although their validity cannot be considered beyond doubt. On the other hand, traditional approaches seem to ignore the wide prevalence of exponential distributions (rather than Gaussians), and have not led to any rational models for turbulent flow yet. For many decades now the scientific initiative has remained with experiment, and extensive work on coherent structures in a variety of flows has suggested novel methods of turbulence management and possible new candidates for "molecules" of turbulence that may enable progress in theory. While the initiative may now be passing to computers at low Reynolds numbers, a "reconciliation" between the notions of ordered motion and statistical theory is considered likely in the near future, and will probably emerge from greater understanding of the apparently random occurrence of coherent events through a dynamical-systems approach.

Keywords : Turbulence, traditional approaches, coherent structures, modelling, chaos, exponential distributions, working rules

1   What are the "traditional" approaches?

This Conference is going to discuss turbulent flows in separate sessions on coherent structures, dynamical chaos, numerical solutions, computer simulations and mathematical modelling. This suggests an operational definition of traditional approaches for the purposes of the present discussion as whatever is left out from succeeding sessions. However, modelling as a whole cannot be termed non-traditional, as it goes back at least to Boussinesq's introduction of the eddy viscosity idea in 1877. There will also be some question about how non-traditional the idea of coherent structures is (see e.g. Liepmann 1979, Lumley 1981). Comments about these two subjects in this position paper therefore seem essential, so I will not define traditional approaches entirely by exclusion.

Having said this, what more specifically *are* the traditional approaches? We begin inevitably with Reynolds (1883), who introduced a decomposition of flow-field variables into means and fluctuations, leading to what are now called the Reynolds-averaged equations (see Note /1/). I do not foresee that interest in the means is ever going entirely to disappear -- if only because there will always be applications in which long-time averages convey useful information (e.g. the drag of the aircraft during the long journeys I have made to come here). But if male is +1 and female -1, it does seem that to say that the average gender in the world is approximately neutral misses the point altogether (in spite of the fact that the mean in my country being 0.035 and not zero has been considered extremely significant by many sociologists). The Reynolds equations are of course not wrong, in the sense that any solution of the Navier-Stokes equations will also be a solution of the Reynolds equations. But the converse is not in general true if by "solutions" of the Reynolds equations (which are *not* closed) we mean those effected by truncation and forced closure at moments of some finite order, as the ad hoc procedures adopted for doing so are not rigorous consequences of the Navier-Stokes equations. There is no guarantee that in a strongly nonlinear problem means can be successfully calculated by equations containing only means.

One method of describing the fluctuations themselves -- there can be others, as we shall discuss later -- is the spectral approach pioneered by G I Taylor and extensively used during the four decades from the 1930s to the 1970s. The Navier-Stokes equations can be written in spectral language, but of course this does not solve the problem of closure -- it now appears in a different guise, that is all. However, the approach has given us some valuable clues, as we shall discuss in Section 2.

Using the spectral description, several lines of attack on the problem of turbulence have been attempted. Among the most striking are:

- quasi-normal theories,
- the direct-interaction approximation and related techniques, and
- universal equilibrium concepts,

the first two having been extensively used especially in homogeneous turbulence. It has long been known (Ogura 1962, Orszag 1970) that quasi-normal theories (going back to Millionshtchikov 1939), which assume that all cumulants above some given order greater than 2 vanish, lead to negative energy spectra: this appears to be a general consequence of a theorem due to Marcinkiewicz (see Rajagopal & Sudarshan 1974, who show that such cumulant-truncation will violate the basic positive definiteness of the probability distribution function itself). Orszag (1970) attempted to tackle the problem by the use of an arbitrary damping term in the equation for the third-order moment, but it turns out that it is necessary to introduce the additional hypothesis of Markovianisation to ensure a positive spectrum; and this requires an assumption about time scales that is questionable in the energy-containing range (see e.g. Lesieur 1987). Direct inter-action theories (Kraichnan 1959, Leslie 1973, Kraichnan & Herring 1978, Nakano 1988) make a variety of assumptions whose validity is difficult to assess in physical terms, and will not be considered further here. We shall return later to equilibrium theories in general.

In attempting to treat turbulent *shear* flows, which are far more common than the homogeneous variety so often the object of attention in fundamental theories, a body of serviceable ideas has emerged from the traditional approaches. These ideas have been exploited effectively by Townsend (1956, 1976) and others in discussing flow development, and will be considered in Section 3. This body of argument and experience, with all its undoubted inadequacies, should be a lasting contribution to our knowledge of turbulent flows.

2   The major triumphs

The traditional approaches can in my view claim two major triumphs which altered our perception of the problem, and perhaps several minor ones whose cumulative effect is not negligible. Each worker in the field will probably have his own candidates for such triumphs.

The first concerns an approach to the phenomenon of "eddy diffusion", or more precisely of dispersion of passive scalars in turbulent flow. Taylor (1921) introduced what is now familiar as the Lagrangian auto-correlation function, and showed how it may plausibly be connected with such eddy diffusion in homogeneous turbulent flow. (He significantly noted that these connections were incomplete, as they assumed convergence of series that he had to truncate.) This study also led to an expression for an effective diffusivity at large times. As we shall note later, the notion of gradient transport has severe limitations, and has unfortunately been widely used as a modelling tool in situations where it is not valid. But Taylor's introduction of the correlation led slowly but inevitably to *a* (often unjustifiably considered *the*) statistical theory of turbulence, with the further introduction of the spectrum (Taylor 1938) and dynamics (Karman & Howarth 1938). None of these studies *solved* anything, of course, as the problem of closure remains; but they did establish connections and offer "transforms" of the problem, so to speak.

The second major triumph is the equilibrium theory of Kolmogorov (1941, 1962) and others. These well-known arguments led to the prediction that the spectrum in the "inertial subrange" -- or more appropriately in the overlap domain between viscous and large-eddy scales (as the "dissipative sub-range" would be an equally valid description if we start from the large eddies) -- followed a $k^{-5/3}$ law. There has been strong evidence in favour of the law (e.g. Grant, Stewart & Moilliet 1962), although the fine-scale motion does not necessarily possess the isotropy that was one of the properties postulated by Kolmogorov, and the proportionality constants cannot yet be deduced in any convincing and rigorous way by theory (despite the valiant attempts of Kraichnan and others). There has also been the criticism that the original Kolmogorov arguments took no account of the intermittency of the dissipative scales. This criticism has been countered by various refinements of the theory (Kolmogorov 1962, Oboukhov 1962, Monin & Yaglom 1975), which are however not without their difficulties (see Note /2/).

Actually, this triumph must be coupled with the name of Millikan (1939), who exploited much earlier the same "matchability" idea that informs Kolmogorov's argument to deduce, without appealing to any specific turbulence model, the log law for the mean velocity distribution in channel flow. One may formally introduce the language of asymptotic expansions (Yajnik 1970, Mellor 1972) and postulate a 'Millikan-Kolmogorov Rule' (see Section 3 below) that requires that solutions over inertial and viscous scales match; the rule is most generally seen (Afzal & Narasimha 1976) as leading to the formulation of a *functional* equation whose solution achieves the required matching. The rule has been widely exploited (often only implicitly) in a variety of flows: e.g. two-dimensional turbulence (Batchelor 1950), boundary layers (separating, Townsend 1960, Kader & Yaglom 1978; axisymmetric, Afzal & Narasimha 1976, 1985), etc.

As will be clear, these are incomplete successes, but they are not the less significant for that reason. Their main contribution has been in introducing and exploiting plausible physical arguments through dimensional and similarity analyses, i.e. through a form of group theory (as discussed e.g. by Birkhoff 1960).

3   Some general principles?

Inability to achieve a rational closure of the Reynolds equations has encouraged on the one hand the invention of a variety of models (which we shall consider later), and on the other the evolution of certain general principles that are intended to help us to surmise the broad features of turbulent flows without appealing to detailed models of questionable validity. Of course, as these general principles have not yet been deduced from the basic laws of fluid motion, they remain open to doubt; they are therefore no more than hypotheses or working rules, and their value must be judged by their usefulness. Some effort at determining the existence of such Rules seems essential if we are to prevent the total 'botanisation' of turbulence research that we have been witnessing for some time now. (This incidentally suggests that it is high-time that some consolidation is attempted by adopting a well-known technique from botany, namely the creation of 'catalogues' of carefully observed flows.)

Some years ago (Narasimha 1983, 1984), it was suggested that experience with various hypotheses of this kind may be summarised in the form of the *five basic working rules* listed in *Table 1*. Of these, WR4 is a very serviceable rule, being often implicitly assumed by engineers (and by research workers in designing the plumbing of their apparatus, although not in measuring the test flow): it is the basis of all handbook charts and correlations, such as e.g. the Moody diagram for pipe losses. WR5 is the Millikan-Kolmogorov Rule already mentioned.

There is nothing really new in these Working Rules: they are what come naturally to workers in turbulence whenever they encounter a totally unfamiliar flow, and have often been used without explicit recognition of their application (reminding one of Moliere's Bourgeois Gentleman who discovered he had been making prose all his life). The point about highlighting them is to openly exhibit them, to help us in assessing how valid they are, and in concentrating on violations (if there are any). Furthermore, it would seem that any turbulence models would have to start from some position on the matters touched upon by the Rules (that models do not do this is in my view one of their serious deficiencies).

Table 1 : The Five Basic Working Rules

(1) As the Reynolds number of any turbulent flow tends to infinity, the fraction of energy contained in the length and time scales directly affected by the viscosity of the fluid becomes vanishingly small; so do the scales themselves, compared to those accounting for the energy.

(2) Any turbulent flow subjected to constant boundary conditions evolves asymptotically to a state independent of all details of its generation save those demanded by overall mass, momentum and energy conservation.

(3) If the equations and boundary conditions governing a turbulent flow admit a self-preserving (or equilibrium) solution, the flow asymptotically tends towards that solution.

(4) A turbulent flow may, before reaching equilibrium, attain a 'mature' state in which the different energetic parameters characterising the flow obey internal relationships, irrespective of the detailed initial conditions as in Rule 2.

(5) Between the viscous and the energetic scales in any turbulent flow exists an overlap domain over which the solutions characterising the flow in the two corresponding limits must match as Reynolds number tends to infinity.

A detailed discussion of these Rules will not be repeated here, but we recall that at the time that they were first presented (1983/84) the assessment was as follows: "They are obviously very useful, being close to reality; but they cannot still be elevated to the status of scientific laws, because the small departures noted from them cannot be dismissed as experimental error, and seem to indicate that the principles are strictly valid only under certain as-yet unstated conditions which would not always be easily obtained." The hypothesis which has been most frequently challenged in recent years concerns the postulated independence from initial conditions in WR2 and 4.

What is the evidence? There is no doubt whatever that initial conditions strongly influence near-field flow development, and that flow configurations need not be unique for given boundary conditions (e.g. separated flow in a plane diffuser

can be in either of two distinct states that correspond to attachment to either wall).
Lee & Reynolds (1985) have shown how jets can 'bifurcate' and 'bloom' when
excited by certain special combinations of axial and orbital excitations: the flow
can be spectacularly different from that in a standard jet, although it is not yet
certain that such abnormal behaviour will persist "for ever". In excited mixing
layers (e.g. Oster & Wygnanski 1982, Roberts & Roshko 1985) the growth rate can
be completely suppressed over a certain region of the flow, but appears to resume
its 'normal' value further downstream.

Wygnanski, Champagne & Murasli (1986) have recently reported that the
development of a two-dimensional turbulent wake, even in the far-field, depends
strongly on initial conditions (*Figure 1*). By comparing wakes of the same
momentum defect generated by different bodies (which included wire screens of
different solidity), they found that at equivalent distances the distribution of the
Reynolds stresses was not identical. George (1988) has argued, with a more
general self-preservation analysis, that there exist a multiplicity of equilibrium
states, each uniquely determined by its initial conditions. On the other hand,
Louchez, Kawall & Keffer (1985) found from extensive experiments, also on a
variety of bodies including strips of screens, that a universal self-preserving state
was attained in all cases except those where the screen wires had a Reynolds

Figure 1
Data on wake velocity defects
from Wygnanski et al (1986).
Abscissa is maximum defect
velocity as fraction of free-
stream velocity, ordinate is the
non-dimensional parameter introduced by Narasimha &
Prabhu (1972). Open points
include no allowance for a virtual origin. The diagram has
been extended upto $u_o/U = 0$,
where the equilibrium value W^*
of W as deduced by Sreenivasan
& Narasimha (1982) is shown
as a star. The question is
whether flow development is
towards W^* or not.

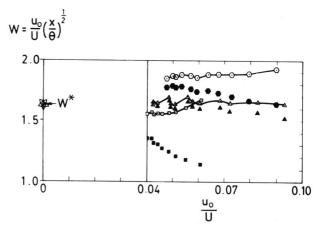

$$W = \frac{u_o}{U}\left(\frac{x}{\theta}\right)^{\frac{1}{2}}$$

□–■ airfoil ; ▲–▲ 70% solidity screen ; ○–● solid strip

number less than about 100. Similar discrepancies in a mixing layer appear to have been resolved by the data of Kleis & Hussain (1979), who found (*Figure 2*) that while flow development for some distance downstream of the splitter plate depended strongly on the state of the boundary layer (especially on whether it was laminar or turbulent) at the trailing edge, these initial conditions were eventually forgotten -- after a distance in their case of more than 4 ft (in a mixing layer 12 ft long). In the turbulent wake of a circular cylinder, Cimbala et al (1988) find an exponential decay of velocity fluctuations at the Karman vortex-street frequency, with a half-life of about 12 diameters, and a life-cycle of growth and decay of large eddies not inconsistent with the equilibrium hypothesis of Townsend (1956).

Conclusions on the existence or otherwise of universal equilibrium states independent of initial conditions are often clouded by an inexplicable reluctance on the part of experimenters to use ideas of asymptotic analysis; it seems as if there is a feeling that this may be relevant for mathematics but not for experiment. The importance of taking an asymptotic view of experimental data has been empha-sised several times (Narasimha 1984, Sreenivasan & Narasimha 1982). For example, if the Townsend expansion of the far wake (1956; p137) is valid, the concept of a unique virtual origin is invalid; and we know this for certain in laminar wakes, from asymptotic solutions derived by Goldstein (1933) long ago, as well as

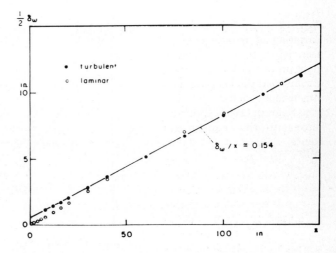

Figure 2
Flow development in a long mixing layer, with boundary layer at trailing edge respec-tively laminar and turbulent, showing that initial conditions appear to be forgotten beyond about 50 in. (From Kleis & Hussain 1979.)

from more recent studies (Wygnanski 1967; see discussion in Narasimha 1984). Thus it is hard to draw the conclusion from data like those in Figure 1 (based on Wygnanski et al 1986) that there is *no* tendency towards a unique equilibrium state; more definitive evidence is necessary before WR2 and 4 can be abandoned. Especially in cases where relaxation times are very long, as in the turbulent wake (Narasimha & Prabhu 1972), the approach to equilibrium can be bizarre. The data of Sharma (1987) make this point forcefully (*Figure 3*). The wake development shown in *Figure 4* (from Prabhu & Narasimha 1972) is, believe it or not, relaxing towards a (moving) equilibrium state *all the time* ; the reason for the apparently odd behaviour of the flow (with the wake thinning down after having reached a maximum thickness; there is a strong favourable pressure gradient) is just that the flow is too slow in tracking a fast-moving equilibrium state.

There is complete agreement that the memory of turbulent flows can be very long; the question is whether it can be infinite. If the answer is yes the implication is that (borrowing terminology from the theory of dynamical systems) there are no attractors in the flow -- strange or otherwise; and this would be hard to swallow.

Returning to the five rules, it must be emphasised that while they do make a lasting contribution to our understanding of turbulent flows, they do not by any

Figure 3
Variation of non-dimensional parameters in wake development in the ΔW plane (non-dimensional thickness vs. non-dimensional defect velocity), from Sharma (1987). The stars represent the equilibrium points.

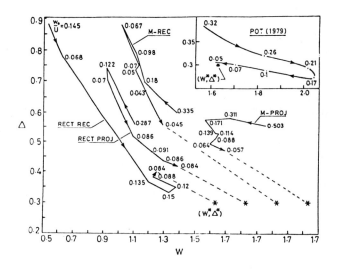

means solve the problem. First of all their scope is limited -- e.g. the flow in the near-field, or outside the overlap domains of the Millikan-Kolmogorov principle (WR5), can not only be interesting but of crucial importance. Secondly, even within their admitted scope, no quantitative results are provided by the rules -- at best they are helpful for organising and assessing experimental data and turbulence models (although this help is not too often taken!). Finally, more precise conditions under which the rules are valid need to be determined and stated. The conflicting conclusions drawn from different experiments suggests that there may be multiple, metastable equilibrium states. There could well be a fundamental difference in behaviour between what one may term weakly stable flows (e.g. constant-pressure boundary layers) and basically unstable flows (e.g. mixing layers and wakes) -- as indeed there does appear to be in their manageability (Liepmann & Narasimha 1987). It is most desirable that the latter class of flows are given the same meticulous scrutiny that Coles & Hirst (1969) bestowed on boundary layers. Continuing the botanical metaphor, we need to ask if the different flow-species can be grouped into broader genera, possibly based on their stability characteristics (such as, when unstable, whether convectively or absolutely).

Figure 4
Development of a wake in a favourable pressure gradient, from Prabhu & Narasimha (1972). Full lines in the diagram show prediction from a model postulating a slow approach to a (local) equilibrium state. So as the wake first grows and then thins down, it is always tracking an equilibrium state, but one that keeps changing.

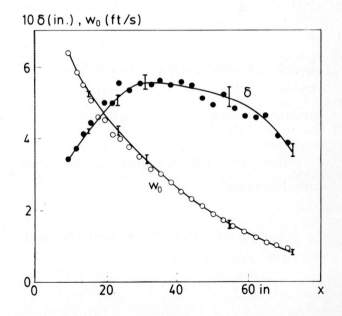

Perhaps at the present stage a Sixth Rule needs to be added:

(6) "Do not be surprised if in some particular flow the above Rules seem to be violated."

4 The failures

Apart from what has been discussed above, the traditional approaches have mainly been a story of failures. We shall list a few major ones here.

4.1 Philosophical

The philosophical issue of how statistical theories may be used for what is basically a deterministic set of equations could never be squarely faced till recently. Turbulent motions were just defined as those where "the velocity takes random values which are not determined by the ostensible, or controllable, or 'macroscopic', data of the flow" (Batchelor 1953).

Recent developments in the theory of dynamical chaos may be considered to have solved the philosophical problem: at least it can now be demonstrated that there are nonlinear dynamical systems whose solutions, because of sensitive dependence on initial conditions, exhibit a weak causality and hence are apparently random. Einstein's objections (in a different context) that God does not play dice can now be countered firmly: He need not, it is enough that He is nonlinear.

But is there a more quantitative connection between dynamical chaos and turbulent flows?

In the first flush of the exciting studies by Ruelle, Feigenbaum and others in the 1970s, extending a line of thought pioneered by Lorenz (1963) more than a decade earlier, it looked to some as if the answers were very close. A decade later, however, the situation is not so clear, in particular in open flows (ducts, boundary layers, etc); Libchaber (1988) says "physicists come and go, and the problem of turbulence remains". None of the routes to chaos till now identified seems directly relevant, e.g., to transition in a boundary layer as we know it. The breakdown observed by Klebanoff, with the relatively abrupt appearance of high-frequency oscillations (Klebanoff, Tidstrom & Sargent 1962), seems as yet to have nothing in common with the slow chaos observed in low-dimensional dynamical systems (Narasimha 1987a, Morkovin et al 1987; see also Spiegel 1985). In real open flows, transition Reynolds numbers are not unique for a flow-type, but seem to increase (indefinitely, as far as one can tell; see Narasimha 1985, Govindarajan & Narasimha 1989) as external disturbances go down; and chaos, once it has set in, persists at higher Reynolds numbers, instead of disappearing as in the Lorenz system. But the chaologists have raised some very interesting questions; e.g., are there only certain fundamental modes of transition to chaos (irrespective of the physical nature of the system), just as there are (or used to be) only a few fundamental particles that constitute all matter? Based on some recent work on a three-dimensional dynamical system that mimics some of the gross characteristics of open-flow transitions, it seems, however, that dynamical chaos could possess the power to explain the phenomenon of turbulence. It is certainly possible to devise a low-dimensional system (Narasimha & Bhat 1988; Bhat, Narasimha & Wiggins 1989) in which the onset of chaos shows the right kind of dependence on external disturbance and Reynolds number: in this view, unforced, nearly inviscid flow is like a homoclinic orbit, which will break into chaos at the smallest perturbation (*Figure 5*). The recent study by Aubry, Holmes, Lumley & Stone (1988) on the dynamics of coherent structures in the wall region of a turbulent boundary layer, discussed at length by Holmes (1989), also provides a link between chaos in relatively low-dimensional systems and turbulence in open flows, using the lowest modes in an expansion of the field in empirical orthogonal functions derived from correlation measurements using the decomposition method of Lumley (1970).

These studies are highly suggestive, but still a long way from being predictive.

4.2 The prevalence of exponentials

Much of the traditional theory takes the Gaussian as the norm. It has already been pointed out that quasi-normal approximations in homogeneous turbulence eventually lead to physically absurd results such as negative energy spectra. In spite of this result, however, the normal distribution appears quite often in theory. If we look at observations, however, it is surprising how often we encounter exponentials, by which I shall mean distributions with tails like $\exp - |x|$ rather than $\exp -x^2$ in the random variable. Let us list some examples.

(a) The distribution of velocity derivatives. The squared derivative is often thought to obey a log-normal distribution (e.g. Monin & Yaglom 1975, following Kolmogorov 1962), but this is inconsistent with the observed non-singular behaviour of the distribution for small values of the derivative (see /2/). *Figure 6* shows experimental data from Van Atta & Chen (1970), and a fit to it which is a Gaussian core with an exponential tail, of the form

Figure 5
Chaotic regime in a simple dynamical system (from Bhat, Narasimha & Wiggins 1989), showing how the transition Reynolds number depends on external disturbance level (lower curve). Note how chaos persists as Reynolds number Re --> ∞, the required disturbance level simultaneously going to zero.

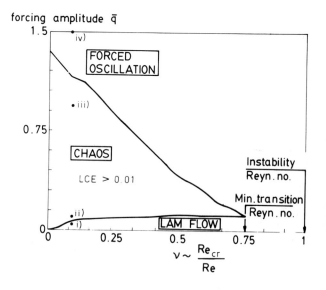

26

$$p(x) = a \exp(-x^2/\sigma^2) + b \exp{-\alpha|x|}. \qquad (1)$$

The implication is clearly that large deviations are vastly more frequent than a Gaussian would indicate. The atmospheric data of Wyngaard & Tennekes (1979) indicate that the weight b of the exponential above increases with the Reynolds number. The recent measurements of Dowling (1988), on concentration derivatives in a jet, show a probability density similar to (1). The total dissipation, which involves the sum of the squared derivatives of velocity components in different directions, does however tend to log-normality -- a point that has been clarified in numerical simulations kindly analysed by Rogers & Moin (1987, private communication; see /2/ and *Figure 7*).

(b) The occurrence of gusts in the atmosphere. In an interesting discussion Jones (1980) shows that measured velocity increments are non-Gaussian and fitted very well by exponential distributions (*Figure 8*).

Figure 6
Probability distribution of velocity derivatives, from Van Atta & Chen (1970). Curve 1 is Gaussian; curve 2 is a fit to equation (1), with a Gaussian core and exponential tail.

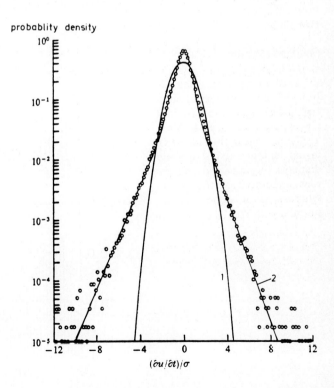

(c) Vorticity. Yamamoto & Hosokawa (1988) show that in their (direct) numerical solution of the Navier-Stokes equations for decaying isotropic turbulence (at a macroscale Reynolds number of 500), the magnitude of the vorticity is distributed exponentially over more than five decades in the cumulative probability.

(d) The intervals between successive zero-crossings of a turbulent signal. In a variety of flows this distribution has an exponential tail (Sreenivasan, Narasimha & Prabhu 1983, *Figure 9*).

(e) Time interval between ejections in a boundary layer. Bogard & Tiederman (1986) find from their flow visualisation experiments that inter-event times follow closely an exponential distribution.

Figure 7

Probability density of distribution of

(a) $(\partial u/\partial x)^2$, and

(b) total dissipation ε, as obtained from a numerical solution for homogeneous turbulent shear flow (Rogers & Moin 1987). It is clear that $(\partial u/\partial x)^2$ is not log-normal but instead displays the expected square-root singularity at the origin, whereas ε is close to log-normal.

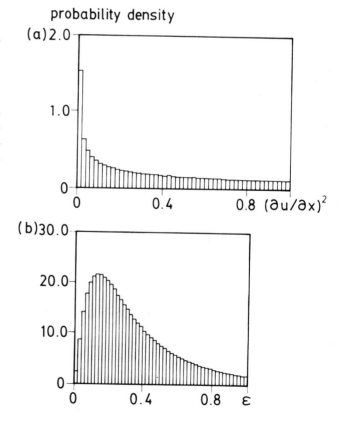

An exponential distribution in time, as in (d) and (e) above, suggests a Poisson stream of events -- a point to which we shall shortly return.

4.3 The absence of rational models

Turbulence modelling of a certain kind is now a widespread industry: models proliferate in the market-place, and continually keep getting more complex and sophisticated. Despite certain fundamental deficiencies in the concepts underlying these models (-- to which we shall return shortly), some of them have been cleverly exploited in a class of applications, highlighted, for example, by the aerofoils designed and tested by Liebeck (1978, see *Figure 10*), yielding the remarkable lift:drag ratio of 220 at a Reynolds number of 10^6 (Lissaman 1983 has

Figure 8
Rate of occurrence of peaks of aircraft response to atmospheric gusts, showing similarity and an exponential distribution that is in part a consequence of a similar exponential distribution of gust intensites. (From Jones 1980.)

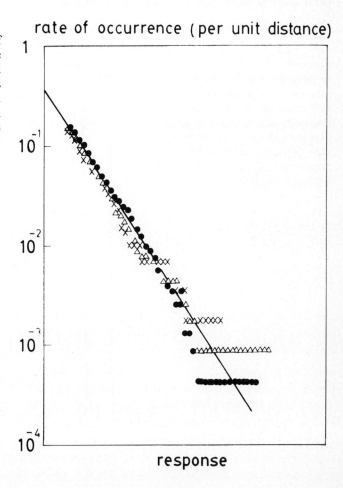

surveyed other interesting work in this area). However, no model in this class has yet emerged that can be said to be rational or dynamically correct; so one criticism of the effort to-date (although it may seem somewhat unfair) must be that they have turned out to be no more than models. When Planck tentatively fitted his famous curve to the spectrum of black-body radiation, he did not realise that the fit was going to turn out to be such an accurate reflection of truth. Unfortunately, no such luck has crowned any of the modellers.

But as long as the turbulence problem remains unsolved (and perhaps even then), models of one kind or another will continue to be invented and explored, and it is necessary to distinguish between different kinds before passing judgement. There appear to be at least four classes of models in general, with the objectives (and examples) listed in *Table 2* : there are many hybrids as well. With this classification in mind, we can re-assert that there are no rational models yet for

Figure 9
Probability distribution of the interval l between successive zero-crossings in a turbulent boundary layer, showing long exponential tail (Sreenivasan et al 1983). Inset shows that the slope of the tail varies little with Reynolds number.

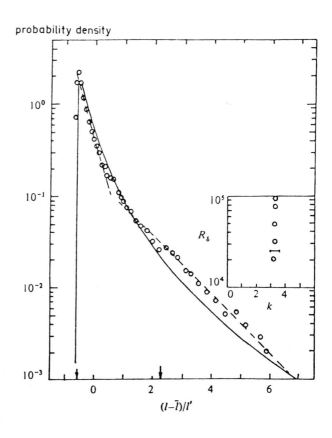

turbulent flows in general, and only a few that are impressionistic or physical; the models that feed industry are all ad hoc.

To amplify this point, consider first gradient transport models (mixing length, eddy viscosity, K-ε etc). That the foundations of these models are shaky has been known for a long time; thus, Batchelor (1950), examining turbulent wakes, said that it was "difficult, if not impossible, to account for [the observed] relation ... on the basis of gradient type of transfer ... [the] past usefulness [of mixing length

Figure 10
An aerofoil designed by Liebeck (1978), giving a lift: drag ratio of 220 at a Reynolds number of $2x10^6$.

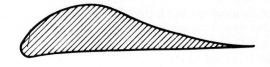

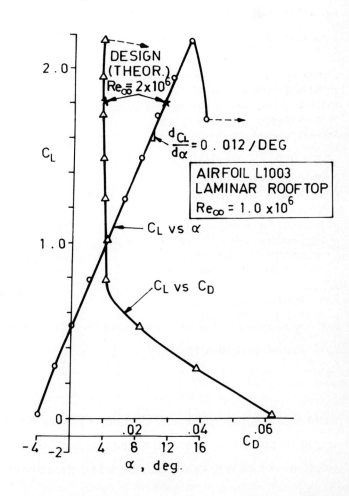

Table 2 : Models Classified

Objective	Examples
1. *Impressionistic*	
To gain insight into structure of problem and solution, without any claim to quantitative accuracy in prediction	Burgers (1948), for turbulence Lorenz (1963), for weather prediction BGK (1954), for kinetic theory of gases
2. *Physical*	
To predict quantities of interest based on assumptions not inconsistent with observed or understood physics, appealing to experimental data for model parameters when necessary	Emmons (1951), for boundary-layer transition Kutta-Joukowski for airfoil lift[a] Kolmogorov (1941), for spectrum of turbulence[b] Mean free-path theory, for molecular transport
3. *Rational*	
To investigate nature of problem and solution through simpler models derivable from more complete system by some limiting process[c]	Burgers (1948), for weak shocks[d] Newtonian model for hypersonic flows ($\gamma \to 1$) (Rapid) distortion of turbulence (Prandtl, Taylor, Batchelor & Proudman 1954)
4. *Ad hoc*	
To provide estimates of quantities of interest, e.g. in engineering applications, without insistence that all assumptions be physically or mathematically justified in detail	Boussinesq's eddy viscosity and Prandtl's mixing length, for turbulent fluxes Miner (1945), rule for cumulative damage in fatigue

[a] Widely suspected to be rational.
[b] Possibly rational in part.
[c] Without necessarily claiming to solve exactly any real-life problem; e.g. Kirchoff free-streamline theory.
[d] As shown by Lighthill (1956).

theory] is slightly spurious". Corrsin (1974) noted that "nearly all traditional transport problems violate [this] requirement" for the validity of a gradient-flux relation -- which is basically that the interaction distance (mixing length in turbulence, mean free path in the kinetic theory of gases) must be *small* compared to the scale of the inhomogeneity. As the two are comparable in most turbulent flows, the problem is clearly analogous to the kinetic theory of strong shock waves, which nobody would now seriously attempt to tackle using gradient-flux (i.e. Chapman-Enskog or Navier-Stokes) relations. A more detailed criticism of gradient transport models based on recent observations of the structure of turbulent flows has been made by Broadwell & Breidenthal (1982).

Stress transport and higher-moment closure schemes (e.g. Bradshaw, Ferris & Atwell 1967, Launder 1989, Narasimha & Prabhu 1972) do not relate stresses directly to velocity gradients, and so are less open to the above criticism; however gradient diffusion usually appears in modelling higher order terms.

There are other 'global' difficulties with models that claim to handle a wide variety of flows. First of all, they do not respect counter-examples (which in hard science would be considered sufficient to dismiss an argument). As one instance, take once again the widespread use of eddy diffusivities, despite the presence of many flow situations where diffusion is against the gradient. A wall-jet provides the most familiar example (*Figure 11*). (Other examples will be found in Sreenivasan, Tavoularis & Corrsin 1981.) It is worth emphasising that counter-gradient diffusion is not uncommon, and one does not have to look for it in more complex stratified or rotating flows of the kind familiar in geophysical situations. It is

Figure 11
Mean velocity and Reynolds shear stress profile in a wall jet (data from S P Parthasarathy 1964, reproduced from Narasimha 1984), showing counter-gradient momentum diffusion over the region 0.5<η<1. The stress at the point of zero gradient is comparable to the wall stress but opposite in sign.

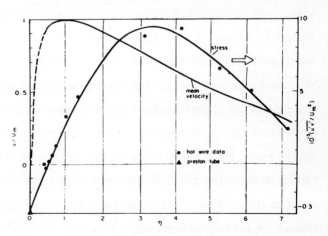

sometimes argued that the counter-gradient diffusion seen e.g. in the wall jet is not 'important', presumably in the sense that some flow parameters (e.g. skin friction) can be predicted reasonably well without modelling the flux accurately (an argument which incidentally justifies calling the model ad hoc). What is disturbing is that the modelling does not take account of a situation that is easily understood physically: in this case the dominant influence ("long arm") of the large eddies in the jet region in producing counter-gradient stresses in the wall region of the flow. It is a matter of serious doubt whether better models will be devised by ignoring such simple counter-examples while continuing to build elaborate super-structures.

A second problem is that models shy away from taking a stand on general principles -- e.g. on the Working Rules listed in Section 3 (with some rare and notable exceptions such as Melnik & Rubel 1983). *Do* turbulent flows eventually forget their initial conditions as postulated in WR2 or not? *Do* the models allow for the Millikan-Kolmogorov Principle or don't they? *When* (if at all) is counter-gradient diffusion possible? (The criticism about counter-gradient diffusion, made during a personal conversation, was taken seriously by Lykossov (1989), who has shown that it is possible to simulate the observed stress distribution in a wall jet in a turbulence model he has developed and used.)

The third problem is that all models in current use, to the best of my knowledge, seem incapable of assessing (let alone predicting) whether, and if so under what conditions, the many interesting and novel methods of turbulence management (even the passive ones) now being investigated may be expected to succeed. Will the so-called large-eddy break-up devices (or blade manipulators) work? (See the conflicting experimental evidence presented by Corke, Guezennec & Nagib (1980) and Yajnik, Sundaram & Sivaram (1987) on the one hand, and by Prabhu, Kailas Nath, Kulkarni & Narasimha (1987) and Sahlin, Johansson & Alfredsson (1988) on the other.) Why do splitter plates behind cylinders (Roshko 1955) and in front of them (Viswanath, Srinivas, Prabhu & Rajasekar 1989, see *Figure 12*) raise base pressures and reduce drag?

Although current models contain a wealth of refinement, it is not clear that they embody any *conceptual* advance over the Prandtl-Kolmogorov models described in the 1940s. They do tend, however, to include many factors characterising complex flows (additional rates of strain, curvature, mass addition, buoyancy, free-stream disturbances etc), but as none of them can be trusted beyond the experimental data against which they have been compared, they are really very elaborate interpolation procedures, and cannot be relied on to make genuine predictions in totally unfamiliar situations. It may however be worth repeating that interpolation procedures can be useful, and are often the bedrock of confident design. Perhaps the time has come to leave their development and refinement largely to industry, to whom in any case it is second nature.

Against this background, the appearance of new 'physical' models for some flows is to be welcomed. It is interesting that these are being devised to handle flows with mixing and combustion. Broadwell & Breidenthal (1982, for mixing layers) and Broadwell (1987, for jets) base their model on a physical treatment of newly entrained flow in two stages, involving a cascade to fine scales and mixing by molecular transport. Kerstein (1988, 1989) makes a stochastic simulation of turbulent transport or what he calls "convective stirring" through randomly occurring "block inversions" in the flowfield; each of these inversions results in a rearrangement of the distribution of composition representing the action of an eddy. These models are interesting in particular because they completely avoid the Reynolds decomposition altogether (on which the more traditional models are

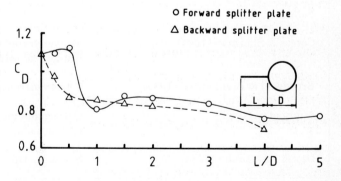

Figure 12
Variation in total drag of a circular cylinder with a splitter plate at a Reynolds number of 6×10^4 based on the cylinder diameter. (Viswanath et al 1989.)

based: see e.g. Bilger 1989), but they need some information on the flow-field structure as an empirical input.

As it is prohibitive in general to start with the (closed) Navier-Stokes equations, solve them completely and average later, it has generally been assumed that the only feasible alternative is to tackle the Reynolds equations, the prescription being: "average first and close later". However, as Corcos & Sherman (1984) have pointed out, there may be intermediate approaches. For example, we may first close *in part*, to solve for an isolated eddy or structure or mode or whatever -- let us call the entity an appropriate 'molecule' of turbulence -- play these molecules off against each other on the computer (preferably with a dynamically derived interaction mechanism), and average last. (In the recent study of Aubry et al 1988, the first partial closure is achieved by appeal to experimental data.) Such approaches should provide considerable insight into the structure of turbulent flows, and appear to have the potential of answering the three criticisms made above of current modelling efforts. (Indeed all dynamical approaches to-date, both traditional and modern, have followed some similar course, and, as the list in *Table 3* indicates, the search for the right 'molecule of turbulence' continues unabated.)

Table 3 : Possible 'molecules' of turbulence	
Spectral	exp $i(\mathbf{k.x}-\omega t)$
Synge & Lin (1943)	Hill's spherical vortices
Townsend (1956)	Distorted eddies
Saffman (1968)	Random array of vortex tubes, sheets
Corcos & Sherman (1984)	T-layer vortex structure for mixing layer (numerical solution with periodic b.c.)
Broadwell & Breidenthal (1982)	(Observed) coherent structure for mixing layer
Aubry et al (1988)	Orthogonal modes derived from observed correlations
Dey & Narasimha (1988)	Spots, for transitional boundary layers

As a parting thought on ad hoc models, I would like to say a good word for the humble integral method: it is cheap, effective (-- it did as well as the fancier ones in the second Stanford competition of 1980-81: see Kline, Cantwell & Lilley 1981), and so unpretentious that it is in no danger whatever of claiming to do too much: all virtues to be prized.

4.4 Turbulence management and coherent structures

Traditional approaches have failed to reveal the many possibilities for manipulation and management of turbulence that have recently been widely investigated (e.g. Liepmann & Narasimha 1987). It is fair to say that these new methods have emerged entirely from an appreciation of the presence of ordered structures in turbulent flow. The realisation that turbulence was not structure-less is not entirely new. Liepmann commented in 1952 that "the importance of the existence of a secondary, large-scale structure in turbulent shear flow has become apparent"; Townsend (1956) inferred structure in several flows from correlation data (of the kind shown in *Figure 13a*). With hindsight, we can now see that the very failure of closure schemes involving truncation at low-order moments indicates at once the importance of higher order correlations and, equivalently, of 'structure' in the flow. Nevertheless, either because these structures were thought to be too weak or because oil was too cheap, the possibilities of managing turbulence now being investigated (mentioned in Section 4.3) did not emerge before the coherent structure idea caught the imagination of fluid dynamicists, provoked by the spectacular pictures of Brown & Roshko (1974) and the work of Kline et al (1969). The possibility of a grooved surface (with a vastly larger surface area) having actually a smaller drag, or of suppressing the growth of a mixing layer by forcing it into oscillation or irradiating it with sound, was not something that appeared a logical consequence of early ideas about structure.

This is not ununderstandable. Townsend's conclusions about structure were often based on measurements of only the 'principal' correlation functions

$$R_{ii}(r_i\mathbf{e}_i) \quad \text{(no sum on i, i = 1,2,3)}$$

-- i.e. the 'normal' correlations with separation *only* along the corresponding axes (magnitude r_i, unit vector $\mathbf{e}_i$). (In some cases, e.g. Tritton (1967) in a boundary layer, there were supporting cross-correlations like $R_{13}(r_1, 0, r_3)$, r_1 being streamwise separation along the plate and r_3 normal to it.) By comparison with results expected when turbulence is rapidly distorted (the only case of a turbulent flow

Figure 13
(a) Townsend's roller eddies in the cylinder wake, as inferred from Grant's correlation measurements by Payne & Lumley (1967).

(b) A conditionally-averaged flow (at constant phase) in the near-wake behind a circular cylinder at a Reynolds number of 140,000 (from Cantwell & Coles 1983).

(c) Sketch of possible coherent structures in two-dimensional wake (Hussain & Hayakawa 1987).

(a)

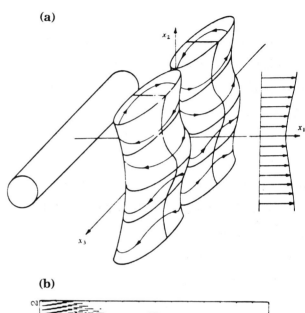

(b)

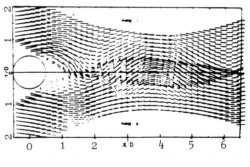

(c)

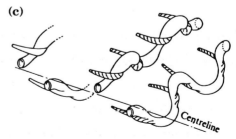

that is not totally unrealistic even with the neglect of nonlinear *and* viscous terms), Townsend was able to sketch the shape of possible structures in the flow. However, as he himself noted (Townsend 1976, p118), "Even with a complete knowledge of the correlation function, the form and distribution of the eddies cannot be calculated without some ambiguity"; and of course that complete knowledge is not available. It is in retrospect remarkable how much insight was obtained by these weak and difficult tools.

Lumley (1970) has shown how the procedure of inferring structure from correlation may be made more objective, by the use of empirically derived orthogonal functions. If higher moments are measured, some information about the phase differences between the structures may also be obtained. However, this information is still incomplete. And the relation to structures derived from conditional sampling is not clear (candidates for the coherent structure in a wake are compared in *Figure 13*, which uses data from Cantwell & Coles 1983, Payne & Lumley 1967, Hussain & Hayakawa 1987; see also Roshko 1976). Thus, the (averaged) 'roller' eddies inferred from correlations as stretching across the whole wake (Figure 13a) may at no instant be connected across the centre line, and may at each instant be linked to each other at the ends: facts that may be of great importance in attempts to manage a wake. Conditional sampling can itself be a treacherous technique, as Lumley (1969, 1970) and Sato (1983) ("in the limit we can find any pattern [we like] in a perfectly random field") have pointed out. However, if there is an independently measured index of the contribution of the structure to some significant quantity such as e.g. the momentum flux (Cantwell & Coles 1983, Hussain 1983, Kailas & Narasimha 1988), the chances of getting misled by conditional statistics are small.

The notion of coherent structures does call into question the appropriateness (and offers a plausible explanation for the ineffectiveness) of a statistical theory based on a generalised harmonic analysis of the field containing waves that extend to infinity in all directions, and on low-order moments that erase phase information. Coherent structures are of finite extent in space and have finite lives (e.g. Cimbala, Nagib & Roshko 1988), and embody extensive information on internal or relative phases. But they too obey their own statistics (e.g. Bernal 1988), and so do not necessarily invalidate *all* statistical approaches. The question is not really about a conflict between one view that is deterministic and another that is statistical at all (this conflict having in any case been resolved by recent developments in the theory of nonlinear dynamical systems, as already discussed) -- it is entirely about the most suitable basis for expansion of the velocity or vorticity field, or (if one wishes to put it that way) about what we earlier called the right 'molecule' of turbulence is (the 'right' molecule being by definition one whose interactions with other similar molecules are simple or weak or easy to exploit). Each such molecule would presumably have a Fourier decomposition itself, with appropriate phase relations among the waves constituting it: it would be a selective sum in the usual harmonic expansion of the field. The fact that the structures do not all have exactly the same shape or size makes matters more difficult: unlike their counterparts in kinetic theory, these turbulence molecules are flabby and mortal. We have however no demonstration yet that any of the coherent structures described to-date in the variety of flows that have been experimentally investigated are fruitful entities for theory, but they have clearly been most fruitful in suggesting new methods of turbulence management.

A different way of posing the problem is to seek the description of turbulence in terms of events rather than waves: the contrast is between 'harmonic' and 'episodic' views of turbulence (Narasimha 1988). Brown & Roshko (1974) have provided an event diagram for mixing layers, and Kailas & Narasimha (1988) have studied 'chronicles' of events in an atmospheric boundary layer (*Figure 14a,b*). The latter diagram in particular points to the feasibility of describing turbulence as a 'point process' (defined as a random collection of point occurrences, in time or space; see Cox & Isham 1980), with events of different magnitude and duration occurring at various instants of time, rather than as a superposition of harmonic waves. An episodic view, embodying the "random" occurrence of coherent events, is completely consistent with a dynamical-systems approach to turbulence, and could be the great reconciliation that is awaiting us in the next decade.

5 End of the road for tradition?

Are there any big successes round the corner using traditional approaches? The answer depends once again on where tradition is supposed to stop.

Take, for example, vortex dynamics. Calculations involving vortex elements have a long history (see e.g. Saffman & Baker 1977, Leonard 1980); in recent decades the use of computers has enabled us to get considerable insight into the structure of many turbulent flows. For example, the calculations of Ashurst (1979) with point vortices show remarkable similarities to the coherent structures in a mixing layer observed by Brown & Roshko.

Point vortex models have their problems: the singularity at the vortex leads to unduly high fluctuations. The situation can be relieved to some extent by smearing the point vortex to make a blob, but this introduces additional ad hoc parameters. A vortex sheet is a logical structure to analyse (see also Brachet, Meiron, Orszag, Nickel, Morf & Frisch 1983), especially in high Reynolds number

Figure 14
(a) "World lines" of coherent structures in mixing layer (Brown & Roshko 1974).

(b) A 'chronicle' of momentum flux events in the atmospheric boundary layer (Kailas & Narasimha 1988) at two different values of averaging time t_{av} and threshold k in a VITA procedure. Each mark indicates the occurrence of an event, the bolder marks signifying a super-event that occurs from a merger of sub-events at a lower value of t_{av} (circled). Also shown are 'correlated' ordinary events marked by arrows, and 'uncorrelated' ordinary events in dashed lines.

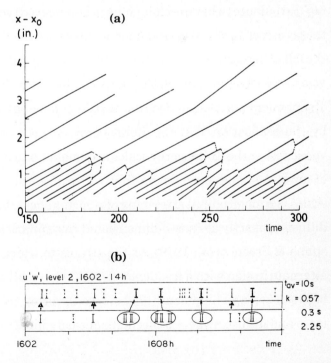

flows (-- this may be as good a place as any to say how odd it surely is to compute numerical solutions of the *full* Navier-Stokes equations to solve high-Re turbulent flows -- one would think it more logical to start with the nearly-inviscid limit and 'smear in' viscosity). The trouble with a vortex sheet however is that after a finite time it develops a singularity, in the form of a cusp in the vorticity distribution along the arc (Moore 1979). It is possible to argue however that this singularity is smeared out by *longitudinal* viscous diffusion -- acting *along* the sheet -- leaving the rest of the sheet to behave pretty much as it would without the singularity or viscosity (cf. shock waves). On this premise a vortex sheet element technique can be devised to simulate the flow (Basu, Narasimha & Prabhu 1989).

The Helmholtz decomposition shows that any velocity field can be split into a potential and a rotational component. If the latter can be handled satisfactorily the former is no problem, and an effective computational tool may be possible.

Kuechemann's characterisation of vortices as the "sinews and muscles of fluid motion" may be even truer for turbulence. It would seem that "understanding" turbulence (whenever it is going to be achieved) will proceed by breaking down the processes involved into simpler ones whose dynamics can be grasped. Consider the following scenario (Narasimha 1987b). A large eddy of size L and characteristic velocity U (say r.m.s. fluctuation) develops boundary layers of size $\lambda = (\upsilon L/U)^{1/2}$ (the Taylor micro-scale). How does the next step in the cascade occur? A clue is provided by the realisation that the Kolmogorov *vorticity* is identical with the Taylor vorticity, but the Kolmogorov scales are smaller and have a local Reynolds number of unity. A simple mechanism that will transform Taylor scales to Kolmogorov scales is a longitudinal stretching of a vortex sheet, limited eventually by viscous diffusion, exactly as in two-dimensional relaminarisation by acceleration (Narasimha & Sreenivasan 1979). As the strains to which a vortex sheet is subjected alternate in sign with a frequency of $O(U/\lambda)$ (-- recall that λ is the average spacing between zeros as Liepmann 1952 showed), vortex sheets will alternately concentrate and diffuse away -- a mechanism easily illustrated (Govindarajan & Narasimha 1988) using a solution of Leonard (1987) for the strained vortex sheet. If these vortex sheets wind in a spiral they can lead to a $k^{-5/3}$ spectrum (Moffatt 1984) -- and if the Reynolds numbers are not very large one won't be able to tell what may in actual fact be a scroll from a rod or filament (viscosity will smear the former into the latter), making it difficult to distinguish between them experimentally, and accounting for the conflicting reports in the literature (e.g. Kuo & Corrsin 1972, Badri Narayanan, Rajagopalan & Narasimha 1977, Saffman 1968).

Is this a traditional approach or a modern one?

It is difficult however to see a future for classical spectral dynamics. The Direct Interaction Approximation and the many improvements that have been suggested may seem a possibility, but, apart from the difficulties already mentioned, at the present time in this commercial age the price-to-earnings ratio (intellectual as well as financial) for this formidable approach seems too high. As for the Reynolds-averaged equations, it is already clear that none of the fresh insights into the turbulence problem that have come during the last two decades (coherent structures, connections with dynamical chaos, new methods of turbulence management) can be attributed to (or could have been derived from) an approach preoccupied with means. We do appear to be close to the end of these roads.

The scientific initiative has since the 1950s remained with experiment, because what is being observed is way ahead of what can be predicted; at low Reynolds numbers it may now be passing to the computer, as the cost of making a calculation continues to fall steeply. But neither experiment nor computation is theory. When (if?) theoretical understanding comes, it will probably use some traditional ideas at least -- perhaps from vortex dynamics. In any case, if traditional approaches have failed, it is not that modern approaches have succeeded -- at least not yet. There is still quite some way to go towards the grand reconciliation I mentioned earlier; the end of the tunnel is not yet in sight.

Notes:

/1/ In some engineering literature the term "Navier-Stokes equations" is often used when only the Reynolds-averaged equations are meant. We always distinguish between the two here, so that "solutions" of the Navier-Stokes equations do not have to be qualified by adjectives like 'true', 'full', 'direct' etc.

/2/ If the probability density of a derivative ξ_1 (spatial or temporal) of a fluctuating quantity (velocity, concentration etc) is $f(\xi_1)$, that of its square $\varepsilon_1 = \xi_1^2$ is $[f(+\sqrt{\varepsilon_1}) + f(-\sqrt{\varepsilon_1})] / 2\,\varepsilon_1^{1/2}$, which is singular at $\varepsilon_1 = 0$ as $f(0) \neq 0$ (see e.g. the measurements of Van Atta & Chen 1970, Figure 6). Now a log-normal distribution for ε_1 does not have this singularity at zero, and hence cannot be correct for low values of any squared derivative. This is also confirmed by the recent data of Dowling (1988) on concentration derivatives in a jet. However, the viscous dissipation of energy (say ε) is a sum of nine squared derivatives of the velocity components in different directions, and of additional cross products; so even if one derivative is zero the others need not be. Rogers & Moin (1987) have undertaken an examination of their numerical solution which shows that while the distribution of ε_1 is not log-normal, that of ε tends to be so (Figure 7). This argument suggests clearly that a scalar dissipation (the sum of only three squared derivatives) may tend less convincingly to log-normality than the energy dissipation.

It would thus appear that the nearly log-normal distribution of the dissipation is not just a consequence of large eddies being crushed into smaller ones like lumps of coal; the fact that it is the sum of a large number of essentially positive terms is also important.

While the views expressed here are my own, I acknowledge with gratitude the influence of work done over many years with colleagues in Bangalore and at Caltech. In particular, I have pleasure in dedicating this paper to Hans Wolfgang Liepmann on the occasion of his 75th birthday, asking him to accept this inadequate return for the privilege I have enjoyed for over thirty years of a free trade in fact and fancy, theory and speculation, insight and prejudice -- and not always only on turbulence!

The preparation of this paper was supported in part by a grant from the Department of Science and Technology. I am grateful to Mr K Hanumantha, Dr S Bhogle and Mr A S Rajasekar for their valuable assistance in putting the manuscript together.

REFERENCES

N Afzal, R Narasimha 1976 *J Fluid Mech* , **74**:113-128

N Afzal, R Narasimha 1985 *AIAA J* , **23**:963-965

W T Ashurst 1979 In: *Turbulent Shear Flows* , (Ed. Durst et al) **1**:402-413

N Aubry, P Holmes, J L Lumley, E Stone 1988 *J Fluid Mech* , **192**:115-173

M A Badri Narayanan, S Rajagopalan, R Narasimha 1977 *J Fluid Mech* , **80**:237-257

A Basu, R Narasimha, A Prabhu 1989 IISc Report 89 FM 1

G K Batchelor 1950 *J Aero Sci* , **17**:441-445

G K Batchelor 1953 *The theory of homogeneous turbulence* , Cambridge Univ Press

G K Batchelor, I Proudman 1954 *Qly J Mech Appl Math* , **7**:83-103

L P Bernal 1988 *Phys Fluids* , **31**:2533-2543

G S Bhat, R Narasimha, S Wiggins 1989, NAL Report PD DU 8901

P L Bhatnagar, E P Gross, M Krook 1954 *Phys Rev* , **94**:511-525

R W Bilger 1989 *Ann Rev Fluid Mech* , **21**:101-135

G Birkhoff 1960 *Hydrodynamics* , Princeton Univ Press

D G Bogard, W G Tiederman 1986 *J Fluid Mech* , **162**:389-413

M E Brachet, D I Meiron, S A Orszag, B G Nickel, R H Morf, U Frisch 1983 *J Fluid Mech* , **130**:411-452

P Bradshaw, D H Ferris, N P Atwell 1967 *J Fluid Mech* , **28**:593-616

J E Broadwell, R E Breidenthal 1982 *J Fluid Mech* , **125**:397-410

J E Broadwell 1987 In: *Turbulent Reactive Flows* , (Ed. Borghi, Murthy) 257-277, Springer-Verlag

G L Brown, A Roshko 1974 *J Fluid Mech* , **64**:775-816

J M Burgers 1948 *Adv Appl Mech* , **1**:171-199

B J Cantwell, D E Coles 1983 *J Fluid Mech* , **136**:321-374

J H Cimbala, H M Nagib, A Roshko 1988 *J Fluid Mech* , **190**:265-298

D E Coles, E A Hirst 1969 (Ed.) *Computation of turbulent boundary layers* , Proc 1968 AFOSR-IFP Stanford Conf II, Stanford Univ

G M Corcos, F S Sherman 1984 *J Fluid Mech* , **139**:29-65

T C Corke, Y G Guezennec, H M Nagib 1980 *Prog Astro Aero* , **72**:128-143

S Corrsin 1974 *Adv Geophys* , **18A**:25-71

D R Cox, V Isham 1980 *Point processes* , Chapman & Hall

J Dey, R Narasimha 1988 IISc Report 88 FM 7

D R Dowling 1988 Mixing in gas-phase turbulent jets, GALCIT Ph.D. thesis

H W Emmons 1951 *J Aero Sci* , **18**:490-498

W K George 1988 In: *Advances in Turbulence* , (Ed. W K George, R Arendt) 39-73; Hemisphere Publishing Corp

S Goldstein 1933 *Proc Roy Soc* , **A142**:545-562

R Govindarajan, R Narasimha 1988 NAL Report PD DU 8803

R Govindarajan, R Narasimha 1989 NAL Report TM DU 8901

H L Grant, R W Stewart, A Moilliet 1962 *J Fluid Mech* , **12**:241-268

P Holmes 1989, *this volume*

A K M F Hussain 1983 *Phys Fluids* , **26**:2816-2850

A K M F Hussain, M Hayakawa 1987 *J Fluid Mech* , **180**:193-229

J G Jones 1980 *Proc Ind Acad Sci (Engg Sci)* , **C3**:1-30

B A Kader, A M Yaglom 1978 *J Fluid Mech* , **89**:305-342

S V Kailas, R Narasimha 1988 The structure of turbulence in a neutrally stable atmospheric boundary layer, Report 88 FM 6, Indian Institute of Science

A R Kerstein 1988 *Comb Sci Tech* , **60**:391-421

A R Kerstein 1989 *Comb Flame* , **75**:397-413

P S Klebanoff, K D Tidstrom, L M Sargent 1962 *J Fluid Mech* , **12**:1-34

S J Kleis, A K M F Hussain 1979 *Bull Amer Phys Soc* , **24**:1132

S J Kline, W C Reynolds, F A Schraub, P W Runstadler 1967 *J Fluid Mech* , **30**:741-773

S J Kline, M V Morkovin, G Sovran, D J Cockrell 1969 (Ed.) *Computation of turbulent boundary layers* , Stanford Univ

S J Kline, B J Cantwell, G M Lilley 1981 AFOSR-HTTM-Stanford Conference on Turbulent Flows, Stanford Univ

A N Kolmogorov 1941 *Doklady AN SSSR* , **30**:299-303, **32**:19-21

A N Kolmogorov 1962 *J Fluid Mech* , **13**:82-85

R H Kraichnan 1959 *J Fluid Mech* , **5**:497-543

R H Kraichnan, J R Herring 1978 *J Fluid Mech* , **88**:355-367

A S Kuo, S Corrsin 1972 *J Fluid Mech* , **56**:447-479

B E Launder 1989, *this volume*

M J Lee, W C Reynolds 1985 Dept of Mech Engg Report TF-24, Stanford Univ, Stanford, CA

A Leonard 1980 *J Comp Phys* , **37**:289-335

A Leonard 1987, *private communication*

M Lesieur 1987 *Turbulence in Fluids* , Martinus Nijhoff

D C Leslie 1973 *Developments in the theory of turbulence* , Clarendon Press, Oxford

A Libchaber 1988 In: *Perspectives in Fluid Mech* (Ed. D E Coles). Lecutre Notes in Physics 320, Springer-Verlag

R H Liebeck 1978 *J Aircraft* , **15**:547-61

H W Liepmann 1952 *ZaMP* , **3**:321-426

H W Liepmann 1979 *Amer Sci* , **67**:221-228

H W Liepmann, R Narasimha 1987 (Ed.) *Turbulence Management and Relaminarisation* , Proc IUTAM Symp, Bangalore, Springer-Verlag

M J Lighthill 1956 In: *Surveys in Mechanics* , (Ed. Batchelor & Davies), Cambridge Univ Press

P B S Lissaman 1983 *Ann Rev Fluid Mech* , **15**:223-39

E N Lorenz 1963 *J Atmos Sci* , **20**:130-141

P R Louchez, J G Kawall, J F Keffer 1985 In: *Turbulent Shear Flows* , (Ed. Durst et al) **5**:99-109, Springer-Verlag

J L Lumley 1969 *Ann Rev Fluid Mech* , **1**:367

J L Lumley 1970 *Stochastic tools in turbulence* , Acad Press, New York

J L Lumley 1981 In: *Transition and Turbulence* , (Ed. R E Meyer) 215-242, Acad Press, New York

V N Lykossov 1989, *private communication*

G Mellor 1972 *Int J Engg Sci* , **10**:851-873

R E Melnik, A Rubel 1983 Proc 2nd Symp Numerical & Physical Aspects of Aerodynamic Flows, Long Beach, CA

C B Millikan 1939 Proc 5th Int Cong Appl Mech, 396-

M Millionshtchikov 1939 *Doklady AN SSSR* , **22**:236-240

M A Miner 1945 *J Appl Mech* , **12**:159-164

H K Moffat 1984 In: *Turbulent and chaotic behaviour in fluids* , (Ed. T Tatsumi) 223-230, Proc IUTAM Symp, Kyoto, North-Holland

A S Monin, A M Yaglom 1975 *Statistical hydrodynamics* , MIT Press, Cambridge

D W Moore 1979 *Proc Roy Soc* , **A365**:105-119

M V Morkovin et al 1987 *Amer Sci* , **75**:119

T Nakano 1988 *Phys Fluids* , **31**:1420-1430

R Narasimha, A Prabhu 1972 *J Fluid Mech* , **54**:1-17

R Narasimha, K R Sreenivasan 1979 *Adv Appl Mech* , **19**:221-309

R Narasimha 1983 *J IISc* , **64A**:1-59

R Narasimha 1984 *The Turbulence Problem* , GALCIT Report FM 84-01

R Narasimha 1985 *Prog Aero Sci* , **22**:29-80

R Narasimha 1987a *Curr Sci* , **56**:629-645

R Narasimha 1987b Preliminary notes on turbulent mixing, Report 87 FM 9, Indian Institute of Science

R Narasimha 1988 NAL Proj Doc PD DU 8808

R Narasimha, G S Bhat 1988 *Curr Sci* , **57**:697-702

A M Oboukhov 1962 *J Fluid Mech* , **13**:77-81

Y Ogura 1962 *J Fluid Mech* , **16**:33-40

S A Orszag 1970 *J Fluid Mech* , **41**:363-386

D Oster, I Wygnanski 1982 *J Fluid Mech* , **123**:91-130

S P Parthasarathy 1964 IISc Associateship thesis

F R Payne, J L Lumley 1967 *Phys Fluids* , **10**:S194-S196

A Prabhu, R Narasimha 1972 *J Fluid Mech* , **54**:19-38

A Prabhu, P Kailas Nath, R S Kulkarni, R Narasimha, 1987 In: Liepmann & Narasimha 1987:97-107

A K Rajagopal, E C G Sudarshan 1974 *Phys Rev* , **A10**:1852-1857

O Reynolds 1883 *Phil Trans Roy Soc* , **174**: 935-982

F A Roberts, A Roshko 1985 *AIAA Paper 85-0570*

M Rogers & P Moin 1987, *private communication*

A Roshko 1955 *J Aero Sci* , **22**:124-132

A Roshko 1974 *J Fluid Mech* , **64**:775-816

A Roshko 1976 *AIAA J* , **14**:1349-1357

D Ruelle, F Takens 1971 *Commun Math Phys* , **20**:167

P G Saffman 1968 In: *Topics in Nonlinear Physics* , (Ed. Zabusky), Springer-Verlag

P G Saffman, G R Baker 1979 *Ann Rev Fluid Mech* , **11**:95-122

A Sahlin, A V Johansson, P H Alfredsson 1988 *Phys Fluids* , **31**:2814-2820

H Sato 1983 *Cognition and description of patterns in turbulent flows* , Proc 2nd Asian Cong Fluid Mech, Science Press, Beijing

S D Sharma 1987 *Phys Fluids* , **30**:357-363

E A Spiegel 1985 In: *Theoretical Approaches to Turbulence* , (Ed. D L Dwoyer, M Y Hussaini, R G Vogt) 303-336, Springer-Verlag

K R Sreenivasan, S Tavoularis, S Corrsin 1981 In: *Turbulent Shear Flows 3* 96-112, (Ed. L O S Bradbury et al), Springer-Verlag

K R Sreenivasan, R Narasimha 1982 *J Fluids Engg* , **104**:167-170

K R Sreenivasan, R Narasimha, A Prabhu 1983 *J Fluid Mech* , **137**:251-272

J L Synge, C C Lin 1943 *Trans Roy Soc Canada* , **37**:45

G I Taylor 1921 *Proc London Math Soc* , **A20**:196-212

G I Taylor 1938 *Proc Roy Soc* , **A164**:446-468

A A Townsend 1956 *The structure of turbulent shear flow* , Cambridge Univ Press

A A Townsend 1960 *J Fluid Mech* , **8**:143-155

A A Townsend 1976 *The structure of turbulent shear flow (2nd edn)* , Cambridge Univ Press

D J Tritton 1967 *J Fluid Mech* , **28**:439-462

C A Van Atta, W Y Chen 1970 *J Fluid Mech* , **44**:145-159

T von Karman, L Howarth 1938 *Proc Roy Soc* , **A164**:192-

P R Viswanath, T Srinivas, A Prabhu, R Rajasekar 1989 NAL TM, *to be published*

I J Wygnanski 1967 *J Fluid Mech* , **27**:431-443

I J Wygnanski, F H Champagne, B Murasli 1986 *J Fluid Mech* , **168**:31-71

J C Wyngaard, H Tennekes 1979 *Phys Fluids* , **13**:1962-69

K S Yajnik 1970 *J Fluid Mech* , **42**:411-427

K S Yajnik, S Sundaram, R Sivaram 1987 In: Liepmann & Narasimha 1987:63-68

K Yamamoto, I Hosokawa 1988 *J Phys Soc Japan* , **57**:1532-1535

The Utility and Drawbacks of Traditional Approaches.

Comment 1.

J. Lumley

Sibley School of Mechanical
  and Aerospace Engineering
Cornell University
Ithaca, NY 14853 USA

Professor Narasimha has produced a very civilized (a kinder and gentler?) paper. In a moment I will give my attention to some of the specific points he raises. First, however, I would like to make some general remarks about recent events that affect us deeply.

During the past several decades part of the turbulence community has advocated a turning away from the approaches with which we were familiar. This turning away is, at least partly, documented in Liepmann (1979).

We are all very close to this situation. Turbulence is, after all, what we do for a living, and it is hard to be objective about events that affect our livelihood. I will try here to describe this situation from as much perspective as I can gain. We have a situation here in which a faction in the community perceives that a condition exists in our field, and decides to take action to change the condition, while other factions of the community react to the action. Phrased this way, it sounds very benign. The perceived condition was a kind of stasis. Liepmann (1979) sketches the process, but does not discuss the mechanism: new ideas are suggested relatively infrequently in a field; as each new idea is suggested, all its implications are explored and it is worked (like a vein of ore) for everything it has to contribute to the field. As time goes on, the return from the idea falls off, because the simpler and more direct consequences have been explored. The field is more-or-less in stasis. The idea is no longer an efficient producer. It is time for a new idea.

The perceived condition in the turbulence field was a stasis induced by too long and too great a dependence on the statistical approach to turbulence. Probably, in fact, the progress possible using a statistical approach, without the introduction of a new idea, had fallen to a very low level. It is difficult to say at this point whether the new idea which came to the fore was introduced intentionally in an effort to move the field out of its stasis, or whether it occurred spontaneously.

The existence of coherent structures in turbulent flows emerged as the new idea, and it was suggested that statistical approaches were not only no longer productive, but were undesirable; that they obscured the physics and made certain questions unapproachable. Attempting to turn the turbulence community away from traditional statistical approaches is equivalent to changing the minds of a large group of people. That is politics. As we know from experience in other areas, swaying a constituency requires clear, simple, perhaps even simplistic, statements, that can be readily grasped. The constituency here is made up of highly trained scientists, but we are as human as any other electorate, we listen as poorly, and are as prejudiced. Many simplistic statements were made in an effort to move the turbulence community away from traditional statistical approaches. Elsewhere (Lumley, 1989) I have commented on some of these statements. What we have here is an unstable situation; situations of this kind tend to polarize, and the protagonists are driven to extreme positions. It is difficult to maintain advocacy of a moderate policy; it is much easier to advocate a clean sweep. This scenario is a familiar one.

I would like to quote a few sentences from *The Structure of Scientific Revolutions*, by Thomas S. Kuhn (1970). Kuhn is drawing what he calls "...[a] genetic...parallel between political and scientific development...". He begins by describing political revolution, and then explores the parallel at length. I have taken the liberty of substituting "scientific" for "political" in the original, and have made a few other similar changes. "[Scientific] revolutions aim to change [scientific] institutions in ways that those institutions themselves prohibit. Their success therefore necessitates the partial relinquishment of one set of institutions in favor of another, and in the interim [the scientific community] is not fully governed by institutions at all. ...In increasing numbers individuals become increasingly estranged from [scientific] life and behave more and more eccentrically within it. Then, as the crisis deepens, many of these individuals commit themselves to some concrete proposal for the reconstruction of society in a new institutional framework. At that point the [scientific community] is divided into competing camps or parties, one seeking to defend the old institutional constellation, the other seeking to institute some new one. Once that polarization has occurred, *[scientific] recourse fails*. Because they differ about the institutional matrix within which [scientific] change is to be achieved and evaluated, because they acknowledge no supra-institutional framework for the adjudication of revolutionary difference, the parties to a revolutionary conflict must finally resort to the techniques of mass persuasion, often including force...."

In the present case, the existence of coherent structures in turbulent flows was not a new idea. J. T. C. Liu (1988) has documented the first appearance of this idea sometime around the outbreak of the second world war. It was probably first articulated by Liepmann, 1952, and was thoroughly exploited by Townsend, 1956, but it did not require the abandoning of traditional statistical approaches. Another factor was probably involved in the move away from statistical approaches, which I have mentioned elsewhere (Lumley 1989). The scientific mind is not comfortable with ambiguity, and a statistical approach is almost an institutionalization of ambiguity. Many scientists are not comfortable with statistical approaches, and have never liked them. The proposal of an alternate approach immediately drew strong support from this group.

Whether or not the idea was new, perhaps it was "an idea whose time had come", that is, relatively neglected when first proposed because it did not fit in with the prevailing view, but now ripe for exploitation because the times had changed. The idea was thirty years old, but had not made much of a splash when first proposed. Exploitation had begun, and the threads of that early exploitation are present in some of the more interesting current work (Aubry *et al,* 1988). From a graph - figure 1 - of the probability of citation (in the Aerospace Database) of a paper containing "turbulence" or "turbulent" in combination with "structure", "eddy" or "organized motion" (including plurals), we see that there was an enormous surge of interest in this area up out of a constant noise at about 1970, roughly the time of first appearance at meetings of the work of Brown and Roshko (appearing in print in 1971). By the end of the decade the probability of citation had increased an order of magnitude.

The question of whether it was necessary or desirable to abandon traditional statistical approaches remains. The presence of coherent structures in a flow is not in any sense incompatible with a classical statistical approach. In fact, one classical statistical approach was devised to meet exactly this need. The coherent structures that appear in turbulent flows are statistical entities themselves, and their size, strength, phase and even orientation are subject to fluctuations. Different turbulent flows have different proportions of organized and disorganized motion (something which is difficult to define precisely), and as the level of disorganization increases, the utility of the statistical approach increases. While there are surely things to be gained by considering the dynamics of individual coherent structures, or collections of coherent structures, without averaging of any sort (see the recent work of Aubry *etal*, 1988), there are equally interesting things that can be done with a statistical approach.

It is often claimed that the presence of coherent structures *must* be recognized in a statistical approach, as an aftermath of the political controversy. It is felt that we need a statistical approach, perhaps a modeling approach, that acknowledges explicitly the presence of coherent structures; that only in this way will we achieve results that more clearly resemble reality. However, it is demonstrable (Shih *et al* 1987) that even some flows containing dominant coherent structures can be satisfactorily modeled by statistical approaches that take no account of the presence of coherent structures. The fact seems to be, that if the structures scale in the same way as the rest of the flow, they can be lumped in with the rest of the disorganized motions, and the prediction will be satisfactory. After all, when we are thinking of a second order statistical model, we are considering only the crudest statistical measure, that takes into account very little of the physical properties of the motions present in the flow. At the level of second order statistics, it is irrelevant whether the motions are disorganized or organized. Structural questions really are not much reflected until third and higher order. Transport (in a second moment equation) is a third order statistical quantity, and we might expect transport to be affected by the presence or absence of coherent structures in the flow.

Enough of historical and philosophical comment. Let me turn now to Professor Narasimha's paper. There is not a great deal that I disagree with, although I have some quibbles that I will bring up later. I am going to bypass several of the areas he mentioned, and concentrate my remarks in one area. Moreover, since he chose not to mention some of the more extreme statements that have been made, I would like to examine them for the record.

Let me pick first on the question of whether statistics hides physics. Professor Narasimha's charming example certainly illustrates how statistics can hide physics. If the proponents of this point of view claimed only that statistical approaches improperly used could obscure the physics, and that in any given situation a statistical approach might not be the most useful one, I would heartily endorse the position. However, this is more than an academic, semantic, quibble. One hears statements like "Reynolds averaging never got us anywhere" from respected people on national platforms, and at least one Department of Defense Agency takes the position that it will no longer support work starting from Reynolds averaging. Thus, the situation is more serious. There is an implication in the statements of the opponents that Reynolds averaging *inevitably* hides the physics; that no physical progress can be made if one starts with Reynolds averaging, and it is this that has been picked up by the funding agencies, among others.

I certainly do not wish to be in the position of pushing Reynolds averaging to the exclusion of other techniques which may, in a given situation be more suitable. I even think it is quite appropriate that we give much of our attention now to exploring other techniques, without abandoning statistical approaches. The problem is that the antagonists of Reynolds averaging make statements that are not true in trying to make their point. These are the extreme statements made in the heat of a political debate. I understand the point they are trying to make, but when I dispute the factual truth of their statements, they suggest that I have missed their point. I have not, but I fail to see why I should help them make it.

So far as whether averaging *inevitably* hides the physics, we can point to a number of representation theorems. For example, the proper orthogonal decomposition theorem (Loève, 1955) says that the *instantaneous* realization of a random function can be represented as a series of deterministic functions (obtained in a certain way) with random coefficients, with convergence defined in a certain way. We may also mention Fourier-Stieltjes integrals (Batchelor, 1953), which permit a similar thing. We use such representations constantly (the usual non-statistical Fourier representation, for example) and never question whether the use of such representations obscures the physics, whether there is any part of the function so-represented that is somehow left out by the representation. In the case of the statistical representation, some sophisticated questions can be asked about the definition of convergence, but generally speaking there is not much doubt that the instantaneous form of the individual realization can be reproduced. We use this concept every time we use Fourier analysis on realizations of a random process on the computer. We know that we can synthesize the realization of the random process from the computed Fourier coefficients and the trigonometric functions.

If that is so, then we must believe that all information about the evolution of the realization is contained somewhere in the statistical treatment. Hence, the statistical treatment obscures nothing *a priori*. Now, there are other, less subtle questions: does a statistical treatment that stops at nth order (second order, for example) obscure the physics? Of course, just as a Fourier representation that omits all Fourier modes above the second leaves out much of interest in the function. I have shown elsewhere (Lumley, 1981) that it is possible to recapture phase information from third order statistics, for example, and other information from fourth order statistics. Professor Narasimha has referred to this in such a way as to suggest that it is still not possible to recapture all information from the statistics. I hope I have convinced you that this is not the case.

Stan Corrsin (private communication) tried very hard to get across the point that we must stop concentarting on the second order statistics, and measure and think about higher order statistics, which contain the information missing from the second order statistics. He never made much progress with the majority of the community, because the higher order statistics are unfamiliar, do not have immediately apparent physical significance, and are hard to measure. These are all problems with which I can have great sympathy, as long as they do not lead one to take untenable positions. With regard to the last aspect, the difficulty of measuring, statistical quantities are difficult to measure if the events contributing to them occur rarely. It is dangerous to draw conclusions about any event that occurs rarely without some statistical measure of its importance in the overall ensemble.

Turbulence is almost by definition disordered, or has the appearance of disorder. It does not matter whether the system is deterministic but appears disordered due to non-linearity, or is "really" disordered, whatever that means. This is reminiscent of controversies in quantum mechanics. A statistical approach is a way of dealing with real or apparent disorder, of lumping together things that are not understood, or with which one does not want to deal. Even very ordered turbulent flows, having much apparent structure, retain a certain disorder in the phase of occurrence, size and intensity of the coherent structures. It is natural to use a statistical approach, something like the shot-effect expansion, to represent such a flow.

Guckenheimer (1986) has suggested that deterministic dynamical systems having a strange attractor, if they have a high Lyapunov dimension, will be too complex to compute, and will probably best be approached statistically. In fact, it seems natural to use a statistical approach to describing the structure of strange attractors of high dimension, since one is certainly not interested in the details, but in global statements regarding mean recurrence times, bounds and so forth. This may be one of the most fruitful avenues for near term research (Ristorcelli, private communication).

Now I would like to take exception to some statements made by Professor Narasimha. He states (section 4.3) that current models ignore the existence of counter-gradient flux. This is not the case. Since the buoyancy-driven atmospheric surface layer is run almost entirely by counter-gradient flux, a great deal of effort went into the development of a rational model displaying this phenomenon. A total of forty-nine papers were published in the open literature by members of my group between 1970 and 1987 (references can be found in Lumley (1983)) describing the development of this model. The relation between the third and second moments in this model is obtained almost entirely from first principles; that is, it might better be described as an approximation, than as a model. It predicts the values of the measured third moments within experimental error. It is not an "elaborate super-structure" "ignoring such simple counter-examples".

Professor Narasimha also referrs to the quasi-normal theories, pointing out that they lead to negative energies. This is something that we have all known for nearly thirty years. Much more to the point is the demonstration by Orszag (1970) that the *eddy-damped quasi-normal* assumption was relatively safe in this regard. Using the eddy-damping concept, Lumley (1978) showed that it was possible to develop a perturbation analysis about the homogeneous state, leading in the case of weak inhomogeneity to cumulant discard as a conclusion, rather than an assumption.

Finally, Professor Narasimha states that the turbulence models have not predicted anything. That, of course, is true. However, I believe it is foolhardy to expect them to. These models are simply an embodyment of experience; they are something constructed to behave like turbulence, in situations where it has been observed, to be used as a design tool. A model cannot, except by accident, contain more than is put into it. One tries, of course, to make models as universal as possible, building into them behavior designed to avoid violation of as many commandments as one can manage. In this sense, a good model can be reasonably forgiving, continuing to produce reasonable looking results far beyond the parameter range for which it was constructed. That does not mean the results are right. If, in this parameter range, a physical phenomenon is important that was not built into the model, the results will be wrong. You should never expect a model to predict something you did not forsee. Use it to get a better numerical value for something you can already estimate on the back of an envelope.

To sum up: statistical approaches are not wrong, or a dead end; they simply have their limitations, like all other approaches, and must be used judiciously. Some of their bad press has been politically motivated. It is not possible to make rational blanket statements such as " don't ever use statistics on turbulence" or "we will support no further such work". At the same time, it must be said that we will probably not make much progress by continuing to apply the old techniques in the old ways. We must have either new techniques or new ways. If statistical approaches are to be used, they must be used creatively. But that was always true.

As for modeling, it is both better than Professor Narasimha knows, and worse than he hopes.

Acknowledgments

I would like to thank Jane Lumley for helpful discussions on the history and philosophy of science, and Sid Leibovich and Zellman Warhaft for generous moral support. The help of John Saylor of the Cornell Engineering Library in preparing the figure is gratefully acknowledged.

References

Aubry, N., Holmes, P., Lumley, J.L. and Stone, E. 1987. The dynamics of coherent structures in the wall region of a turbulent boundary layer. *J. Fluid Mech.* 192: 115-173.

Batchelor, G. K. (1953) *The Theory of Homogeneous Turbulence.* Cambridge UK: The University Press.

Brown, G. & Roshko, A. (1971) The effect of density difference on the turbulent mixing layer. In *A.G.A.R.D. Conference on Turbulent Shear Flows.* Conf. Proc. No. 93, pp. 23/1-23/12.

Guckenheimer, J. (1986) Strange attractors in fluids: another view. *Ann/. Rev. Fluid Mech.* 18:15-31.

Kuhn, T. S. (1970) *The Structure of Scientific Revolutions.* International Encyyclopedia of Unified Science, Volume 2, No. 2. Chicago: The University of Chicago Press.

Liepmann, H. W. (1952) Aspects of the turbulence problem. Part II. *Z. Angew. Math. Phys.* 3: 407-426.

Liepmann, H. W. (1979) The rise and fall of ideas in turbulence. *The American Scientist* 67: 221-228.

Liu, J. T. C. (1988) Contributions to the understanding of large scale coherent structures in developing free turbulent shear flows. In *Advances in Applied Mechanics,* 26: 183-309.

Loève, M. (1955) *Probability Theory.* New York: Van Nostrand.

Lumley, J. L. 1978. Computational modeling of turbulent flows. In *Advances in Applied Mechanics* 18, ed. C. S. Yih, pp. 123-176. New York: Academic Press.

Lumley, J. L. 1981. Coherent structures in turbulence. In *Transition and Turbulence,* ed. R. Meyer, pp. 215-242. New York: Academic.

Lumley, J. L. 1983. Atmospheric Modeling. The Institution of Engineers, Australia: *Mechancial Engineering Transactions.* ME8: 153-159.

Lumley, J. L. 1989 The state of turbulence research. In *Advances in Turbulence,* eds. W. K. George and R. Arndt, pp. 1-10. Washington DC: Hemisphere.

Orszag, S. A. (1970) Analytical theories of turbulence. *J. Fluid Mech.* 41: 363-386.

Shih, T.-H., Lumley, J. L. & Janicka, J. 1987. *Second order modeling of a variable density mixing layer. J. Fluid Mech.* 180: 93-116.

Townsend, A. A. (1956) *The Structure of Turbulent Shear Flow.* Cambridge, UK: The University Press.

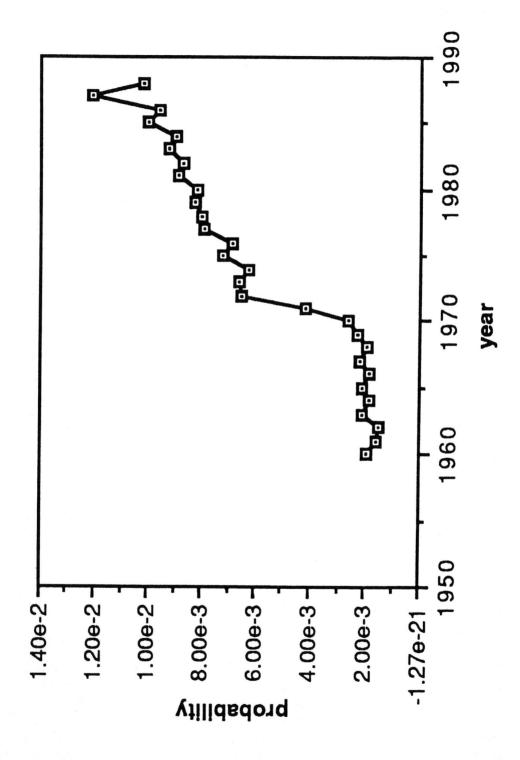

The Utility and Drawbacks of Traditional Approaches.

Comment 2.

M. Lesieur

Institut de Mécanique de Grenoble
BP 53 X
38041 Grenoble-Cedex
France

1. Introduction

Before starting this discussion, I would like to stress that I do not feel much inclined to fight about what is the best approach to turbulence, since I do not know of any which is really satisfactory.

The first question to be asked is of course: how can we define the *traditional approaches*? Is it Taylor's diffusion theory or Kolmogorov kinetic energy cascade, mentioned by *Narasimha*'s [1] position paper? Or Prandtl's mixing-length theory? Certainly, the one-point closure modelling theories are traditional also. *Narasimha* [1] agrees that it is also true for Millionshtchikov's 1941 Quasi-Normal theory and for two-point closures such as Kraichnan's DIA or Orszag's Eddy-Damped Quasi-Normal Markovian theory (EDQNM). I think the discussion is somewhat biased in trying to assess globally all the various statistical modelling approaches of turbulent flows. You can compare and evaluate only things which are comparable: in the same way as you do not expect a bus to fly, you do not ask your favourite one-point closure model to predict a two-point quantity such as a $k^{-5/3}$ Kolmogorov kinetic energy cascade. Conversely, you do not ask DIA or EDQNM to predict how much petrol you will save by putting riblets on your bus.

2. Two-point closures

I will spend some time discussing the possibilities and limits of a traditional approach I happen to have studied quite extensively, that is, the EDQNM approach. This statistical theory assumes that, in the moments hierarchy for the velocity, the fourth-order cumulants relax linearly the third-order moments, through an eddy-damping which is generally dimensionally chosen in order to be compatible with a $k^{-5/3}$ Kolmogorov cascade in isotropic turbulence (see e.g. [2] for a review). The markovianization procedure ensures the positivity of various spectra, contrary to the Quasi-Normal approximation, as mentioned by *Narasimha* [1]. There is, for the velocity field, only one adjustable constant, which fixes the value of the Kolmogorov constant. At variance with the statement on *the absence of rational models* stressed in [1], I consider the EDQNM as a very nice rational model for isotropic turbulence, forced or unforced. It shows in the latter case how the Kolmogorov cascade builds up at a finite time t_*, when starting initially from a kinetic energy spectrum peaking in the large scales. This time t_*, which is of the order of $5\ D(0)^{-1/2}$ (where $D(0)$ is the initial enstrophy $(1/2) < \vec{\omega}^2 >$), is in good agreement with large-eddy simulations done in [3]. This phenomenon could correspond to a blow up of the enstrophy in the Euler equations, and hence to a singularity. In this case (that is with these particular initial conditions), the closure predicts that the kinetic energy is conserved before t_*, and dissipated at a finite rate ϵ afterwards. This is, why not, a candidate to explain the origin of turbulence, and it provides to Mathematicians tracks to explore the possibility for singularities to appear in the unforced Euler equation: the behaviour of the *skewness factor*, a statistical quantity, is essential within this context.

3. Unpredictability

Another very important quality of these two-point closures is their ability to handle the unpredictability property, which is not only characteristic of chaotic dynamical systems: closures allow to write an evolution equation for the *error spectrum*, that is, the spectral kinetic energy density of the difference between two realizations of the turbulent velocity field. If the error is initially very small, it will grow and eventually contaminate the whole scales of motion, in a finite time in three dimensions as shown in [4] (see also Fig. 1, taken from [5]). To me, *Lorenz*'s 1969 [6] paper about the two-dimensional turbulence predictability (studied with the Quasi-Normal approximation), is may be more important than his more celebrated 1963 one [7], which introduces the so-called Lorenz attractor: indeed, the former involves an infinite number of spatial scales, and allows to show in particular how error in the small scales can contaminate the large scales. This kind of non-local spatial information is of course impossible to obtain from the dynamical systems theory, where you freeze a very small number of spatial degrees of freedom.

I would like also to point out that it is from the statistical results predicting the growth of unpredictability in two-dimensional isotropic turbulence, that we conjectured, for the coherent structures of a two-dimensional plane mixing layer, the same unpredictable character. We verified this conjecture using direct numerical simulations [8]. In the same paper, this result was used to study the growth of three-dimensional instabilities in the following manner: a one-mode

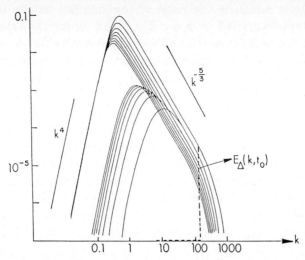

Fig. 1. EDQNM calculation showing the inverse cascade of error from small to large scales in decaying isotropic three-dimensional turbulence. Notice also the appearance of the infrared k^4 and ultraviolet $k^{-5/3}$ ranges in the kinetic energy spectrum (from [5])

spanwise spectral truncation of the three-dimensional flow allowed to identify the average kinetic energy of the spanwise wave to the error energy in the two-dimensional predictability problem. Hence, growth of the 2D error corresponds to a process of three-dimensionalization.

4. Inertial ranges

Closures work pretty well to predict the direction of various velocity or scalar cascades in 3D or 2D turbulence: it is by using these closures that *Kraichnan* [9] could check the validity of his 1967 [10] phenomenological predictions concerning the enstrophy and the inverse energy cascade in two-dimensional turbulence. These closures display also all the various ranges (inertial-convective, inertial-diffusive, viscous convective) predicted on phenomenological grounds for the diffusion of a passive scalar in 3D and 2D turbulence ([11], [12]).

5. The theory of non-local interactions

A very nice possibility given by the closures is to enable to compute analytically *local interactions*, by expansion of the statistical transfers with respect to a small parameter which is the ratio of the smallest to the largest wave number in the interacting triad k, p, q. The result is, for the velocity field in three dimensions, the prediction of a k^4 positive transfer for $k \to 0$. Such a transfer is dominant at low wave numbers for decaying turbulence with a sharply peaked initial spectrum ([13], see also Fig. 1), and is confirmed by direct or large-eddy simulations, such as in [3] (see Fig. 2). This k^4 infrared transfer is not, to my knowledge, predicted by the Renormalization Group techniques (RNG, see e.g. [14], [15], [16]), whose original formalism is unable to handle such a decaying situation, and needs for the turbulence to be sustained by a forcing $F(k) \propto k^r$: for $r = -1$, the resulting stationary kinetic energy spectrum happens to be a $k^{-5/3}$ Kolmogorov spectrum, except that the convergence of the procedure becomes then questionable.

6. Kinetic energy and passive scalar decay

Thus, EDQNM closures give extremely satisfactory results for problems which are controlled by the large-scale dynamics, such as the kinetic energy decay problem in 3D isotropic turbulence at high Reynolds number: for an initially sharply peaked spectrum, the kinetic energy eventually decays like $t^{-1.38}$, as shown in [13]. Large-eddy simulations [17] tend to confirm this result. Notice also, as stressed by *Herring* [18], that this isotropic kinetic energy decay problem admits self-similar solutions which depend upon the infrared exponent s of the initial kinetic energy spectrum $E(k) \propto k^s$ for $k \to 0$ (for $s \le 4$). This is at variance with the principles stated in [1], about self-similar states independant of the initial conditions.

I think all these statistical results had some importance for our present understanding of turbulence, even though they may not all be correct in reality. Hence, it is a little bit excessive to dump them in *a story of failures*, as done in [1]. But, to be honest, some failures exist: I agree that closures do not properly predict structure functions of high order, and are generally inadequate to take into account effects of intermittency such as exponential probability distributions. This is discussed extensively in [18]. Other interrogations which can be raised concern the ability of the closures to model the passive scalar decay problem in 3D isotropic turbulence: indeed, there is a lot of evidence from direct numerical simulations of Navier-Stokes or Euler[1] equations, or large-eddy simulations (see e.g. [3]), that the scalar cascades quicker to high wave numbers

---

[1] When starting initially with kinetic energy or scalar spectra peaked at $k_i < k_m$, where k_m is the maximum wave number permitted by the simulation, an inviscid calculation will be valid during the initial stage of the cascade to higher wave numbers, until spectra reach k_m.

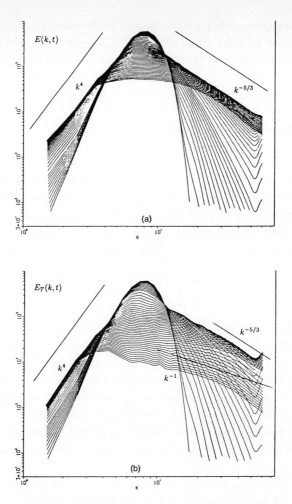

Fig. 2. Spectral large-eddy simulations of decaying 3D isotropic turbulence (128^3 modes): a) kinetic energy spectrum; b) passive scalar spectrum. Both spectra have build up a k^4 range in low wave numbers. The scalar spectrum depletes in the large scales, tending to form a k^{-1} range (from [3])

than the velocity. This might be due to the absence of pressure term in the passive scalar equation (Rogallo, 1988, private communication). A consequence seems to be, at high Reynolds number and molecular Prandtl number of the order of one, the formation of a scalar spectrum close to k^{-1} in the kinetic energy containing range ([3], [17], see Fig. 2 b). We have checked in two independant calculations, done at NASA-Ames and Grenoble, that the scalar spectrum in this range is of the form

$$E_T(k) = 0.1 \, \eta \, \frac{< \vec{u}^2 >}{\epsilon} \, k^{-1} \quad . \tag{1}$$

Another consequence is that the scalar variance decays faster ($\propto t^{-1.85}$) than the kinetic energy, even though the scalar and velocity integral scales are initially equal[2]. Up to now, this behaviour has not been obtained within the closures. The latter seem to fail here, but also the standart phenomenology.

7. Spectral large-eddy simulations

The concept of spectral eddy-viscosity in three-dimensional isotropic turbulence was introduced in [20]. For large-eddy simulation purposes, it can be formulated in the following way (see [2]): if k_c is a cutoff wave number, $-2\nu_t(k|k_c)k^2E(k)$ corresponds to the kinetic energy transfer from k across k_c. The EDQNM theory shows that, for $k \ll k_c$, $\nu_t(k|k_c)$ becomes independant of k and equal to:

$$\nu_t(k|k_c) = \frac{1}{15} \int_{k_c}^{\infty} \theta_{0pp}[5E(p) + p\frac{\partial E}{\partial p}] \, dp \quad , \tag{2}$$

where θ_{kpq} is the EDQNM relaxation time for triple correlations. Here, the positive k^4 transfer is negligible as soon as $E(k) \geq E(k_c)$. Since (2) expresses the action of very small scales $\sim k_c^{-1}$ on structures $\sim k^{-1}$, it is feasible that the same eddy-viscosity will enable to describe the action of subgrid-scale modes $p > k_c$ on $k \ll k_c$ in the momentum equation. Any random backscatter on k would act on the phase, not the energy, of the complex velocity $\hat{u}(\vec{k})$: but such an effect does not propagate instantaneously, and requires a finite time comparable to the time necessary for the unpredictability to cascade from k_c to k.

If k_c lies in a wide $k^{-5/3}$ inertial range, ν_t, calculated using the EDQNM theory, is equal to its asymptotic value $\sim [E(k_c)/k_c]^{1/2}$ up to $k = k_c/3$. In the neighbourhood of k_c, it displays a strong rising cusp due to local interactions. The same behaviour has been found in [21] for the spectral eddy-diffusivity defined in the same manner. When using these eddy-coefficients as subgrid-scale models in spectral large-eddy simulations, and recalculating explicitly the eddy-viscosity from the simulation itself (see [3], [17]), the latter is found to behave in the way predicted by the closures, that is with a plateau and the rising cusp behaviour[3]. When the eddy-diffusivity is recalculated in the same fashion, the plateau turns out to be replaced by a logarithmically decreasing region (in k), which is evidently related to the above mentioned k^{-1} scalar

---

[2] This can be only a high Reynolds number effect, since moderate Reynolds number experiments show unambiguously [19] that the scalar and the velocity decay following approximately the same law. Things are different if the scalar is injected initially at scales smaller than the velocity [19].

[3] Remark that, when analogous quantities are computed using RNG techniques, a decreasing cusp is found.

range, and disagrees with the plateau closure prediction (see Fig. 3). It remains however to confirm experimentally the existence of the anomalous large-scale k^{-1} range, in order to be sure that it is not due to numerical artefacts.

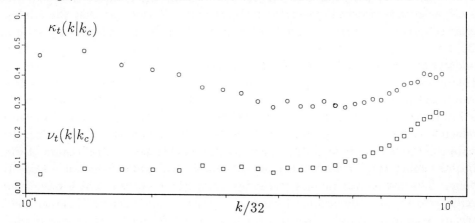

Fig. 3. Same calculation as in Fig. 2: spectral eddy-viscosity (squares) and eddy-diffusivity (dots), calculated from interactions across a fictitious cutoff wave number $k'_c = k_c/2$, where $k_c = 64$ is the maximum wave number (from [3])

Notwithstanding these difficulties, these spectral eddy-coefficients, derived from the closures, are very good tools to be used as subgrid-scale parameterizations for large-eddy simulations in spectral space[4]. They have been used in particular in [22] for stably-stratified Boussinesq homogeneous turbulence[5], and in [24] for a three-dimensional temporal mixing layer. In the latter case, the large-eddy simulation allows to recover the longitudinal vortices found experimentally in [25]. However, when applied to a flow which is strongly intermittent in physical space, the spectral eddy-viscosity has to be defined locally in this space, in regions where exists a strong small-scale three-dimensional turbulence. Hence, the concept of local kinetic energy spectrum (in physical space) must replace $E(k_c)$ in the definition of the spectral eddy-viscosity.

8. Coherent structures

It was proposed in [2] that coherent structures, being unpredictable in position, orientation and wave length, were an integral part of the turbulence. The same point of view is advocated by *Lumley* [26]. It has been found for instance in the above quoted numerical simulations of the two-dimensional mixing layer

---

[4]   when it is possible to use spectral methods

[5]   where the temperature, coupled with the velocity, loses its anomalous character regarding the eddy-diffusivity [23]

[8], that a continuous spatial spectrum of slope intermediate between k^{-3} and k^{-4} forms when the fundamental eddies pair. This is reminiscent of *Kraichnan*'s enstrophy cascade [9], [10], and favours the point of view that this is a special case of two-dimensional turbulence [8]. Notice also that the mixing-layer coherent structures resemble very much the coherent eddies displayed in [28] for two-dimensional isotropic turbulence. Fig. 4 shows, in a mixing-layer calculation involving initially 8 fundamental vortices, a comparison between the distributions of vorticity (top) and of a passive scalar having the same initial profile as the basic velocity (from [27]). In such a calculation, where randomness is introduced by superposing a tiny white noise perturbation onto the basic velocity profile, various pairings involve generally eddies of different size: for instance, the two pairs of eddies, finishing to merge in the central region of Fig. 4 a, are going to pair with the smaller fundamental eddies which are close to them. The result will be structures involving three eddies, which have been found also in two-dimensional isotropic turbulence simulations (Legras, 1988, private communication). Notice also that, if k_a is the fundamental mode[6], the energy is transferred successively to $k_a/2$ (first pairing), $k_a/3$ (second pairing), $k_a/6$ (third pairing between the two $k_a/3$ eddies). This is evidently not a spatial period-doubling process, and poses questions about the possibility of interpreting the mixing layer in terms of chaotic dynamical systems.

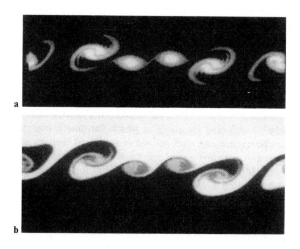

a

b

Fig. 4. Temporal mixing-layer calculation involving initially 8 fundamental vortices: a) vorticity; b) passive scalar initially identical to the basic velocity (from [27])

The more we will explore the turbulent medium, either experimentally or numerically, the more we will discover coherent structures of various type and shapes, associated with global or local instabilities of the velocity. One might even ask the provocative question: *may be turbulence is only composed of*

---

[6] the most amplified mode in the linear stability theory

coherent structures? As stressed by *Lumley* [26], this might not be contradictory with the use of statistical models. Following Einstein and *Narasimha* [1], I agree that God does not play dice, is certainly non linear, chaotic, unpredictable and coherent. But why shouldn't He be allowed to use statistics if He finds it usefull?

9. Perspectives

I will not try to draw general conclusions, nor to decide *who is at the end of the road?*, or *who is traditional and who is not?*. Two centuries after 1789, and more than 20 years after May 1968, the French know too well how tradition may blow up into revolution, and regenerate from it. May be the road has no end any way, and winds chaotically around some fractal attractor.

Speaking more seriously, experiments are certainly a very useful tool for the future, but they have also severe limitations, in particular when precise measurements[7] and three-dimensional visualizations are needed.

Dynamical systems might prove to be of some utility to understand the large-scale dynamics, but I do not believe they will ever be able to simulate any sort of Kolmogorov broad-band spectrum in isotropic turbulence.

Two-point closures were an invaluable tool to understand the phenomenology of isotropic turbulence, in three or two dimensions. They gave us a qualitative understanding which is still largely valid, and was of great help when direct or large-eddy numerical simulation codes were built. I do not think they can be much improved in the isotropic case. But they are also valid in anisotropic or inhomogeneous situations, although quite heavy to handle then: one possible development is to derive from them models simple enough to improve and give more fundamental basis to one-point closure models.

Renormalization Group techniques pose problems of convergence when they are used in a $k^{-5/3}$ Kolmogorov energy cascade, and are unable to handle large-scale dynamics in decaying situations. I do not understand how they can be properly applied to inhomogeneous turbulence, or for large-eddy simulation purposes.

Personally, I have been involved a lot these last years in the utilization of isotropic two-point closures as subgrid scale models for large-eddy simulations: this seems to be at hand when the subgrid-scale turbulence is close to isotropy, but progresses have to be done for more realistic flows, constrained by boundaries in particular. Finally, it is obvious that the extraordinary progresses done by computers, both in power and in fast visualizations of the results, give unprecedented opportunities for computational fluid dynamics and turbulence: a large number of statistical-theory predictions can now be checked by direct or large-eddy simulations, and, if valid, used in order to improve the latter.

---

[7] of vorticity, for instance

A brief closing comment: I have worked on turbulence for about 20 years, and I do not know if my contributions have helped turbulence understanding to progress much. But at least I have the feeling I understand it better.

Aknowledgements: this work was sponsored by DRET under contract 87/238.

References

1 . Narasimha, R., 1989, this volume.

2 . Lesieur, M., 1987, *Turbulence in Fluids*, Nijhoff Publishers, Dordrecht, Holland.

3 . Lesieur, M. and Rogallo, R., 1989, Phys. Fluids A, **1**, pp 718-722.

4 . Leith, C.E. and Kraichnan, R.H., 1972, J. Atmos. Sci., **29**, pp 1041-1058.

5 . Métais, O. and Lesieur, M., 1986, J. Atmos. Sci., **43**, pp 857-870.

6 . Lorenz, E.N., 1969, Tellus, **21**, pp 289-307.

7 . Lorenz, E.N., 1963, J. Atmos. Sci., **20**, pp 130-141.

8 . Lesieur, M., Staquet, C. Le Roy, P. and Comte, P., 1988, J. Fluid Mech., **192**, pp 511-534.

9 . Kraichnan, R.H., 1971, J. Fluid Mech., **47**, pp 525-535.

10 . Kraichnan, R.H., 1967, Phys. Fluids, **10**, pp 1417-1423.

11 . Herring, J.R., Schertzer, D., Lesieur, M., Newman, G.R., Chollet, J.P. and Larchevêque, M., 1982, J. Fluid Mech., **124**, pp 411-437.

12 . Lesieur, M. and Herring, J.R., 1985, J. Fluid Mech., **161**, pp 77-95.

13 . Lesieur, M. and Schertzer, D., 1978, J. Mécanique, **17**, pp 609-646.

14 . Forster, D., Nelson, D.R. and Stephen, M.J., 1977, Phys. Rev. A, **16**, pp 732-749.

15 . Fournier, J.D., 1977, Thèse de Spécialité, Université de Nice.

16 . Yakhot, V. and Orszag, S.A., 1986, J. Sci. Comput., **1**, p 3.

17 . Lesieur, M., Métais, O. and Rogallo, R., 1989, C. R. Acad. Sci. Paris, **308**, Ser II, pp 1395-1400.

18 . Herring, J.R., 1989, this volume.

19 . Warhaft, Z. and Lumley, J.L., 1978, J. Fluid Mech., **88**, pp 659-684.

20 . Kraichnan, R.H., 1976, J. Atmos. Sci., **33**, pp 1521-1536.

21 . Chollet, J.P., 1983, Thèse de Doctorat d'Etat, Grenoble.

22 . Métais, O. and Chollet, J.P., 1989, in *Turbulent Shear Flows VI*, Springer-Verlag, pp 398-416.

23 . Métais, O. and Lesieur, M., 1989, in *European Turbulence Conference, Berlin, 1988*, Springer-Verlag. Berlin

24 . Comte, P., 1989, Thèse de l'I.N.P.G., Grenoble.

25 . Bernal, L.P. and Roshko, A., 1986, J. Fluid Mech., **170**, pp 499-525.

26 . Lumley, J. L., 1989, this volume.

27 . Comte, P., Lesieur, M., Laroche, H. and Normand, X., 1989, in *Turbulent Shear Flows VI*, Springer-Verlag, pp 360-380.

28 . McWilliams, J., 1984, J. Fluid Mech., **146**, pp 21-43.

The Utility and Drawbacks of Traditional Approaches.

Comment 3.

Jackson Herring

National Center for Atmospheric Research
P.O. Box 3000
Boulder, CO 80307

Narasimha's paper has touched on many important issues relative to the utility of the statistical approach to turbulence. I shall use his organization of topics, but will respond to only a few issues he raises. My own comments stress—perhaps more than his—two-point closures, and their possible contributions and defects.

1. What Are Traditional Approaches?

I take the definition of the "traditional approach" as primarily statistical. This by and large is in agreement with Narasimha's definition. It also focuses discussion. Further, I exclude large eddy simulations (*LES*), a hybrid method which is part statistical and part deductive. *LES* is essential for practical problems but it will be adequately discussed elsewhere. As for the statistical theory of turbulence, it seems to me that such methods are grounded in the traditions of non-equilibrium statistical mechanics, perhaps a wider perspective than Narasimha would need . In that case, the direct interaction approximation (*DIA*) (Kraichnan, 1959) is perhaps the most comprehensive of the "traditional" approximations, albeit not one well justified as a convergent perturbation approximation. Narasimha notes—correctly—that the *DIA* contains arbitrary assumptions, whose validity is "difficult to access."

To amplify on this point, we note that the *DIA* was originally proposed in the context of a perturbation approximation, but one which was subsequently found to be either divergent or at best asymptotic. This means that the method has no internal error estimates—perhaps another way of voicing Narasimha's objections. Efforts to extend perturbation analysis have not been profitable, to my knowledge. Recently, however, Kraichnan (1985) has succeeded in making an *in principle* error estimate for approximations, such as the *DIA*, by a new technique he calls a decimation scheme or statistical interpolation. This method incorporates statistical information (in the form of moment realizability constraints) and amplitude equations, at a sample (decimated) set of wave-number points to produce a system of equations that minimizes an error norm. The *DIA* is retrieved from this procedure in the perturbative limit. There is no space here for adequate discussion of these matters.

To balance the "arbitrary assumption" point let us list some of the more important statistical constraints preserved by the DIA.

(a) Full realization for $\langle u(r,t)u(r',t')\rangle$; i.e., positivity for energy spectra $E(k)$, not only for homogeneous problems, but also for any inhomogeneous problems as well.

(b) Representation independence; i.e., the equations may be derived in any representation (Fourier modes, etc.) and transformed to any other without altering their physical content.

(c) Ability to distinguish "spacially turbulent" systems; i.e., systems that evolve complicated spacial structures, but without much temporal fluctuations.

(d) Exact solutions of a model system (the random coupling model, [Kraichnan, 1961]), which has non-Gaussian components. If Navier-Stokes is written as

$$\partial x_n/\partial t = \sum_{ml} M_{nml} x_m x_l - \nu_n x_n \ , \tag{1}$$

then the random coupling model is the $N \to \infty$ limit of

$$\partial x_n^\alpha/\partial t = N^{-1} \sum_{ml\beta\gamma} C_{nml}^{\alpha\beta\gamma} M_{nml} x_m^\beta x_l^\gamma - \nu_n x_n^\alpha \tag{2}$$

where $C_{nml} = \pm 1$, randomly, and Greek superscripts label the N-ensemble members.

(e) Generalizes to inhomogeneous problems; a starting point for further (realizable) approximations of the second order sort. We will discuss this point in more detail later. (See Leslie, 1973; Herring, 1973; Yoshizawa, 1983.)

(f) Fully compatible with the inviscid, (wave number) truncated, Euler dynamics: This is the "thermal equilibrium" solution $E(k) \sim k^{n-1}$, (with n the dimensionality of the problem). For this state, the multivariate distribution function for $u(k)$ is Gaussian, and the two-time statistics satisfy the fluctuation dissipation theorem (Kraichnan, 1958; Leith, 1975).

We may hope that there is a sufficient list of constraints which would guarantee a satisfactory theory at the covariance level. There are important missing elements in the above list, such as invariance to *random* Galilean transformations, which poses a serious problem for the Eulerian DIA. However, Kraichnan (1965) and Kaneda (1978) have proposed Lagrangian methods that obviate this problem. In my comments, I shall stress two-point closure (and in particular the DIA) as a "traditional method." By the same token, I would also include much of RNG, but the efficacy of this method is addressed elsewhere (Yakhot and Orszag, 1986).

2. Major Triumphs

Narasimha states major triumphs as: (1) Eddy viscosity concepts, and (2) $k^{-5/3}$. The concept of eddy viscosity and eddy conductivity flow from the DIA in a natural way, as described in the response of Lesieur (discussant #2) (see Lesieur, 1988 for more details). In fact, the heart of this sort of approximation is that it models the nonlinear terms as a random stirrer, plus a compensating eddy viscosity. The methodology is such that the usual coefficient arbitrariness is avoided. Narasimha notes correctly, in my opinion, that efforts to determine the Kolmogorov constant (and we should add the scalar Batchelor and Corrsin-Oboukhov constants) are only partially successful. A more serious issue is the correct prediction of all these constants from a central theory which is free from arbitrary coefficients. To do this, we may have to address the inertial range intermittency, at least for the scalar. We should note the recent RNG claims in this connection (Yakhot and Orszag, 1986).

3. Some General Principles?

Narasimha lists five basic working rules which serve as guides in thinking about turbulence, if you are a traditionalist. These are interesting items to check, particularly numbers (2) and (3). However, I think we get a more satisfactory perspective if we examine these rules from the two-point or spectral point of view. I would argue that if we do this, the two-point versions may violate the single-point version, but this is to be expected. After all, linearity (and other simple things) in the spectral domain translates into non-linearity for single-point quantities; consider pure viscous decay. Before discussing the two-point perspective let us simply state some examples in which the single-point rules are inadequate.

Rule (2)—that turbulent flows evolve asymptotically to a state independent of initial conditions—is contradicted by decay experiments in stratified flows (Métais and Herring, 1989), and rule (3) (the approach towards self-similarity) is contradicted by experiments (which do not decay as $E(t) \sim t^{-1}$, as would be expected from complete self-similarity; see Lesieur's comments). In spectral form perhaps we could supplement (3), with the idea that turbulence seeks ("asymptotically tends toward") the highest level of statistical symmetry consistent with forcing, boundary conditions, and energy-like constraints. This is a kind of Parkinson's law; turbulence fills-up the (phase) space available to it, each available space receiving equal amounts. It is just an expression of point (f) of section Sec 1. As a corollary, we may assume that if there is no external forcing in a range of scales, then the flow in this range will seek the full symmetry of inviscid equilibrium, provided viscous effects are remote to the range of scales considered. Such a principle was used profitably in quasi-geostrophic turbulence, in predicting the partitioning of kinetic and potential energies at various vertical levels of the flow (Salmon *et al.,* 1976). Of more relevance to our present discussion, is the distribution of the largest scales of turbulence, and how their behavior enters the law of decay of total energy. This is one of the items discussed by Lesieur, (Lesieur and Schertzer, 1978), and I shall touch on a couple of these issues here.

First, a straightforward application suggests that $E(k) \to k^2$, and hence by Lesieur's arguments, $E(t) \sim t^{-6/5}$; Saffman's (1967) law. Suppose that we say simply that the tendency should be toward equipartitioning. Then, given that eddy viscosity acts on these large scales (a triumph remember), we would write, $\partial E(k,t)/\partial t \to Ak^4 - \nu_{eddy} k^2 E(k,t)$. If $E(k,0) \to k^n, n > 4$ this equation says that $E(k,t) \sim k^4$ a short time t later and hence the initial value sensitivity noted by Lesieur. The case $E(k,0) \to k$ is the only completely self-similar solution, but if $E(k,0) \sim k^n, n > 4$, then "you can't get there from here" (unless you start there). What about the Saffman result? It could certainly be that $E(k,t) \sim k^4$, at an intermediate phase, but $\sim k^2$ at very long times. This requires detailed calculations to settle, but the answer seems to be that this cannot happen either.

4. Failures

4.1 Philosophical

There is a suggestion here that statistical theories somehow are at variance with the determinism imbedded in Navier-Stokes. However, in statistical mechanics, the posing of a statistical (random) initial value problem for a *deterministic* system is clear. One finds it useful to pose such a question if the system considered is thought to be unstable, in that small, initial differences among ensemble members become, for practical purposes, untraceable as time goes on. The fluctuation field takes on the aspect of a random field for two reasons: first, because the initial ensemble is chosen Gaussian, and second, because the effects of non-linearities are usually very complex. One may do statistical theories even if the second reason is removed, but the economy of the statistical initial value problem is gone.

Narasimha's discussion here of dynamical system theory touches on the predictability problem; how fast do neighboring points in solution phase space diverge? We should recall that this problem has been examined with more traditional techniques (DNS) and two-point closures for over 25 years (see Thompson, 1984 for a review). We should stress that it has also been a subject to which two-point closures have been involved with at least partial success. In this connection the meteorological community has long used results from two-point closure (Leith and Kraichnan, 1972) to judge the quality of numerical predictions of large-scale forecasts. The question here is, to what extent are closure results at variance with dynamical system theory? We know that the *RCM* expresses the exact dynamics of the *DIA*. Does the *RCM* have dynamical system aspects? The answer, I suspect, is yes: but it seems doubtful that the statistical theory of turbulence (in its present form) can be of much use here.

4.2 The Prevalence of Exponentials

Narasimha notes that in an effort to ascertain the full distribution of certain aspects of Navier-Stokes, there have been many measurements (and numerical experiments) of quantities such as velocity, temperature, and vorticity. These turn—he notes— out to be more exponential than Gaussian.

I would like at this point to consider the distribution for $\partial u/\partial z$, and $\partial T/\partial z$, as computed from direct simulation (128^3) for homogeneous turbulence at $R_\lambda \sim 20$. These are as depicted in Fig. 1(a,b) showing fairly good exponentials (see also Kida and Murakami, 1988). But for both energy and scalar variance spectra, the *DIA* at this R_λ works rather well. These are shown in Fig. 2 (a,b). We show also the *TFM*

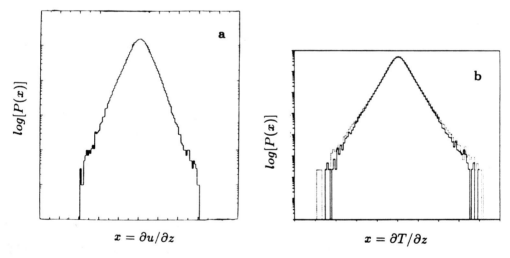

Fig. 1. a: Distribution function $P(\partial u/\partial z)$ for direct numerical simulation of isotropic turbulence at $R_\lambda \simeq 20$ after Métais and Herring (1989). Resolution of the *DNS* is 128^3. b: Distribution function for $\partial T/\partial z$ for the same conditions as in a: dashed lines indicate $Pr = 1$, solid lines, $Pr = .7$. From Métais, private communication. Same conditions as in a, except code is Shear3 (Orszag). Calculations were made on the Cray2 in Paris.

results in Fig. 2 (c,d) in order to measure the effects of non-invariance to random Galilean transformation. Since the DIA is based on a sort of quasi-normality, how can we understand the rather good comparisons for spectra? The answer may lie in the fact that the DIA is fairly good in the energy-containing range, but fails progressively as $k \to \infty$ (the dissipation range). The latter is where the "structures" and their associated exponential effects are concentrated. At least this picture is verified by low R_λ DNS. The key question here is whether intermittency moves from the small-scale dissipation range of k's into the inertial range in a serious way as R_λ increases. By serious, I mean, do the inertial range structures represent strong departures from Gaussianity? The DIA's forecast for $\langle | \partial u/\partial t |^2 \rangle$ (as shown in Fig. 3) is not bad. That the curves drop significantly below unity indicates a degree of self organization, or "Eulerization" in the flow, where the nonlinearities are supressed.

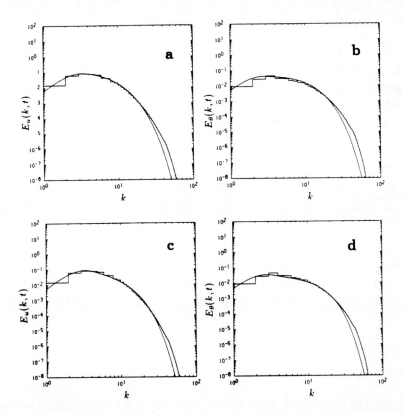

Fig. 2. a: Energy spectrum $E(k,t)$ at one eddy circulation time for roughly the same initial conditions as in Fig. 1 Smooth curve DIA, histogram DNS. From Herring and Kerr (1989). b: Scalar variance spectrum $E_\theta(k,t)$ for $Pr = 1$. From Herring and Kerr (1989). (c, d) are the same as (a, b), except for TFM instead of DIA.

(For a discussion of the dynamics of this issue, see the paper of Moffatt in the present volume). The evaluation of $\langle |\, \partial u(t)/\partial t\, |^2\rangle$ requires fourth moments of u. But all this says nothing about the distribution function of $\partial u/\partial t$, since such would require the infinite sequence of moments $\langle |\, \partial u(t)/\partial t\, |^n\rangle$. These could, in principle, be obtained from the *RCM*. I see *no* indication that such would be Gaussian. The *RCM* evaluation of the fourth moment $\omega = \nabla \times u$ is however Gaussian (see Chen *et al.*, 1989).

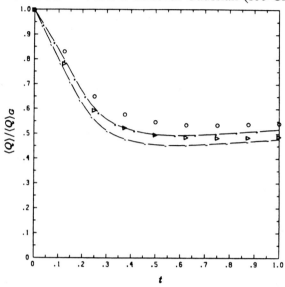

Fig. 3 Comparison of *DIA* values for $\langle |\, w\, |^2\rangle/\langle |\, w\, |^2\rangle_G$, with w the nonlinear force for Navier-Stokes, and $\langle |\, \partial u/\partial t\, |^2\rangle(t)/\langle\cdot\rangle_G$ with the *DNS* for approximately the same conditions as in Figs. 1 and 2. Here, the subscript G denotes evaluation with the Gaussian part of u. Solid lines with $(+,-)$ are *DIA* values, and circles and triangles are *DNS* values. After Chen *et al* (1989).

With respect to the question of strong non-Gaussian features in the inertial range, we know that such a situation may arise in two-dimensional turbulence (McWilliams, 1984; Herring and McWilliams, 1984), and in two-dimensional MHD (Frisch *et al*, 1978).

4.3 The Absence of Rational Models

Much of the criticism Narasimha makes here is, in my view, well founded. One must take seriously counter gradient transport and a rational model must predict under what circumstances this happens. However, to paraphrase a recent comment of Lumley (1989), it is easy to underestimate the complexity of turbulence. For this reason we should not shy away from models, which on the surface are correspondingly complex. Take again the *DIA genre* two-point covariance equations. As noted by Yoshizawa (1983), these equations may be reduced to a sort of second-order model (of the Hanjalic-Launder, 1972 variety) by the *einsatz*, that dependencies on $r_1 - r_2$ are much sharper than those on $r_1 + r_2$. (The same sort of approximation used to crank out the rapid distortion terms.) One gets the full panoply of terms; production, diffusion, dissipation, buoyancy, etc., except those which had been arbitrary coefficients now contain no arbitrary elements (they are spectral integrals over the Fourier transform which conjugate to the variable $r_1 - r_2$). Moreover, the approximation is guaranteed realizable, if the technical approximations made on the way to the single-point reductions are valid. Is this a rational model of turbulence? It is based on a divergent perturbation approximation, whose basic structure is to some, intuitively appealing; random forcing by nonlinearities, plus eddy-viscosity, but no arbitrary coefficients. Higher-order statistics are presumed not too far from Gaussian. I would argue that it is a rational program, and as a set of evolution equations for Reynolds stress that the closure has a validity for short times (and even long times, if Lesieur's arguments are correct). Thus, from my perspective, it is a rational model, but it clearly omits structures and hence, may diverge from reality. However, to account for all the relevant physics, should we not, to paraphrase Woody Allen, promote it from a "model" to a theory?

Narasimha states that current models furnish elaborate equations that really embody little by way of new concepts above the Prandtl-Kolmogorov models. This statement is, in essence, correct. In my view, closures are computational tools, not methods to produce deep insights into the dynamics of turbulence. However, do we really know what new concepts are needed?

5. End of the Road for Traditional Approaches?

Finally, Narasimha raises the issue of cost effectiveness of spectral (two-point) closures similar to the *DIA*. This to me is an interesting issue. First, historically, we should note that the design of efficient computational algorithms was a definite motivational consideration in advancing this approximation, despite the fact that the closure is much more complicated on the surface than Navier-Stokes. The point is, of course, that whereas the degrees of freedom of $u(k, t)$ evolved in a complicated, chaotic manner in both k and t, the correlation functions $\langle u(k, t)u(k', t')\rangle$, are smooth functions of their arguments, and thus should be easier to approximate with either numerical algorithms or analytic means. The *should be* here means that it is up to our ingenuity to design such algorithms. Thus, whereas direct simulations of u generally use a cubic lattice of k's (N^3 points) for the statistical theory, one may use logarithmic spacing of scalar k's

($\leq N$), at least for problems having a high degree of statistical symmetry (isotropic turbulence, being the simplest case). For problems of less statistical symmetry, the relative computational load on the closure goes up. Of course one can deduce analytically certain properties from the closures, such as the inertial range shape, the decay laws for large R_λ flows (Lesieur and Schertzer, 1978), and rates of approach to isotropy of Reynolds stress, heat flux, and scalar variance. We should note that the DIA results presented earlier take about one minute, whereas the DNS results to which they were compared take about one hour. Simpler closures (such as the $EDQNM$) should take a few seconds. The situation is even more favorable to closures if we recall that with judicious distribution of computational nodes $\{k_i\}$ and time-stepping algorithms (quasi-linearization) (Herring, 1984), the time needed to do evolution calculations at arbitrarily high R_λ does not increase much. Of course we must add here that what emerges from such closure calculations are very simple statistical information (spectra, some fourth-moment information, various derivative skewness information, and estimates of singularity times mentioned by Lesieur). There is also the issue that the qualitative accuracy of this information is uncertain; there are the serious problems in two-dimensional MHD, noted above. There may be even more serious problems at very high R_λ associated with possible "Cantor-set" features of the inertial range (Argoul et al, 1989).

Homogeneous, anisotropic problems (including stratification and rotation) are probably in good shape although final results on the closure applications are just now becoming available (Cambon, 1989). For simple anisotropic flows, we have the extensive studies of the group at Lyon already (Cambon et al, 1981), which time limitations prevent any discussion. We should at least note the satisfactory agreement of the closure with the experiment reported at the Stanford convergence (Cambon al., 1985), as well as the closure-prediction of the $k^{-7/3}$ law for the Reynolds stress spectrum.

For problems with less statistical symmetry and with boundary layers, such as shear flow or convection, we must admit that insufficient resources have been devoted to analytic and asymptotic methods to make closure methods competitive with DNS and LES. We must note, however, the studies of Dannevik (1984), which effectively reduce the (t, t') dependencies (in the DIA) to a single time problem by using Padé tables. Dannevik's work aside, the reason for this "algorithm deficiency" is perhaps that the next step requires a few years' effort in the development phase. Yet the possible end result of such an application may yield mixed results, as noted in the preceding paragraph. In this circumstance, who is willing to invest the effort? We seem to lack the means for long-term, high-risk investments. Yet surely a certain level of such investments is prudent for the long-term health of our enterprise.

References

Argoul,F., A. Arnéodo, G. Grasseau, Y. Gagne, E. J. Hopfinger, and U. Frisch, 1989: Wavelet analysis of turbulence data reveals the multifractal nature of the Richardson cascade. Preprint.

Bertoglio, J. P., 1981: A model of three dimensional transfer in non-isotropic homogeneous turbulence. *Third International Symposium on Turbulent Shear Flow*. Davis, Calif., Sept. 1982.

Cambon, C. and L. Jacquin, L. 1989: Spectral approach to non-isotropic turbulence subjected to rotation. To appear in *J. Fluid Mech.*.

Chen, H. D., J. R. Herring, R. M. Kerr, and R. H. Kraichnan, 1989: Non-Gaussian statistics in isotropic turbulence. Preprint, submitted to *Physics of Fluids A*.

Dannevik, W., 1984: Two-point closure study of covariance budgets for turbulent Rayleigh-Benard convection. Ph. D. Thesis, St. Louis University, St. Louis Mo.

Frisch, U., A. Pouquet, P. L.Sulem, and M. Meneguzzi, 1983: The dynamics of two-dimensional ideal magneto-hydrodynamics, in *J. Mécanique Théor. Appl. Suppl.*, R. Moreau, ed., 191–216.

Hanjalic, K., and B. E. Launder, 1972: A Reynolds stress model of turbulence and its application to thin shear flows. *J. Fluid Mech.*, **52**, 609–638.

Herring, J. R., and J. C. McWilliams, 1985: Comparison of direct numerical simulation of two-dimensional turbulence with two-point closure: The effects of intermittency. *J. Fluid Mech.*, **158**, 229–242.

Herring, J. R., 1973: Statistical turbulence theory and turbulence phenomenology. *Proc. Langley Working Conference on Free Turbulent Shear Flows*, NASA SP321, Langley Research Center, Va., 41 pp.

Herring, J. R., 1984: Some contributions of two-point closure to turbulence, in *Frontiers in Fluid Mechanics*. S. H. Davis and J. L. Lumley, eds., Springer-Verlag, pp. 68–86.

Herring, J. R., and R. M. Kerr, 1988: Numerical simulation of turbulence, in *Proceedings of the International Union of Theoretical and Applied Mechanics*, Springer.

Kaneda, 1985: Attempts at Statistical Theories of Turbulence in *Recent Studies on Turbulent Phenomena*, T. Tatsumi, H. Maruo, and H. Takami *Eds*. Association for Science Documents Information, Tokyo. pp. 79-90.

Kida, S., and Y. Murakami, 1989: Statistics of velocity gradients in turbulence at moderate Reynolds numbers. *Fluid Dynamics Research*, **4**, in press.

Kraichnan, R. H., 1959: Classical Fluctuation–Relaxation Theorem., *Phys. Rev.*,**113**, 1181.

Kraichnan, R. H., 1959: The structure of isotropic turbulence at very high Reynolds numbers. *J. Fluid Mech.* **5**, 497.

Kraichnan, R. H., 1961: The dynamics of nonlinear systems. *J. Math. Phys.*, **2**, 124.

Kraichnan, R. H., 1965: Lagrangian–History Closure Approximation for Turbulence. *Phys. Fluids*, **8**, 575–598.

Kraichnan, R. H., 1985: Decimated amplitude equations in turbulence dynamics. *Theoretical Approaches to Turbulence, Appl. Math. Sci.*, **58**, (Springer-Verlag), New York, Berlin, Heidelberg, Tokyo. D. L. Dwoyer, Y. Y. Hussaini, and R. G. Voigt, Eds., pp.91-136.

Leith, C. E. and R. H. Kraichnan, 1972: Predictability of turbulent flows,*J. Atmos. Sci.*,**29**, 1041-1058.

Leith, C. E., 1975: Climate Response and Fluctuation Dissipation, *J. Atmos. Sci.*, **32**, 2022–2026.

Lesieur, M., and D. Schertzer, 1978: Amortissement auto similarité d'une turbulence a grand nombre de Reynolds. *J. de Mécanique*, **17**, 609–646.

Lesieur M. 1987: *Turbulence in Fluids*, Martinus Nijhoff Publishers, Dordrecht/Boston/Lancaster.

Leslie, D. C., 1973: *Developments in the Theory of Turbulence*, Oxford.

Lumley, J.L. 1988: Review of Turbulence in Fluids by M. Lesieur. *AIAA Journal*, **26**, 1288.

McWilliams, J. C., 1984: The emergence of isolated coherent vortices in Turbulent Flow. *J. Fluid Mech.*, **146**, 21–43.

Metais, O., and J. R. Herring, 1989: Numerical studies of freely decaying homogeneous stratified turbulence. *J. Fluid Mech.*, **202**, 117–148.

Saffman, P.G. 1967: Note on the decay of homogeneous turbulence. *Phys Fluids*, **10**, 1349.

Salmon, R., G. Holloway, and M. Hendershott 1976: The equilibrium statistical mechanics of a simple quasi-geostrophic model. *J. Fluid Mech.* **75**, 691–703.

Thompson, Ph. D. 1984: A review of the predictability problem, in *Predictability of Fluid Motions* LaJolla Institute-1983, G. Holloway and B. West, eds. American Institute of Physics, New York.

Yakhot, V., and S. A. Orszag, 1986: Renormalization group analysis of turbulence, I: Basic Theory. *J. Scientific Computing*, **1**, 3–51.

Yoshizawa, A. 1983: A Statistical Theory of Thermally–Driven Turbulent Shear Flow, with the Derivation of a Subgrid Model. *J. Phys. Soc. Japan*, **52**, 1194–1205.

Discussion of "Utility and Drawbacks of Traditional Approaches"

Reporter J. G. Brasseur

Dept. of Mechanical Engineering
223 Hallowell Building
The Pennsylvania State University
University Park, PA 16802

Uriel Frisch:

I have three comments on the topics which were touched upon this morning by other people concerning the predictive power of closure and other models. I think the situation is slightly better than suggested by various people. There is one particularly good success. You know that the inverse cascade which takes place in two-dimensional incompressible flow turbulence was predicted first on the basis of absolute equilibrium models, and DIA and DIA-related models by Bob Kraichnan [1], and before there was any serious evidence that it existed. The evidence came later in numerical form [2,3,4]. I'm sure one can find other successful closure-based predictions but this one is especially prominent.

The second point is probably a recurrent theme of this workshop—it concerns the duality between the statistical approaches and those by coherent structures. There are things that are interesting that we can learn by looking at other branches of physics. For example, the study of phase transition which has, at least superficially, some relation to turbulence. Everybody knows that sometime ago, now it's nearly twenty years in this very place, Kenneth Wilson developed a renormalization group for studying phase transitions [5]. He was particularly successful with a certain kind of phase transition for critical phenomena in ferromagnets governed by models like the Ising model, and he developed a tool which certainly one can describe as statistical. It involved expansions around rather simple states which would be the equivalent of viscous-dominated homogeneous turbulence, and the theory was very successful. A couple years later there was another problem in statistical mechanics which was solved by J.M. Kosterlitz and D.J. Thouless [6]. It is a variant of the Ising model which is called the X-Y model, when you have spins which have two components in the plane. If one tries to solve that model by using just the same sort of tools as done for the ordinary Ising ferromagnet, one gets nowhere—until one recognizes that there are what <u>we</u> would call coherent structures. These are actually vortices. The spins organize themselves in the form of vortices. Once these vortices have been identified, if one goes to suitable collective variables and redescribes the interaction in terms of those vortices, then rather simple statistical arguments, with renormalization group having the flavor of what Wilson did on the other problem, are successful. Therefore, you see, the message of that is that it is okay to do statistics provided you do it on the right animals.

My third point concerning successes and shortcomings of closure has to do with the fact that in general, second-order closure tends to overestimate the strengths of nonlinearities. As many know, real flows, incompressible flows, tend to form extremely flat structures—pancakes, ribbons, and things like that. Heuristically it's rather well understood why they are formed; you have strain action and you get very flattened structures. Now, such structures have strongly reduced nonlinearities because the dependence is mostly on one coordinate, the coordinate transverse to the pancake; and if it were strictly one-dimensional, you would have vanishing nonlinearity in incompressible flow. What is happening is that there are cases, there is at least one well-documented case [7], where a closure has a wrong prediction of what is happening there. It is a variant of two-dimensional turbulence in a conducting flow, an <u>ideal</u> conducting flow, where vorticity is no longer conserved because the Lorentz force can enhance vorticity. If you apply closure to that problem you predict that there is blowup after finite time. It is not a mathematical proof, but you know the closure equations can be integrated at extremely large Reynolds numbers. The evidence is massive. Now, if you study the problem by direct numerical simulation, which you can also do at very high resolution, you find that there are sheets forming and these sheets have random orientation. One place it goes like this, another place it goes like that, and you have strongly reduced nonlinearities; and actually the growth isn't catastrophic, it's just exponential. So that you don't get finite-time singularity, and you can show that this spurious effect, I mean that this flattening effect, cannot possibly be captured if you limit yourself to second-order moments. You would have to go to fourth-order moments in order to correctly describe the statistics by which various vectors, all randomly distributed, get aligned. This is not something that can be captured with second-order statistics.

Now, the question which still remains open is of three-dimensional Euler flows where closure also predicts finite time singularity, as Marcel [Lesieur] has shown in his talk, and where we are still waiting for the evidence to point in one way or the other. At the moment we don't have any clear evidence that the Euler equations blow up, and it could possibly behave like the case I just mentioned before—namely, remain smooth forever through depletion of nonlinearities. Of course, Marcel doesn't want to believe this and I am not conjecturing one way or the other. I'm just saying that one has to be careful there, that there is no doubt that there are instances where closure over-estimates nonlinearities.

Actually, the very concept of the Reynolds number overestimates nonlinearity. We tend to describe the Reynolds number as something measuring the strength of nonlinearities, forgetting that flows tend to organize themselves very often in a way in which the nonlinearities are almost vanishing. If you then try to estimate the strength of nonlinearity by dimensional analysis you can be off by orders of magnitude.

References

1. R.H. Kraichnan: *Phys. Fluids* **10**, 1467 (1967)
2. D.K. Lilley: *Phys. Fluids*, Suppl. II **12**, 240 (1969)
3. E.D. Siggia, H. Aref: *Phys. Fluids* **24**, 171 (1981)
4. U. Frisch, P.-L. Sulem: *Phys. Fluids* **27**, 1921 (1984)
5. K.G. Wilson: *Rev. Mod. Phys.* **47**, 773 (1975)
6. J.M. Kosterlitz, D.J. Thouless: *J. Phys.* **C6**, 1181 (1973)
7. U. Frisch, A.Pouquet, P.-L. Sulem, M. Meneguzzi: *J. Méc. Théor. Appli.* (Paris), special issue on 2D turbulence, 191 (1983)

Ravi Sudan:

This is largely an audience concerned with normal fluids; however, there are some of us who deal with fluids that are ionized. In addition, these fluids are magnetized so that they are not describable by the normal Navier-Stokes equation. Nevertheless, the problems of turbulence are important—experimentally in high temperature plasmas, observationally in space, and in astrophysics. To some extent we rely on the methods developed by the fluid dynamicist because you have had access to fluids and have built up an intuition over many years, whereas our knowledge of plasmas is only recent so that intuition does not always guide us there. Now, the only methods that would be successfully extrapolated to these more esoteric situations are those that are grounded on first principles and are not model-dependent. That is, in cases where a particular experiment is modeled by a particular set of equations, even though you may get a very satisfactory agreement between theory and experiment, the methods applied are not those that would be useful in situations that are orders of magnitude larger in scale, and otherwise depart radically from the behavior of normal fluids. For this reason, the methods that have been applied are based on the DIA approximation or its equivalent, simply because physicists can understand DIA, appreciate some of its limitations, and therefore claim to be on somewhat firmer ground. What I would like to see is that in the discussions and perspective developed here you keep in mind the community that is outside normal fluid dynamics, because they too depend very much on the theories and concepts developed by fluid dynamicists.

With respect to coherent structures, I would comment that there is an example in high frequency plasma turbulence, *viz.* Langmuir turbulence, where the appropriate fundamental 'basis function,' so to speak, is not the sine wave term, but the soliton. A picture of turbulence can be developed using the soliton, or more accurately, the caviton. The caviton is a physical entity which, created at some scales, collapses continuously inward to a very small size until kinetic effects, such as Landau damping, destroy it. The theory of turbulence applied to this particular phenomenon has been developed by using the physical picture of the collapsing caviton.

H. Keith Moffatt:

I would like to put in a word for the Reynolds equation before it is buried altogether. As Professor Sudan has just mentioned, one of the great fields in the application of turbulence theory has been in the wider domain of magneto-hydrodynamics—in the diffusion context, in astrophysics, and so on. I think to the list of great achievements of traditional methods over the last fifty years should be added the discovery of the alpha effect in magneto-hydrodynamics. The alpha effect was really based on very simple arguments involving straightforward averaging of the equations of magneto-hydrodynamics coupled with some rather inspired, but quite rigorous, dimensional analysis; and coming to the very remarkable conclusion that, as an effect of turbulence which lacks reflectional symmetry (and that is generally the case for turbulence in a rotating medium), an electric current can be driven parallel to a magnetic field. That discovery in the 1960's by Steenbeck, Krause, and Radler [1] has really revolutionized the whole subject of turbulence in the astrophysics context.

References

M. Steenbeck, F. Krause, K.-H. Radler: Z. *Naturforsch.* **21**, 369 (1966)

William K. George:

This is a three-minute comment, so don't panic. I want to focus for a moment on an aspect that has not been discussed, namely Dr. Narasimha's discussion on rules for turbulence—particularly the question of self-preservation. To illustrate the point, I raise the question "Can you have self-preservation in states that are still dependent on the initial conditions?" Figures 1 and 2 show spectra measured in a homogeneous shear flow by Chuck Van Atta and his co-workers in a water channel in San Diego [1]. Notice that the spectra collapse very nicely with a single length scale (in particular, the Taylor microscale) over the entire range of measurement, at least at scales for which you can assume the flow is homogeneous. Now what is interesting is that this behavior falls right out of a self-preservation analysis of the dynamical equations for homogeneous shear flow, which predicts that the Taylor microscale should approach a constant while the energy increases exponentially, and all sorts of other things that are observed [2]. What is particularly interesting is that the predicted self-preserving state is not universal, but is in fact uniquely determined by the initial conditions. This is also observed in the experiments.

Now, with this example as a background let me make a few quick comments. A situation like this would seem to be consistent with proposition number three of Narasimha—namely, that if you have a self-preserving state, the flow will relax to it. But this comes into direct conflict with rules number one and five, which include Kolmogorov similarity—the assumed local similarity that we have all come to believe in. The same problem also arises in grid turbulence where one can carry out a similar similarity analysis, and discover that the experiments do, in fact, collapse with a single length scale and with energy decay rates determined by the initial conditions [3]. Thus, we discover things that seem inconsistent with our rather entrenched ideas. Long and Chen [4] made a comment in a paper that was published (with editor's protest) in *JFM* that is perhaps very relevant in this context. He said something like, "no amount of experimental evidence can overturn an idea which has been around long enough, however insufficient the evidence on which the idea came to be accepted in the first place." It's a very interesting comment.

The second question which I ask leads into the discussions of tomorrow: "Is self-preservation, whether dependent on initial conditions or not, a property of the 'strange attractors' for these turbulent shear flows?" One of the things that you often hear discussed as a property of strange attractors in dynamical systems theory is similarity, often described in terms of fractals. Is the self-preservation of turbulent flows a dynamical property that is associated with the 'strange attractors' of the Navier Stokes equations? I suspect the answer is probably "yes!".

The last question is: "Do coherent structures, and the manner in which they arise and are sustained by the flow, provide a way for the flow to remember its initial conditions?". There are a number of mechanisms which have been suggested as to how coherent structures arise in particular situations and simply keep replicating themselves downstream (*e.g.*, [5]). It's not hard to imagine a situation where one could start a flow in different ways and entirely different series of structures are replicated downstream, so that the flow relaxes to different self- preserving states— states determined by the initial conditions.

Well, these are just some points that I wanted to bring to your attention—reinforcing the point that self-preservation determined by the initial conditions does appear to be possible, that it might have a relation to coherent structures, and in fact that it might also be related to strange attractors.

Let me just make a comment in closing. I think it would probably be useful in our discussions if we distinguish between our attempts to *close* turbulence (or to model turbulence) and our attempts to *understand the physics* of turbulence. Sometimes there is certainly an overlap—in the large eddy simulation and the direct simulation— but I think very often a lot of our controversy is because we are not saying the same thing. Closing the turbulence problem, or providing engineering models, certainly is not the same as understanding the physics. And clearly, as engineers, which many of us are, we have to do both.

References

1. J.J. Rohr, E.C. Itsweire, K.N. Helland and C.W. Van Atta: *J. Fluid Mech.* **187**, 1.(1988)
2. W.K. George, M.M. Gibson: *Proc. 7th Symp. Turb Shear Flows*, Stanford Univ. (1989)
3. W.K. George: in M. Hirata, N. Kasagi (Eds.) *Transport Phenomena in Turbulent Flows*, Hemisphere Publ. Co., New York (1987).
4. R.R. Long, T.C. Chen: *J. Fluid Mech.* **105**, 19 (1981)
5. M.N. Glauser, W.K. George: *Proc. 6th Symp. on Turb. Shear Flows*, Toulouse, France, p. 93.1 (1987)

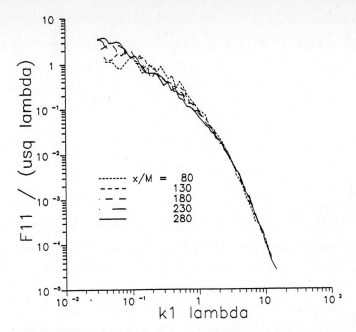

Figure 1. Normalized longitudinal velocity spectra of Rohr *et al.* [1], taken from George & Gibson [2].

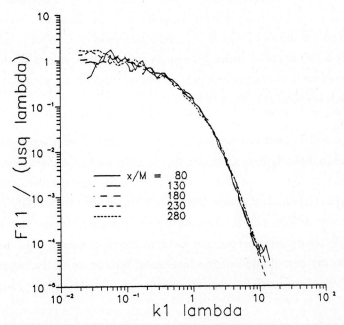

Figure 2. Normalized vertical velocity spectra of Rohr *et al.* [1], taken from George & Gibson [2].

Fazle Hussain:

I would like to address the question of whether coherent structures and statistical approaches to turbulence are in conflict with each other. Actually, it turns out that my comments are related to what Bill [George] just said, although they were initially meant to be a response to the points brought up by Roddam, Marcel, and John. I wish to discuss this issue using a particular example (Fig. 1), the evolution of the centerline turbulence intensity u'_c normalized by the local centerline mean velocity U_c of jets of various geometries—circular, plane, and elliptic—with different initial conditions. These data were taken in our laboratory and are discussed in a forthcoming paper [1].

First, let us examine elliptic jets. In Fig. 1a we plot the axial variations of the half-widths of elliptic jets of various aspect-ratio along two mutually perpendicular planes. Since the elliptic structures undergo deformation, primarily due to curvature-dependent self-induction, the jet cross-section switches major and minor axes. However, with increasing axial distance the ellipticity of structures decreases and the jet eventually becomes axisymmetric. Axial variations of normalized centerline turbulence intensity for elliptic jets (initially laminar; sharp orifice and contoured nozzles), circular jets (initially laminar and turbulent), and plane jets (initially laminar and turbulent) are shown in Figs 1b, 1c, and 1d, respectively. What is surprising is that for this wide variety of orifice geometries and initial conditions, the normalized centerline turbulence intensities collapse to what one would call a self-preserving state with a turbulence level of 0.245 ± 0.005.

Clearly, the initial conditions and initial structures are different among the various cases shown in these figures. Compare, for example, the high-aspect-ratio plane jets and the circular jets. Even though we have not documented the details of the organized structures in all the cases, there is little doubt that they are different. These differences are reflected in the near-field local turbulence intensities which are dominated by the dynamics of the characteristic structures in each flow. In the near-field region the contribution of the incoherent turbulence to the time-averaged turbulence intensity is much less than the contribution from the coherent fluctuations.

The essential point of this discussion is that, with increasing distance, structures undergo multiple stages of interactions—like pairing, tearing, shredding, and cut-and-connect. We do not claim that the structures in the far field are identical; most probably they are not. So far we have studied the coherent structure in the self-preserving region of circular jets only [2]. I suspect that the coherent structures are different in the fully-developed regions of plane and circular jets. For example, Mumford's observation of double-roller structures in a plane jet [3] are quite different from what we observed in a circular jet (axisymmetric, helical, and bihelical structures). It is possible, therefore, to have similar statistics with different structures. This is a point for both structures people and statisticians to address.

References
1. F. Hussain, H.S. Husain: *J. Fluid Mech.* (in press)
2. J. Tso, F. Hussain: *J. Fluid Mech.* **203**, 225 (1989)
3. J.C. Mumford: *J. Fluid Mech.* **118**, 241 (1982)

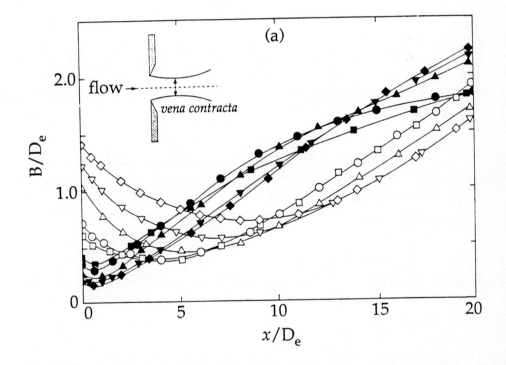

Figure 1a.   Axial variations in jet half-widths (B) of elliptic jets of various aspect ratios. ☐ 3:2 jet; ○ 2:1 jet; △ 4:1 jet; ▽ 6:1 jet; ◇ 8:1 jet. Open and solid symbols refer to jet half-widths on the major-axis and minor-axis of the jet exit planes, respectively. Jet exit velocity $U_e = 29.26$ ms^{-1}. Effective exit diameter $D_e = 2.54$ cm, where $D_e = 2(ab)^{1/2}$. U_e and D_e are the same for all jets.

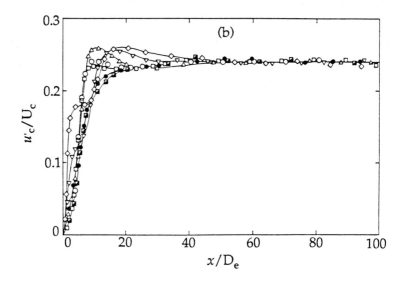

Figure 1b. Axial variations of u'_c/U_c in circular and plane jets, where u'_c and U_c are the local centerline turbulence intensity and mean velocity, respectively. □ 3:2 jet; ○ 2:1 jet; △ 4:1 jet; ▽ 6:1 jet; ◇ 8:1 jet; ▨ circular orifice jet (D = 2.54 cm); ● contoured nozzle, circular jet (D = 3 cm). For all elliptic jets De = 2.54 cm and U_e = 29.26 ms⁻¹.

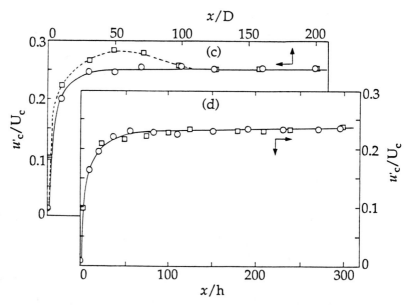

Figure 1c,d. Axial variations of u'_c/U_c in circular (c) and plane (d) jets. (c) Initially laminar and turbulent circular jets: D=3 cm, U_c=65 ms⁻¹. (d) plane jet: slit width h = 1.12 cm, U_e = 45 ms⁻¹ for the initially turbulent jet. ○ initially laminar; ◻ initially turbulent.

Jay P. Boris:

In the first position paper today a description was given of the nonlinear dynamics of coherent structures. The particular point was made that there are situations which relax in different ways or at different rates to the Kolmogorov spectrum scaling, such as the velocity field. In this regard, simple nonlinear dynamics models based on ordinary differential equations [ODE's] can be extended to model a cascade in which the coefficients connecting a number of different dynamic scales have group renormalization properties and produce a Kolmogorov cascade. Perhaps this approach can only be applied to one-point or homogeneous models, but these types of models are worth looking at, not only because one might learn something about the temporal cascade of instabilities in turbulence, but also because they could be used (in abbreviated form) in two- or three-equation models, as the basis of a subgrid model in large eddy simulations, or in other practical contexts.

A few years ago I applied one of these coherent-structure dynamics models to a few very simple interactions among vortices of different scale [1,2], building up a set of stiff coupled ODE's spanning about five orders of magnitude in scale. This model showed that, while the Kolmogorov cascade was reproduced very nicely as a dimensional necessity, there are also an infinite number of nondimensional features of turbulence associated with non-self-similar behavior on the different scales. For example, we found that two weak vortices can have the same affect on the cascade as one strong vortex. The result is something which is demonstrably non-self-similar geometrically, but not dimensionally. In this case the non-dimensional numbers were packing fractions, the number of vortices of a given size per unit volume, which don't appear at all in dimensional analysis. In fact, in that case, in contra-distinction to the case mentioned this morning where the scalar converged faster than the velocity, the packing fractions converged much more slowly than the turbulent velocity spectrum.

This is a situation where the energetics and the dimensional aspects of the problem look very much like a cascade, so you would think that a renormalization-group or self-similar theory should apply. In reality, however, the geometry of the different scales in this very simplistic model took orders of magnitude longer to relax to a self-similar condition than the energy spectrum. I think these kinds of models and effects need to be looked at, because dynamic models for unresolved, or averaged turbulence scales may depend very much on the kinds of assumptions that are made.

References

1. J.P. Boris, E.S. Oran: *Proceedings of the International Colloquium on Gas Dynamics of Explosions and Reactive Systems*, AIAA, Vol 76, 187 (1981)
2. J.P. Boris, E.S. Oran, M.J. Fritts, J.M. Picone: *Bull. Amer. Phys. Soc.* **27** (9), 1163 (1982)

Julian C. R. Hunt:

I'd like to have a conversation—after all, this is a conversation. In particular, I want to address something to John [Lumley]. He actually said in his last talk, "modeling equals second-order moments." Does he really mean that? I hope that the subject of the conference is, in fact, a broader concept of turbulence. For example, there are particular practical problems where one really needs to know the spectrum; for example, where there is a building that is sensitive to a particular part of the frequency range—or you may have bubbles and particles responsive to certain eddy structures, and so on. 'Modeling', for many people, is for many people, is just a matter of modeling the mean flow, for others just the Reynolds stress. 'Modeling' really depends on the particular application you have in mind, and I hope that when people talk about modeling at this conference they will say what they mean. He said what he meant and I know what I mean by modeling, and they are different from each other and from other definitions!

Stephen J. Kline:

I want to make a philosophical comment. The question was raised about overthrowing paradigms, as Tom Kuhn called them (and John Lumley called them by some other name). The existence of contradictory data does not overthrow paradigms. Kuhn said that in his book, *The Structure of Scientific Revolutions* [1] (if you don't know that book, it is well worth looking at). Why doesn't it happen? It doesn't happen because you and I have to have some kind of model to think about anything. We do not think about nothing very well. So we don't throw out a paradigm (which is a high-level model) until we get a better one. I can give you lots of examples of this, but I won't right now. More importantly, I agree with everything John [Lumley] said, and he put it very well, concerning what I call the "guideline for scholarly controversy." This guideline comes from looking at arguments in scientific fields, not only in physical science but in many other disciplines, where you have the following situation:

You have legitimate scholars—not people doing mysticism, or saying they believe something because, say, Aristotle said it, or the Pope, or The Buddha, or whoever you want—but are legitimately looking at the problem in the modern scientific mode. And you have two (or more) camps—here, coherent structures and statistics—and they go on arguing with each other for a long time. Now if you look at the probabilities of that situation, what you'll find is that as time goes on the probability more and more favors the idea that both camps have some of the truth, and neither of them have all of the truth—otherwise somebody would get carried out on the shield, as John put it. So what that means, and I'm delighted to hear that happening here, is people saying— yes, if you get the statistics of the right animal, which means the right coherent structure, then of course you are doing the right things. So that guideline may help us. It is clear that there is some truth in coherent structures and some truth in statistics, and in the end they ought to be consistent with each other. That's what I want to say.

References

1. Kuhn, Thomas S. 1970 *The Structure of Scientific Revolutions*. Chicago: The University of Chicago Press.

Session Two

The Role of Coherent Structures

Discussion Leader: J. C. R. Hunt, Cambridge University

Future Directions in Turbulence Research and the Role of Organized Motion

Brian Cantwell

Stanford University
Stanford, CA

1. Introduction

Before focusing on the role of organized motion I believe it is appropriate to respond to the overall premise of this meeting as set forth in the green flyer which advertises it. In this vein a general discussion follows which includes remarks about the other areas of the meeting. Then general discussion provides a background for remarks on the role of the organized motion. Since this paper will be read by persons who have not attended or been involved with the meeting, the first paragraph of this flyer is reproduced here.

"Turbulence is rent by factionalism. Traditional approaches in the field are under attack, and one hears intemperate statements against long time averaging, Reynolds decomposition, and so forth. Some of these are reminiscent of the Einstein-Heisenberg controversy over quantum mechanics, and smack of a mistrust of any statistical approach. Coherent structure people sound like *The Emperor's New Clothes* when they say that *all* turbulent flows consist primarily of coherent structures, in the face of visual evidence to the contrary. Dynamical systems theory people are sure that turbulence is chaos. Simulators have convinced many that we will be able to compute *anything* within a decade. Modeling is thus attacked as unnecessary or irrelevant because it starts with Reynolds averaging or ignores coherent structures. The card-carrying physicists dismiss everything that has been done on turbulence from Osborne Reynolds until the last decade. Cellular Automata were hailed on their appearance as the answer to a maidens prayer, so far as turbulence was concerned. It is no wonder that funding agencies are confused."

We have been brought to discuss issues implied by the premise that:

modern approaches to turbulence have been overpromoted by their proponents attracting inappropriate levels of funding at the expense of traditional mean flow turbulence modeling.

Intemperate claims by mean flow turbulence modelers seem not to have occurred! Where are these sentiments coming from? What is meant by "classical approaches" and "mean flow turbulence modeling"? With a little bit of digging I was eventually able to satisfy myself that I understood the motivation for the meeting and I will try to summarize those findings here. I will take "mean flow turbulence modeling" to mean, as it is implied in the green flyer, modeling based on the time-independent Reynolds equations without consideration of the time-dependent organized part of the motion. This definition is consistent with the approach of

Shih, Lumley and Janicka (1987) who use a carefully developed algebraic stress model to solve for the mean properties of a variable density mixing layer of the type studied by Rebollo (1973) and Konrad (1976). Modeling in this case is put forth, not just as a method for solving a practical problem, but as a means of obtaining a better physical understanding of the flow. In particular the model is used to argue that, at high Reynolds numbers and low Schmidt numbers, density is transported like a passive scalar and the turbulence behaves like constant density turbulence. It is pointed out that significant details of the mean flow are reproduced by the model even though there has been no attempt to explicitly include the large eddies. This is contrary to views such as those expressed by Broadwell and Dimotakis (1986) who warn that models which do not explicitly or implicitly include the organized structure cannot capture subtle features of the mean flow particularly the asymmetry in the entrainment ratio on the two sides of the layer. It clear from the concluding remarks of Shih et al. that their work is motivated by the desire to squelch such claims although no reference is given.

My position on this particular issue is that the entrainment ratio is a consequence of the basic geometrical asymmetry of the spatially developing mixing layer and the inviscid character of the large eddies which, on dimensional grounds, assures a linear growth rate for the layer. Any model that does a decent job of reproducing the mean velocity profile (ie, the correct spreading rate, maximum velocity gradient and position of the dividing streamline) should get the right entrainment ratio. The same holds for the maximum Reynolds shear stress. By the same token this is not an especially stringent test of a model. The ability to reproduce the mean density profile and mass fraction fluctuation profile is the most impressive feature of the model. It is interesting and perhaps significant in thinking about the current state of turbulence modeling to note that, except for the streamwise normal stress, the simplified model equations in Shih et al. agree more closely with the experimental data than the full equations. This is reminiscent of something which was noted in the 1980-81 AFOSR - Stanford Conference On Complex Turbulent Flows (Kline, Cantwell and Lilley 1981); model effectiveness did not necessarily increase with increasing model complexity. From the standpoint of modeling, the more significant issue raised by Broadwell and Dimotakis is the persistence of Reynolds number effects beyond the point where the mixing layer would be considered fully developed.

Additional background for the meeting can be found in Lumley (1981) and Lumley (1989). The sentiment expressed in Lumley (1981) is that organized structure is a characteristic of low Reynolds number turbulence and probably not important at high Reynolds numbers where conventional modeling approaches are satisfactory. Much of the paper is devoted to a discussion of statistical approaches, particularly the use of proper orthogonal decomposition as an unbiased method to extract organized structure from an inhomogeneous random field. The 1989 paper reiterates and updates these ideas with notions about the role of chaos. Reference is made to the review by Guckenheimer (1986) who discusses the difficulties inherent in treating dynamical systems characterized by a high dimensional attractor and the possible need for a probabilistic rather than a geometrical approach. In the context of quoting unnamed colleagues Lumley (1989) presents visual evidence of the absence of organized motion in the form of a schleiren photograph of a round jet with a turbulent exit flow. The difficulties inherent in the interpretation of such images are not discussed. Throughout both of these papers there is a high degree of ambivalence toward organized structure; it is important, it is not; it is practical, it is academic. Methods for finding organized structure are put forward while at the same time the significance of the organized motion is called into question.

II LOOKING FORWARD

The green flyer lays down a challenge to identify the role of organized motion in the technology of turbulence. By this I mean the ability to solve practical engineering problems using methods founded on fundamental physical principles. In order to make an honest stab at responding to this challenge in the context of a debate over the basic premise described above I think it is both necessary and useful to try to imagine where turbulence will be a decade or so from now. Only in this way can we come up with a realistic set of expectations for progress and a reasonable set of recommendations for future funding priorities. The goals are to further basic understanding of turbulent phenomena, promote the efficient development of methods for solving engineering problems and to insure the general intellectual health of the field.

It is true that the movement of ideas from research to application in turbulence has been agonizingly slow and this has strained the patience of funding agencies and the credibility of basic research which has told us much about the nature of turbulence without generating new methods for solving engineering problems in turbulence. This was one of the concluding remarks in my 1981 Annual Reviews article and it still holds true today. This is not to say that there hasn't been any progress or that we haven't made significant inroads into certain areas. It is just that new methods of general applicability have not emerged. Looking forward to the world of the 1990's which promises rapid growth in the technologies of computer science, new materials, biology and the like, the technology of turbulence appears to be a backwater, stunted by the lack of progress in using new approaches to create new engineering methods.

II.1. Computation

Will the technology of turbulence still be a backwater ten years from now in 1999? I think not. I think the rate of advancement of supercomputing and especially personal computing technology will have a very significant impact on the way we approach engineering problems in turbulence ten years from now. To a certain extent I am with the simulators, mentioned in the green flyer, in that I believe that when one looks at the broad spectrum of problems where turbulence plays a significant role; from flows in turbomachines and combustors to chemical and biological processing to atmospheric and oceanic flows; many will be solved either by direct numerical simulation or by large eddy simulation. Mean flow turbulence modeling will remain a useful tool for certain cases however the range of problems requiring this approach seems likely to decrease.

Computer scientists are talking seriously, and I am told by my colleagues reliably, about teraflop (10^{12} floating point operations/sec) computing capabilities by the latter part of the next decade. With a relatively modest investment (< 40 K) one can buy desktop computing capability exceeding 20 megaflops today. Sun Microsystems has committed itself to doubling the speed of its processors every year for the next five years. Intel recently announced a new 64 bit RISC microprocessor which runs at 150 million instructions per second. It is probably not unrealistic to envision desktop supercomputing capability approaching 2 - 10 gigaflops by the late 1990's. We are likely to see similar advancements in memory capacity and high volume, local storage capabilities as well. Stunning high resolution imaging of data will be a routine tool. Significant advances in CFD software availability will also occur as the interest in "desktop engineering" takes on commercial dimensions. Moderately capable packages are already available. What this means is that the capability we see concentrated at a few centers today will, in the relatively near future, be available to most researchers and industry engineers while the capability of those centers will be greatly enhanced. Today we have a direct simulation of an incompressible turbulent boundary layer at $R_\theta = 1410$ (Spalart 1988) with a barely discernible logarithmic layer. Ten years from now, with the increased power predicted above, we should expect to have simulations at $R_\theta > 10,000$ corresponding to the upper end of available experimental data. Direct numerical simulations with a well developed turbulent spectrum with widely separated scales will exist for an increasing number of elementary flows including cases with compressibility and chemical reactions and most workers will have the capability of storing the information from these flows locally as well as the ability to carry out moderate Reynolds number simulations of their own. It is important to note that to a large degree the field of turbulent simulation has been motivated and supported by experimental observations of organized structure.

However simulation alone is not enough; it generates too much information. Simulations need to be accompanied by systematic methods for flow interpretation which can reduce complex flow patterns to basic elements in order to identify, summarize and relate significant features of the data.

II.2. Experiment

In the next decade numerical simulation will be increasingly used as a laboratory tool. A researcher doing fundamental experiments may well have available data from a high Reynolds number direct simulation of the flow under study and will also have the capability of carrying out numerical simulations in the laboratory. The ability to couple experimental measurements of organized structure with visualization of simulation - generated data for difficult to measure variables will be of great use in identifying new effects and developing explanations of underlying physical mechanisms. The use of the computer in the laboratory for data acquisition _and_ flow simulation will greatly broaden the prospects for basic experimental research which leads to significant technological applications as well as improved understanding. For example, direct simulations will be used to develop control strategies based on the sensitivity of the large eddies and their interactions to perturbations applied to the transition region. The same computer generated information could then be used as a command function to control the laboratory flow. Flow control based on modifications of the organized motion will probably occupy an increasingly important place in turbulence research. The researcher will be concerned with using basic understanding to generate methods for improving flow behavior in a coordinated program of computation and experiment. Flow control is recognized today as an important subject for basic research and there is no doubt that interest in this area will grow.

Broadly speaking current experimental turbulence research can be broken down into two categories. There are experiments directed at identifying and exploring new physical phenomena and there are experiments directed at confirming computations. Unfortunately support for experiments of discovery has been on the wane while support for experiments of code verification, particularly NASA support, has increased. True enough, measurements of this type serve in the development of improved standards of computation; an issue which will grow in significance with the democratization of computational power. However at present the coordination between experiments and computations is less than ideal often involving an experimenter who mistrusts the code and an analyst who doesn't understand the intricacies of measurement. I have been shocked at meetings to witness presentations of computational results with comparisons to experimental data with no acknowledgement of the source of the data. I consider this an abberation of the current dichotomy between computation and experiment. The sharp intellectual and social divisions which exist today will have no place in the laboratory of the future. Closely coordinated experiments and computations will be the hallmark of the best research and I believe that this will hold the key to advancements, not only in understanding, but also in the technology of turbulence. In an integrated research environment code verification will be a benchmark in any project. In this instance the emphasis will shift away from experiments where code verification is a primary objective and toward innovative high risk experiments directed at using basic understanding to develop methods for improving flow behavior.

The tensions which exist between computation and experiment arise partly from a lack of confidence. For various reasons, universities, including Stanford, do a poor job of instilling in their fluid mechanics PhD graduates the professional confidence required to move comfortably between theory, computation and experiment. There has been a considerable expansion of the curriculum to include CFD, hypersonics and other important new areas at the expense of an equally considerable drop in the time devoted to graduate level laboratory education. Associated with this has been a considerable amount of fragmentation in the way we teach fluid mechanics and the associated mathematics. Finally we have to deal with a natural human tendency to develop narrow interests in response to an increasingly diverse world. It is a little hard to predict where, in the future, advancements in mathematical and computational methods will be developed. The centers of computation which exist today will still exist with greatly enhanced capability in the late 1990's. However it seems to me that if necessity is the mother of invention then an increasing source of such methods will be the laboratory where computations and experiments are being carried out side by side by individuals versed in experimentation, mathematics and numerical analysis. Today's PhD graduate in fluid mechanics has a considerable amount of self education to do if he or she is to be effective in the fully integrated research environment of the late 1990's.

II.3. Modeling

There are many technologically significant problems where turbulence modeling can be used to generate useful engineering solutions and a broad range of examples can be found in the Proceedings of the AFOSR -Stanford 1980-81 Conference on Complex Turbulent Flows. Indeed mean flow turbulence modeling is presently the only method of solving certain practical problems and the problems of today can't wait for new approaches to gain practicality. There is a need today for improved turbulence models just as there was in 1981 (cf. the position paper by Bradshaw, Cantwell, Ferziger and Kline in Volume I of the 1980-81 proceedings). If this were 1981 with the future described above eighteen years off I would probably support the idea of putting more research money into improved mean flow turbulence modeling if only to resolve some of the issues raised during this conference concerning the mixing of numerical errors with modeling errors. Because of these problems we didn't get a very accurate picture of the real capabilities of turbulence models. But it is now 1989 and progress in simulation has moved rapidly forward. Has mean flow turbulence modeling made similar progress ? Can we expect that, with a large infusion of research funding, progress in mean flow turbulence modeling will outpace the progress in computation driven by advancements in computer hardware? I would put my money on the hardware.

In many circumstances modeling works; the difficulty I see is that when it works we usually don't know why and when it fails we don't have the basic physical understanding required to correct the problem. But will this remain so ? By the mid 1990's a design engineer needing to solve a turbulent flow may well have the capability of carrying out a direct or large eddy simulation of the flow. Even in large scale engineering projects involving high Reynolds numbers and complex geometries it seems likely that some large eddy simulations will be computed for at least a few cases. One could envision using a limited number of simulations to develop a time-averaged, flow - specific turbulence model tailored to the flow in question. Lilley (1983) has pointed out the need for mean flow turbulence models which reflect the changing character of the large eddy structure from flow to flow. The single flow model would be used in cases where repetitive calculations are needed. When a regime is encountered where the turbulence model fails, simulations can be carried out to develop a physical understanding of why the failure occurred and to suggest adjustments for the model. If such a capability is developed then it reduces the need for developing universal mean flow turbulence models. It may well be that Reynolds averaged turbulence models of the late 1990's will not be significantly different from the models of today (which are not significantly different from the models of 1981). What will have changed dramatically will be our ability to understand why models work when they do and why they fail when they don't. Model failure is not such a disconcerting thing if one can determine where and why the failure occurred and has the tools available to correct it.

Mean flow turbulence models will probably always enjoy some degree of success in predicting simple as well as complex turbulent flows. Mean flow fields are smooth and, in the absence of shocks, continuously differentiable to a high order. Any carefully drawn, dimensionally-consistent, mean flow turbulence model which conserves mass, momentum and energy and is invariant under basic groups of translation, rotation and stretching should give reasonable answers in circumstances where it is applicable. Therefore the question of whether mean flow turbulence modeling is useful or not will probably never have a clear cut answer. There will always be situations where it is useful and others where it is not and so it will always remain as one among a number of tools for solving problems in turbulence. The real question from the standpoint of this meeting is whether the range of applications of mean flow turbulence modeling will grow or diminish in the future. If the advances described above come to pass it seems certain that simulation techniques which incorporate the organized structure will continue to encroach upon the domain of mean flow turbulence modeling. One exception may be in the area of supersonic boundary layers where there will be an increased interest in turbulence research across the board.

I think mean flow turbulence modeling will always be a tool for a few specialists whereas simulation is likely to have broader appeal. There are several reasons for this. The process of model construction is complex and lacking in physical - intuitive concepts for guiding the uninitiated; it is very hard to get into. When I study equations typical of mean flow turbulence models the first question that comes to my mind is: what physical picture can I form which will help me understand the various terms in these equations? Nothing comes. Direct simulation techniques are not for the fainthearted either but at least one has confidence that the basic equations one is trying to solve are correct. Large eddy simulations probably lie somewhere in between although significant improvements in sub-grid-scale models are needed. The main reason why turbulence models are under attack from some quarters today, why they divide the community so deeply and why they will probably never gain wide acceptance in the future is simply their sheer lack of a sound physical and theoretical foundation. The fact that they work in certain circumstances is not enough. When one looks at the breadth of problems which need to be solved and, as we shall do shortly, at the list of cases which likely cannot be handled it is hard to envision the widespread acceptance of mean flow turbulence models by engineers and scientists of the late 1990's especially when more attractive and powerful methods are available. The complexity and absence of a sound theory are fundamental drawbacks of mean flow turbulence modeling which should not be underestimated.

There are currently a number of efforts to include organized structure in models of turbulence. Closure schemes based on the use of a superposition of inviscid hairpin vortices have been developed by Perry (1987), Perry, Li and Marusic (1988) and Perry, Li, Henbest and Marusic (1988). The method has been used to model wakes, mixing layers and wall turbulence. In the case of wall turbulence the model has shown the connection between properties of the organized structure and classical eddy viscosity models. The recent review of Liu (1988) discusses models of coherent structure as nonlinear instability waves

II.4. Toward a Theory of Turbulence

By a "theory of turbulence" I mean a theory which tells us about the nature of solutions of the Navier-Stokes equations at high Reynolds number. I do not mean a theory which is disconnected from the Navier-Stokes equations although it may turn out that the equations will be viewed as imbedded in a more general form.

Chaos theory provides a current example in turbulence research where inadequate education impedes progress and understanding. If you were to ask a physicist working in chaos to discuss the state of our understanding of shear layer mixing, or an engineer working on shear layers to explain the notion of strange attractors it is likely that neither response would be particularly edifying. Circular-Couette flow, and Bernard cell experiments aside, the connection between chaos theory and the general problem of turbulent shear flow has not yet been made although I think it eventually will be. Lack of understanding in this area is one of the reasons why chaos theory remains largely a curiosity to the average engineer. The situation is not helped by the fact that chaos has found its way into the popular literature where the technical treatment of turbulence is often superficial and misleading. See for example the discussion of turbulence in the otherwise interesting book by Gleick (1987).

Yet here is a field which has a significant potential for improving our understanding of the nature of nonlinearity and for providing a setting within which unresolved issues can be precisely stated. If a theory of turbulence is developed in the next decade my guess is that the wellspring of this theory will be a conjunction of three current areas of active research; research concerned with the topology of unsteady flow solutions in coefficient space (chaos), research concerned with the topology of unsteady flow patterns in physical space (organized motion) and research concerned with mathematical and numerical techniques which can be used to relate the two. The recent work of Aubry, Holmes, Lumley and Stone (1988) is an effort in this direction. Such a theory, if it is developed, would tell us much about the nature of solutions of the Navier-Stokes equations as the Reynolds number goes to infinity; it would be known whether the solutions remain regular or whether singularities could develop; constants in the Kolmogorov theory and the law of the wall and their possible dependence on Reynolds number would be determined, the existence or nonexistence of asymptotic growth rates for elementary shear flows would be known and rates of growth would be predicted. In a number of areas, empirical knowledge would receive a theoretical underpinning.

We might ask what contribution would such a theory make to the technology of turbulence ? Would it be of direct use in the design of, say, a wing-body junction on an airplane ? My guess is that for large scale complex flows the existence of a theory of turbulence would probably not have a very large direct impact on engineering practice. The situation would be somewhat analogous to the situation today in quantum mechanics. We feel very comfortable about our understanding of solutions of the Schrodinger equation and for very simple atomic and molecular systems we can solve for energy levels and transition probabilities from first principles. But for complex systems one still has to resort to numerical analysis coupled with empirical models. The impact of the theory is that it provides a guide for building the models. Similarly I think the main contribution of a theory of turbulence would be to provide fundamental principles which would guide the development of models. It would greatly enhance our understanding of why models work and why they fail. The limitations of a given model would be known a priori instead of a posteriori.

III. THE ROLE OF ORGANIZED MOTION

A number of reviews related to organized motion in turbulence are available including Willmarth (1975), Roshko (1976), Cantwell (1981), Ho and Heurre (1984), and Liu (1988). In addition the role of organized motion in turbulence has been discussed by Hussain (1981) and Coles (1981) and someone wishing further information should consult these references. In the context of this meeting I have chosen to frame this paper not as a review but as both a response to the basic premise of the meeting and an attempt to articulate my views of where the field of turbulence is headed. The previous sections did this in general terms, touching slightly on all the areas of the meeting. I would now like to turn to the role of organized motion.

III.1. The Role of Organized Motion in Transport

Townsend (1956) emphasized the importance of the large eddies in controlling turbulent transport and recognized that the eddies ought to have a quasi-deterministic form. Over the past thirty years numerous studies have have revealed aspects of the structure of the organized motion in a wide variety of flows. The topology and range of scales of the organized part of the motion varies widely from flow to flow. For example in the plane mixing layer the most apparent part of the organized motion appears to be a double structure consisting of strongly interacting two-dimensional rollers which span the layer plus a relatively well organized streamwise superstructure of smaller scale associated with regions of high strain which occur between the rollers. In the turbulent boundary layer the presence of a wall leads to a structure which is strongly three-dimensional, highly intermittent and in general much more complex than in free shear flows. The most intense motions are at a scale which is small compared to the width of the boundary layer.

The wide variation in organized structure from flow to flow is not suprising in that one would expect the motion to be the saturated nonlinear result of a basic instability driven by the boundary conditions and forces which define the overall flow. This notion underlies the so-called principal of marginal stability used by Lessen (1978) to account for trends in the turbulent Reynolds numbers of simple free shear flows. In this approach the organized structure is viewed as the result of linear instability of the mean velocity profile. It is essentially a heuristic in that the velocity fluctuations of the organized motion are not small and the mean profile is rarely, if ever, realized instantaneously nor is the instantaneous profile slowly varying. Nevertheless it is a useful idea which accounts for differences in flow geometry and forcing and can be used to argue for the regeneration of the large eddies in fully developed turbulence. The method was used effectively by Marasli, Champagne and Wygnanski (1989) to explain features of a turbulent wake in terms of the interaction of basic instability modes derived from a linear stability analysis of the mean profile. Aspects of the turbulent structure of the wake were accounted for although growth rates of various modes did not match the experiments.

There are also wide variations in the energy and stress associated with various scales of the organized part of the motion. In general studies show that eddies at the largest scale of the flow only account for a modest fraction of the stress (Hussain 1981). The most significant role of these eddies is that they differentiate the flow into regions of strongly varying strain and rotation and thus provide the setting for the first stage of coupling to finer scales. In this respect they control the overall transport. The near wake of a circular cylinder is a case where the large eddies are extremely well defined and would be expected to provide a substantial fraction of the stress. The measurements of Cantwell and Coles (1983) show that the periodic part of the motion associated with the large eddies accounts for only about fifty percent of the Reynolds stress. Hussain (1981) has attributed this to jitter in the position of the organized structure. However the observation holds within the first few diameters of the wake where such effects are not important as evidenced by relatively small values of background turbulent kinetic energy in regions where high gradients in the periodic part of the motion occur. In a frame of reference moving with the eddies the flow field is apparent as a moving pattern of centers and saddles. The saddle points are found to be regions of high shear stress and high production of turbulent kinetic energy. When the correlation coefficient of the background turbulence is formed it is found to vary widely with values near the saddle as high as 0.5 and as low as 0.1 at the centers of the vortices. Measurements in the plane mixing layer (Hussain 1980 , 1981) also indicate a peak in the production of turbulent kinetic energy at the saddles between the two-dimensional rollers. The mechanism in each case appears to involve stretching and organizing of three-dimensional vorticity by the strain field of the saddle. The stretched vorticity is aligned with the diverging separatrix of the saddle (ie. the positive direction of strain) forming an array of counter rotating vortices similar to those observed by Breidenthal (1981) and modeled by Corcos and Lin (1984). There really are no experimental studies with enough detail or simulations at high enough Reynolds number to tell us the scale at which organized structure ceases in a typical fully developed shear flow. We cannot, at this point, say definitively what fraction of the mean Reynolds stress could be regarded as contributed by motions which, by some measure, could be considered ordered. It is probable that the answer to this question, as with so many questions in turbulence, varies from flow to flow.

III.2. The Description of Organized Structure

I think it is fair to say that the observations of organized structure over the past four decades have been the motivation for virtually all new approaches to the field which have occurred during this time. From early studies of the intermittent nature of turbulence by Corrsin (1943) and Townsend (1947), to numerical simulations, to modal decompositions of flow solutions, to the theory of chaos (which is in fact the theory of *order - in - chaos*); one way or another, all have been motivated by the notion that the large scale motion in turbulent shear flows is characterized by a certain degree of order. While there is general agreement that order exists, there is a huge variety of methods for defining that part of the motion which is ordered and the subject as a whole remains controversial.

Now that simulations of turbulent flows are available, a major problem is one of understanding a vast amount of information. Simulation can compute the flow but it cannot interprete the flow solution. Despite the success of simulation there is a feeling that a true understanding of turbulence still eludes us. The problem solving engineer of the late 1990's will be faced with the same difficulty. To an important degree the usefulness of simulation for problem solving and new understanding will rest on the ability to synthesize complex flow information. The key to this will be to find an appropriate framework for describing and summarizing the organized part of the motion. A satisfactory methodology for the interpretation of complex three-dimensional data must involve more than simply an improvement in the technology of displaying the data but requires a systematic method for reducing complex flow patterns to basic elements in order to summarize and draw relationships between significant features of the data. An important point to keep in mind in thinking about this issue is that there is a vast variety of problems in turbulence and a flow interpretation scheme which satisfies one set of needs may be quite unsatisfactory for another. In this instance there are good technical imperatives for a pluralistic approach. A second point is that the structure of turbulence can be described on a number of different levels ranging from the mean or ensemble averaged flow at the highest level to the instantaneous flow at the lowest level and methods of flow field interpretation are needed at every level.

Representations of organized structure tend to fall into three catagories which could be roughly described as statistical, phenomenological, and topological with a good deal of overlap between the three. Statistical approaches have played a very important role in the experimental determination of the organized structure of turbulence. The history of this subject is one where, at each stage of development, the amount of information derived has been limited by the available techniques of the day. The evolution of statistical knowledge has moved from spatial correlation information derived from a few thousand single or two point hot-wire velocity measurements to elaborate computer controlled conditional sampling experiments involving millions of measurements over a field. Antonia (1981) has given a comprehensive review of conditional sampling techniques and the various approaches used to identify organized motions experimentally. A recurring point is the difficulty of relating Eulerian and Lagrangian information. Hussain (1981) discusses analytical tools for decomposing the flow field, various methods for characterizing coherent structures and experimental techniques for finding them in boundary layers and free shear flows. Hussain (1981, 1986) defines a coherent structure as a turbulent fluid mass connected by a phase correlated vorticity. Beside the somewhat circular nature of this definition I am not comfortable with the idea of trying to assign any sort of strict nounal definition to what is essentially an adjectival concept; ie descriptive of that part of turbulence which is ordered. I think to tie down the concept of organized motion too closely would be to cause it to lose its usefullness. For this reason I dont like the phrase "coherent structures" because it predisposes us to the idea that the organized part of the turbulent motion consists of "things"; like soft mushy rotating billiard balls and tends to ignore the elliptic field nature of turbulence which I believe must be faced head on if we are to reach an improved understanding of the subject.

The availability of direct simulations has enabled conditional averaging techniques to be applied to complete three-dimensional flow fields. Moin (1984) used the technique of proper orthogonal decomposition, suggested by Lumley (1981) as an unambiguous way of identifying organized structure, to study a simulation of turbulent channel flow. Recently Adrian and Moin (1988) have developed a rapid estimation technique which enables simulations of homogeneous shear flow to be used to study average motions conditioned on the velocity vector and deformation tensor; a study which would be impossible without simulations. They were able to identify the topology of correlated flow events which contribute to the Reynolds shear stress.

Phenomenological approaches look directly at the flow field and attempt to use visual information to identify key features of the motion and to identify the role of these features in the generation of turbulent transport. Experimental studies of this sort have been limited by the experimental techniques at hand and hampered by the conflicts which arise in the attempt to relate the instantaneous velocity and vorticity fields to the results of flow visualization which usually involved the time integrated effect of the flow on a tracer. The ambiguities of Lagrangian visualization techniques are nicely illustrated by the far wake studies of Cimbala, Nagib and Roshko (1988). Shariff (1989) studied timelines in the flow about a vortex pair subjected to an oscillating strain field. Extremely complex patterns were produced even though the underlying velocity field was quite smooth and Shariff discusses the difficulties of relating Lagrangian and Eulerian turbulence. Simulations have helped to clarify some of these problems. The recent survey of near wall structure by Kline and Robinson (1988) is an attempt to reach a consensus on a variety of disparate pictures which currently exist in the literature and to begin piecing together a single, consistent physical model of the flow. This is another example of an undertaking which would not be possible without the kind of simulation data which is now available. Using the computations of Spalart (1988), Robinson, Kline and Spalart (1988) have studied graphical images of various quantities in the turbulent boundary layer including the pressure field, stress fields, velocity vector field and vorticity. A great deal of information is emerging from these studies which are still at a relatively early stage. The physics of the motion is found to be even more complex than was previously thought and it appears that a given visual observation can have more than one underlying mechanism (Kline, private communication).

Topological methods are useful in the description of fields. They can be used in the multidimensional space of coefficients of a modal decomposition of the equations of motion, where the object of study is the region of attraction to which the solution tends at large time, or they can be used in three-dimensional physical space where the object of study is the critical behavior of the unsteady flow field. The former is the setting for theories of chaos, the latter is the setting for a topological description of organized motion. Perry and Chong (1988) have recently reviewed the use of critical point analysis in the description of unsteady flow fields. They emphasize the fact that the method provides a wealth of topological language particularly well suited to the description of fluid - flow patterns. Perry and his co-workers (Perry and Lim 1978; Perry, Lim and Chong 1980; Perry and Chong 1987) have made extensive use of critical point theory to describe complex flow patterns in steady and unsteady three-dimensional flows. Topological methods have recently been used by Lewis, Cantwell, Vandsburger and Bowman (1988) to describe the kinematics of flame breakup in an unsteady diffusion flame. This method focuses on the problem of connecting vortex structures together to complete the flow field. However there are significant conceptual problems involved in interpreting the unsteady streamline patterns as they relate to entrainment since streamlines can move across pathlines and the pattern of streamlines depends on the frame of reference of the observer.

Certain time dependent flows can be reduced to a self-similar form including the class of flows referred to below as one-parameter shear flows. In this case the topology of the flow can be described in terms of particle paths in similarity coordinates. This procedure was used by Cantwell, Coles and Dimotakis (1978) to describe the self-similar flow in the plane of symmetry of a turbulent spot. Experimental data was collapsed onto (x/t, y/t) coordinates and the phase portrait of particle paths was used to determine the rate at which fluid was entrained into various regions on the centerline of the spot. Glezer (1981) used $(x / t^{1/4}, y / t^{1/4})$ coordinates to freeze the large-scale structure of a turbulent vortex ring. In both of these cases, an assumption of Reynolds number invariance was invoked and non-self-similar motions following a viscous timescale were averaged out through the use of the large eddy timescale for assigning phase information to the velocity data. The entrainment diagrams generated by this procedure have the useful property that they are independent of the observer. Prospects for using this approach in modeling the organized structure of turbulent shear flows are discussed by Coles (1981) who proposes an eddy viscosity model for the propagation of the turbulent-non-turbulent interface with a diffusivity proportional to the background turbulent kinetic energy. Griffiths (1986) has used a drifting Stokes flow solution in similarity coordinates to describe entrainment by a buoyant thermal. The model is used to explain laboratory observations of the distortion of dyed fluid blobs at low intermediate and high values of the Rayleigh number. Critical Reynolds numbers in the starting process for a class of impulsive jets were determined by Cantwell (1981, 1987). A complete picture of the evolution of the flow with increasing Reynolds number was deduced just from considerations of boundary conditions, integrals of the motion and the invariance properties of the governing equations. It was found that all of the significant topological properties of the solution could be conveniently represented by trajectories of the critical points in the space of invariants of the local deformation tensor. This scheme of flow representation was first used by Cantwell (1979, 1981) to classify the topological properties of various turbulent shear flows. The method has several attractive features for concisely summarizing flow fields. In an incompressible flow the first invariant of the deformation tensor is zero and therefore the trajectories of the critical points are restricted by continuity to lie in the plane of the second and third invariants even though the flow field may be three-dimensional. Thus the complete topological history of a three-dimensional unsteady flow can be represented in a plane.

Recently Chong, Perry and Cantwell (1988) have described a generalized approach to the classification of elementary three-dimensional flow patterns in compressible and incompressible flow. Although the attention in this paper is on the topology of the velocity field as determined by its associated deformation tensor, the method can be applied to any smooth vector field and efforts are currently under way to determine the topology of the vorticity and pressure gradient vector fields in the compressible wake computations of Chen, Cantwell and Mansour (1989). The vorticity field is interesting because the first invariant of the vorticity deformation tensor is zero for both compressible and incompressible flow. The pressure gradient field is interesting because the deformation tensor of this field always has real eigenvalues with orthogonal eigenvectors.

III.3. The Effects of Reynolds Number

The dynamical significance of the mean flow is often ignored. Not only is the mean field an expression of the forces and boundary conditions which define the flow, but it exhibits the most important dynamical property of turbulence; Reynolds number invariance. This is the well known property that, once the Reynolds number is large enough for turbulence to occur, the overall properties of the flow away from walls are observed experimentally to be almost independent of the Reynolds number. Most models of turbulence begin with something equivalent to an assumption of Reynolds number invariance although it does not have a strong theoretical foundation. Probably the clearest way to think about this is to imagine that the kinematic viscosity of the fluid is varied while all forces and boundary conditions are held constant. For example in a high Reynolds number jet a factor of 10 decrease in the kinematic viscosity will have relatively little effect on the mean velocity and Reynolds stress fields. The velocity fluctuation levels scale with the characteristic velocity of the flow $U' \sim U_0$ and tend to be independent of ν. The main effect is that the spectral content of the velocity widens to contain more high frequency components. The reason this is dynamically significant is that it implies that the rate of dissipation of turbulent kinetic energy is also independent of viscosity. The dissipation is linearly proportional to ν, yet observed to be independent of ν when the Reynolds number is large.

The simplest flows to consider are one-parameter turbulent shear flows. These are flows governed by a single invariant of the motion or global parameter; a momentum flux in the case of jets, a velocity difference in the case of mixing layers, a buoyancy flux in the case of plumes, a drag per unit volume flux in the case of wakes, hydrodynamic impulse in the case of vortex rings, etc. (See Cantwell 1981 for a more detailed enumeration). Once the assumption of Reynolds number invariance is invoked and viscosity is removed from the problem the solution depends only on the global parameter. In this case the existence of a similarity solution for the ensemble - averaged flow is assured. If the flow is governed by more than one parameter with units that are incommensurable the symmetry of the problem is broken and a global similarity solution does not exist although there may be regions in the flow where local similarity holds.

There probably does not exist a real flow which is completely governed by only a single global parameter. Virtually all flows involve a variety of length and velocity scales related to the effects of transition, geometry of the apparatus, presence of free stream turbulence, noise, etc. Once the flow is fully developed to a point where Reynolds number invariance can be invoked, the global parameter of the motion dictates the power of time or space with which the overall velocity and length scales of the flow develop nevertheless the effects of local parameters of the problem can creep in through modifications of the rates of growth or decay. Although one-parameter flows are relatively simple they are an important class in that they form the basis for much of what we know about turbulence and a tremendous amount of research has been devoted to their study. Their geometrical simplicity helps to focus basic unsolved issues of turbulence. Historically they were the genesis of simple zero and one equation models of turbulence and with the input of a single empirically determined constant (a spreading rate or mean velocity decay rate constant) useful engineering solutions were produced.

However modeling shed no light on how this constant could be determined and it is still so today that we do not have a theory which will enable us to solve, from first principles, the simplest conceivable turbulent flow with the simplest conceivable boundary conditions: Why ? In a sense Reynolds number invariance is both a simplification, because in so far as the mean is concerned viscosity can be approximately ignored, and a source of great complexity because the role of viscosity is subtle and cannot be ignored completely. Viscosity plays a central role in the time evolution of the flow through instability. With the observations of organized structure and the recognition that the time-dependent motion needs to be included in models of the flow, mean flow turbulence models were replaced by other approaches. Vortex methods, for example, do a reasonable job of simulating the spreading rate of turbulent shear flows but the solutions depend on how the vortices are defined and on how the vorticity is introduced into the flow. Among all the possible solutions which the Euler equations might admit for a given flow geometry and a given source of vorticity, viscosity limits the possibilities in ways we are only beginning to understand.

III.3.a Transition

In recent years beginning with the work of Bradshaw (1966) we have come to appreciate the importance of the transition region in determining downstream flow behavior. The spectral content of the initial region and the relative phases of various modes have a very strong effect on the interactions of the developing large eddies in turbulent shear flows and thus on the way the mean flow develops, even though the initial amplitudes may be very small. The question has been raised as to whether the high Reynolds number growth rate of the plane mixing layer is unique. This issue has been studied carefully by Browand and Latigo (1979) who used a large facility to study a plane mixing layer with laminar and turbulent splitter plate boundary layers. The turbulent case was found to grow more slowly at first, due the time required for the turbulence to adjust from a boundary layer structure to that of a free shear layer, but eventually approached the same spreading rate as the laminar case. Recent measurements by Wygnanski and Weisbrot (1987) show that coherent forcing of the mixing layer can increase the distance required for the flow to approach the asymptotic spreading rate. Nonuniqueness in the similarity structure of fully developed wakes has been observed by Wygnanski, Champagne and Marasli (1986) who studied the far wakes of various shaped two-dimensional bodies which were designed to have the same drag. They found that, while the wakes they studied were self similar when normalized by their own velocity and length scale, the evolution of the characteristic velocities and lengths depended on the geometry of the wake generator. Different bodies generated different dimensionless wakes.

III.3.b Mixing and Combustion

Viscosity plays a particularly subtle role in the problem of scalar mixing. Beyond the usual velocity transition region lies a so-called mixing transition associated with the onset of three-dimensionality (Breidenthal 1981). A conceptual model of this three-dimensional motion has been described by Bernal and Roshko (1986) and a theoretical model has been developed by Corcos and Lin (1984). The detailed structure of three-dimensional disturbances upstream of the mixing layer have been shown by Lasheras Cho and Maxworthy (1986) to influence the subsequent development of streamwise vorticity. In the case of a chemical reaction, the dynamics of the reaction are most strongly coupled to the flow at scales where the scalar gradients and strain are the largest; scales where viscosity dominates. Chemically reacting flows push the limits of full simulations but a few studies in relatively simple geometries are beginning to appear (McMurtry, Jou, Riley and Metcalfe 1985, Jou and Riley 1987, Rutland and Ferziger 1989, Mahalingham, Cantwell and Ferziger 1989). A fast numerical method for computing large eddies in reacting flow fields is described by Oran, E. S. and Boris, J. P. 1987. This method relies on numerical dissipation to stabilize the computation of solutions of the Euler equations on a coarse grid. While the method is incapable of treating viscous effects except by analogy, basic features of the large eddies and their interactions are reproduced rather faithfully. Models of shear layer mixing with chemical reaction which implicitly include the organized structure have been developed by Broadwell and Breidenthal (1982), Dimotakis (1989) and Broadwell and Mungal (1988). These models account for the effects of Reynolds, Schmidt and Damkohler number on the overall reaction rate. These studies are motivated by the recognition that an understanding of the effects of viscosity, scalar diffusion and chemistry on combustion requires an understanding of how the flow responds to the straining and rotational motions induced by the large eddies. One of the important results of this experimental work which has implications for modeling is the observation that significant molecular diffusion effects persist to high Reynolds numbers at which the flow would ordinarily be regarded as Reynolds number independent.

III.3.c Jet Noise

The role of organized motion and the possible effects of viscosity comes up in connection with the problem of jet noise. Inviscid models of vortex ring dynamics produce a large variety of solutions and instability modes only a few of which are observed in experiments. In the case of the azimuthal instability of vortex rings for which an inviscid theory has been provided by Widnall and Tsai (1977) two modes have nearly identical growth rates (within two-tenths of a percent) but, apparently due to viscous effects, only one of them is experimentally observed (Shariff - private communication). Kambe (1986) has studied the head-on collision between two vortex rings. He points out the importance of viscous effects in the noise produced in the late stages of the collision when the cores come into contact and their radii increase rapidly. Experiments by Hussain (1983) with controlled excitation revealed that vortex pairing can be an important source of sound but it is pointed out that in the natural jet clean vortex pairing events are relatively rare and other sound generation mechanisms must be sought. Vortex structures in the near field of the jet undergo azimuthal breakdown starting at about two diameters from the exit and this led Bridges and Hussain (1987) to suggest that vortex filament cut-and-connect processes, which occur during vortex breakdown, are a more important sound generation mechanism than vortex pairing in practical turbulent jets. On the other hand Michalke (1983) has demonstrated that azimuthal coherence is necessary for sound production and only low-order azimuthal modes can radiate efficiently. Michalke notes that the sound field of a jet depends significantly on the axial and azimuthal source coherence and that the coherence length scales of the sound radiating turbulence increase when the turbulence is excited artificially. These studies suggest that phase relationships between the large eddies and coupling to finer scale motions through instabilities governed by viscosity may play a significant role in the generation of noise.

III.3.d Separated Flows

Mean flow turbulence models are often concerned with complex geometries where viscous effects of the type described above are likely to be particularly difficult to handle. Regions where separation occurs at low Reynolds number may diminish or disappear altogether at high Reynolds number. Should we therefore expect mean flow models to work better as the Reynolds number becomes very large? Can we ever expect to be free of the effects of transition? The classic example of flow past a circular cylinder illustrates well the persistence of transition to be expected in the flow about any smooth bluff shape. Even with fixed separation points, bluff body flows show the effects of transition to Reynolds numbers well beyond the point where we expect the flow to be fully developed (Roshko and Fizsdon 1969). Recently Schewe (1986) has studied jumps and hysteresis effects in the flow about a circular cylinder in the critical Reynolds number range 300,000 - 400,000. Sudden changes in flow structure triggered by small disturbances in the surface boundary layer can cause asymmetric lift with lift to drag ratios as large as two. Recently Williamson and Roshko (1988) have documented a whole range of resonances, jump phenomena and hysteresis in the vortex wake of a circular cylinder subjected to controlled oscillations. Similar complex phenomena were observed by Nakamura and Nakashima (1986) in the wakes of self excited bluff prisms. Although both sets of experiments were at relatively low Reynolds numbers it is not unreasonable to expect that aspects of the same phenomena will occur in the fully turbulent case. The general class of problems with large scale overall unsteadiness would appear to be one which is clearly outside the scope of mean flow turbulence modeling but amenable to simulation. The problem of determining the lift and drag of a bluff body for truly high Reynolds numbers ($Re > 10^{10}$) will continue to be beyond the power of simulation, and for that matter experiment, for decades to come. The problem studied by Schewe may be just barely in range of simulation by the end of the next decade.

III.3.e Flows with buoyancy

Zeman (1981) reviews the status of turbulence modeling of planetary boundary layers and discusses the use of second-order closure schemes. The complexities of modeling buoyancy driven turbulence which can support countergradient transport and the lack of information on the contributions of buoyancy to the pressure-velocity terms in the Reynolds stress transport equations are discussed. Zeman points out the importance and difficulties of modeling molecular diffusion terms and similar comments on the modeling of these terms can be found in Shih et al (1987). The viscous terms involve correlations of fluctuating strain rates which receive their largest contribution from the fine scale motions. The hope is that these motions will exhibit universal character. Antonia, Anselmet and Chambers (1986) have studied the local isotropy of various fields in a heated plane jet at moderate Reynolds numbers and they review the general state of knowledge of local isotropy. They point out that available data indicates that mean square derivatives of velocity and temperature are anisotropic and suggest the need to include this in models.

Recent research in buoyancy driven flames has focused on the coupling between the flow field and the reaction field and facets of this issue have been studied by a number of investigators. The effect of pressure variation on the structure of a low speed, co-flowing jet diffusion flame was studied by Strawa and Cantwell (1989). This type of flame is subject to a classical flickering instability manifested as strong, self-excited, longitudinal oscillations driven by, but not solely dependent upon, buoyancy. The flame was found to be extremely sensitive to the frequency of velocity perturbations applied to the jet exit flow. When the frequency of excitation was close to the flickering frequency the flame was seen to break up into a sequence of turbulent flamelets which exhibited an extremely repeatable three-dimensional structure. In an effort to understand some of these effects in a simpler flow configuration with approximately the same density ratio, Subbarao (1987) carried out an extensive experimental study of a co-flowing jet of helium into air subject to self-excited oscillations similar to the flickering oscillations of the low speed flame. In a buoyancy dominated range of Richardson numbers above one the helium jet was also found to exhibit an extremely regular and repeatable structure over a wide range of scales at Reynolds numbers where the jet would ordinarily be considered turbulent. However the helium jet was very insensitive to perturbations of the jet exit flow. The conclusion drawn from these studies is that the downstream development of the flow is strongly dependent on the details of how buoyancy is released. In the flame buoyancy is produced in a spiky fashion in the flame sheet near the jet exit. This flame sheet is very sensitive to small fluctuations of the jet exit velocity. In the helium jet the buoyancy flux is produced across the entire jet exit and the flow is much less sensitive.

III.4 Flow Control

In the integrated research environment described in Section II the use of simulation in conjunction with experiments to accomplish improvements in flow behavior will be of increasing interest. In one of the earliest examples of flow control Roshko (1954) demonstrated that a splitter plate placed in the wake of a bluff body caused a significant increase in the base pressure and consequently a reduction in drag. The splitter plate interferes with the transverse flow in the near wake. The strength of the vortices is reduced and they are forced to form further downstream relieving the base of the body from being subjected to the very low pressures associated with unencumbered vortex formation. The basic idea of interfering with or modifying the organized structure of the flow to achieve a more desirable flow state forms the basis for a whole variety of experiments directed at flow control. A summary of recent work in this fast growing field may be found in Liepmann and Narasimha (1987). A recent review of turbulence control in wall-bounded flows is given by Bushnell and McGinley (1989). Most of the attention in these references is on open loop flow control in which there is no attempt to use feedback.

From the body of research which has been carried out thus far, most of which is devoted to the open loop response of flows to external forcing, it seems clear that the controllability of a flow is intimately connected to its stability. For this reason attention has been focused on the transitional region of the flow and its role in determining the downstream development. In the few cases where feedback and control has been attempted such as the work of Liepmann and Nosenchuck (1982) this has been the case. In their work feedback was used to suppress a pure harmonic in the linearly unstable region of a flat plate boundary layer. In a more complex situation involving say mixing in a free shear layer, one can conceive of using combinations of unstable modes to achieve a certain desired effect on overall flow behavior in the presence of external disturbances.

Research on flow control will increasingly involve coordinated experimental and computational efforts. In this approach experiments are used to search for and identify physical mechanisms which can then be examined in detail using simulations. The recent thesis work of Mittlemans (1989) is an effort to use flow simulation to control the wind tunnel response of a delta wing at high angles of attack. Experiments will need to access more realistic flow conditions which are not achievable in simulations. It is critical to be able to establish whether the mechanisms identified for control at, say, low Reynolds number still play a role at high Reynolds number and future research in active flow control has to be capable of addressing this question.

III.5 Modeling

In the last decade or so studies of organized motion have led to a recognition that viscous and molecular diffusion effects persist to higher Reynolds numbers than was previously thought and play an important role in determining the properties of fully developed turbulent flows. These effects are felt directly through the influence of viscosity on three-dimensional breakdown and scalar transport processes and indirectly through the influence on transition and the phase relationships between the developing large eddies which affect downstream flow development. These are the common threads which run through the examples discussed above. The list of flows known to be subject to viscous effects grows longer as our fundamental understanding of the nature of turbulence improves. It is likely to grow longer in the future as flow control becomes a central theme of turbulence research. How will mean flow turbulence models incorporate these effects? Will they do it before the power of simulation overtakes them?

IV. WHAT IS NEEDED FOR THE FUTURE

IV.1. Pluralism

The field of turbulence is described in the green flyer as "rent by factionalism". But what some may term factionalism I would call pluralism. It is true that turbulence is being studied along a number of different lines of approach and in recent years there has been an influx of physicists to the field which has been traditionally dominated by engineers. Physicists bring a somewhat different point of view to bear; more inquiring of fundamental questions, less interested in practical applications, more critical of models which lack underlying theoretical justification. The problem of turbulence is presently being approached along a broad front and in my view this is very appropriate given the importance of the subject, the breadth of problems which need to be solved, and the current lack of basic understanding. It seems to me that so far, funding agencies have been fairly enlightened about their willingness to fund new approaches and this should continue. Support for research into cellular automata as a means of flow simulation is in this spirit. The computationally intensive nature of these calculations presents a significant challenge which needs to be overcome and it is probably too early to tell how this approach will contribute to the problem of turbulence simulation. It may gain in significance with the advent of massively parallel machines. In any case I believe a pluralistic approach is essential for the future progress of the field and that it will remain so into the 1990's and beyond. By this token continued funding should be available for mean flow turbulence modeling as long as it does not represent a significant shift away from more promising areas of research.

IV.2. The Democratization of Supercomputing

By the late 1990's many workers will have desktop computing capability comparable to the largest supercomputers of today. In order for such a vision to be realized, an immense capital investment will be needed by universities, industry and the funding agencies. Advances in technology have a way of forcing themselves on us and I expect that one way or another the funds will be found to achieve this but *real increases are needed in funding for turbulence research if the technology of turbulence is to develop as it should.* The merits of widespread supercomputing are described above; but there are challenges too. The number of workers in CFD today is very large and growing and complaints about the quality and reliability of what is being done are common. If the future described above is realized the problem will be completely out of hand unless some sort of standards exist. Reliable workers in the field routinely do this today; full simulations are checked to demonstrate consistency with the results of linear stability theory, global conservation laws are checked, etc. In the case of compressible simulations the acoustic transmission properties of the grid should be carefully documented. Anisotropy of the acoustic speed or wave reflections caused by the grid can feed erroneous perturbations into the flow throwing off calculations of viscous unstable flows. I am unaware of anyone who does this rigorously at present.

The issue of maintaining computational standards is an especially important and difficult one for mean flow turbulence modeling since the problem is exacerbated by the fact that as the constitutive equations grow in complexity the numerical schemes needed to solve them also become more complex and the ability to make numerical checks with established theories diminishes. This came up in the 1980-81 AFOSR-Stanford Conference on Complex Turbulent Flows (Kline, Cantwell, Lilley Vol III 1981) where numerical problems severely limited the ability to make meaningful comparisons. One of the challenges of the 1990's will be to develop a set of standards of computation which workers in the field will adhere to and which will insure that published results are reliable while not discouraging innovation.

By the end of the 1990's high Reynolds number direct simulations of a number of elementary flows will exist and the results of these simulations will need to be made available to researchers who will have the capability of storing computed flow solutions locally or running codes to generate additional data. This will have the beneficial effect of forcing error checks and uncertainty analyses to be carried out on the numerical data similar to checks of experimental data which are released for general use today.

Software which generates software for turbulence simulation will be needed to relieve the average researcher from having to re-invent the wheel. The software packages which accomplish this in the future will resemble highly evolved versions of typical scientific software packages available today. The main difference is that they will need to be designed to allow a great deal of user intervention to try out new ideas and approaches.

IV.3. Support for Integrated Research

The funding patterns of today don't particularly encourage the kind of closely coordinated experiments and simulations which I have described above simply because the capability for such research is only beginning to occur. As the power of flow simulation moves into the laboratory, increased funding will be required to support research efforts which involve fully integrated experimental and computational investigations of turbulence. There is a number of possible ways that turbulence research might benefit from this approach.

(1) Perhaps the greatest benefit of having flow simulation capability in the lab is the possibility for rapid interpretation of experimental observations which would be used to guide the next round of experiments. The position of the experimenter would be rather like that of an amatuer sport fisherman who is suddenly handed a sonar imaging system for finding schools of fish. There is likely to be a sort of positive feedback effect where better experiments lead to better simulations which lead to better experiments and so on. Rapid turn around between the experiment and the simulation is essential for this synergism to work effectively.

(2) In - the - lab - simulation will afford flexibility in the rapid display of difficult to measure variables to understand how they relate to observables of the flow. This will aid in the efficient and unambiguous interpretation of flow visualization data. Experimentation will be made more efficient by permitting the numerical study of a large number of flow conditions punctuated by a few well chosen measurements.

(3) Advanced methods of experimental flow control are likely to be highly dependent on simulations for the generation of command and feedback information for the flow. Ultimately the simulation will be part of the control system.

IV.4. Funding Priorities

What will the funding for turbulence research look like in the late 1990's? If the last decade is any indicator it is not likely that turbulence research will receive a significant increase in real dollars. The fact is that funding for turbulence research is too low and we have taken an awful beating in recent years; the funding agencies just got tired of hearing that the solution to the turbulence problem is just around the corner. Perhaps my description of the future will be passed off as just another call for patience. I hope not. I believe today we can see the future more clearly. We have all witnessed the amazing progress in computing. There has also been significant progress in analysis and in the development of numerical methods so that today we have fairly mature methods of direct numerical simulation in hand for a limited number of flows. It is not too much of a leap to suggest that flow simulation will become an engineering tool in the future.

Turbulence should not be regarded as just another engineering discipline. Progress in almost any technical endeavor involving movement at sea or in the air or involving industrial processing of a fluid is nearly always limited in one way or another by the problem of turbulence. We have not been effective at getting this message across to the funding agencies and as a result contract monitors in fluid mechanics have not been able to compete effectively with their counterparts in other disciplines. Part of the reason is that too many people are chasing too few dollars; the inevitable result of that situation is that congress and the funding agencies do not get unbiased advice. People working in different areas tend to disagree with one another in rather uninformed ways and funding agencies dealing with fixed budgets hear a cacophony of inconsistent voices from researchers with disparate points of view. In spite of this it not my impression that the funding agencies are confused. They are just forced to work with limited funds.

For the purposes of this meeting, our discussion of the future of turbulence should probably be predicated on the assumption that any significant increase in funding in one area will have to be at the expense of another. The real question is: *How should we view funds for basic research and what is the appropriate balance between short term and long term goals?* Should basic research money be used to fund the development of methods which don't advance our understanding but do advance our capability of solving practical problems even though the actual employment of such methods in engineering practice may never occur? Or should these funds be used to develop new fundamental knowledge of lasting value? At the present time the emphasis is on fundamentals and the field is advancing rather rapidly. Given the applications that this basic knowledge will have in the future this does not seem to be the time to move toward mean flow turbulence modeling if it is at the cost of reduced funding for research in simulation, modeling via simulation, studies of organized structure, or other areas of fundamental research.

I don't think we should regard current funding levels as acceptable. Increased funding is needed. As a community we need to think seriously about trying to develop a consensus on a reasonable set of priorities and to get our message across to Congress and the funding agencies in a unified way. Physicists have done this effectively for many years. Our job is more difficult in that the list of priorities must satisfy both scientific and engineering needs.

V. CLOSING REMARKS

In this position paper I have tried to express my view of where the field of turbulence is headed and to describe some of the forces which shape current research.

(1) Research on organized structure has led to an increased awareness of the importance of viscous effects which persist to high Reynolds number.

(2) Aspects of this work form the impetus for an increasing emphasis on flow control through the use of coherent forcing to modify the complex interactions of the large eddies.

(3) At the same time powerful simulation methods are appearing which have the potential for handling these complexities.

(4) There is rapid progress in the hardware needed to use simulation methods. Things are moving vertically toward greater computational power and horizontally toward increased availability.

The future of this field is just beginning to be realized and there are exciting prospects for combining theory, experimentation and simulation to rapidly advance the technology of turbulence. Meanwhile basic questions about the physical and theoretical foundations of mean flow turbulence models remain unanswered and the overall progress of modeling has been disappointing. Although there will continue to be a need for improved mean flow turbulence models, the combined effects of the forces which drive current turbulence research seem likely to reduce the range of problems treated by these models in the future.

VI. REFERENCES

Adrian, R. J. and Moin, P. 1988 Stochastic estimation of organized turbulent structure: homogeneous shear flow. J. Fluid Mech. 190 : 531 - 559.

Antonia, R. 1981 Conditional sampling in turbulence measurement. Ann Rev. Fluid Mech. 13 : 131 - 156.

Antonia, R. A., Anselmet, F. and Chambers , A. J. 1986 Assessment of local isotropy using measurements in a turbulent plane jet. J. Fluid Mech. 163 pp. 365 - 391.

Aubry, N., Holmes, P., Lumley, J. and Stone E. 1988 The dynamics of coherent structures in the wall region of a turbulent boundary layer. J. Fluid Mech. 192 pp. 115 - 173.

Kline, S. J., Cantwell, B. J. and Lilley, G. M. 1981 Proceedings of the 1980 - 81 AFOSR - HTTM - Stanford Conference on Complex Turbulent Flows.

Bernal, L. P. and Roshko, A. 1986 Streamwise vortex structure in plane mixing layers. J. Fluid Mech. 170 pp. 499 - 525.

Bradshaw, P. 1966 The effect of initial conditions on the development of a free shear layer. J. Fluid Mech. 26 (2): 225 - 236.

Bridges, J. E. and Hussain, A. K. M. F. 1987 Roles of initial condition and vortex pairing in jet noise. J. Sound and Vib. 117 (2) : 289 - 311.

Breidenthal, R. E. 1981 Structure in turbulent mixing layers and wakes using a chemical reaction. J. Fluid Mech. 109 : 1 - 24

Broadwell, J. E. and Breidenthal, R. E. 1982 A simple model of mixing and chemical reaction in a turbulent shear layer. J. Fluid Mech. 125 pp. 397 - 410.

Broadwell, J. E. and Dimotakis, P. E. 1986 Implications of recent experimental results for modeling reactions in turbulent flows. AIAA J. 24 (6): 885 - 888.

Broadwell, J. E. and Mungal, G. M. 1988 Molecular mixing and chemical reactions in turbulent shear layers. 22nd International Symposium on Combustion, Seattle. The Combustion Institute.

Browand, F. K. and Latigo, B. O. 1979 Growth of the two-dimensional mixing layer from a turbulent and non-turbulent boundary layer. Phys. Fluids 22 (6) : 1011 - 1019.

Bushnell, D. M. and McGinley, C. B. 1989 Turbulence control in wall flows. Ann. Rev. of Fluid Mech. 21 : 1 - 20.

Cantwell, B. J. 1979 Coherent turbulent structures as critical points in unsteady flow. Arch. Mech. Stosow. 31 : 707 - 21.

Cantwell, B. J. 1981 Organized motion in turbulent flow. Ann. Rev. Fluid Mech. 13 : 457 - 515.

Cantwell, B. J. 1987 Viscous starting jets. J. Fluid Mech. 173 pp. 159 - 189.

Cantwell, B. J. and Coles D. E. 1983 An experimental study of entrainment and transport in the turbulent near wake of a circular cylinder. J. Fluid Mech. 136 pp. 321 - 374.

Cantwell, B. J., Coles, D. and Dimotakis, P. 1978 Structure and entrainment in the plane of symmetry of a turbulent spot. J. Fluid Mech. 87: 641 - 72

Coles, D. E. 1981 Prospects for useful research on Coherent structure in turbulent shear flow. In Surveys in Fluid Mechanics eds. Narasimha and Deshpande. Indian Academy of Sciences.

Corcos, G. M. and Lin, S. J. 1984 The mixing layer : deterministic models of the turbulent flow. Part 2. The origin of the three-dimensional motion. J. Fluid Mech. 139 pp. 67 - 95.

Corrsin, S. 1943 Investigations of flow in an axially symmetric heated jet of air. NACA Adv. Conf. Rep. 3123.

Chen, J. Cantwell, B. and Mansour, N. 1989 The effect of Mach number on the stability of a plane supersonic wake. AIAA paper 89 - 0285, Reno.

Chong, M. S., Perry, A. E. and Cantwell, B. J. 1988 A general classification of three-dimensional flow patterns or a traveler's guide to (P, Q, R) space. Stanford University SUDAAR 572.

Cimbala, J. M., Nagib, H. M and Roshko, A. 1988 Large structure in the far wakes of two-dimensional bluff bodies. J. Fluid Mech. 190 pp. 265 - 298.

Dimotakis, P. E. 1989 Turbulent free shear layer mixing AIAA paper 89 - 0262. 27th Aerospace Sciences Meeting, Reno, Nevada

Gleick, J. 1987 Chaos: making a new science. Viking. pp. 121 - 125.

Glezer, A. 1981 An experimental study of a turbulent vortex ring. PhD thesis, GALCIT.

Griffiths, R. W. 1986 Particle motions induced by spherical convective elements in Stokes flow. J. Fluid Mech. 166 pp. 139 - 159.

Guckenheimer, U. 1986 Strange attractors in fluids: Another view. Ann. Rev. Fluid Mech. 18 : 15 - 31.

Ho, C. M. and Huerre, P. 1984 Perturbed free shear layers. Ann Rev. Fluid Mech. 16 : 365 - 424.

Jou, W.H. and Riley, J. J. 1987 On direct numerical simulations of turbulent reacting flows. AIAA - 87 - 1324.

Hussain , A. K. M. F. 1980 Coherent structures and studies of perturbed and unperturbed jets. In The Role of Coherent Structures in Modeling Turbulence and Mixing (ed. J. Jimenez). Lecture Notes in Physics

Hussain, A. K. M. F. 1981 Role of coherent structures in turbulent shear flows. In Surveys in Fluid Mechanics, Narasimha and Deshpande (eds.).

Hussain, A. K. M. F. 1983 Coherent structures reality and myth. Phys. of Fluids 26: 2816 - 2850.

Hussain, A. K. M. F. 1986 Coherent structures and turbulence. J. Fluid Mech. 173 pp. 303 - 356.

McMurtry, P. A., Jou, W.H., Riley, J.J. and Metcalfe, R. W. 1985 Direct numerical simulations of a reacting mixing layer with chemical heat release. AIAA - 85 - 0143.

Kambe, T. 1986 Acoustic emissions by vortex motions. J. Fluid Mech. 173 : 643 - 666.

Kline, S. J., Cantwell, B. J. and Lilley, G. M. Proceedings of the 1980 - 81 AFOSR - HTTM Conference on Complex Turbulent Flows.

Kline, S. J. and Robinson S. 1988 Quasi - coherent structures in the turbulent boundary layer: Part I: Status report on a cumminity wide summary of the data. Zorin Zaric Memorial International Seminar on Near - Wall Turbulence, Dubrovnik, Yugoslavia 16 20 May, Hemisphere Publishing.

Konrad, J. H. 1976 An experimental investigation of mixing in two-dimensional turbulent shear flows with applications to diffusion-limited chemical reactions. PhD thesis, California Institute of Technology (also Project SQUID Technical Report CIT - 8 - PU).

Lasheras, C., Cho, J. S. and Maxworthy, T. 1986 On the origin and evolution of streamwise vortical structures in a plane free shear - layer. J. Fluid Mech. 172 pp. 231 - 258.

Lessen, M. 1978 On the power laws for turbulent jets, wakes and shearing layers and their relationships to the principal of marginal stability. J. Fluid Mech. 88 : 535 - 540.

Lewis, G. S., Cantwell, B. J., Vandsburger U. and Bowman C.T. 1988 An investigation of the structure of a laminar non-premixed flame in an unsteady vortical flow. Twenty second International Symposium on Combustion, Seattle.

Liepmann, H. W. and Narasimha, R. 1987 editors, Turbulence management and relaminarisation. IUTAM symposium Bangalore, India.

Liepmann, H. W. and Nosenchuck, D. M. 1982 Active control of laminar - turbulent transition. J. Fluid Mech. 118 : 201 - 204

Lilley, G. M. 1983 Vortices and turbulence J. Roy. Aeronautical Soc.

Liu, J. 1989 Coherent structures in transitional and turbulent free shear flows. Ann Rev. Fluid Mech. 21 : 285 - 315.

Lumley, J. 1981 Coherent structures in turbulence. In Transition and Turbulence , R. Meyer (ed.). Academic Press.

Lumley, J. 1989 The state of turbulence research. In Advances in Turbulence, W. K. George and R. Arndt (eds.). Hemisphere.

Marasli, B., Champagne F., and Wygnanski I. 1989 Modal decomposition of velocity signals in a plane turbulent wake. J. Fluid Mech. 198, pp. 255 - 273.

Mahalingham, S., Cantwell, B. and Ferziger, J. 1989 Effects of Heat Release on the Structure and Stability of a Coflowing, Chemically Reacting Jet. AIAA paper 89 - 0661, Reno, Nevada.

Michalke, A. 1983 Some remarks on source coherence affecting jet noise. J. Sound and Vib. 87 (1): 1 - 17.

Mittlemans, Z. 1989 Unsteady aerodynamics and control of delta wings with leading edge blowing. PhD thesis Department of Aeronautics and Astronautics, Stanford University.

Moin, P. 1984 Probing turbulence via large eddy simulation. AIAA paper 84 - 0174.

Nakamura, Y. and Nakashima, M. 1986 J. Fluid Mech. 163 pp. 149 - 170.

Oran, E. S. and Boris, J. P. 1987 Numerical Simulations of Reactive Flows, Elsevier.

Perry, A. and Chong, M. S. 1987 A description of eddying motions and flow patterns using critical - point concepts. Ann. Rev. Fluid Mech. 19 : 125 - 55.

Perry, A. E. and Lim, T. T. 1978 Coherent structures in coflowing jets and wakes. J. Fluid Mech. 88 : 451 - 64.

Perry, A. E. Lim, T. T. and Chong, M. S. 1980 Instantaneous vector fields in coflowing jets and wakes. J. Fluid Mech. 101 : 243 - 256.

Perry, A. E. 1987 Turbulence modeling using coherent structures in wakes, plane mixing layers and wall turbulence. In Perspectives in Turbulence (ed. Meier H. U. and Bradshaw, P.). Springer - Verlag 115 - 153.

Perry, A. E., Li, J. D. and Marusic, I. 1988 Novel methods of modeling wall turbulence. AIAA - 88 - 0219, Reno.

Perry, A. E., Li, J. D., Henbest, S. M. and Marusic, I. 1988 The attached eddy hypothesis in wall turbulence. Zorin Zaric Memorial International Seminar On Near-Wall Turbulence, Dubrovnik, Yugoslavia 16 - 20 May, Hemisphere Publishing.

Rebollo, R. M. 1973 Analytical and experimental investigation of a turbulent mixing layer of different gases in a pressure gradient. PhD thesis, California Institute of Technology.

Robinson, S. K., Kline, S. J. and Spalart, P. R. 1988 Quasi - coherent structures in the turbulent boundary layer: Part II. Verification and new information from a numerically simulated flat - plate layer. Zorin Zaric Memorial International Seminar On Near-Wall Turbulence, Dubrovnik, Yugoslavia 16 - 20 May, Hemisphere Publishing.

Roshko, A. 1954 On the drag and shedding frequency of two-dimensional bluff bodies. NACA TN 3169.

Roshko, A. 1976 Structure of turbulent shear flows: a new look. Dryden Research Lecture. AIAA J. 14 : 1349 - 57.

Roshko, A. and Fiszdon, W. 1969 On the persistence of transition in the near wake. Problems of Hydrodynamics and Continuum Mechanics, SIAM.

Rutland, C. J. and Ferziger, J. 1989 Interaction of a vortex and a premixed flame. AIAA paper 89 - 0127, Reno, Nevada.

Schewe, G. 1986 Sensitivity of transition phenomena to small perturbations in flow around a circular cylinder. J. Fluid Mech. 172 pp. 33 - 46.

Shariff, K. 1989 Dynamics of a class of vortex rings PhD thesis Stanford University.

Shih, T.-H. Lumley, J. L. and Janicka, J. 1987 Second order modeling of a variable density mixing layer. J. Fluid Mech. 180 pp. 93 - 116.

Spalart, P. 1988 Direct simulation of a turbulent boundary layer up to $R_\theta = 1410$. J. Fluid Mech. 187: 61 - 98.

Strawa A. W. and Cantwell, B. J. 1989 Investigation of an excited jet diffusion flame at elevated pressure. J. Fluid Mech. 200 : 309 - 336.

Subbarao, E. R. 1987 An experimental investigation of the effects of Reynolds number and Richardson number on the structure of a co - flowing buoyant jet. PhD thesis, Stanford University. SUDAAR 563.

Townsend, A. A. 1947 Measurements in the turbulent wake of a cylinder. Proc. R, Soc. London Ser. A 190 : 551 - 61.

Townsend, A. A. 1956 The structure of turbulent shear flow. Cambridge Univ. Press. 1st ed.

Widnall, S. E. and Tsai, C.-Y. 1977 The instability of the thin vortex ring of constant vorticity. Phil. Trans. Roy. Soc. London A287 : 273 - 305.

Williamson, C. H. K. and Roshko, A. 1988 Vortex formation in the wake of an oscillating cylinder. Journal of Fluids and Structures 2, 355 - 381.

Willmarth, W. W. 1975 Pressure fluctuations beneath turbulent boundary layers. Ann. Rev. Fluid Mech. 7 : 13 - 37.

Wygnanski, I., Champagne F., and Marasli B. (1986) On the large-scale structures in two-dimensional, small-deficit, turbulent wakes. J. Fluid Mech. 168, pp. 31 - 71.

Wygnanski, I. and Weisbrot, I. (1987) On the pairing process in an excited, plane, turbulent mixing layer. In IUTAM Symposium on Turbulence Management and Relaminarisation, Springer - Verlag, Liepmann, H.W. and Narasimha, R. (eds.)

Zeman, O. (1981) Progress in the Modeling of Planetary Boundary Layers. Ann. Rev. Fluid Mech. 13 : 253-272.

Whither Coherent Structures?

Comment 1.

James Bridges, Hyder S. Husain and Fazle Hussain

Department of Mechanical Engineering
University of Houston
Houston, TX 77204-4792

1. Role of Coherent Structures in Turbulence

Currently, the state of research in coherent structures (CS) is not unlike the well-known story of the five blind men and the elephant. But the problem runs much deeper than failing to agree on the form of the beast -- we must also understand how to harness and ride it! This meeting on "Turbulence at the Crossroads" was convened to ask representatives of several groups not only to defend their chosen approaches to turbulence research but also to explain how their approaches will be used in engineering applications. This is a difficult task for those who study CS, as there is yet no consensus on the definition and role of CS, let alone a formalism for incorporating CS into predictive models. Research toward understanding the topology and dynamics of time-evolving, three-dimensional CS is still in its infancy, and thus differing perspectives from different researchers are not only natural and unavoidable, but even desirable. We are in full agreement with Cantwell (1989)[†] that CS research, as does turbulence research, needs a diversity of approaches, justified not only by the current state of inadequate knowledge about CS, but also by their richness and the prospects for their control in a variety of technological applications. However, we are not totally ambivalent about what constitutes the best manner of pursuit of CS: the analogy of the elephant ends quickly because, in our view, not all definitions of CS are equally relevant or even effective in capturing the essential physics. Although the term 'coherent structure' is widely recognized as meaning the large-scale organized motions of a flow, we have definite ideas as to what specific definition of CS directly addresses most critical basic and technological questions.

We will resist the temptation of simply reviewing our own results in CS, as this has been done previously (Hussain 1981, 1983, 1986; referenced as H81, H83, H86, respectively). Nor will we add another encompassing review to the literature; competent reviews have already been written by, among others, Davies & Yule (1975), Willmarth (1975), Roshko (1976), Kovasznay (1977), Yule (1978), Cantwell (1981), Lumley (1981), Coles (1981, 1985), Antonia (1981), Laufer (1983), Ho & Huerre (1984), Hunt (1987), Fiedler (1988a), Blackwelder (1988) Reynolds (1988), Kim (1988) and Liu (1989). Instead, we will address what we consider to be the issues of the day for CS. Chief among these are (i) the need for a mathematical framework in which to describe the organization of the flow, (ii) ways in which the CS concept has been and will be useful in turbulence, and (iii) the need for advanced experimental and numerical tools to further CS research.

Fluid mechanics, like most other physical sciences, uses a blend of physical observation and mathematical tools. We see the CS approach as merely an evolutionary extension of traditional approaches to turbulence. Both are driven by experimental observation: traditional exper-

---

[†]This paper is a good example of the plurality of views on CS of which Cantwell speaks. Being unsure as to exactly what his paper would finally contain, we elected to write a second position paper on this topic.

imental devices such as pitot tubes, hot-wires and analog computational circuits have been effectively employed to obtain time-average information in the form of single and joint statistics, while CS have become the focus of basic researchers since the advent of high speed imaging systems, multiple hot-wire rakes, computerized data acquisition and direct numerical simulation (DNS). Both approaches are trying to satisfy the demands being made of researchers in turbulence: traditional approaches are most useful in predicting time-average quantities such as drag, lift and spread rate, while CS methods seek more detailed insight into instantaneous mechanisms and flow physics, such as mixing mechanisms, active separation control for aircraft supermaneuverability, flow noise and wake signature reduction. With our interest, abilities and needs going beyond time-average information, the CS approach becomes the direct successor to traditional approaches and seeks to incorporate the time-dependent information of dynamically significant events in the flow. However, it has not yet begun to fulfill the needs of engineering science and will require much more work before it can.

What troubles us most is our inability to embody the information gleaned from the experimental studies of CS into a mathematical framework. We feel that providing a mathematical basis for the CS concept will be the topic of considerable effort for years to come and will bring about a much better *understanding* of turbulence.

Let us enumerate what we mean by 'understand' as just used. Science is expected to deliver a theory/model for physical systems which can be used to (1) explain, (2) predict and, in technological applications, (3) control the systems. These three aspects of a theory are what we mean by understanding a system or a phenomenon. In the following, we will first address how we see CS as contributing to turbulence research and then give specific references and examples illustrating how CS are currently aiding efforts to understand turbulence.

1.1. Example of Explanation Using CS

The concept of CS has helped explain many important features of flows, as well as peculiarities which are not at all discernible from time-average data. Early successes of the CS approach include explanations of mixing and noise production by two-dimensional pairing (Browand & Weidman 1976; Petersen *et al.* 1974). Two-dimensional studies also gave explanations for negative production (H83) and turbulence suppression in mixing layers (Zaman & Hussain 1981). More recently, consideration of the three-dimensional CS of mixing layers, in particular, stretching of streamwise ribs by the nominally spanwise three-dimensional roller structures, has shed considerable light on the mechanisms of entrainment, mixing, production, dissipation, helicity generation, etc. (H86), as well as mechanisms of mixing layer chemical reaction and flame shortening (Metcalfe & Hussain 1989). As our knowledge of the three-dimensional structure increases, more detailed explanations of turbulent phenomena undoubtedly will be found.

As an example, time-average velocity fields of elliptic jets show that their cross-section switches orientation (major and minor axes of the jet cross-section interchange) at axial locations which depend upon the aspect ratio of the nozzle and the initial condition (Husain & Hussain 1983). Under preferred mode excitation, high aspect ratio (≥ 4:1) elliptic jets show only one such switchover, while jets of lower aspect ratios continue this switchover behavior for tens of diameters. Consideration of the time-average fields gives no indication as to the cause of the axis switching and the effect of aspect ratio. The first clue came from flow visualization and from consideration of vortex dynamics as applied to the elliptic CS in the jets.

Elliptic vortex rings have complicated motions due to curvature-dependent self-advection, which cause their aspect ratio to change as they evolve; this is shown qualitatively in figure 1(a). Furthermore, visualization revealed that the rings in high-aspect-ratio jets switch axes only once because each ring bifurcates into two side-by-side rings, as shown in figures 1(b,c). These observations have since been confirmed in air jets by phase-locked hot-wire measurements which clearly show the two separate rings retaining their distinct identities for several diameters. As a side note, that the low-aspect-ratio jets continue to switch axes in the far field indicates that elliptic CS rings are important even tens of diameters away from the origin (Hussain & Husain 1989).

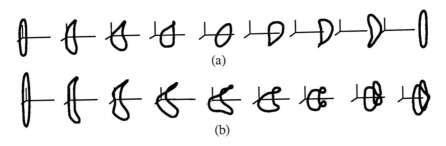

(a)

(b)

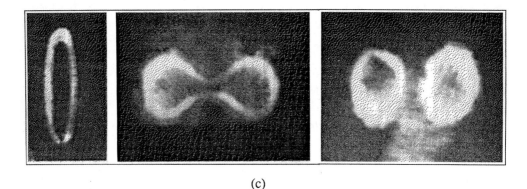

(c)

Figure 1. (a) Evolution of low aspect ratio (2:1) elliptic vortex ring as given by vortex dynamics and as observed in experiments. (b) Evolution of high aspect ratio (4:1) elliptic vortex ring as observed in experiments. (c) Flow visualization of three phases in the evolution of a vortex ring in a high aspect ratio jet showing the original orientation of the ring, the ring just before and just after bifurcation (Hussain & Husain 1989).

1.2. Example of Prediction Using CS

For most present-day engineering applications, time-average quantities, such as moments of the probability distribution and single- and two-point correlations, are all that need to be predicted. This is partly because the evaluation of performance, such as lift, drag, thrust, mixing rate and spread rate, is in these terms and partly because the additional information cannot be readily measured and has not been considered important. For applications requiring this level of information, the CS concept may not have much impact. It is not until one wishes either to make fine improvements on performance, such as reducing instantaneous dips which lower time average values or removing 'hot spots', or to understand local changes which make global differences -- such as the onset of turbulence in the entrance of a pipe, flame shortening or quenching in combustors, or local shear layer dynamics in flow separation -- that one needs to consider detailed, dynamic events within the flow. In these cases, more understanding is gained by considering the time-evolving measures of dynamically significant events rather than a set of time-average measures.

Prediction in engineering science usually entails the evaluation of formulas. The CS concept is not yet developed to the point of providing such quantitative answers. Rather, with CS we are capable of 'soft' prediction. As will be mentioned later, if we specify that vorticity is to be the governing property of CS (as we insist that it is), then vortex dynamics gives intuitive answers as to the behavior of the CS. Given the topology of the vortical CS we can say roughly how it will evolve and interact with other CS. This is an advantage not held by other specific definitions of CS. Vortex dynamics gives local, short-term predictability of the dynamically important aspects of the flow.

As an example, in early exploration of CS in various flows, it was found that elliptic jet geometries produced greater mass and momentum transports than circular jets (Husain & Hussain 1983; Ho & Gutmark 1987), and that three-dimensional deformation of the elliptic ring CS was responsible for the enhanced transports. By increasing the strength and regularity of the CS in elliptic jets using excitation, the turbulence could be enhanced even further. In a separate study of whistler jets with circular nozzles (Hasan & Hussain 1982), we discovered that the effect of the whistler configuration on CS was the same as applying a massive amount of external forcing. The advantage of the whistler configuration is that the flow develops geometry-dependent resonances, making it *self-excited* and requiring no external power input. Only a few moments' thought was required to predict that the enhancement mechanisms found in unexcited elliptic and circular whistler jets could be combined in an elliptic whistler jet which would give greater mixing than either the circular whistler jet or the unexcited elliptic jet. A study in our laboratory of the mass entrainment in the different jets has confirmed this prediction. Figure 2 shows the amount of mass entrained into a jet as a function of downstream distance for unexcited circular and 2:1 elliptic jets, and for circular and 2:1 elliptic whistler jets under the conditions (*i.e.* the selection of collar length) of maximum exit turbulence intensity. We see that the elliptic whistler jet has the greatest entrainment, especially around 10 diameters downstream, the location of maximum turbulence intensity.

1.3. Example of Control Using CS

The CS approach promises not only to add insight into mechanisms of turbulence, but also to allow these mechanisms to be controlled. The dynamical significance of CS in the transports of heat, mass and momentum, which in turn affect mixing, chemical reaction, drag and aerodynamic noise, has been aggressively studied by many researchers. It has also been noted that CS generation, evolution and interaction can be modified by controlling the instabilities which create these structures. It follows that turbulence phenomena can be selectively augmented or suppressed by controlling CS. We call this *turbulence management*.

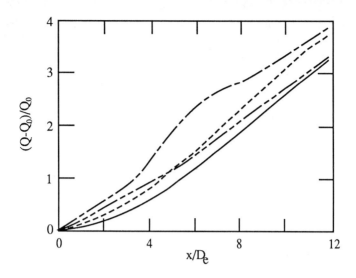

Figure 2. Amount of mass entrained in jet cross-section in different jets as a function of downstream distance. ———, unexcited circular jet; — – – —, unexcited 2:1 elliptic jet; – – – –, self-excited (whistler) circular jet; — – —, self-excited (whistler) 2:1 elliptic jet.

We will limit our discussion to turbulence management in free shear flows and give examples of *active* (requiring input of external power) and *passive* (requiring no external power) control which affect the CS through the following:

i) Boundary condition: Changing the physical shapes of boundaries of the flow system. Examples include using different shapes of nozzles, *e.g.* circular, plane, elliptic, square, rectangular, triangular, and indeterminate origins, and corrugating trailing edges of splitter plates of mixing layers. This type of passive manipulation of CS takes advantage of the fact that the self-induced velocity of a segment of vortical structures depends on its local curvature (as well as core radius and strength) to produce systematic deformation of advecting structures and thus large-scale, three-dimensional motions which enhance transports of mass and momentum. Nozzles with sharp corners, such as triangles, produce structures with very high curvature and this region of the structure undergoes such rapid distortion (rapid coiling of vortex lines, local axial flow and localized breakdown) that it quickly produces small scale mixing, complementing the large-scale engulfment caused by the overall noncircular geometry (Toyoda 1987).

ii) Initial condition : Changing the inflow condition (*e.g.* mean and fluctuating velocity profiles, spectral content, etc.). Examples of various initial conditions are laminar or turbulent boundary layers of moderate thickness using contoured nozzles with appropriate boundary layer suction and tripping, extremely thin boundary layers using sharp-edged orifice nozzles, fully developed laminar velocity profile resulting from a long approach channel, and a turbulent freestream from grids upstream. These methods primarily affect CS through the instabilities which generate them by changing the strength and spacing of the CS.

iii) Excitation: Perturbation at selected frequencies. Examples include external sources such as acoustic drivers and oscillating flaps or ribbons, and internal sources set up by flow resonance such as a whistler jet or pulsed combustor. Controlled excitation can generate a particular mode of CS, either enhancing or attenuating turbulence (Zaman & Hussain 1980). We have observed four basic modes of structures in the transitional region of shear layers and jets and their driving frequencies (in terms of $St_L \equiv fL/U$) are: *shear layer stable pairing* ($St_{\theta e} \cong 0.012$), *shear layer suppression mode* ($St_{\theta e} \cong 0.017$), *jet column preferred mode* ($St_D \cong 0.4$) and *jet column mode of stable pairing* ($St_D \cong 0.85$), where θe and D are the exit momentum thickness and jet diameter. Merger/pairing of vortices plays an important role in large-scale engulfment and subsequent mixing (Browand & Laufer 1975; Zaman & Hussain 1980), while prevention of pairing, either via excitation (Zaman & Hussain 1981) or in an accelerated mixing layer (Fiedler 1988b), decreases turbulent mixing. By controlling the phase and amplitudes of two harmonically related perturbation frequencies (*e.g.* f and f/2), the growth of the subharmonic (associated with pairing of vortices) can be augmented or attenuated over a wide range of Strouhal numbers (Husain *et al.* 1988). Subharmonic resonance (Kelly 1965; Monkewitz 1988) has been extended recently to the jet column domain by Mankbadi (1985) and Raman & Rice (1989).

Elliptic jets. An excited elliptic jet is an interesting example of a combination of passive and active control. Elliptic CS undergo three-dimensional deformation due to curvature-dependent self-induction, and, when excitation is applied, the CS are strengthened, which augments the deformation process.

Wall jet. In a wall jet -- a mixing layer juxtaposed with a boundary layer -- shear layer structures induce the boundary layer to separate at the wall, a process which can be either augmented or suppressed via proper excitation of the shear layer. Suppression of separation has important applications in aircraft control surfaces, while augmentation enhances heat transfer from the wall and oxidation of wall-fed fuel in combustors.

Noise suppression. Inasmuch as laminar shear layers can produce loud aeroacoustically generated sound from the interaction of strong laminar shear layer vortices, the removal of these structures, either by active suppression by excitation or by passively introducing turbulence into the shear layer, can significantly reduce the overall sound of a jet (Hussain & Hasan 1985; Bridges & Hussain 1987).

Other examples. Many examples exist in the literature where active excitation has been used to modify the CS in a flow and alter the flow. Lee & Reynolds (1982) and Parekh, Leonard & Reynolds (1988) have shown that using dual-mode (axial and helical), dual-frequency forcing, the spreading of a circular jet is drastically increased because of the ejection of rolled-up vortices in different directions. Another method of changing the spanwise symmetry of the mean velocity cross-section of a jet, starting from a circular nozzle, was demonstrated by Cohen & Wygnanski (1987) who used circumferential mode excitation at the nozzle lip to give the jet a variable (*e.g.* square) mean velocity cross section.

In other studies of shear flows, excitation has been used to obtain favorable results, presumably by controlling the dynamics of the CS of the flow, even though no direct study of the CS was made (*e.g.* Oster & Wygnanski 1982). Sigurdson and Roshko (1989) observed that drag on a bluff body can be minimized significantly via excitation at about four times the shedding frequency of separation bubbles. Physically, controlled excitation increases entrainment in the early part of the shear layer, producing reduced reattachment length, bubble height, and drag on the body surface. External acoustic excitation has also been used to control flow separation on an airfoil (Ahuja & Burrin 1984; Zaman *et al.* 1989). On a stalled airfoil, excitation produced a narrower wake with a smaller velocity defect, causing a substantial improvement of the aerodynamic efficiency (lift/drag ratio). These improvements appear to be the direct result of manipulation of the CS in the separating shear layer.

2.   Importance of Vortex Dynamics in CS and Turbulence

Early quantitative measurements of noninteracting CS taken at selected phases of their evolution were helpful in explaining some aspects of the role of their different parts in entrainment and production. More important, however, is the effect of CS interactions on the flow. CS studies have begun to classify and differentiate structure interactions and show how these differences affect turbulent phenomena (a good example is turbulence suppression (Zaman & Hussain 1981)). It is crucial for a candidate theory of CS to be able to predict the dynamics and interactions of CS. *If CS are defined by vorticity, their evolution and interactions are directly connected to their topology through vortex dynamics.* This is why it is important to categorize CS morphologically; the study of their forms is immediately related to the study of their dynamics, as their shape affects their evolution through both self- (local and nonlocal) and mutual-inductions. This is the primary reason why we believe that vorticity is the most useful quantity to define CS. *Vortex dynamics is the missing mathematical framework for the study of CS.*

Turbulence has often been defined as flow with three-dimensional, random vorticity and has been regarded as a tangle of vortex filaments (Tennekes & Lumley 1972). Given this predisposition toward vorticity as the prime quantity of interest in turbulence, the use of vorticity in defining CS is natural. Not only is vorticity Galilean invariant, it remains independent of the frame even in unsteady linear translation, allowing CS to be studied in whatever frame is convenient (typically the one tracking the CS). Using vorticity to define CS also allows us to predict flow evolution in complicated flow situations using intuition (developed from local and nonlocal induction concepts) instead of having to resort to direct calculation; one cannot make such direct use of a coherent velocity or energy (see Aubry *et al.* (1988) for an example of obtaining a dynamical system representation of a coherent velocity field using Fourier modes in homogeneous directions and Karhunen-Loeve expansions of the velocity autocorrelation tensor in inhomogeneous directions). Of course, the main drawback to using vorticity to define CS is the difficulty in measuring it. This partially accounts for the popularity of CS definitions which are based on velocity, pressure, temperature or other simpler observables.

2.1.   Vorticity Topology and Turbulence

With our present understanding, we have begun to classify CS interactions by the topology of the structures involved. For example, the differences between various types of two-dimensional interactions involving vorticity which is aligned parallel (pairing, tearing, shredding; circulation-preserving) and antiparallel (collision, initial stages of reconnection; circulation-destroying) are important to the mixing and entrainment resulting from these interactions. We have only begun to explore three-dimensional CS interactions as the experimental and numerical tools being developed have allowed. Uncovering the rib/roll interaction mechanism of production in the braid region has shown that turbulent mixing can be enhanced by the proper augmentation of streamwise vorticity in the mixing layer. Discovery of the importance of ribs in mixing has also given a detailed view of combustion processes in diffusion flames and of flame short-ening.

The use of vortex dynamics to model CS is not limited to incompressible and inviscid flows. Vortex filaments, which do have such limitations, can be useful approximations to CS in some cases, but this is only one aspect of vortex dynamics (which might be renamed *vorticity dynamics*). We are just beginning to learn about viscous, compressible vortex dynamics, pri-marily through numerical simulations as the details are not yet accessible experimentally because of limitations of measurement technology. Some notable progress has been made in compress-ible vortex dynamics and reconnection mechanisms in an effort which has just begun (Kerr *et al.* 1989). Let us only consider incompressible flows in this paper and concentrate on how the equations of motion are affected by the ensemble averaging process. We will then explore un-der what conditions vortex filaments can be used to locally model CS, and finally look at an ex-ample where the limitations of filament methods are being circumvented through the use of DNS to study coherent structures in flow situations of far-ranging importance to turbulence.

Using a decomposition of the field variables of interest, *i.e.* velocity, vorticity, etc., given by $f = \langle f \rangle + f_r$, where f is the field variable and $\langle \, \rangle$ denotes the conditional ensemble averaging procedure used to separate the coherent $\langle f \rangle$ and incoherent f_r components (discussed in detail in section 3), we can derive an incompressible coherent vorticity transport equation,

$$\frac{\hat{D}}{Dt}\langle \vec{\omega} \rangle = \langle \vec{\omega} \rangle \cdot \nabla \langle \vec{u} \rangle + \frac{1}{Re}\nabla^2 \langle \vec{\omega} \rangle + \langle \vec{\omega}_r \cdot \nabla \vec{u}_r \rangle - \langle \vec{u}_r \cdot \nabla \vec{\omega}_r \rangle \tag{1}$$

where $\hat{D}/Dt = \partial/\partial t + \langle u_j \rangle \, \partial/\partial x_j$. We note that (1) is simply the vorticity transport equation with two additional coherent terms arising from incoherent turbulence: the third term of the RHS is the stretching of incoherent vorticity fluctuations by incoherent velocities, while the fourth is the advection of incoherent vorticity by incoherent velocity fluctuations, both of which have been organized by the coherent field so as to affect the coherent field itself. Evaluation of these two nonlinear source/sink terms is not straightforward and constitutes the prime limitation on our ability to predict the evolution of CS.

The two incoherent contribution terms in (1) must be modeled, even if that model is to ig-nore them for short-time evolution. The analysis of H86 (appendix B) indicates that the impor-tance of these terms depends upon the scales of the incoherent components. If the incoherent component is mostly small-scale, as it is when there is a single dominant mode, then the inco-herent terms contribute heavily; intuitively, we would model such a situation by some CS ver-sion of an eddy viscosity. If the incoherent component is of a larger scale then the terms may be of less importance but require a more sophisticated modeling. In handling the incoherent terms, one must consider that CS do not so much *die*, but rather *interact* with other CS such that they then take a *different form*. For this reason, basis function representions of CS (discussed later) may not be able to describe an evolving CS for as long as representations which do not require the structure to retain its form.

In the incoherent component we have stored our ignorance about the differences between in-stantaneous vortical patches. Such differences are caused by their origins and by their environment and will cause each instantaneous realization to depart at some time from the ensemble average. In some cases, we can reduce the incoherent component by using different or additional eduction criteria (dicussed in section 3) which will distinguish these differences and either allow better alignment of the instantaneous vortical patches or let us find more than one distinct CS mode. Having done either of these, the individual structures will follow the ensemble average for a longer time and reduce the importance of the incoherent terms in (1).

Let us now turn to models of vortex dynamics which can be used to predict the evolution of CS under the restrictions just discussed.

2.2. Filament Models

Many researchers use a modified Biot-Savart equation to explain and describe motions that they encounter in their studies of CS in incompressible flows. Filament models, such as those of various ring topologies (Dhanak & DeBernardinis 1981; Bridges & Hussain 1988; Ashurst 1983; Inoue 1988), give insight into CS dynamics in jet flows (the examples of elliptic jets discussed earlier). Other examples of filament models either of CS themselves or of their dynamics are the filament simulations of the blooming jet (Parekh, Leonard & Reynolds 1988), the study of a hairpin vortex in a boundary layer (Moin, Leonard & Kim 1986), and the multifilament simulation of the three-dimensional plane mixing layer (Ashurst & Meiburg 1988) and plane wake (Meiburg & Lasheras 1988).

As the Biot-Savart equation is derived solely from consideration of incompressible continuity equation and the definition of vorticity, it is actually only a model for the vorticity transport equation, and when modified to become an evolutionary equation for vortex filaments (Leonard 1985), it has three main assumptions or restrictions: (i) the flow is inviscid so that vorticity lines are material lines, (ii) any velocity superimposed by sources not included in the Biot-Savart integral is irrotational (all vorticity is calculated for explicitly), and (iii) filaments do not come into close proximity with other filaments or themselves.

When CS are modeled by filaments, the filament model is approximately solving (1), ignoring the last three terms. The solution is only valid within the restrictions of the filament model and the restrictions discussed in section 2.2 concerning the unknown incoherent contributions to the coherent vorticity. Assumption (i) of the filament method is satisfied for CS by virtue of their large Reynolds number which makes them nearly inviscid (of course, Re is not very large when effective viscosity due to fine-scale turbulence is considered). The inviscid assumption also carries with it the theorems of Helmholtz and Kelvin, principally the statement that vortex lines cannot be broken. This (along with the third assumption) rules out the use of filament models in the study of vortex reconnection. The second assumption requires that we either (a) include all the coherent vorticity in the field explicitly -- which entails specifying the spatial relationships between CS -- or (b) resign ourselves to validity for the short time during which the induced velocity produced by CS excluded from the computation has little effect. If we know the relationships among a few CS, we may elect to include these in the computation and use option (b) for the rest, especially if the CS being calculated explicitly are interacting strongly with each other and not so strongly with the excluded CS. Assumption (iii) requires either that we take care in choosing our model, such as using multiple filament bundles instead of single filaments to model merging of CS, or that we abandon the simulation if the filaments come too close to each other for accurate solution. Since single filament representations of CS cannot properly model their interactions, especially in three-dimensional CS, multiple filament models are often required.

2.3. Vorticity Dynamics in Viscous Flow

Despite their limitations, filament models give great insight into the physics of many flow situations encountered in CS studies. However, there are a wealth of vortex interactions which are inaccessible by this type of model, such as vortex reconnection, in which two vortex filaments crosslink at their point of closest approach forming two new vortices each consisting of parts of the old vortices. It is becoming apparent (Melander & Hussain 1988; hereafter referred to as MH88) that vortex reconnection, also called *cut-and-connect*, occurs whenever anti-parallel vortices approach one another. Furthermore, filament studies by Siggia (1985) have shown that whenever two vortices at arbitrary orientations interact, they locally become anti-parallel, thus likely undergoing a reconnection. Reconnection is important in turbulence because it is a mechanism for isotropization (by causing vortexlines to be created orthogonal to their original orientation), production, dissipation, aerodynamic noise generation, and energy and enstrophy cascades. This vortical event has been observed in the bifurcating elliptic jet (see figure 1c) and the Crow instability (Crow 1970), and we believe it to be a fundamental part of the breakup of the CS at the end of the potential core in jets. This motion may also be found in boundary layers when hairpins pinch off to form 'typical eddies' (Falco 1979; Moin *et al.* 1986).

Since the apparent cutting of vortex filaments observed during reconnection is directly contrary to Kelvin's theorems for inviscid filaments, reconnection requires viscosity to occur (Takaki & Hussain 1985). To study vorticity dynamics in viscous and compressible flow, we have to resort to direct numerical simulation.

We often study prototypical CS interactions by means of direct numerical simulations of relatively simple vorticity fields, such as those in recent studies of vortex reconnection (Ashurst & Meiron 1987; Kerr & Hussain 1989; Kida, Takaoka & Hussain 1989; MH88). Full field simulations of turbulent flows with present-day computers disallows adequate resolution to study this event, and experimental study by successively firing vortex rings (Schatzle 1986) cannot produce exactly repeatable reconnection details on a scale fine enough to allow precise phase-locked measurements; hence the requirement of using an idealized initial condition and DNS. As an indication of how the physics of the interactions are masked by lack of resolution, every new study of reconnection which increased the resolution of the simulation gave totally new interpretations of the resulting flow. (Some studies of this problem did not suffer from inadequate resolution but used complicated, unphysical initial conditions, such as the sheathed vortices of Meiron *et al.* 1988.) This apparent disagreement has been more or less settled now as the required resolution to fully resolve the scales of motion in the interaction has presumably been reached in the MH88 simulation.

Figure 3a shows a schematic of the initial vortices and coordinates used by MH88 who studied the details of a vortex reconnection event using DNS of crosslinking of two antiparallel (viscous) vortex tubes of initial Re ($=\Gamma/\nu$) up to 1000. Dealiased (2/9 k-space truncation), pseudo-spectral (Galerkin) method with fourth-order, predictor-corrector time stepping was employed to solve the Navier-Stokes equations in a 64^3 domain with periodic boundary conditions. The vortex tubes of circular cross-section had an initial Gaussian vorticity distribution with compact support such that the vorticity field was divergence free and the box was circulation free. The vortices are initially given small sinusoidal perturbations to hasten the early stages of the process and to ensure that they are pressed against each other as viscous annihilation continues. In the figure, t denotes time t* nondimensionalized by the initial peak vorticity $|\vec{\omega}_m(0)|$ such that $t = t^* |\vec{\omega}_m(0)|/20$. Figures 3(a-e) show wire plots of the vorticity norm $|\vec{\omega}|$ surface at t=0, 3, 3.75, 4.5 and 6, the surface being at a level of 0.3 $|\vec{\omega}_m(0)|$. The mechanism has been explained to be the consequence of mutual vorticity annihilation by viscous cross-diffusion. The remainders of the annihilated vortexlines are cross-linked, stretched and advected by the vortex swirl and collected at the forward stagnation point of the vortex dipole as two vortex bundles or *bridges* (figure 3c). Induction by the bridges reverses the curvature of the dipole vortices and stretches them into slowly decaying *threads* as the bridges pull apart under self-induction (figure 3e). This shows a detailed mechanism by which small scales are produced from large-scale structures and vorticity lines become reoriented.

One interesting observation concerns the vortexlines in the bridges. As a result of the non-concentric vorticity distribution, the vortexlines will be twisted, resulting in an induced axial flow and vortex compression. This constitutes an inviscid mechanism, distinct from diffusion, for smoothing the vorticity along the reconnected vortex lines. This phenomenon seems to be related to the axial flow smoothing which was predicted by Moore & Saffman 1972.

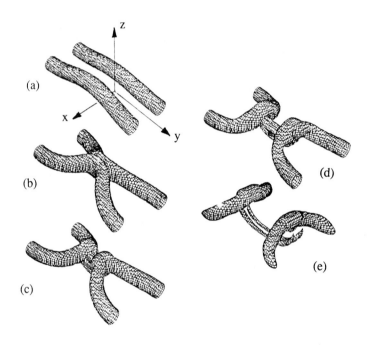

Figure 3. Simulation of vortex reconnection given at times t = 0, 3, 3.75, 4, 5 and 6. Surface shown is a constant enstrophy surface, $|\omega| = 0.3\ |\omega_{max}|$, where ω_{max} is the peak vorticity at time t = 0 (Melander & Hussain 1988).

2.4. Fallacy of Flow Visualization Exemplified

A by-product of the MH88 study was the illustration of a limitation of flow visualization apparently overlooked by many researchers. While most visualization involves large Schmidt number (Sc), suggesting different boundaries of vortical and marker domains, the difference in domains found in the reconnection problem occurs even at Sc = 1. Consider equations for vorticity $\vec{\omega}$ and scalar c in incompressible flow:

$$\frac{D\vec{\omega}}{Dt} = \vec{\omega} \cdot \nabla \vec{u} + \nu \nabla^2 \vec{\omega}, \tag{2}$$

$$\frac{Dc}{Dt} = \mathcal{D}\nabla^2 c. \tag{3}$$

Thus even at Sc = $\nu/\mathcal{D}$ = 1, unless the flow is two-dimensional (when the vorticity and scalar boundaries would be identical), the boundaries of $|\vec{\omega}|$ and c are different: vorticity undergoes self-augmentation (due to vortex stretching) but the scalar does not. As a result, markers are depleted away from locations of vorticity augmentation; that is, the domains of intensified vorticity lose markers and gradually become invisible while markers accumulate in domains of decreased vorticity, viz. domains of reduced dynamical significance. If we agree that vorticity is the principal characteristic of turbulent flows, it follows that flow visualization will miss dynamically significant domains of turbulent flow. This is compounded by the fact that visualization markers are usually not introduced locally (as is often possible by smoke wires and hydrogen bubbles where they capture the local dynamics adequately for short distances or times from the point of introduction) but typically are introduced upstream, say at the cylinder in case of a wake, splitter plate in a mixing layer and nozzle lip in jets. Thus, there is an integration effect of flow markers, progressively deviating from domains of dynamically significant events. The contrast between scalar and vorticity boundaries has been more clearly demonstrated recently by Melander & Hussain (1989).

3. Definitions of CS

One wonders if history will record all of the discordances which have accompanied the topic "definitions of CS" over the past 15 years. The earliest perceptions of CS were heavily biased by the researcher's method of observation. Based on flow visualization alone, researchers argued about how to characterize the organization that they saw. The need for quantitative data pushed them into single-point measurements in transitional regions of shear layers, where CS were often described in terms of waves using a nonlinear extension of instability waves; others have inferred CS from spectra, correlation, signals (single-point) with characteristic features or fronts. More recently, researchers engaging in fundamental study of CS are searching for descriptions which can be incorporated into a mathematical framework, with an eye towards application and a degree of universality. Given that any spatial structure we observe in the flow is a property of the mathematical solution to equations governing the flow, it is now more to the point that we find the most convenient approximations and variables in which to study the flow and its structures rather than argue in loose terms about what CS are.

A popular notion of the past few years has been that the dynamics of CS are given by some form of a low-dimensional, nonlinear dynamical system which would embody enough of the physics to model most aspects of turbulence -- a drastic reduction in complicacy! Keefe (1987) has made a more direct connection between CS and nonlinear dynamical systems by starting with the statement that the flow is a trajectory in phase space and that the conditional measurements made in physical space correspond to sampling the state of the system as its trajectory passes through a low-dimensional window in phase space. One of the more successful attempts to construct a nonlinear dynamical system from a CS definition has been the work of Aubry et al., (1988) (see also Holmes 1989; Sirovich 1989) which uses the proper-orthogonal-mode decomposition definition of CS, retaining the dominant modes and maximizing time-average kinetic energy in successive spatial modes (Lumley 1967). Another definition (Adrian 1979) uses conditional events (a 'conditional eddy') as the initial conditions whose evolutionary equations can be derived from the Navier-Stokes equations after their dependence upon the conditioning event is obtained (the two-point correlation tensor or a model thereof; Adrian & Moin 1988). Our vortical CS, defined by conditional averaging of the vorticity field, produces vortical entities underlying the random vortical turbulent field, the CS being defined by measures of their topology such as shape, strength and size, and are not constrained to a basis function representation. This last definition being our own, we will expound on it in the subsequent section, hopefully clarifying its basis, execution, and value in studying the physics of turbulence.

3.1. Conceptual Basis of CS Eduction

Instead of trying to refine a definition of CS down to a few words, we will state the progressive steps with which our CS is conceived, giving the method and philosophy of acquiring (or educing) the CS of a flow.

It is profitable to consider turbulent flow as a collection of spatially discrete vortical structures. This is the conceptual premise of CS study, based upon observations of fluid motion. By studying the behavior of large-scale organized motions within the flow, we expect to better understand the overall behavior of the flow. In essence, we are assuming that the flow can be decomposed into spatially disparate parts whose dynamics can be described by self- and mutual-interactions.

The structures which occur within a region of the flow are classified by their topology so that 'representatives' of each class can be studied. In section 2, we have explained why the topology of the vortical CS is important to its dynamics. We do not expect all flow structures to have the same topology; in fact, while there are universal features of CS in different flows, we generally expect to find many structures whose differences in topology are important to their evolution. Consistent with our definition, the underlying common feature among various flow realizations is the CS. An instantaneous flow realization consists of superposition of CS and incoherent turbulence. We focus on some dominant CS modes and educe them (iteratively) through ensemble averaging of properly aligned realizations.

In the past, we have used such topological measures as orientation, location, size, strength and aspect ratio, although much more complicated descriptors can be used when there is enough resolution in the data (Jeong, Hussain & Kim 1987). At this point we only want to choose measures of the structures' topology which are going to be used to distinguish or detect the events; values of these characteristics are set after we have obtained data from the flow to guide the choice of values. An important point to mention is that the limited number of sensors in previous laboratory eduction experiments has not allowed a large number of measures to be used.

Probability distributions of the topological measures determine the classes of structures. The flow data, in whatever form available, is searched for spatial patterns, usually using some broad topological criteria such as peak vorticity magnitude. The selected topological measures of the structures are taken and their joint probability distributions calculated. From these distributions, the structures will be classified by searching for local maxima in the distributions.

The detail with which we wish to know the flow determines the number of classes of events we define. Since we will be studying an *average* structure from each class, there will always be some aspects of the instantaneous realizations which are not represented. In double decomposition, these unaccounted-for details make up the incoherent component and contribute to the nonlinear terms in the coherent vorticity equation (1). We have only recently tackled situations in which the coherent part consisted of different modes (with the exception of Tso & Hussain 1989; see also, Jeong, Hussain & Kim 1987), as previously we studied excited flows containing only one dominant type of CS. In post-processing, the coherent, incoherent, and time-average Reynolds stresses (see H83 for exact definitions) can be used as an indication of the dynamical significance of the CS educed in a particular flow. For example, the CS found in the wake of a cylinder had these peak values of coherent, incoherent, and time-average Reynolds stresses (Hayakawa & Hussain 1985):

x/D	$\dfrac{\langle u_c v_c \rangle}{U_\infty^2}$	$\dfrac{\langle u_r v_r \rangle}{U_\infty^2}$	$\dfrac{\overline{uv}}{U_\infty^2}$
10	-0.039, 0.031	0.013	~ 0.005
20	-0.014, 0.015	0.006	~ 0.003
30	-0.006, 0.007	0.003	~ 0.001
40	-0.003, 0.004	0.002	~ 0.001

The CS educed are dynamically significant immediately behind the cylinder as to be expected. Further downstream they are still significant, but weaker, and do not overshadow the incoherent turbulence (H83).

The representative of a class of structure is the coherent event. Obviously, instead of trying to study each and every structure that catches our eye, we do better by defining and studying a representative of the structures which draw our attention. The representative is the CS and is the expected value of all the members of the class, obtained by ensemble averaging the structures of a given class together. Since we consider these structures to be similar in the measures specified, we can normalize them in these measures, such as spatially shifting structures (normalizing them by location) to bring them into better alignment as determined by their cross-correlation with the ensemble average. The eduction scheme is intrinsically *iterative*, with different measures being used to find the ones with the best separation of classes and with the normalization of structures within each class being adjusted to bring about the best alignment with the ensemble average.

The incoherent turbulence is the departure (typically rms value) of each realization from the final ensemble average. In the end, everything is accounted for; nothing is filtered away or left out.

The finer details of the topology of the coherent event which were not specified by the classification are the most important part of the coherent event. Our method picks out the most statistically and dynamically significant structures based on loose topological measures, such as overall size and strength, leaving the details about the shape and vorticity distribution of the CS unconstrained and determined only by the flow. The eduction brings out the underlying common structure with all its detailed features. This is what distinguishes our definition and operation from those which "set a template" and find structures to match it (Mumford 1983).

So what keeps the method of classification from becoming just an exercise in finding whatever you want to find?

If a set of measures cannot be found whose joint probability distribution have a limited number of peaks, then the CS approach will not be fruitful in that flow. By this we mean that if there is uniform probability of finding whatever we look for no matter what we want, then

Lumley's criticism (Lumley 1981) is correct. With an eye toward application, we might add that if a flow does not have a limited number of classes which can be represented by a limited number of entities -- the concept of preferred modes -- then the study of specific structures and topologies will be too time-consuming a course to follow. In most flows where the CS has been educed, there were obviously a limited number of classes, or modes, so the actual construction of distribution functions was not necessary.

4. Future Road of CS

The primary goal of CS research today, as we see it, is to provide interpretion of turbulence mechanisms, not necessarily to construct global quantitative models. The insight gained from CS concepts may lead to such models, however; when properly applied, large eddy simulation (LES) is conceptually similar to a coherent structure simulation with the large scales (the CS) being calculated explicitly while the small scales (typically involving incoherent motions) and their contributions to the large scales are modeled statistically. From simulations of this kind there will arise an even greater need for approaches like CS to interpret the complicated motions observed in LES or even DNS, as well as in physical turbulent flows; this may lead to improved intuition, prediction, and management methods. The future goal of CS will be to interpret turbulent field data (obtained both experimentally and numerically) that our technology provides and requires.

4.1. Importance of Advanced Experimental Techniques

As discussed in section 2.4, flow visualization does not record dynamically important vortical events. The statement* that flow visualization provides "visual evidence . . . contrary" to the existence of CS exemplifies the need for experimental techniques which will let us "see" the vorticity field. Then the methods of eduction can be used to extract the important large-scale organized motion which is often the objective of flow visualization studies. In such studies, the visualization technique itself is often being used (erroneously) to make the extraction. Measurement techniques which can adequately resolve the vorticity field (or even the velocity field if the spatial resolution is adequate to allow computation of circulations over a small enough region to approximate large-scale vorticity) in space and time will bring about a major advance in defining and understanding CS and turbulence.

In a way, the concept of CS is out of step with experimental techniques. Most experimental techniques in fluid mechanics measure single-point velocities with extremely fine temporal resolution. The study of CS needs simultaneous measures at many points without as much temporal resolution. Only recently have techniques which address this need been developed, and all existing techniques suffer to some extent from the same two problems: *effective sensor density* and *information recording rate*. Only holographic particle velocimetry has shown the potential for instantaneously capturing three-dimensional flow fields at a reasonable Re; they can be converted into digital form off-line, alleviating the problem of information recording rate. However, accurate conversion of the information available in the hologram has not yet been

---

* From the infamous green flyer announcing this conference; see Cantwell's paper in these proceedings.

rigorously demonstrated, so the absolute effective sensor density of this technique cannot be quantified at this time. A three-dimensional time-evolving flow event is not likely to be captured by holography or other techniques in the immediate future.

Cantwell (1989) expresses a belief in a brighter future to be brought about by continued growth in computational power and DNS capabilities. We are not persuaded that full Navier-Stokes simulations at high Re will be possible in the near future, even at the current rate of increase in computing power. In addition, the problem of specifying absorbing outflow boundary conditions in spatially evolving simulations will not be solved strictly by increased computational power and may remain a problem for years to come. These reservations require that we develop advanced measurement techniques so that the appropriate data for CS can be acquired in physical experiments. On the positive side, the projected increase in computational capability at lower cost, to be found in desktop or laboratory computers and graphics workstations, may well prove to be as beneficial as high-power DNS in advancing the CS concept. Their availability will allow recording, storage and analysis of more flow data than has been possible before, making it easier for researchers to manipulate and visualize data from all sources.

4.2. Approaches to Two- & Three-dimensional Measurements of Velocity

In the following, we briefly review several current and proposed methods of obtaining three-dimensional, time-resolved velocity field data in turbulent flows which might be used in CS study.

Multipoint extensions of present sensors *(currently used; moderately expensive; poor spatial resolution)*
Rakes of hot-wires and LDA's, scanning LDA, and line or sheet LDA all fall under this category of ways to extend single-point measurements to multipoint spatial measurements. The limitations, chief among them being resolution and cost (and interference for hot-wires), are obvious, and have been covered extensively elsewhere (H83).

Image-displacement *(expensive; medium to high spatial resolution; computationally intensive; 3D by holography)*
Several groups are working on techniques which estimate the velocity field by measuring displacement of flow-tagging particles, a technique which can in principle be applied in two or three dimensions (using holography). Generally, there is a problem with the range of velocities which can be measured; particles must move enough between frames that the displacement can be discerned, but not so much as to be confused with other nearby particles. When ciné film is used (as opposed to multiple exposures on a single photographic plate), this problem can be solved by measuring displacement between different numbers of frames, such that the Δt part of $u \cong \Delta x/\Delta t$ can be varied as required to compensate for lack of range in Δx. Also, the flow facility must optimize the relationships between particle size, flow geometry, velocity gradients, velocity ranges and displacement times. The problem of obtaining displacements from images is often underestimated; the amount of processing required to obtain the velocity or vorticity fields from holographic cinéfilm requires computational hardware the equivalent of today's supercomputers. For example, assuming that the holographic recording of flow particle positions were made on 35mm film, there would be approximately 10^{13} bytes of 8 bit/voxel information per frame and hundreds of frames would have to be analyzed. FFT-based image correlation methods will have to be employed (Adrian & Landreth 1988) using dedicated hardware to evaluate displacements between frames as the hologram is being digitized. With today's technol-

ogy, this method can deliver velocity fields comparable in resolution to those being generated by DNS and a velocity transduction rate of several three-dimensional velocity vectors per second.

Doppler shift *(expensive; many implementations limited to 2D (or slow 3D); low-velocity cutoff; can often give other quantities, such as pressure or concentrations)*
Another group of velocity measurement techniques extracts the velocity of the fluid by observing the Doppler shift in radiation from the fluid. Acoustic Doppler techniques are used to monitor wind conditions near airports and to measure laminar flows in biomedical applications, but suffer badly from refraction, making them unsuitable for use in multielement arrays. In techniques which measure the fluid velocity relative to the speed of light (*i.e.*, planar laser-induced fluorescence velocimetry: Paul *et al.* 1989), the velocities of the flow must be rather high. Since in many flows there are points where the instantaneous velocity is zero, this lower threshold may cause problems. Also, these techniques can measure only one component of velocity at a time (the component parallel to illumination) and are intrinsically two-dimensional because they use two-dimensional recording techniques (usually an intensified CCD array). Any extension to three dimensions and multiple velocity components must be done by sequential planes at the expense of data rate, which causes a tradeoff between the information recording rate and the time resolution. The bonus of these techniques is their potential to provide additional information about the flow, such as species concentration, which is especially important in combustion studies. Simultaneous study of CS and their effect on mixing and combustion rates is very appealing.

A subset of the Doppler techniques utilizes molecular resonance and tomography to measure the velocity in a fluid volume, but has yet to be realized outside of limited medical research applications in steady flows (Wedeen *et al.* 1986). Application of magnetic resonance and electron spin resonance imaging requires that the velocity field be reconstructed from an integrated signal collected as moving magnetic gradients select different slices of the flow field. The advantages and limitations of the technique (especially in turbulent flows) are unknown at this time except that the velocity field is only one of several fields which can potentially be obtained and that the flow facilities will have special restrictions allowing the required electric and magnetic fields to be imposed on the flow.

5. Summary

The CS concept is the outcome of our search for 'order in the disorder' of turbulence. Even though it has been the focus of much recent effort, CS research is still in the information-gathering stage. In fact, we are just beginning to move beyond the stage of looking for CS in flows and toward the development of concepts and models which address their evolution and dynamical significance.

While a variety of methods, mostly flow visualization, have been employed to discern CS in different flows, a discussion of their dynamical role must involve quantitative data; flow visualization is only qualitative and frequently can be grossly misleading. Of the variety of quantitative approaches, the ones based on vorticity appear to be the most appealing. Vortex dynamics provides a powerful tool to understand and control CS and serves as the starting point for efforts aimed at placing CS in a quantitative mathematical framework.

While CS are dominant in transitional flows, they are not the whole story in fully turbulent or self-preserving regions of flows where incoherent turbulence is also significant. Here, either a statistical model for the effect of incoherent turbulence must be developed or the motions currently viewed as incoherent must be differentiated into organized modes of motion.

While DNS seems to be extremely valuable for discerning local flow in more detail than is possible experimentally, experiments are unavoidable for flows at higher Re, in complicated geometries and in many spatially evolving flows. Often we must combine the powers of both methods to reveal the flow details that either method alone is inadequate to provide. The need for advanced experimental techniques to instantaneously measure three-dimensional flow fields is great, as is the need for studies which synthesize and interpret data collected from both numerical and experimental methods.

Preparation of this manuscript has been supported by ONR Grant N00014-87-K-0126 and DOE Grant DE-F605-88ER13839. Many thanks to Mr. George Broze for his discussions and proofing.

6. References

Adrian, R. J. 1979, *Phys. Fluids* **22**, 2065.
Adrian, R. J. & Moin, P. 1988, *J. Fluid Mech.* **190**, 531.
Adrian R. J. & Landreth, C. 1988, *Laser Topics* (Spring 1988), 10.
Ahuja, K. & Burrin, R. H. 1984, AIAA-84-2298.
Antonia, R. A. 1981, *Ann. Rev. Fluid Mech.* **13**, 131.
Ashurst, W. T. 1983, AIAA-83-1879-CP.
Ashurst, W. T. & Meiburg, E. 1988, *J. Fluid Mech.* **189**, 87.
Ashurst, W. T. & Meiron, D. T. 1987, Phys. Rev. Lett. **58**, 1632.
Aubry, N., Holmes, P., Lumley, J. L. & Stone, E. 1988, *J. Fluid Mech.* **192**, 115.
Blackwelder, R. F. 1988 in *Transport Phenomena in Turbulent Flows*, (eds. M. Hirata & N. Kasagi), Hemisphere.
Bridges, J. E. & Hussain, A. K. M. F. 1987, *J. Sound Vib.* **117**, 289.
Bridges, J. & Hussain, F. 1988, IMACS 12th World Congress, Paris.
Browand, F. K. & Wiedman, P. D. 1976, *J. Fluid Mech.* **76**, 127.
Cantwell, B. 1981, *Ann. Rev. Fluid Mech.* **13**, 457.
Cantwell, B. 1989, in *Whither Turbulence: Turbulence at the Crossroads*, Cornell U., Springer.
Cohen, J. & Wygnanski, I. 1987, *J. Fluid Mech.* **176**, 221.
Coles, D. 1981, *Proc. Indian Acad. Sci.* **4**, 111.
Coles, D. 1985, AIAA Dryden Lecture.
Crow, S. C. 1970, *AIAAJ* **8**, 2172.
Davies, P. O. A. L. & Yule, A. J. 1975, *J. Fluid Mech.* **69**, 513.
Dhanak, M. R. & Bernardinis, B. 1981, *J. Fluid Mech.* **109**, 189.
Falco, R. 1979 *Symp. on Turbulence* (Rolla: Univ. of Missouri).
Fiedler, H.E. 1988a, *Prog. Aerospace Sci.* **25**, 231.
Fiedler, H.E. 1988b, in *Current Trends in Turbulence Research* (eds. H. Branover, M. Mond & Y. Unger) **112**, 100.
Hasan, M.A.Z. & Hussain, A. K. M. F. 1982, *J. Fluid Mech.* **115**, 59.
Hayakawa, M. & Hussain, A. K. M. F. 1985, *Turbulent Shear Flows V*, 4.33.
Ho, C. M. & Huerre, P. 1984, *Ann. Rev. Fluid Mech* **16**, 365.

Ho, C. M. & Gutmark, E. 1987, *J. Fluid Mech.* **129**, 383.

Holmes, P. 1989, in *Whither Turbulence: Turbulence at the Crossroads*, Cornell U, Springer.

Hunt, J. C. R. 1987, *Trans. Can. Soc. Mech. Eng., 11, 21.*

Husain, H. S. & Hussain, A. K. M. F. 1983, *Phys. Fluids* **26,** 2763.

Husain, H. S., Bridges, J. E. & Hussain, A. K. M. F. 1988, in *Transport Phenomena in Turbulent Flows* (eds. M. Hirata & N. Kasagi) **111**, Hemisphere.

Hussain, A. K. M. F. 1981, *Proc. Indian Acad. Sci.* (Eng. Sci.) **4**, 129.

Hussain, A. K. M. F. 1983, *Phys. of Fluids* **26**, 2816.

Hussain, A. K. M. F. 1986, *J. Fluid Mech.* **173**, 303.

Hussain, F. & Husain, H. S. 1989, *J. Fluid Mech.* (in press).

Inoue, O. 1988, personal communication.

Jeong, J., Hussain, A. K. M. F. & Kim, J., 1987, NASA report CTR-S87, 273.

Keefe, L. R. 1987, personal communication (APS/DFD 1987, paper FC4)

Kelly, R. E. 1967, *J. Fluid Mech.* **27**, 667.

Kerr, R. M. & Hussain, F. 1989, *Physica D*, (in press).

Kerr, R. M., Virk, D. S. P. & Hussain, F. 1989, *IUTAM Symposium on Topological Fluid Mechanics*, Cambridge, UK.

Kida, S., Takaoka, M. & Hussain, F. 1989 *Phys. Fluids* **A1**, 630.

Kim, J. 1988, in *Transport Phenomena in Turbulent Flows* (eds. M. Hirata & N. Kasagi) **111**, Hemisphere.

Kiya, M. & Ishii, H. 1988, *Fluid Dynamics Research* **3**, 197.

Kovasznay, L.S.G. 1977, *Symp. on Turbulence* (Rolla: Univ. of Missouri), 1.

Laufer, J. 1983, *J. Appl. Mech.* **50**, 1079.

Lee, M. & Reynolds, W. C., 1982 *Bull. Amer. Phys. Soc.* **27**, 1585.

Leonard, A. 1985, *Ann. Rev. Fluid Mech.* **17**, 523.

Liu, J. T. C. 1989, *Ann. Rev. Fluid Mech.* **21**.

Lumley, J. L. 1967, *Atmospheric Turbulence and Radio Wave Propagation* (eds. Yaglom & Tatarsky), Moscow, NAUKA.

Lumley, J. L. 1981, *Transition and Turbulence* (ed. R. E. Meyer), **215**, Academic.

Mankbadi, R. R. 1985, *J. Fluid Mech.* **160**, 385.

Meiburg, E. & Lasheras, J. C. 1988, *J. Fluid Mech.* **190**, 1.

Meiron, D. I., Shelley, M. J., Ashurst, W. T. & Orszag, S. A. 1988, *Mathematical Aspects of Vortex Dynamics* (ed. R. E. Caflisch), SIAM, 183.

Melander, M. V. & Hussain, F. 1988, NASA report CTR-S88, 257.

Melander, M. V. & Hussain, F. 1989, *IUTAM Symposium on Topological Fluid Mechanics*, Cambridge, UK.

Metcalfe, R. W. & Hussain, F. 1989, *IUTAM Symposium on Topological Fluid Mechanics*, Cambridge, UK.

Moin, P., Leonard, A. & Kim, J. 1986, *Phys. Fluids* **29**, 955.

Monkewitz, P.A. 1988, *J. Fluid Mech.* **188**, 223.

Moore, D. W. & Saffman, P. G. 1972, *Phil. Trans. Royal Soc. Lond. A* **272**, 403.

Mumford, J. C. 1983, *J. Fluid Mech.* **137**, 447.

Oster, D, & Wygnansi, I. 1982, *J. Fluid Mech.* **123**, 91.

Parekh, D. E., Leonard, A. & Reynolds, W. C. 1988, Report TF-35, Stanford U.

Paul, P. H., Lee, M. P. & Hanson, R. K. 1989, *Optics Letters* **14**, 417.

Peterson, R. A., Kaplan, R. E. & Laufer, J. 1974, NASA CR-134733.

Raman, G. & Rice, E. J. 1989, NASA TM-101946.

Reynolds, W. C. 1988, in *Transport Phenomena in Turbulent Flows* (eds. M. Hirata & N. Kasagi) **111**, Hemisphere.

Roshko, A.1976, *AIAAJ* **14**, 1344.

Schatzle, P.R. 1987, PhD thesis, California Institute of Technology.

Siggia, E. D. 1985, *Phys. Fluids* **28**, 794.

Sigurdson, L.W. & Roshko, A. 1988, in *Turbulence Management and Relaminarization* (eds. H.W. Liepman & R. Narasimha), 497, Springer.

Sirovich, L. 1989, *Physica D.* (in press).

Takaki, R. & Hussain, A. K. M. F. 1985, *Turbulent Shear Flows V*, 3.19.

Tennekes, H. & Lumley, J. L. 1972, *A First Course in Turbulence*, MIT Press, Cambridge.

Toyoda, K. 1987, Univ. Houston Aero. & Turb. Lab. Internal Report.

Tso, J. & Hussain, F. 1989, *J. Fluid Mech,* **23**, 425.

Yule, A.J. 1978, *J. Fluid Mech.* **89**, 413.

Wedeen, V. J., Rosen, B. R., Buxton, R. & Brady, T. J. 1986, *Mag. Res. Medicine* **3**, 226.

Willmarth, W. W. 1975, *Advances in Applied Mechanics* **15**, 159, Academic Press.

Zaman, K.B.M.Q. & Hussain, A. K. M. F. 1980, *J Fluid Mech.* **101**, 449.

Zaman, K.B.M.Q. & Hussain, A. K. M. F. 1981, *J Fluid Mech.* **103**, 133.

Zaman, K.B.M.Q., Bar-Sever, A. & Mangalam, S. M. 1989, *J. Fluid Mech.* **182**, 127.

The Role of Coherent Structures – Comment no. 2

Peter Bradshaw

Mech. Engg. Dept., Stanford University, Stanford, CA 94305, USA

1. Introduction – Chaos and Cosmos

The "organization" of turbulence comes from its obedience to Newton's simple and nominally linear second law of motion, as embodied in the Navier-Stokes equations. These equations may not be all that simple but Newton's law produces a nice degree of non-linearity, enough to generate solutions with millions of degrees of freedom but not enough to constrain the solutions to low-dimensional chaos. (The Greek word chaos is the antithesis of "cosmos", which originally meant "order" or "arrangement".) The memory requirements of full turbulence simulations for low Reynolds numbers demonstrate that millions of degrees of freedom really are needed for a complete description, and even high-Reynolds-number large-eddy simulations, which at best reproduce only the energy-containing motion, seem to need storage of the same order.) Whether 'degrees of freedom" always approximates to "attractor dimension" is not clear, but it is certainly not proven that a system with millions of degrees of freedom can be approximated, to engineering accuracy, by a low-dimensional attractor. I am much less confident than Brian Cantwell that any quantitatively useful connection exists between turbulence and the chaotic behaviour of systems with only a few degrees of freedom.

The unlimited divergence of turbulence solutions with infinitesimally different initial conditions is often quoted as a link between Navier-Stokes solutions and strange-attractor chaos: but it seems a perfectly natural result of nearly-random diffusion, like a drunkard's walk – the miracle would be if independent solutions somehow reduced their initial differences. It is also fairly obvious that the rate at which solutions diverge will increase with the number of degrees of freedom in the system.

One must not judge the usefulness of chaos studies or other modern mathematical approaches by some of the intemperate claims that have been made. Ten years or so ago ago we had many intemperate claims by the mean-flow (Reynolds-averaged) modellers: their hopeful statements about the universality of their methods were shot down at the 1980-81 Stanford meeting on Computation of Complex Turbulent Flows, and Reynolds-averaged modelling has remained under a cloud – unfairly, I believe. Many of the more recent theoretical ideas (cellular automata, RNG, strange attractors) are attempts to introduce principles not contained in the Navier-Stokes equations – but if the N-S equations are a complete description of turbulence, any principles not deducible from them must be empirical, and therefore deserving of no more respect than empirical correlations of results from experiments and N-S simulations. Brian Cantwell's definition of "theory" at the start of section II.4 hits the nail on the head – but even basic mathematical results like existence and uniqueness are pretty rare. I agree with his conclusion that theoretical studies of the N-S equations would probably not contribute much to engineering prediction methods, mainly because they are likely to give order-of-magnitude results like attractor dimensions rather than quantitative information of engineering accuracy.

Sections 2 to 4 refer, approximately, to the same sections of the Position Paper.

2. Modelling

By referring, in the last paragraph, to "results from experiments and N-S simulations" in the same breath, I am accepting most of the comments in sections II.1 and II.2 of the position paper. A combination of simulations for geometrically-simple flows (illustrating complex phenomena) and experiments for more realistic flows is the only obvious way forward in development of turbulence models for engineers and earth scientists. Brian Cantwell's breakdown of experimental work into "experiments of discovery" – the search for new physical phenomena – and "experiments of code verification", directed at *confirming* computations, is too simplistic. It ignores experiments aimed at contributing to turbulence modelling! All too many "experiments of discovery" produce pretty pictures without contributing at all to the development of improved prediction methods: "code verification" too often means mean-flow measurements alone, sufficient to show if a model is right or wrong but not sufficient to show why.

I wish I could agree with the statement in II.3 that FTS or LES calculations for engineering flows are likely to be available by the mid-1990s (qualified further down the page). This seems very optimistic – storing or evaluating coordinate-transformation metrics at 128^3 points is a bad enough prospect, but even in LES, 128^3 points is a very small number for coverage of, say, a complete swept wing. Current large-eddy simulations do not resolve the whole of the energy-containing range, so that a significant fraction of the Reynolds stress – especially near a solid surface – is obtained from the very crude eddy-viscosity formula of the sub-grid-scale model. Thus, current LES techniques would not be reliable in flows where the law of the wall breaks down – which of course is one of the places where Reynolds-averaged models run into trouble. To increase the Reynolds number of a full simulation by a factor of 10 (section II.1) implies increasing the number of grid points by a factor of the order of 1000.

For the next decade, simulations are likely to equate to lab. experiments in idealized geometries, rather than configuration tests on engineering hardware. That is, for the next decade at least, engineers are going to be stuck with mean-flow (i.e. Reynolds-averaged, specifically time-averaged) models. "Very-large-eddy simulations", and other attempts to link a quasi-deterministic or time-dependent solution of the organized motion to a time-averaged model for the smaller-scale motion, have not yet shown any promise of better accuracy or universality than Reynolds-averaged models, and it seems inescapable that time-dependent models will need computing times an order of magnitude larger than Reynolds-averaged models. Remember that engineers are not yet convinced that anything better than an eddy-viscosity method (Baldwin-Lomax, Cebeci-Smith or k, ϵ) is cost-effective!

The "zonal modelling" concept seems to be the only alternative to the ideal of a universal model – in fact this is a truism. The catch about the "local model" scenario in section II.3 of the position paper is that it is difficult to find out where a turbulence model fails unless one has quite detailed experimental data including turbulence statistics. I heartily agree that ability to "determine where and why failure occurred" is essential to convert turbulence modelling from an art to a science – or at least a branch of technology. Virtually every piece of engineering hardware, from a domestic appliance to an aircraft, has qualitative or quantitative restrictions in the instruction manual or in hardware interlocks: any piece of software should contain the equivalents of the boil-dry cutout in the coffee maker or the stick pusher in the aircraft. Belief in the possibility of a universal model (other than the N-S equations) has been responsible for a lot of fuzzy thinking, both in the use of data from homogeneous decaying turbulence or rapid-distortion theory to calibrate methods intended for shear layers, and in the failure to print out warning messages when a program reaches the envelope of its empirical data.

3. Organized Motion

In section III.1, Brian Cantwell quite rightly points out that the organized structure in turbulence (which, except for cases like the mixing layer or Taylor-Couette flow which are dominated by quasi-inviscid instability, really means no more than Townsend's large eddies) varies from flow to flow. There may be common qualitative features: perhaps the large eddies are all trying to become horseshoe vortices, with a degree of success that varies from flow to flow. However the quantitative differences surely imply that the empirical coefficients in any turbulence model based on organized-structure ideas will also vary from flow to flow – in other words, study of organized structure will not lead to a universal model. One hopes that this study will lead to a better understanding of basic processes like scrambling of Reynolds stress by pressure-strain interaction, but so far it has not even led to an improvement on the eddy diffusivity concept used, *even in Reynolds-stress transport models*, for turbulent transport of stress and energy (and that's a "disorganized-structure" concept if ever there was one!).

Reynolds-number invariance, discussed in section III.3, is one of those simple principles which have surprisingly restricted validity. For example, virtually every laboratory experiment that includes Reynolds-stress spectra shows that the ratio of shear-stress (uv) cross-spectral density to energy spectral density is significant even at the highest wave numbers (frequencies): the shear-stress-containing and dissipating ranges overlap. Locally-isotropic inertial subranges are likely to be found only at geophysical Reynolds numbers. Combustion is a particularly sad case because of the increase in viscosity with temperature –

note the absence of small-scale structure in any laboratory flame. The resolution of the "viscous superlayer" in FTS is probably very poor: granted that the superlayer is probably thicker than the sublayer, but the grid can be refined in the sublayer whereas we cannot afford a fine grid over the whole of the intermittent region. The same criticism is bound to apply to internal scalar interfaces which are so important in diffusion-limited reactions. Thus, simulations with limited resolution can over- or underestimate effects of molecular diffusion.

4. Conclusions

Despite the "Closing Remarks", I feel that engineers and earth scientists (and therefore funding agencies) will see Reynolds-averaged mean-flow modelling as the only quantitative contribution that turbulence research will make to the real world in the next decade or so. This is not to decry qualitative understanding from basic experiments or simulations, but both experimenters and simulators ought to ask themselves what they are doing, directly or indirectly, short-term or long-term, to help engineers. So far, the large amount of research on organized motion has really not contributed much, with the exception of the light shed on diffusion flames by the study of organized structure and "mixing transitions" in mixing layers. I believe that there is plenty of mileage left in mean-flow Reynolds-stress modelling, providing that we accept that different species or zones of turbulent flow will require different empirical coefficients. If the model is well-founded, the differences will be small enough for interpolation between different zones to be non-critical: but the computer program must identify the zones for itself, and stop when it enters uncharted territory. A better understanding of organized motion (the traditional "large eddies" or the unusually-orderly structures found in a few special flows) should help the modellers, but let us not replace the unfulfilled belief in a universal turbulence model with the equally hopeful (and closely related) notion of universal organized motion.

The Role of Coherent Structures

Comment 3.

R.A. Antonia

Department of Mechanical Engineering
University of Newcastle
N.S.W. 2308
Australia

ABSTRACT

This paper is partly a discussion of some of the issues raised by Brian Cantwell in Section III of his paper. It also briefly considers the importance of the organised motion in terms of contributions to the Reynolds stresses and the average lateral heat flux, the effect of Reynolds number on this motion and some implications for local isotropy.

TOPOLOGY OF TURBULENT SHEAR FLOWS

As Brian notes, significant use has already been made of the concept of critical points for describing the topology of turbulent shear flows. When the velocity field is viewed in a frame of reference which translates at an appropriate convection velocity (U_c say), critical points, or points where the velocity is zero, can be identified. The spatial arrangement of these points often gives an indication of the degree of organisation of the flow. This is illustrated in Figure 1 for the far-wake of a cylinder and in Figure 2 for a turbulent boundary layer.

The sectional streamlines shown in these figures are everywhere tangent to the velocity vector in either the (x,y) plane (x is in the streamline direction and y is in the direction of main shear) or the (x,z) plane (z is in the spanwise direction). The velocity fluctuations in these planes were obtained with a linear array of X-probes providing data for velocity fluctuations u,v in the (x,y) plane and u,w in the (x,z) plane.

Comparison of Figure 1 and Figure 2a shows that the spatial arrangement of critical points (the streamline slope is indeterminate at these points) is much more systematic in the wake than in the boundary layer. In the wake, the arrangement is suggestive of a predominatly antisymmetric disposition of spanwise vortices about the centreline. In contrast with Figure 1, the patterns in Figure 2a suggest a high degree of complexity, which is reflected, inter alia, by the wide range of distances from the wall at which critical points are found.

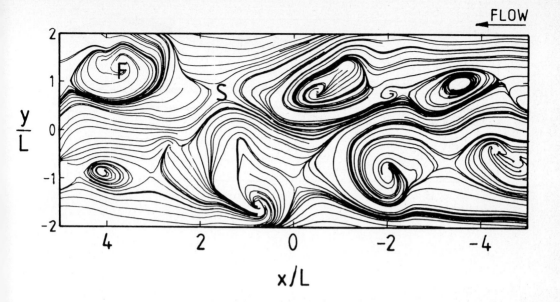

Figure 1 Instantaneous sectional streamlines in the (x,y) plane of the far-wake of a circular cylinder. The Reynolds number R_d, based on the free stream velocity U_1 and cylinder diameter d, is 1170. Streamlines are observed in a frame of reference which travels in the flow direction at $U_c = 0.95U_1$. The origin for x is arbitrary. L is the mean velocity half-width of the wake. Examples of a focus and a saddle point are denoted by F and S respectively.

An obvious disadvantage of the presentations in Figures 1 and 2 is the dependence of the streamline patterns on U_c (the dependence is less severe for Figure 1 than Figure 2, since the maximum mean velocity variation across the wake is about 5% of U_1). This difficulty can be circumvented by considering quantities which are independent of U_c. Two such quantities are the vorticity and the strain rate. Arrays of X-probes aligned in either the y or z directions permit approximations to smoothed z or y vorticity components, the degree of smoothing depending, in the first instance, on the separation between X-probes. This approach has been used by Hussain and Hayakawa [1], Antonia et al. [2] and Bisset et al. [3]. Our perception has been that vorticity contours, while extremely important in the context of identifying three-dimensional structures, do not by themselves provide a sufficiently complete description of the flow topology. Typically we find that vorticity patterns need to be

complemented by, inter alia, contours of the strain rate and quantities such as uv and $v\theta$ (θ is the temperature fluctuation) to yield sufficient insight into the physics and importance of the organised motion.

The association of critical points with various physical quantities is important from the point of view of providing a complete description of the physics of the organised motion. Computer simulations are proving useful in this context : for example, Robinson et al. [4] observe that some of the near-wall shear layers roll up into transverse vortices. At the tip of the vortex, the spanwise

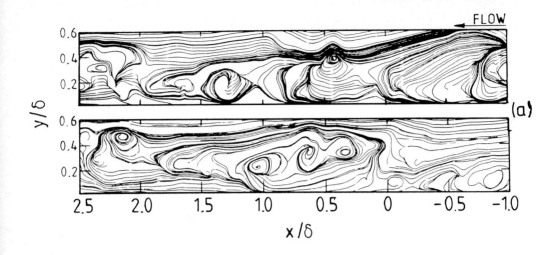

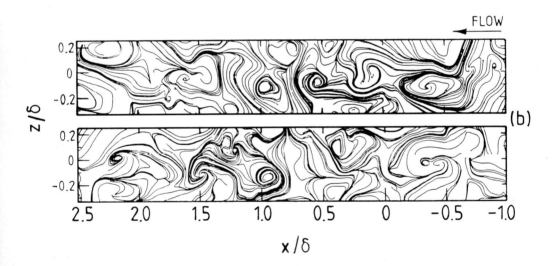

Figure 2   Instantaneous sectional streamlines in the (x,y) and (x,z) planes of a zero pressure gradient turbulent boundary layer ($R_\theta = 2200$). Streamlines are observed in a frame of reference which travels in the flow direction at $U_c = 0.8U_1$. Parts (a) and (b) are chosen from the subset of detections defined in Figure 3. The origin for x is arbitrary. [a] (x,y) plane;   [b] (x,z) plane.

vorticity is high and the pressure is low. The spanwise vorticity is equally high almost all along the shear layer.

DETECTION OF ORGANISED MOTION AND CONDITIONAL AVERAGING

Implicit in the requirement that the organised motion should survive ensemble averaging (e.g. Coles [5]; Hussain [6,7]) is the expectation that adequate criteria can be designed for detecting this motion. The appropriateness of these criteria should of course depend on the particular definition of the motion. Also, different aspects of the organised motion (e.g. features associated with different scales or different structures), may require the implementation of different criteria for their detections. In this context, a number of observations can be made :

1. If the conditionally averaged topology of a particular flow implies the existence of a unique spatial relationship between different critical points, it is reasonable to expect that criteria which detect different critical points (different features of the topology) should lead to the same description of the organised motion. In the context of Figure 1, it is not surprising that ensemble averaged streamlines obtained in the far-wake ([3,8]) are essentially the same, irrespective of whether the detection focuses on a saddle point, a diverging separatrix or a focus. In this context, I endorse Brian's view that the concept of organised motion should not be tied down too closely. This seems especially true in the case of a more complex flow, such as the boundary layer where the organisation can be described in terms of several features. In this flow, I doubt that the eight categories of quasi-coherent structures (see Kline [9] for a useful documentation) could be captured by a single, precise definition of the organised motion.

Note that the classification of flow patterns using the critical-point theory approach (Perry and Chong [10]) or Hunt et al.'s [11] procedure for describing different zones allows features to be defined precisely without explicit commitment to the overall definition of the organised motion of which the features are aspects.

2. The remarks in (1) suggest that, from an experimental point of view, much can be learnt by using relatively simple detection schemes. It is almost axiomatic that schemes that are based on information measured at several points in space should yield more insight into the spatial organisation of the motion than those that are based on one point detections. In the sense that organisation often implies a spatio-temporal relationship of structures or events, it is useful to examine the histogram or the probability density function of the time between consecutive detections obtained with one point criteria. Figure 2 shows the probability density function (pdf) of the time interval between detections obtained

at one point ($y/\delta = 0.27$) in a boundary layer with the window average gradient (WAG) approach (e.g. [12]) which focuses on the δ-scale u discontinuity. A possible strategy is to select a subset from the original set of detections consisting of only those detections which straddle the pdf peak. This was the procedure followed by Browne et al. [13] in the wake and Antonia et al. [2] in the boundary layer. Arguably, this approach reduces detection jitter and sharpens the organisation under study. The instantaneous patterns in Figure 1 correspond to the detection subset defined in Figure 2 while Figure

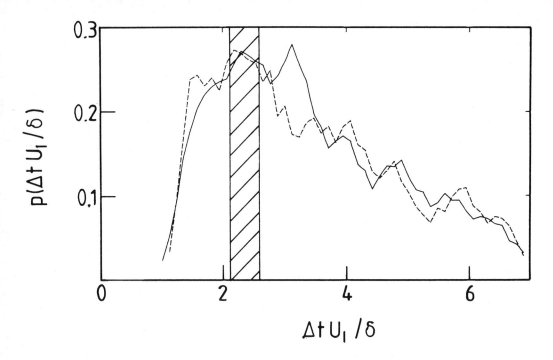

Figure 3 Probability density function of the time between consecutive WAG detections, based on the velocity u at $y/\delta = 0.27$ in a zero pressure gradient turbulent boundary layer ($R_\theta = 2200$). The hatched region defines the subset of detections used in Figures 2 and 4. The solid and broken curves correspond to detections in the (x,y) and (x,z) planes respectively.

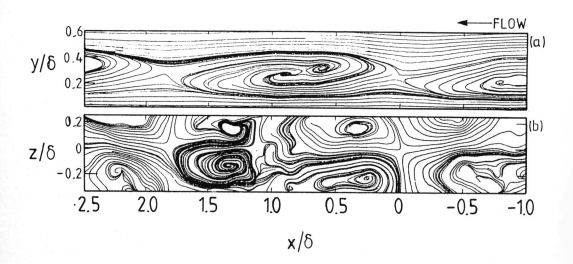

Figure 4 Ensemble averaged sectional streamlines in the (x,y) and (x,z) planes for a zero pressure gradient turbulent boundary layer ($R_\theta = 2200$). The ensemble for which the averaging was done is defined in Figure 3. Streamlines are observed in a frame of reference which travels in the flow direction at $U_c = 0.8U_1$. The origin for x is the detection location. [a] (x,y) plane; [b] (x,z) plane.

3 shows streamlines averaged over the total number (about 100) of detections of this subset. Obviously, many other selection strategies - aimed at enhancing the organisation - are possible, e.g. the pattern recognition approach of Wallace et al. [14] and different bootstrapping procedures (e.g. [15]).

Note that the scope and quality of the information can be significantly enhanced when the detection subset is derived from information at several points in space. Such an approach was used to assess the importance of different spatial arrangements of the organised motion on opposite sides of the centreline in the far-wake of a cylinder [3].

3. It is highly desirable that detection procedures are "calibrated", preferably against visually recognised features of the organised motion. This is not always feasible but existing attempts at calibration have greatly improved our ability to assess the performance of different conditional techniques. It is worth mentioning here the attempts by Offen and Kline [16], Bogard and Tiederman [17] and Talmon et al. [18] at calibrating several conditioning techniques by direct comparison with visually detected ejections; such calibrations are restricted to relatively low Reynolds numbers so that the application of these results at higher Reynolds numbers requires validation. There have also been attempts to calibrate conditional techniques against δ-scale discontinuities observed in signals from arrays of cold wires (e.g. [19]) or arrays of hot wires (e.g. [2]).

It is difficult to overemphasise the role flow visualisation has played in identifying various aspects of the organised motion in different flows. This can be illustrated by taking a brief look at the history of attempts at unravelling the organisation of a turbulent spot. The early flow visualisations, e.g. Elder [20], had delineated the planform arrow-like outline of the spot. Conditional techniques which identified the interface between the laminar and turbulent flow regions reproduced the planform arrow-like boundary. However, when the same techniques were applied to obtain ensemble averaged velocity field within the spot, the resulting signature was interpreted to be consistent with the presence of one or two relatively large scale vortices (e.g. [21]). Later flow visualisations (e.g. [22-24]) and observations of signals from arrays of hot wires [25,26] indicated the presence of internal features whose number increased as the spot developed downstream. Subsequently, and not unpredictably, attempts have been made (e.g. [27]) to devise schemes for identifying these features and for educing their velocity signature. Instantaneous sectional streamlines in the (x,y) plane of the spot (Figure 5) clearly show that a description in terms of one or two vortices is grossly inadequate : the number of vortices increases with distance from the spot origin in a manner which accounts for the streamwise growth of the spot [28]. Although the patterns of Figure 5 may indicate a somewhat higher degree of order than those of Figure 2a, there is sufficient jitter in the locations of the internal features to make the eduction of every feature a relatively difficult task [29]. This is the reason why earlier attempts at conditioning on either the downstream or upstream turbulent/laminar interface failed to reveal the internal make-up of the spot. It would be misleading to imply that our knowledge of the spot is now

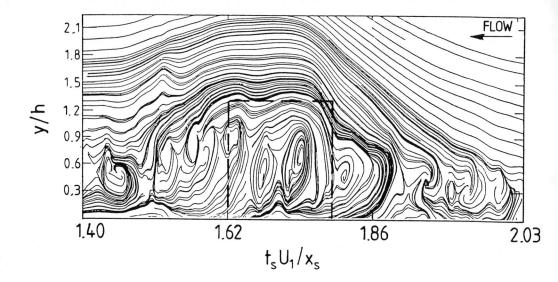

Figure 5   Instantaneous sectional streamlines in the (x,y) plane for a single turbulent spot viewed in a frame of reference which travels in the flow direction at $U_c = 0.84U_1$. xs and ts are the distance and time respectively measured from the spot disturbance; h is the height of the spot. The spot was selected from a population of spots containing the most probable number (5 at this location) of features. The boxed-in section has been investigated in more detail in Sankaran et al. [29].

complete or that we now understand how new structures are formed. Clearly, more information is required on the kinematics and dynamics of these structures. This information may be more easily captured outside the plane of symmetry of the spot where the jitter is less severe than at z = 0.

Brian Cantwell mentions the usefulness of the concept of similarity in time [30] in the context of analysing the development of the organised motion in turbulent shear flows. This concept has been applied, with only partial success, to the turbulent spot. While the concept helps to describe the development of the large scale motion of the spot, it does not focus on the internal features.

4.   Brian points out that the structure of turbulence can be described on a number of different levels, ranging from a Reynolds-averaged level down to an instantaneous level, possibly via conditionally averaged descriptions.

It has been established (e.g. [4,9]) that patterns that emerge from a conditionally averaged topology may not be present instantaneously, in which case the conditionally averaged topology should be interpreted with caution. One reason for this difference is that the particular detection scheme simply fails to capture some of the features observed instantaneously. Another possibility is that the scheme detects features that are aligned differently in space without distinguishing them. In either case, potentially useful information may be lost when conditional averaging is applied (e.g. [31,32]). As one illustration of this loss, only those patterns which are associated with the length scale defined in Figure 3 emerge in Figure 4a. Another illustration is the comparison between instantaneous and ensemble averaged patterns in the (x,z) plane. Instantaneous u patterns in this plane often exhibit strong asymmetries about z which may disappear completely after averaging. Conditionally averaged contours of u must be symmetrical with respect to z when a detection procedure is applied only at z = 0. The detections used to obtain Figure 4b were applied at z = 0; the nearly symmetrical double-roller type patterns of Figure 4b occur only infrequently (see Figure 2b). Conclusions regarding the shape of the organised motion on the basis of such conditionally averaged information may not be completely realistic : whereas the symmetrical horseshoe-type vortex structure is consistent with conditionally averaged data (e.g. [33]), Robinson et al. [4] note that this structure seldom occurs instantaneously in direct simulations of the boundary layer. It would seem that serious thought will have to be given in future on how to incorporate the instantaneous information (and how much of it) into the description of the organised motion.

CONTRIBUTIONS TO REYNOLDS STRESSES AND AVERAGE LATERAL HEAT FLUX

At the end of Section III-1 of his paper, Brian states "We cannot, at this point, say definitively what fraction of the Reynolds stress could be regarded as contributed by motions which, by some measure, could be considered ordered. It is probable that the answer to this question, as with so many other questions on turbulence, varies from flow to flow". While it is difficult to disagree with both these statements, it is useful to provide some estimate, albeit tentative, of the magnitude of the fraction and of its variation in different flow.

The expectation that this fraction can be estimated presupposes that the organised motion can be adequately described. Since the organised motion may be associated with different length scales, geometries, strengths, it would be reasonable to expect estimates of its contribution to momentum or heat transport to be appropriately qualified. Although a comprehensive classification of contributions attributable to different characterising parameters of the organised motion has yet to be attempted, useful qualitative estimates have been obtained in several "classical" flows (plane jets, wakes, boundary layers) by making use of the triple decomposition[*]

$$F = \overline{F} + \tilde{f} + f_r \quad .$$

Here F represents an instantaneous quantity, $\overline{F}$ is its time average, $\tilde{f}$ is the organised motion component and f_r is the so-called random, incoherent or sometimes background motion component. It seems

---

[*]In contrast to the triple decomposition, a double decomposition is needed to yield the patterns in Figure 3.

reasonable to view f_r as including the truly random as well as any co-existing coherent motion of a different kind or scale to the one educed. The organised motion component can be determined by ensemble averaging $F - \overline{F}$, viz. $\tilde{f} \equiv \langle F - \overline{F} \rangle$, (the angular bracket denotes averaging over the ensemble of detections) and depends on τ, the time measured from the detection instant. Introducing another quantity g (here, f and g represent the fluctuations u, v or θ), the equation

$$\langle fg \rangle \;=\; \widetilde{fg} + \langle f_r g_r \rangle$$

can be used to determine $\langle f_r g_r \rangle$. A measure of the contribution of the organised motion to the product $\overline{fg}$ (e.g. the Reynolds shear stress $\overline{uv}$ or the average heat flux $\overline{v\theta}$) is given by the ratio $\widetilde{fg}/\overline{fg}$, where the double overbar denotes an average over a suitable time interval, τ_1 say. Clearly, care is needed in interpreting the numerical values of $\widetilde{\widetilde{fg}}/\overline{fg}$ and also $\overline{\langle f_r g_r \rangle}/\overline{fg}$ since they will depend, inter alia, on the particular conditioning technique, the unavoidable detection jitter and the choice of τ_1. Note that the use of a technique such as WAG with a detection subset, as defined in Figure 2, would ensure uniformity (in terms of a time or length scale) of members of the ensemble. Structure variability and motions with different scales from the one educed are naturally ignored in the coherent contribution and included, by default, in the incoherent component.

Notwithstanding the previous reservations, it is worth drawing attention to the main results that have been obtained to date :

1. The magnitude of $\widetilde{\widetilde{fg}}/\overline{fg}$, although smaller than that of $\overline{\langle f_r g_r \rangle}/\overline{fg}$ leaves little doubt that the contribution from the organised motion cannot be ignored. Typical values of this ratio, shown in the Table below, were obtained using a (saddle-type) temperature based detection.

TABLE TYPICAL AND APPROXIMATE VALUES OF COHERENT CONTRIBUTIONS IN THREE FLOWS

Flow	$\widetilde{u^2}/\overline{u^2}$	$\widetilde{v^2}/\overline{v^2}$	$\widetilde{\theta^2}/\overline{\theta^2}$	$\widetilde{uv}/\overline{uv}$	$\widetilde{u\theta}/\overline{u\theta}$	$\widetilde{v\theta}/\overline{v\theta}$
Plane Jet	0.20	0.20	0.40	0.42	0.40	0.52
Boundary Layer (log region)	0.20	0.08	0.30	0.22	0.32	0.26
Plane Wake	0.09	0.22	0.33	0.28	0.30	0.45

2.   The table indicates that the contribution to $v\theta$ is consistently larger than the contribution to uv.

3.   In the self-preserving part of the flow, the data suggest that similarity should be satisfied, although the contributions may depend on y/δ (for the boundary layer) or y/L (for the wake and the jet, where L is the mean velocity half-width). For example, the coherent contribution to the temperature variance is especially large in the outer part of the wake. In the boundary layer, the burst-sweep sequence is largely responsible for the generation of the Reynolds shear stress and heat flux near the wall.

4.   Contributions from the organised motion are different in different flows. The Table indicates that, with the exception of $\tilde{v}^2/\overline{v^2}$, values of $\overline{\tilde{f}\tilde{g}}/\overline{fg}$ are consistently larger in a plane jet than in a plane wake (also [8]). Also, the contributions to $\overline{u^2}$ and $\overline{u\theta}$ in the boundary layer are much greater than the contributions to $\overline{v^2}$ and $\overline{v\theta}$ respectively, while the situation is reversed in the free shear flows, especially the wake. While the magnitudes of these contributions depend to some extent on the particular conditional technique used, their relative importance is not affected.

Some of the above results should have implications with respect to both turbulence modelling and flow control strategies. With regard to result (4), it is worth noting that the topology of the plane jet appears to have quite similar features to that of the plane wake and yet, there must be sufficient differences between the organised motions in these two flows to account for the difference in the relative values of the coherent and incoherent contributions to the fluxes. If use, in terms of modelling, were to be made of our knowledge of the organised motion, then careful consideration should be given to such differences[*]. It would be of interest to relate these differences to a parameter such as the non-dimensional mean shear used by [34] in comparing homogeneous with non-homogeneous shear flows. With respect to (3), it would clearly be of use to establish the dependence of the topology, in the self-preserving flow region, on initial conditions : if the idea of a multiplicity of self-preserving stages is correct [35], there should be detectable differences in the topology of the self-preserving region.

EFFECT OF REYNOLDS NUMBER ON ORGANISED MOTION

It is important to know whether the organisation changes with the Reynolds number since such information may allow us to relate observations at low Reynolds numbers to what may occur at higher Reynolds numbers. This information may also allow better use to be made of results from direct or large eddy simulations which are carried out at low to moderate Reynolds numbers.

---

[*]Bradshaw [36] has noted that the study of orderly structure is not a route to a universal model, at least at a Reynolds-averaged level.

Here my remarks will be directed to the boundary layer as most of the work appears to have been done for this flow. The influence of Reynolds number on the mean velocity, the skin friction coefficient, various turbulence quantities are well documented in the literature [37-41]. Useful insight into the effect of Reynolds number on the organisation in both the wall and outer regions of a turbulent boundary layer has been provided by a number of flow visualisation experiments (e.g. [42-44]).

Recently, the effect of R_θ on the topology of the large scale motion was studied using an array of X-probes in the plane of main shear [2]. Although the study was not extensive (only four Reynolds numbers were considered), all results at the two largest Reynolds numbers ($R_\theta = 6030$ and 9630) are practically identical, in conformity with the general consensus that the influence of R_θ ceases when the latter exceeds about 5000. Below $R_\theta = 5000$, estimates of the contribution from the organised motion (using a procedure similar to that outlined in the previous section) to all the Reynolds stresses suggest an increase as R_θ decreases. At small R_θ, the region $0 \le y^+ \lesssim 100$ is virtually coincident with $0 \le y \lesssim \delta$ so that a large overlap exists between the largest scales of the flow and the scales which carry most of the Reynolds stresses.

Perhaps of significance from the point of view of modelling is the observation that streamlines and contours of the strain rate and spanwise vorticity are essentially unchanged at all Reynolds numbers. Although details are lacking, the instantaneous topology provides some support for the presence of hairpin vortices extending to various distances from the wall irrespective of R_θ - in contrast with the stronger distinction between low and high R_θ, indicated by Head and Bandyopadhyay's [43] smoke photographs. I expect that experiments will continue to supply data at higher Reynolds numbers than is currently feasible in numerical simulations, although a major limitation in experimental work is the rapid impairment in the spatial spanwise resolution when R_θ increases. This problem may be somewhat alleviated in future due to improvements in computer technology and the miniaturisation of sensors and supports. Brian expects that direct simulations at $R_\theta > 1000$ should be possible in about 10 years from now. Only time will tell whether such an expectation is reasonable!

Low Reynolds number effects must be taken into account when considering the scaling of the average bursting period [45]. This is important since the Reynolds number range is small in laboratory experiments and smaller in computer simulations. In the context of the established Reynolds number dependence of the outer region of the boundary layer, it is perhaps not surprising that the average bursting period - as determined for example by a "calibrated" approach (such as the modified u-level method of [17]) - does not scale on outer variables, even if the outer large scale motion exerts an influence on the wall region (e.g. the direct simulations of [46] up to $R_\theta = 1410$) at low Reynolds numbers. The experimental results of Shah and Antonia [45] for a boundary layer and a fully developed duct flow appear to be consistent with scaling on three different groups of variables, depending on the Reynolds number. Over the Reynolds number range for which viscous effects are observed in the outer region, scaling on wall variables seems appropriate (a result supported by the experiments of Luchik and Tiederman [47], Tiederman [48], and the numerical simulations of Kim and Spalart [49]). At higher Reynolds numbers, scaling on a mixture of wall and outer variables (as introduced by [50]) seems to provide the best agreement with the data, although it should be clearly kept in mind that the detection technique requires calibration at these Reynolds numbers. The possibility of scaling on just outer variables cannot be ruled out for even higher Reynolds numbers. This trend would seem reasonable in terms of reconciling results in the atmospheric surface layer with those in the laboratory boundary layer.

IMPLICATIONS FOR LOCAL ISOTROPY

The existence of a motion that is quasi-organised in space and time has a number of implications with regard to the concept of local isotropy, a concept which has a place of importance in turbulence research. Before the implications can be properly assessed, possible sources of error associated with the measurement of small scale turbulence must be recognised; a review of these difficulties was given by Antonia et al. [51] and will not be discussed here. One other aspect, also covered in the review, relates to the degree of sensitivity of different tests of local isotropy. When the focus is solely on the high wavenumber part of the spectrum, all available data suggest that local isotropy is satisfactorily validated above a certain wavenumber, whose magnitude depends, inter alia, on the flow, the Reynolds number or the particular quantity used in the test. For example, the high wavenumber behaviour of spectra of vorticity and dissipation (either energy or temperture) in a self-preserving turbulent wake are consistent with isotropy [52]. When the test does not emphasise the high wavenumber part of the spectrum, the experimental evidence points away from loal isotropy. For example, relations between mean square values of derivatives with respect to different spatial directions of either velocity or temperature indicate strong departures from local isotropy (e.g. [53,54] for the wake). Consequently, the replacement of the average dissipation or the mean square vorticity by their isotropic values is incorrect, even at large Reynolds numbers; implications for turbulence modelling should be obvious.

CONCLUDING REMARKS

Considerable progress has been made with the experimental identification and description of various features or characteristics of the organised motion in a wide range of turbulent shear flows. More work is needed to provide a complete three-dimensional description of this motion and to sensitise this description to possible changes in initial and boundary conditions. While it is not difficult to acknowledge that direct computer simulations have several advantages over experiments, both with respect to describing the organised motion and guiding the development of turbulence models, I expect that experiments will continue to provide valuable data on both the organised and random motions over a wider Reynolds number range than is currently possible with direct simulations. At this point in time, the main difficulty appears to be how to make the most effective use of the data, irrespective of whether they are generated by computer or by experiment. For the future, Brian Cantwell's suggestion of a laboratory-based coupling between experiments and direct simulations seems attractive.

ACKNOWLEDGEMENTS

The financial support of the Australian Research Council is gratefully acknowledged. I would like to thank Mr. D. K. Bisset for many interesting discussions in connection with this paper.

REFERENCES

1. A. K. M. F. Hussain, M. Hayakawa : *J. Fluid Mech.* **180** 193 (1987).
2. R. A. Antonia, L. W. B. Browne, D. K. Bisset : *Proc. Zaric Memorial International Seminar on Near-Wall Turbulence*, Dubrovnik (1988).
3. D. K. Bisset, R. A. Antonia, L. W. B. Browne : *J. Fluid Mech.* [submitted] (1989).
4. S. K. Robinson, S. J. Kline, P. R. Spalart : *Proc. Zaric Memorial International Seminar on Near-Wall Turbulence*, Dubrovnik (1988).
5. D. Coles : in R. Narasimha and S. M. Deshpande (eds.) *Surveys in Fluid Mechanics*, Bangalore, 17 (1982).
6. A. K. M. F. Hussain : *Phys. Fluids* **26** 2816 (1983).
7. A. K. M. F. Hussain : *J. Fluid Mech.* **173** 303 (1986).
8. R. A. Antonia : in M. Hirata and N. Kasagi (eds.) *Transport Phenomena in Turbulent Flows : Theory, Experiment and Numerical Simulations*, New York, Hemisphere, 91 (1988).
9. S. J. Kline : *Proc. Zaric Memorial International Seminar on Near-Wall Turbulence*, Dubrovnik (1988).
10. A. E. Perry, M. S. Chong : *Ann. Rev. Fluid Mech.* **19** 125 (1987).
11. J. C. R. Hunt, P. Moin, A. A. Wray : *Center for Turbulence Research Report CTR-S88*, 1 (1988).
12. R. A. Antonia, L. Fulachier : *J. Fluid Mech.* **198** 429 (1989),
13. L. W. B. Browne, R. A. Antonia, D. K. Bisset : *Phys. Fluids* **29** 3612 (1986).
14. J. M. Wallace, R. S. Brodkey, H. Eckelmann : *J. Fluid Mech.* **83** 673 (1977).
15. G. L. Brown, A. S. W. Thomas : *Phys. Fluids* **20** S243 (1977).
16. G. R. Offen, S. J. Kline : *J. Fluid Mech.* **70** 205 (1975).
17. D. G. Bogard, W. G. Tiederman : *J. Fluid Mech.* **162** 389 (1986).
18. A. M. Talmon, J. M. F. Kunen, G. Ooms : *J. Fluid Mech.* **163** 459 (1986).
19. C. S. Subramanian, S. Rajagopalan, R. A. Antonia, A. J. Chambers : *J. Fluid Mech.* **123** 335 (1982).
20. J. W. Elder : *J. Fluid Mech.* **9** 235 (1960).
21. B. Cantwell, D. Coles, P. Dimotakis : *J. Fluid Mech.* **87** 641 (1978).
22. T. Matsui : in R. Eppler and H. Fasel (eds.) *Laminer-Turbulent Transition*, Berlin, Springer, 288 (1980).
23. M. Gad-el-Hak, R. F. Blackwelder, J. J. Riley : *J. Fluid Mech.* **110** 73 (1981).
24. A. E. Perry, T. T. Lim, E. W. Teh : *J. Fluid Mech.* **104** 387 (1981).
25. I. Wygnanski, J. H. Haritonidis, R. E. Kaplan : *J. Fluid Mech.* **92** 505 (1979).
26. R. A. Antonia, A. J. Chambers, M. Sokolov, C. W. Van Atta : *J. Fluid Mech.* **108** 317 (1981).
27. I. Wygnanski : *Proc. Seventh Symposium on Turbulence*, University of Missouri-Rolla, 390 (1983).
28. R. Sankaran : Ph.D. Thesis, University of Newcastle, Australia (1988).
29. R. Sankaran, M. Sokolov, R. A. Antonia : *J. Fluid Mech.* **197** 389 (1988).
30. B. J. Cantwell : *Archives of Mechanics (Archivum Mechaniki Stosowanej)* **31** 707 (1979).

31. A. V. Johansson, P. H. Alfredsson, J. Kim : *Center for Turbulence Research Report CTR-S87* 231 (1987).
32. Y. G. Guezennec, U. Piomelli, J. Kim : *Center for Turbulence Research Report CTR-S87*, 263 (1987).
33. J. Kim : *Phys. Fluids* **28** 52 (1985).
34. P. Moin : *Proc. Zaric Memorial International Seminar on Near-Wall Turbulence*, Dubrovnik (1988).
35. W. K. George : in W. K. George and R. Arndt (eds.) *New Horizons in Turbulence*, New York, Hemisphere (1988).
36. P. Bradshaw : *Center for Turbulence Research Report CTR-S87*, 313 (1987).
37. D. Coles : *Rand Corp. Report R-403-PR, ARC 24478* (1962).
38. L. P. Purtell, P. S. Klebanoff, F. T. Buckley : *Phys. Fluids* **24** 802 (1981).
39. J. Murlis, H. M. Tsai, P. Bradshaw : *J. Fluid Mech.* **122** 13 (1982).
40. J. Andreopoulos, F. Durst, Z. Zaric, J. Jovanovic : *Expts. in Fluids* **2** 7 (1984).
41. L. P. Erm., A. J. Smits, P. N. Joubert : in F. Durst, B. E. Launder, J. L. Lumley, F. W. Schmidt and J. H. Whitelaw (eds.) *Turbulent Shear Flows 5*, Berlin, Springer, 186 (1987).
42. R. E. Falco : *AIAA Paper 74-99*, presented at AIAA 12th Aerospace Sciences Meeting, Washington (1974).
43. M. R. Head, P. Bandyopadhyay : *J. Fluid Mech.* **107** 297 (1981).
44. C. R. Smith, S. P. Metzler : *J. Fluid Mech.* **129** 27 (1983).
45. D. A. Shah, R. A. Antonia : *Phys. Fluids A* **1** 318 (1989).
46. P. R. Spalart : *J. Fluid Mech.* **87** 61 (1988).
47. T. S. Luchik, W. G. Tiederman : *J. Fluid Mech.* **174** 529 (1987).
48. W. G. Tiederman : *Proc. Zaric Memorial International Seminar on Near-Wall Turbulence*, Dubrovnik (1988).
49. J. Kim, P. R. Spalart : *Phys. Fluids* **30** 3326 (1987).
50. P. H. Alfredsson, A. V. Johansson : *Phys. Fluids* **27** 1974 (1984).
51. R. A. Antonia, F. Anselmet, A. J. Chambers : *J. Fluid Mech.* **163** 365 (1986).
52. R. A. Antonia, D. A. Shah, L. W. B. Browne : *Phys. Fluids* **31** 1805 (1988).
53. R. A. Antonia, L. W. B. Browne : *J. Fluid Mech.* **163** 393 (1986).
54. L. W. B. Browne, R. A. Antonia, D. A. Shah : *J. Fluid Mech.* **179** 307 (1987).

Discussion of "The Role of Coherent Structures"

Reporter Sherif El Tahry

Fluid Mechanics Dept.
General Motors Research Laboratory
Warren, MI 48090-9055

Julian Hunt:

In order to facilitate our discussion, I would like to focus on the major issues that have been raised, and consider them one at a time. I will even suggest what they are.

It seems appropriate to begin by discussing what different people mean by coherent structures. This afternoon coherent structures were described as motions that make significant turbulence and velocity fields; Fazle Hussain defined them by the vorticity field, Brian Cantwell by the pressure-gradient field, and so on. Presumably, the way we define coherent structures depends on how we intend to use the definition. For example, will the definition be used in a dynamic simulation of the flow ? Or will it just be used in a finger printing operation in order to compare between different flows? We ought to discuss this, and see if we can get any consensus about how research should be going in terms of defining what we mean by coherent structures in different flows. I want to call on Dr. Perry who has some interesting concepts of using mathematical invariance properties. I wonder if he could give us his point of view on how we should define coherent structures.

Anthony Perry:

Basically, I don't know what a coherent structure is. I once asked John Lumley what he thought it is, and he said a coherent structure is a pattern that recurs in the flow. This is probably what it is, a pattern that you recognize and it recurs. To me this recurring pattern does not, necessarily, have an order to it; its scale and velocity are random. However, its orientation is fixed. It has to have some characteristic orientation otherwise it would not be possible to recognize it. I do not know how to define a coherent structure better than that.

Regarding the comment made by Julian Hunt on certain invariances, that has to do with another discussion I had with him on how to define a vortex. i.e., What is a vortex? When is a vortical flow a vortex? To get the best answer to these questions, Brian Cantwell and I have considered the flow as seen following a fluid element. The topology of the flow surrounding such an element can be accurately described by the Jacobian tensor, which is actually the rate of deformation tensor. The rate of deformation tensor is made up of two tensors- the rate of strain tensor which is symmetric, plus a rotation or spin tensor which is skew-symmetric. When the rotation tensor dominates over the rate of strain tensor, the eigenvalues of the rate of deformation tensor are complex, and when you map out the instantaneous streamlines of such a flow, we get streamlines that spiral or wrap around an axis and will spiral in or spiral out. This then is how I'd define a vortex.

William George:

I would like to discuss the definition of coherent structures, and also the question raised by Brian Cantwell regarding whether there was any physics in mean-flow modeling. I suggest that a problem with both the definition of coherent structures and with mean-flow modeling is that we do not really recognize physics when we see it because we have forgotten what physics is. Fazle Hussain, rightly perhaps, suggested that coherent structures should be coherent vorticity; but then, I contend, he has never seen coherent vorticity because fluctuating vorticity cannot be easily measured. One cannot use time derivatives because the convective fields are such that Taylor's hypothesis does not apply. Moreover, due the Nyquist criterion in space, it is not possible to resolve vorticity by taking differences between hot-wires that are finite distances apart, at least at the Reynolds numbers of interest. In fact, I wonder sometimes what Fazle has measured-- it is certainly not vorticity, but must be a highly aliased field.

There is a story in the Book of Genesis which I think is appropriate. It is a story about Adam who has been commissioned to name all the animals in the Garden of Eden. Adam dutifully fulfilled his responsibilities: he identified tigers, lions, birds, foul, fish, and so on. I do not recall anybody suggesting that Adam was a biologist; yet I suggest that we are guilty of just naming the animals and pretending that makes us physicists. We identify chicken necks, blobs, eddies, and everything else under the sun, and because we assign names to them and identify all their features, we think that is physics. I thought that physics meant writing the equations and describing dynamic interactions between structures. That is not what we are doing in the coherent structure business. Characterization is a very important part of science, but let's not confuse it with physics. I suggest that we are not going to be able to define coherent structures until we decide what we want to do with them, and derive equations expressing their evolution. My favorite procedure, that achieves all that, is John Lumley's proper orthogonal decomposition.

Fazle Hussain:

I was hoping to stay out of the discussion on coherent structures (CS), as I felt I have had other opportunities to record my ideas on CS in the literature. However, I feel compelled to respond, especially by the comments of Bill George. The basis for my definition of CS and the scheme we have employed to educe them in different physical and numerical experiments is that underlying the random vortical field that we call turbulence there are domains of spatially correlated instantaneous vorticity patches that we call CS. This correlated part of the vorticity is coherent vorticity; the remainder is incoherent vorticity. similarly follows the definition of coherent and incoherent turbulence. It is not meaningful to ask if one has "seen" CS; who has seen the mean profile in a turbulence jet? CS is not only conceptually precise, but also can be measured accurately as has been well demonstrated by us in a number of flows in the laboratory or in numerically simulated flows: laboratory flows studied include near and far fields of circular, plane and elliptic jets, plane mixing layers and cylinder wake; CS have been educed from simulations of plant mixing layers, a channel flow, a boundary layer and a homogenous shear flow (mostly from NASA-Ames database and some simulated by Metcalfe at University of Houston). The eduction algorithm is based on appropriately aligning flow realizations containing similar structures (identified by correlated vorticity patches of similar size, shape, strength, etc.) and then ensemble averaging (for details see Hussain, A.K.M.F. ,1986, J. Fluid Mech., 173, 303.).

Bill George's concerns regarding vorticity measurements are well known and have been adequately addressed in our works. Since we are concerned with coherent vorticity (associated with large-scale events), the crude spatial spacing of X-wires in a rake is an unavoidable compromise (the time resolution is much finer). As emphasized in our papers, the accuracy of coherent vorticity has been carefully validated through much finer-resolution X-wire measurements using phase-loced measurements in periodically forced flows (e.g., Hussain, A.K.M.F. and Zaman, K.B.M.Q., 1980, J. Fluid Mech., 101, 493.). If one is able to understand the concept of coherent vorticity, one should be able to realize that the measurements are valid. The state-of- the-art measurement technology forces the use of Taylor's hypothesis, with which also we have been concerned; we have even directly addressed the limitations of Taylor's hypothesis when applied to CS and quantified errors associated with different choices of convection velocity (Zaman, K.B.M.Q. and Hussain, A.K.M.F., 1981 J. Fluid Mech., 112, 379.). We have shown that for the cases we are talking about the use of Taylor's hypothesis causes negligible error because we always use the frame of the advecting CS. Bill, perhaps others as well, may have overlooked the fact that the eduction scheme has been applied to DNS database and thus Taylor's hypothesis is not a factor at all in these studies. In the case of comparison I have shown between educed CS in laboratory and numerical mixing layers, the CS details (e.g. peak vorticity, shape, size, orientation, coherent Reynolds stress, production, etc.) show amazing agreement. This is another evidence that the measurements are reliable. As an aside, the agreement of the structure details between experiments (Reynolds number based on vorticity thickness, RedW H of order 106) and simulation (RedW H 103) can perhaps be viewed as a support for the concept of Reynolds number similarity.

Let me also address Bill George's amusing comment about observations, equations and physics. I believe science progresses through observation, synthesis and interpretation. The first step is the development of taxonomy for the observed events. Having developed a definition of parts, we have a cognitive reference so that we know what we mean by certain words, what are the different parts, and what are their roles. Then we begin to investigate how different observations fit together Finally, we develop an analytical or mathematical model for predictive, or even control, purposes. Unfortunately, the science of CS is still in the first step, that of developing a taxonomy of the features that dominate turbulence phenomena such as transports of heat, mass and momentum, chemical reaction, and generation of drag and noise. We are still in the information gathering stage. CS is a very rich field which justifies such an apparently protracted information gathering phase. It is just too early to make any definitive statement regarding how the observations are to be incorporated into an analytical framework.

I. Wygnanski:

One of the questions which arises is how do we tie a definition or a description of coherent structures with the Navier-Stokes equations, and with the boundary conditions that are imposed on a given flow? If we look at different types of flows, we obviously will get different coherent structures. I think the idea that there are universal structures of some kind is inappropriate. I suggest that one possible tie between the definition of coherent structures and the Navier-Stokes equation, is to say that coherent structures are the predominant modes of instability. If you say that, and you do not necessarily assume that the problem is linear, you get a quantitative definition of coherent structures in different mean flows. You can also assign different importance to different coherent structures, and compute their interactions. This would at least give a physical intuition, and would define the range of parameters that you can use in order to control these structures. If you then take the triple decomposition, as suggested, then the incoherent motion could be stated as a measure of our incompetence to define it in any better way. Therefore, the more we advance with that sort of model, the smaller the terms which are incoherent will become. I believe that this is a framework with which we can work, and work efficiently.

Julian Hunt:

That is a very important point you make.

Mark Morkovin:

I believe that there is a great deal in what Wygnanski has said, and I think its significance is evident in the claims that have been made for coherent structures. The results shown by Fazle Hussain in his last Figure demonstrate this; they give us a physical feeling for what it is all about, and show us how we can control coherent structures. However, I have not seen a comparable Figure for wall turbulence. The vast majority of flows that have this type of applicability are in fact inflectional mean flows. I do not know the role of large coherent structures, as Antonia was trying to define them for us, in the mechanics of wall boundary layers.

Wall boundary layers are obviously different; they have different types of instability, and this becomes important when you go to high speeds. At high speeds, vorticity provides insufficient description; you have to go to angular momentum, because the distribution of density and angular momentum (density-weighted vorticity) become the more basic variables. At low speeds, these definitions are one and the same thing, but we know that as we go to Mach five the turbulent coherent structures in the inflectional cases of free shear layers behave very differently (because of lack of mutual interactions outside of Mach cones). In this case there is no tendency for the roll-up type of W_z instabilities that we see in Fazle's figures, whereas the compressible, turbulent wall boundary layer at Mach five goes merrily along. While it may have sufficiently different structure from what it had at low speeds, nevertheless there is a very direct connection with what it is at low speeds. Nearly streamwise structures swept back nearly 90R) may have little sensitivity to Mach number.

At lower speeds, I think I understand quite a bit about what is happening in free shear layers, in terms of the instability ideas that Wygnanski is talking about. But I am asking, what can we say along these lines about the wall cases which are, of course, extremely important?

Stephen Kline:

Fazle has shown a lot of things that are useful and important. However, there is one thing on which I have to disagree with him; this concerns vortices. It needs a preliminary remark: It is quite important for at least half a dozen reasons to distinguish vortices (closed near circular motions in the plane normal to the core observed at core speed) from vorticity lines (lines everywhere parallel to the vorticity vector). Three of these reasons are: vortices are the structures that carry the low pressures through the flow; the vortices relate to all the other structures of importance, and very directly to uv2 and uv4; the vortex field is much less noisy than the field of vorticity field. None of these remarks apply to all the field of vorticity lines. As a result looking at vortices is a particularly good way to organize our understanding; looking at the field of vorticity lines is not. Despite all this, if one considers the structures to be the vortices, only the vortices and nothing but the vortices, then a complete picture of the inner layers of boundary layer, what Mark was asking about, cannot be obtained because there are other forms of structure (e.g. ejections, sweeps, sharp-shear layers...) that also play important roles. Indeed, one cannot even get the picture of how vortices are formed and evolve since the vortex formation most often occurs by rollup of what Steve Robinson and I have called in the new nomenclature list for structures "Near-wall-shear-layers". Thus here also, what I called "The Guideline for Scholarly Controversy" in the discussion of the first session, applies. All the structures people have measured are real. What we need to do now is relate these measured structures to each other and sort out which structures are more, and which are less, important. On the definition of structure, somehow we have to broaden the definition beyond just vortices therefore. As Tony Perry said, the structures must be recurring events, but that is perhaps a bit too broad. Perhaps we need to say, "Structures are those recurring events which are essential to the dynamics" (the production of Re stresses, dissipation of TKE, etc).

Fazle Hussain:

It seems I need to respond to others who have posed questions to me. Mark [Morkovin] has made an interesting observation regarding why is it that we know so much more about CS in free shear flows than in boundary layers. My personal apology is that I have had no boundary layer tunnel, but have had jets and mixing layers (they are cheaper to build). The rich information we have gathered in jets and shear layers (see for example, Hussain, A.K.M.F., 1980, Lect. Notes Phys., 136, 252) piqued my interest in wakes; we converted a mixing layer facility into a stop-gap tunnel for wake studies. All these results have given us important information regarding dynamics of CS.

The corresponding level of information is lacking in boundary layers, for obvious reasons. Nearer the wall, there are progressively smaller structures which are harder to fix or to educe. Also, contrary to the desires of physicists, there is no scaling of structures in turbulent shear flows: fine scales are not scaled down versions of large scales. The outer layer structures have been educed (see Hussain, A.K.M.F., Jeong, J. and Kim, J., 1987, CTR Rept. CTR-s87, pp. 273-290.) and they look similar to mixing layer structures at corresponding shear rates. Only recently we have been able to educe the 3D structures in channel and homogeneous shear flows (using NASA-Ames database and in collaboration with John Kim). First of all, both flows are characterized by hairpins; in the homogeneous shear flow haripain and inverted hairpin structure occur with equal probability. There are amazing similarities in the topological details such as shape, coherent vorticity, Reynolds stress, turbulence and enstrophy productions, pressure transports, etc. The same eduction scheme can now be applied to all scales and all distances from the wall; it will be a lot of work.

I am not persuaded by the comments of Mark regarding high M flows. Visualization may not reveal structures, but I suspect that they are there, presumably of finer scales. One must look at the vorticity field to definitely answer this question.

Steve's [Kline] comment on my definition of CS is a valid one, but also deserves careful comments. First of all, we have argued tenaciously for two decades against relying too heavily on flow visualization for studies of vortex dynamics or CS in turbulence (see also Bridges et al. in this volume). We have also argued why other quantities such as velocity, pressure, temperature, or intermittency cannot be consistently used as detectors of CS. We finally settled on vorticity and have been able to use it successfully to map out details of CS in different laboratory and numerical flows. Is our definition the best? Probably not. But for experimental education we consider vorticity to be the best. We have been searching hard for a mathematically tractable approach to CS, different from Lumley's and Adrain's approaches. It is likely that a topology based definition is the way to proceed. We also feel that structures in a turbulent shear flow do not decay and reform, but undergo evolution through interactions such as cut-and-connect, pairing and tearing, and local and nonlocal inductions.

It is not clear that Perry's and Lumley's suggestion of CS as a recurring pattern without any order is consistent with my idea of CS, which necessarily has order. To me, a pattern without order is not interesting. To be measurable, it must have an order. In my view, saddles and foci are examples of order.

Wygnanski's concept of instability modes is interesting, but I do not support the notion of instability of time-mean flow, as the flow itself is dominated by structures of scales at which instability is sought (Hussain, A.K.M.F., 1983, Phys. Fluids, 26, 2216.). As an example, the turbulent boundary layer mean profile is stable but there are CS in it. While indeed there is no scaling of structures in turbulent shear flows, there are some generic features: examples include saddles, foci, etc. CS in their totality are not universal, but have some universal features.

Javier Jimenez:

I want to disagree with what Wyngnanski said about coherent structures being instabilities. It is not clear that this is the case. Coherent structures are flow patterns that stay in flows which are very non-linear. It is only free shear layers that are so unstable that it is almost unavoidable for an instability to grow into something non-linear. In something like a pipe flow which is linearly stable, coherent structure cannot be tied to the flow instabilities because there aren't any. The fact that coherent structures coincides with instabilities, in free shear layers, seems to be a coincidence.

Parviz Moin:

We have heard today (from Wygnanski and Bradshaw) that the notion of the existence of universal coherent structures in different flows is inappropriate. I would like to point out that this is not necessarily the case. In the last couple of years, results from full numerical simulations have shown that, under the right conditions, coherent structures in different flows can be strikingly similar. For example, take the case of homogenous shear flow. Here is a flow without any boundaries, and the mean flow does not have nay inflection points; it is just a straight line. In this homogeneous flow one can find the same kind of structures as in the flat-plate boundary layer flow. One can find hairpin vortices or streaks, depending on whether the non-dimensional shear parameter is comparable to that in the log-layer or to the shear rate in the vicinity of the wall respectively. You can find elongated streamwise structures if you have high enough shear rates. So, the notion that coherent structures in different flows are totally different, and therefore there is no hope for universal turbulence modeling, may be a pessimistic view.

The challenging problem is to identify the proper dimensionless parameters (such as the shear rate) that govern the flow structure; and the models should be concocted in terms of these parameters.

Julian Hunt:

I want to add a comment to what Jimenez has said. The most interesting things about uniform shear flows is that they have no eigenmodes; they are neutrally stable and yet they develop these very strong structures. In some sense, a point made in

Townsend's 1976 book, was that if you take homogeneous, isotropic turbulence and shear it, you develop some sort of structure. While this structure clearly reflects its initial condition, the flow pattern that comes out has nothing to do with the eigenmodes. I am not disagreeing with Wygnanski's point that is certainly valid, (especially for the largest-scale structures) in many flows. But I think there is another mechanisms for producing typical structures, which is the interactions between a rather disordered body of turbulence and shear. In this case, a streak-like behavior seems to appear in the flow which is characteristic of near-wall turbulence. When Townsend was asked about this, he said he had moved away from the notion of wall turbulence being the same as random turbulence.

Roland Stull:

There is another possible definition of coherent structures. Namely, they are entities that cause transport of momentum or tracers across a finite distance in a non-diffusive way. That definition works not only for structures possessing vorticity, but also for structures like rising warm thermals that don't have so much vorticity, like we see in meteorology. This definition allows the concept of non-local transport. It also answers the question raised by Hunt at the start of this session: how do we tie together these coherent structures with statistics? For example, we can measure the transport between different heights in the atmosphere caused by an ensemble of large coherent structures. This is transport across different finite distances.

We find that transport cannot be described by local first-order closures, or local second-order closures. It should be described by a non-local statistical process. This kind of description of mixing between many different points in space is known, in the meteorological literature, as transilient turbulence. This theory relates an ensemble of coherent structures to the net statistical affect that they have on a flow.

Jackson Herring:

It seems to me that in assessing whether coherent structures destroy Reynolds-average modeling, it is very useful to have a notion as to how intensive these structures are. That is to say, is the vorticity kurtosis four, five or fifty? That is very important, I think, in order to make a distinction between situations where such averaging applies and where it doesn't.

Julian Hunt:

Is this the vorticity over the whole flow field, or just in the coherent structure?

Jackson Herring:

How much is the vorticity within the coherent structure above the mean or background value?

Fazle Hussain:

The vorticity associated with coherent structures is much lower, orders of magnitude lower, than the instantaneous vorticity; so I don't think that the kurtosis of vorticity is a good measure. But coherent vorticity as well as other dynamical measures such as incoherent Reynolds stress, enstrophy production, strain rate, etc., would be a meaningful quantity. Typically the peak incoherent Reynolds stress is of the order of two to six times the time-average background turbulence Reynolds stress. This is why we have asserted that both coherent and incoherent turbulence are important in fully turbulent shear flows.

Uriel Frisch:

I have a provocative question for Fazle Hussain. His eduction procedure involves an averaging; and linear superposition of solutions to the Navier- Stokes equations are not solutions to the Navier-Stokes equations. So, does that mean that the educed structures are well defined, but are not solutions of the Navier-Stokes equations?

Fazle Hussain:

Educing a structure has nothing to do with whether at a particular instant the structure satisfies the Navier-Stokes equation. Is the evolution of that patch of vorticity in that projection, which is the coherent vorticity, a solution of the Navier- Stokes? That would be true only if the corresponding governing equation is linear. It is not. What you left out, the incoherent turbulence, is also organized by the coherent structures, so there will be a non-linear term appearing through the interaction of the incoherent turbulence organized by the coherent structures. If this nonlinear interaction is accounted for, CS evolutions are solutions of N.S. Equations.

Ravi Sudan:

For the motion of coherent structures to be analyzed by an equation, these structures must have a definite lifetime. And so my question to those who have seen them: is the lifetime of the structures much longer than the eddy turnover time, for instance? Because if they diffuse away rapidly, then there is no reason for developing an equation for these particular structures.

U. Schumann:

From large eddy simulations of the atmospheric convective boundary layer, we have identified coherent structures by identifying such things as updrafts and downdrafts. Such motions have lifetimes of the order, or slightly longer, than the eddy turnover time. That is for the main structures. Of course, these structures contain smaller-scale structures which also have coherent characteristics. They have a much shorter lifetime, but they are nonetheless important in transporting heat and other quantities. I want to make one further point, and that is that I don't agree with the statement that mean flow-models have no physical basis, and therefore should not be accepted. I think this overstresses the point. There are many mean flow models which we all use, and they do have a physical basis. One should not be so simplistic in arguing that point. In fact, what was missing today are predictive mean-flow models based on coherent structures. They do exist in the literature. For example, again from the atmospheric field, there is a paper by Chatfield and Brost (J. Geophys. Res. 92, 13263, (1987)) that proposes a two-stream model for the up and down transport in a boundary layer that is based on the concept of coherent structures.

Brian Cantwell:

I made the statement that mean-flow models do not have a sound physical or theoretical basis, but I did not say that they should be rejected out of hand, or that they are not useful. They will continue to be useful.

Julian Hunt:

May I ask you to elaborate on that, because it struck me that in your paper you mentioned that these models satisfy continuity and momentum conservation, and these are theoretically sound physical ideas. In what sense do you mean that?

Brian Cantwell:

The sense in which I mean that they don't have a sound theoretical foundation is basically closure. Although closures are sometimes based on full numerical simulations, they are usually accomplished by ad-hoc assumptions. Also, when these models go wrong, there is no theory that tells why they went wrong, or predicts where they will go wrong. This is the theoretical underpinning that is missing.

Julian Hunt:

So what you are saying is that the fine detail has no theoretical or physical foundation.

Brian Cantwell:

The development of these models is wonderful because they are derived from the Navier-Stokes equations, but the closure schemes are ad-hoc.

I. Wygnanski:

I would like to address the comments made by Jimenez and Morkovin. In the free shear flows we have various strengths of the instability that exists in a given flow. If you compare the mixing layer, wakes and jets, it is strongest in the mixing layer. Therefore, we can identify the coherent structures there most easily. But even there, it is not very definitive because you can see two modes of instability competing, even on a linear basis. In an axi-symmetric jet, the competition is stronger and therefore the flow is apparently less coherent. Now, where is the problem when one considers wall boundary layers and pipe flows? In such flows, there might be two or three modes that are equally unstable, and which compete with one another at the same level; therefore, we see double systems. If one wants to see a wall flow where we can distinguish coherent structures, one should look at a boundary layer at the point of separation. There, you see coherent structures which are very similar to those in the mixing layer. Another example is the wall jet, where you see coherent structures which are very similar to those in the two-dimensional jet, inspite of the fact that there is a secondary interaction with the wall which is not insignificant. Extrapolating that point of view to wall flows gives answers. At this point in time, I am trying to understand a couple of these, primarily the wall jet and the separating boundary layer. The results are not as definitive as in the mixing layer, but I still see the same type of approach working, and giving reasonable results.

Brian Launder:

I was not going to comment on the remarks that have been made about mean-flow closures, but I think Brian Cantwell's reiteration of some of the things in his paper require that I should make an interjection or two. I don't know what he means by ad-hoc. Would he regard local isotropy as an ad-hoc concept to apply to a model? Would be consider realizability as ad-hoc? Many of the models for processes within second-moment closures fit precisely the form of the model; there are not dozens of constants to be checked out and kluged up. He mentioned that the models used at the Stanford Conference and the models used today are not very different. Current second-moment models, as for example those developed by W.C. Reynolds and his students, are substantially different than the closures adopted in the 1981 Stanford Conference.

Brian Cantwell:

Local isotropy is ad-hoc since it is applied to all scales. Realizability sounds reasonable, but can force erroneous answers in practice.

E. Novikov:

I would like to remind the audience of an old idea put forward in the forties by Onsager. According to this idea, there are flow structures that are out of equilibrium-- they cannot fit in the small scale equilibrium. Such structures, I think, are what we call today coherent structures. This definition could be useful in our analysis.

H. K. Moffatt:

Fazle Hussain made the remark that helicity was an "unpopular" concept. I think I know what he means, but it would be interesting for this gathering if I invite him to expand a little on the comment. I will have an opportunity to respond in my paper.

Fazle Hussain:

Helicity is a fascinating property about which much has been speculated and written, but its relevance to hydrodynamic turbulence is unclear. My comments were based on the numerical studies made by Rogers and Moin. Their studies did not show the importance of helicity claimed or suggested by Moffatt and others. However, I feel helicity may be useful in understanding local kinamatics in vortical or turbulent flows. We have found it particularly helpful in interpreting the details of reconnection mechanisms and associated hydrodynamics.

One point should not be left unanswered. Ravi Sudan raised the important question: if structures appear and disappear within a turnover time, why should we worry about them? Here is my answer. If structures spontaneously appear and over one turnover time disappear, I absolutely agree with him. But, that really is not the story, nor is the picture proposed by Townsend that structures form as a result of an instability, then break down, and the flow undergoes a new instability and forms new structures, and so on. The structures are really transient as you see them, they are undergoing continual evolutionary changes through non-linear interactions with other structures. When you make observations you see a structure for only a short time. But they are not disappearing via breakdown; they are interacting with other structures.

Ron Blackwelder:

I would like to comment on what Wygnanski and Morkovin said regarding instabilities. I think instabilities are a very useful way of looking for coherent structures because they are things we can identify. We have equations for the instabilities (whether they are linear or non-linear is not important) and we can follow them.

The Goertler instability and the resulting streamwise vorticies are examples of structures that are very similar to those observed in the wall region of a turbulent boundary layer. As for their eddy turnover time, it depends on how you want to define it. If the vorticies turn 90 degrees in the course of their lifetime that is enough to set up inflectional profiles and secondary instabilities which create the transition to turbulence. If you look at the eddy turn-over time in terms of the RMS values and the length scales, they may turn over several times, but the physical mechanism is a lot different. You have to separate the physics from the statistics that we look at.

If you a make simple calculation for Goertler vortices you find that they don't have to turn over, but only rotate 10 to 20 degrees before they set up a secondary instability that creates turbulence. This secondary instability is an inflectional instability, and it grows an order of magnitude faster than the initial instability. Consequently, you have to look at it on a local scale. The mean flow that the secondary instability sees is a base flow of the Blasius type, plus the Goertler votices which produce inflectional profiles. The time scales of the inflectional instability are so small that these profiles essentially look steady. That is, the growth rate of the inflectional instability is enormous. As an eddy moves one wave- length downstream, its energy essentially increases by a factor of 1000, if you accept the simple linear two-dimensional theory. Obviously, they are not two-dimensional or steady in time-- so they do not grow by a factor of 1000, but by a factor of possibly 100. Whatever it is, it is a very large number. In the case of a uniform shear flow, in which isotropic turbulence is strained (referred to by Julian Hunt) you have to look at the intensity and type of eddies imposed on the mean flow to determine the resulting coherent structures.

The point I am trying to make is that the eddy structure is the mean flow. We cannot separate it out. Triple decomposition gets us a little bit further in terms of understanding, but it is blinding us, possibly, to the fact that the mean flow is concomitant with the coherent structures. If you think from that point of view, you can see how you can get elongated structures in a mean homogeneous shear. If you have an instantaneous shear flow that is unstable, it becomes more unstable resulting in more turbulence downstream.

Stephen Kline:

I want to answer three questions that were asked in this discussion period. All the remarks are based on recent observations of the DNS simulation of Spalart for the flat plate.

First, "What are the strengths of the vortices seen in the near wall region of the boundary layer?" In terms of pressure depression in the core below local, $4\,\rho(U_{-\tau})^2$ is a typical value. Values as high as $16\,\rho(U_{-\tau})^2$ occur in terms of vorticity, values as high as ten times the mean vorticity of the layer occur frequently. So the vortices are strong compared to other motions in the layer; however, the variation from one to the next is large. Distributions are given by Robinson in the paper for the IUTAM Zurich meeting in July, and will appear in his dissertation shortly.

Second, regarding persistence one has to distinguish transverse vortices (heads) which are the common type in the outer layer from tilted, quasi-streamwise vortices (legs) which are the common type in the wall layers. Heads are typically quite persistent; we have seen cases of persistence beyond 5000 wall units for large heads. Legs, on the other hand come and go much more rapidly. In the simulations, one can commonly see a leg grow outward from a head right to the wall, stretch and disappear, and then grow again from the same long-lived head. Details can be found in the videos now coming from the simulations.

Third, I found Wygnanski's remark about the boundary layer intriguing because, in studying the various forms of structure the Spalart flat plate DNS simulation, we find that most of the important actions do occur in two (and sometimes three) ways. This includes uv^2, uv^4, formation of vortices, entrainment, and both formation and lifting of near-wall streaks. This is to me quite surprising; one expects to find in the physics of inert naturally-occuring systems one-to-one cause and effect. But this is not so in the boundary layer.

Julian Hunt:

The aim of this part of our discussion was to focus on how people are analyzing or defining coherent structures. As I expected, this moved on to a discussion of dynamics. Some people are hopefully relating their studies on coherent structures to certain dynamical approaches, and there are two dynamical approaches. One is based on vorticity dynamics, and another one based on eigenmodes (we probably we will hear more about the latter tomorrow). Trying to relate the definition and measurement of these structures to dynamics seems to me to be a hopeful sign. I am sure that it will be necessary.

Other questions that arose during the discussion seemed to revolve around the point raised by Cantwell, about the time structure of coherent structures, the fact that coherent structures vary significantly as they evolve in time. There is too much analysis of time slices of these things. Cantwell talked about defining them in some Lagrangian sense, in terms of particles within them. I wonder if there are any other views about a more Lagrangian or time evolution approach for looking at how structures evolve, in terms of perhaps different Lagrangian or Eulerian definitions.

Roland Stull:

The two approaches mentioned by Julian Hunt excluded the suggestion I made earlier of a non-local Eulerian approach - the transilient approach. That approach gives time information; namely, the amount of mixing between two points in space obviously depends on the time that mixing occurs. This non-local Eulerian approach is useful and can easily be applied (to practical problems compared to Lagrangian approaches which are difficult to apply). It gives you similar information about the non- local nature of transport as the Lagrangian approach. Transilient turbulence is a different approach that perhaps not many people here are considering.

William George:

You have twice used a word that I have not heard before. Can you repeat it?

Roland Stull:

Transilient - it comes from the latin root translire, which means jump over or leap across. It is meant to convey something different than diffusion. It has been used in meteorology for some time (Ebert et al, 1989, J. Atmos, Sci.). It answers Sudan's question: what good is an eddy if it dies after one eddy turnover time? Well, during its lifetime, it causes a non-local transport that cannot be explained by a series of small-eddy diffusions. Its more like advection rather than a diffusion process, and this really is what turbulence is.

Ravi Sudan:

Is the transport greater than what the eddy would furnish at that time?

Roland Stull:

The transport by a large coherent structure can be greater than that by a small eddy, because small-eddy transport is a consequence of diffusion down a local gradient.

Julian Hunt:

One of the other things we need to think about is the following: is the use and study of coherent structures going to change dramatically with new types of experimental methods? This is one of the questions suggested by George, Hussain and others. Are we just going to measure these three-dimensional velocity fields in great detail, or is there likely to be some new quantitative look at coherent structures, with new experimental methods?

William George:

I would like to take a stab at this. But, let me first respond to Fazle Hussain's comments about my comments. I am still comfortable with my statements, although I am impressed by the full-simulation results.

As far as what is coming up in experimental methods, part of it depends on our ability to define a structure in space and time if we are going to deal with tracking it. I think particularly interesting in this context is Mark Glauser's work, where he made many measurements in an axi-symmetric jet and sees pretty clear evidence of an axisymmetric mode hopping into something like modes four, five, or six-- presumably by some kind of secondary instability mechanism. To me, that says that if we can measure at many points simultaneously, we can begin to unravel the dynamics of these structures. In particular, do they really start instabilities in the mean flow? Are there secondary instabilities driving them to the next stage? And so forth.

Laurence Keefe:

What Bill George is simply advocating is being able to look at the entire state space of a flow. I have maintained for some time that, operationally, when people educe coherent structures (and I don't care what scheme it is - VITA, QUADRANT analysis, or any variant of Hussain's technique), they go into a high dimensional state space, described by many coordinates, and look at a very small number of these coordinates to do detection and extraction. So when Wygnanski talked about unstable modes as coherent structures, you have to ask - are we talking about global instabilities (on a large scale, spanning all state space coordinates)? Or are we talking about instabilities which are apparently only local (close to a wall, affecting only a few state coordinates)? Detection and eduction are local operations. In mixing layers all localities see the rollup due to global instability and so detection picks up the instability as the structure. In wall flows the locally detected bursts and streaks are not necessarily parts of some global instability of the flow.

I believe you can specify and find almost any reasonable coherent structure, any sequence of events in the flow, provided the sequence doesn't violate the fundamental equations. Set up a detection scheme for that structure, examine the flow, and I think you will find it, although in may cases its probability may not be large. However, I bet you can find structures, essentially arbitrary, which have equal probability to the ones we have latched onto over the years: bursts, streaks, ... etc. Take variants of the VITA scheme, fiddle with the constants, and you can produce several different kinds of structures, and they all have at least equal probability to the ones we are familiar with.

A. Smits:

I want to address what the experimentalists are up to. I think we are seeing some things happening in experimental methods that will give us three-dimensional, time evolving fields-- not in as much detail as we can get from numerical simulations, but probably good enough to start making some direct comparisons with three-dimensional motions. Also, I think we are often at fault in fluid mechanics for not listening to some of our colleagues in combustion. For example, they have developed very sophisticated techniques for the interrogation of very hostile and complex flows. These can help us a lot when we look at some of our simple flows by providing even more interesting detail. I am very hopeful.

Julian Hunt:

It was suggested to me that the one thing we ought to move towards at the end of these discussions is the issue of the broad strategy of support and funding for research in this general area. Two arguments were put forward with unequal emphasis by different speakers. One was: it is important to study coherent structures because they are a key to the fundamentals of turbulence. The other one was: they are very important for the technology; i.e., the application of turbulence to technological problems. I was wondering if the representatives from the agencies would like to comment on that, or if anybody else would like to comment on how they see the strategic-support element.

S. C. Traugott:

I'm not sure it's a comment, but I have a question. Of the various topics that were brought up as having potential payoff, I heard a new one. How do we define a coherent structure? That is what a lot of the discussion has been about. What should a federal-agency program director reaction be if someone came with a proposal to define coherent structures--a new subject which has just been invented? Should he say, "don't be ridiculous", or say "this is worth funding"?

Julian Hunt:

What nobody has done yet in this field is employ a technique that is widely used in engineering science. A given set of data is passed around between different laboratories, and everybody analyzes the same data in different ways. If this were done in this field, we see whether data from Houston looks like a coherent structure at Caltech or vice-versa. This has not been done. If it had, some of the differences between Hussain and George, that we have seen, might not have existed.

Hassan Nagib:

I would like to address Julian Hunt's question as to which of the two routes should we pursue, or are they both important. I don't think you can pursue the application route without the understanding. Last week there was a meeting on turbulence control, and one theme was that we can have some ad-hoc successes under specific conditions, but in order to succeed under general conditions we have to understand. That is where the emphasis of the funding agencies should be-- on the understanding rather than on just the few cases of applications here and there.

William Reynolds:

I am struck with the thought that a transition spot may have a lot to tell us about coherent structures. To draw a parallel, I am imagining I am a physicist working on the critical phenomenon, and I discover that near the critical point there are these clusters of something or other that moves around in an organized way. Here I am trying to discuss all that, and I have not yet figured what a molecule is! Returning to coherent structures, I am wondering maybe we should give more attention to what the little things are, that are in fact running along together. The number of events in a transition spot probably is what really is important for some things - exactly where they are is what produces the noise we see in signals. Maybe we should give more attention to the little things that go into making up the big things.

Uriel Frisch:

Just one thing. Maxwell worked out his kinetic theory of gases when molecules were not yet understood.

Dennis Bushnell:

I want to address Hassan Nagib's comment. The easiest way, the useful way to generate funds for scientific fields is to produce something useful.

The Utility
of
Dynamical
Systems
Approaches

Discussion Leader: A. E. Perry, Melbourne University

CAN DYNAMICAL SYSTEMS APPROACH TURBULENCE?

Philip Holmes[1]

[1] *Departments of Theoretical and Applied Mechanics and Mathematics, Cornell University*

'This may not be a Theory of Everything:
I hope it is at least a theory of something.'

F. Wilczek — lecture delivered
at Caltech, November 1988

Abstract

I review some ideas and methods from dynamical systems theory and discuss applications, actual and potential, to the study of fully developed turbulent flows in an open system: the wall region of a boundary layer. After a brief account of applications to a closed flow system, the approach I concentrate on attempts a marriage between statistical methods and deterministic dynamical systems, both orderly and chaotic. Specifically, coherent structures are identified with combinations of certain basis functions using the proper orthogonal decomposition. A relatively low dimensional ordinary differential equation describing the dynamical interactions of a set of these spatially organized structures is then derived by Galerkin projection of the Navier-Stokes equations. The resulting system is optimal in the sense that it retains the greatest turbulent kinetic energy, in a time averaged sense, among all projections of the same dimension. The model is analyzed using the methods of dynamical systems and symmetries are found to play a crucial rôle. In particular, structurally and asymptotically stable heteroclinic cycles emerge as a common feature in models of various dimensions and orbits attracted to these cycles lead to solutions exhibiting intermittent, violent "events," which appear to reproduce key features of the bursting process. I speculate on the validity of this approach, the "understanding" of turbulent processes it offers and on how some of the gaps in the procedure might be bridged. I do not suggest that this is the only way in which dynamical systems methods can be used, but it is one which seems worth pursuing.

1. Introduction

I have been asked to present my position on the utility of the dynamical systems approach to turbulence. I had better say immediately that, in spite of recent hyperbole

(Gleick [1987]), I do not believe that there is a single, unified "new science" that can be called "chaos theory" or even, less spectacularly "the dynamical systems approach." Current research in dynamical systems is a rich mixture of rigorous abstract mathematics, formal manipulations, derivation of models, numerical experiment and unbridled conjecture. At the same time, the "problem of turbulence" seems rather to consist of many interconnected facets than one grand question. Liepmann [1979] addresses this latter issue in a delightful and cautionary essay by listing several specific questions to which a good theory of turbulence should provide answers. While empirical laws, phenomenological models, experiments and, more recently, massive numerical simulations, can address and even adequately answer *some* such questions, many fundamental issues remain. As Liepmann describes, several waves have swept over turbulence research, usually promising more than their residue revealed, but each one nonetheless contributing some new ideas and methods. In this essay, I hope to suggest that certain ideas from dynamical systems theory might also leave their mark before the next enthusiasm overwhelms us. To do this, I shall focus on a specific problem and argue that some of the ideas and methods of dynamical systems theory have helped to elucidate it. (This might be easier if the grand claims, referred to above, had never been made.)

While I do not see a single theory of dynamical systems, there is a loose battery of theorems and methods applicable to the study of nonlinear ordinary and partial differential equations (ODE and PDE), largely developed over the past twenty or thirty years, which can be usefully added to "classical" methods such as perturbation and asymptotic analyses and so which can, together with them, be brought to bear on specific *pieces* of the turbulence problem. I will attempt to describe some of these new mathematical ideas and suggest turbulent fragments on which they might appropriately be exercised. My main example will be a *relatively* low dimensional model for the dynamical interaction of longitudinal vortices in the wall region of a fully developed, turbulent boundary layer. However, this open flow system will not be too open (one stays near the wall) and the dimension of the model system will be rather larger that has been usual in dynamical systems (10–20 ODEs).

Many of the ideas of dynamical systems theory have their origins in the work of Poincaré [1880, 1890, 1899], who proposed that the solution of differential equations be approached by asking *qualitative* questions about the *global* behavior of large sets of solutions, rather than seeking particular analytical (or, today, numerical) solutions for special initial data. He developed the basic ideas of invariant manifolds and bifurcation of equilibrium and periodic solutions and recognized the importance of homoclinic orbits in the generation of chaotic motions. While his main stimulus came from the Hamiltonian systems of celestial mechanics, the methods he suggested have been equally successful in the study of dissipative dynamical systems. It is worth noting that, in the process, Poincaré invented topology (*analysis situs*), and that now, largely through the work of Smale [1967], Arnold [1983] and their colleagues and students, powerful topological tools have returned to shed light on differential equations. However, as with any other craft, the tools must be used correctly and with care; as we shall see, metaphorical connections between chaos in dynamical systems (mathematics) and spatio-temporally irregular, turbulent, fluid flows (physics) are easy to come by, but serious understanding of the latter via rational mathematical models is much more difficult. Dynamical systems theory per se applies to differential equations; one must still connect the equations with turbulence.

The current interest among mathematicians in the application of dynamical systems approaches to turbulence stems from a paper by Ruelle and Takens [1971], who, apparently following a suggestion of Arnold, proposed that a mathematical object called an Axiom A strange attractor might exist for the Navier Stokes equations defined on a bounded domain and under appropriate conditions, and that solutions attracted to such an object might correspond to "turbulence." In spite of the abstract flavor of their paper (in which the N.S. equations appear only as "$dx/dt = F(x, \mu)$"), they were in essence developing and commenting on the idea of successive bifurcation to quasi-periodic flows of increasing complexity due to Landau [1959] and Hopf [1948]. Somewhat earlier, Lorenz [1963], who had studied with G. D. Birkhoff (another pioneer in dynamical systems: Birkhoff [1927]), published a paper containing analysis and numerical simulations of a set of three first order differential equations obtained by drastic truncation of the Fourier-Galerkin projection of the equations for the velocity and temperature fields in a two dimensional convecting layer. This work remained unknown to the mathematical community until Jim Yorke discovered it in the early 1970's. It was the first, and is so far the *only*, example of a model, connected however tenuously with a "real" fluid system, which can *almost* be proved to have a strange attractor, albeit only in a physically insane parameter limit (cf. Sparrow [1982], Guckenheimer [1976], Robinson [1989]). (The Lorenz system *does* offer a reasonable model for a highly constrained physical system: the closed loop thermosyphon; see Gorman, et al. [1986], for example.)

The moral implied by this historical tale is instructive: (1) there is a huge gap between specific, low dimensional differential equations and the theorems one can prove about strange attractors and other dynamically interesting sets; (2) there is another gap, at least as large and probably larger, between realistic continuum models of fluid systems, such as the Navier-Stokes equations with appropriate boundary conditions, and the aforementioned low dimensional equations about which one can, in any case, prove very little. Nonetheless, in the remainder of this paper I will attempt to indicate how some parts of these gaps can be bridged, if not in a rigorous manner, at least rationally enough to bear the weight of the bridge builder and a few close friends.

The paper is organized as follows. In Section 2 I give a brief account of some of the ideas and methods of dynamical systems theory. I then recall in Section 3 how these tools have been used to study what might be called "preturbulence": hydrodynamic instabilities in spatially constrained or "closed" systems such as Rayleigh-Bénard convection in a small box or the Taylor-Couette experiment. There is a large literature here and since it falls outside my main theme, I merely sketch some ideas, mainly because closed systems afford rather clean applications of the methods, and so may help introduce them.

The remainder of the paper centers on my main topic of "open" flows and fully developed turbulence. Section 4 turns from mathematics to fluid mechanics and briefly reviews the notion of coherent structures, since they provide the pegs from which the low dimensional models will hang. In Section 5, I recall the proper orthogonal decomposition (or principal factor or Karhunen-Loève decomposition) first proposed in the context of turbulent velocity fields by Lumley [1967, 1970]. It supplies the (more or less) unbiased mechanism by which the coherent structures are identified with the basis of a finite dimensional subspace in which the model will live. Section 6 discusses the Galerkin projection procedure in general and Section 7 contains a summary of the heirarchy of models obtained by this process by Aubry, et al. [1988, 1989]. This work has been extensively reported elsewhere and I merely highlight some aspects of it, in particular pointing out the crucial rôle played by *symmetries* in determining the global behavior of solutions and the resulting physical implications.

At this stage, there are many gaps; the bridge sways and totters. In Section 8, I discuss some ideas which may help close one of the gaps: the important notion of *inertial manifolds*. I speculate on the relation between these objects and the finite dimensional subspaces produced by proper orthogonal decomposition and I raise other issues concerning the applicability of the approach. Finally, I attempt to summarize in Section 9.

Readers wishing for details will have to go to the sources in the bibliography; here I provide only a broad sketch of dynamical systems methods and some of their applications. I have, however, tried to put the new ideas in the context of earlier work, and to reflect on their applicability and what new understanding they might offer.

2. Dynamical Systems

I will sketch some ideas from dynamical systems theory, drawing on sources such as Guckenheimer-Holmes [1983], Arnold [1973, 1983] and Bergé, et al. [1987] which the reader should consult for details. Here I only attempt to convey the general flavor.

One of the main tasks of dynamical systems theory is to study the qualitative behavior of parameterized families of deterministic differential equations. Any ordinary differential equation can be written as a first order system,

$$\dot{\mathbf{x}} = \mathbf{F}(\mathbf{x}, \underset{\sim}{\mu}) \; ; \qquad \mathbf{x} \in M \, , \quad \underset{\sim}{\mu} \in \mathrm{I\!R}^k \, , \tag{2.1}$$

where the variables $\mathbf{x} = (x_1, \ldots, x_n)$ live in the phase or *state space* M, usually a smooth manifold and often Euclidean n-space $\mathrm{I\!R}^n$, and the parameters $\underset{\sim}{\mu} = (\mu_1, \ldots, \mu_k)$ vary in an open set of $\mathrm{I\!R}^k$. The phase space need not be finite dimensional; many partial differential equations, including the Navier-Stokes and reaction diffusion systems and long wave approximations such as the Korteweg deVries and Ginzburg-Landau equations can be cast as infinite dimensional evolution equations like (2.1), where $\mathbf{F}$ is a nonlinear (differential) operator on a suitable Hilbert space. Temam [1988] is a good recent reference in this respect. Thus, the general strategy sketched below can in principle be applied to continuum models of fluid systems, although it should be noted that there are considerable technical problems when dealing with open flow systems having infinite spatial extent. Also, while most of the ideas outlined below do extend to infinite dimensional problems, applying them in this situation, or even to large ($n \approx 10$, say) sets of ODEs, is often very difficult; consequently, the vast majority of dynamical systems applications thus far have been to low (≈ 3 or 4) dimensional problems.

Rather than attempting to solve (2.1) for fixed $\mu = \mu_0$ and specific initial data $x(0) = x_0$, we ask how the typical limiting ($t \to +\infty$) behavior of solutions of (2.1) varies as we change μ; for example, how do the *attracting sets*, which capture *all* nearby initial conditions and so correspond to physically observable behavior, appear, vanish and metamorphose as μ varies? This is the business of bifurcation theory.

Local bifurcation theory addresses the creation of new invariant sets such as fixed points, periodic orbits, invariant tori and exotic "chaotic" sets, in the neighborhood of a degenerate equilibrium or fixed point. An *invariant set* Λ for (2.1) is a (compact) set composed of solutions of Λ. If $x(0) \in \Lambda$ then $x(t) \in \Lambda$ for all t. Invariant (or integral) manifolds play a major rôle in dynamical systems theory (Hale [1969]). If $\mathbf{F}(x_0, \mu_0) = 0$ and the Jacobian derivative $\mathbf{DF}(x_0, \mu_0)$ is *non degenerate* (all its eigenvalues have nonzero real part) then the linear system

$$\dot{\xi} = \mathbf{DF}(x_0, \mu_0)\, \xi \tag{2.2}$$

determines the behavior of solutions of (2.1) in a neighborhood of x_0 and we call the fixed point x_0 *hyperbolic*. This is the basis of traditional linearized stability theory; of course, computing eigenvalues and eigenfunctions of $\mathbf{DF}$ in the infinite dimensional case can be a substantial task (cf. Chandrasekhar [1961] or Drazin and Reid [1981]). If $\mathbf{DF}(x_0, \mu_0)$ is degenerate (x_0 non-hyperbolic), we must augment (2.2) by the addition of nonlinear terms. Determining the behavior of even a quadratic differential equation in three dimensions is very difficult (think of the Lorenz example), but fortunately a reduction principle comes to the rescue. One need only consider behavior along the "neutral" directions, tangent at x_0 to the eigenvectors (eigenfunctions) whose eigenvalues have zero real part. The *center manifold theorem* (Carr [1981], Pliss [1964]) permits one to extract a low dimensional system and the *normal form theorem* then allows one to perform nonlinear, near-identity coordinate changes to "simplify" the Taylor series expansion of $\mathbf{F}$ near x_0, to give a system of the form

$$\dot{\xi} = \mathbf{A}(\mu)\xi + \mathbf{g}(\xi, \mu) + \mathcal{O}(|\xi|^{k+1}) \tag{2.3}$$

in which g is a kth order polynomial, many of whose coefficients are zero. Here $\xi \in \mathbb{R}^d$, and d is the dimension of the neutral or center eigenspace: much lower than n, especially if $n = \infty$. This procedure is closely related to formal methods which yield weakly nonlinear "Landau" or amplitude equations. The importance of this local reduction principle, first enunciated by Pliss [1964] cannot be overemphasized. I shall return to its global manifestation in my discussion of inertial manifolds in Section 8. Center manifold reduction and normal form computations can often be (partially) automated using computer algebra (cf. Rand [1984], Rand and Armbruster [1987]).

One next attempts to classify all possible phase portraits of (2.3) in a neighborhood of the degenerate fixed point x_0 as μ varies near μ_0 and to prove that this behavior persists (qualitatively) on the center manifold for the full system in M with the higher order terms included. Classification proceeds via the notion of *codimension*; roughly speaking, this is the number of parameters which must be varied to *unfold* the degeneracy and reveal *all* the topologically distinct behaviors that can appear. Local codimension one bifurcations of fixed points are the *saddle node*, in which a pair of fixed points coalesces, and the *Hopf*, in which an isolated periodic orbit or *limit cycle*

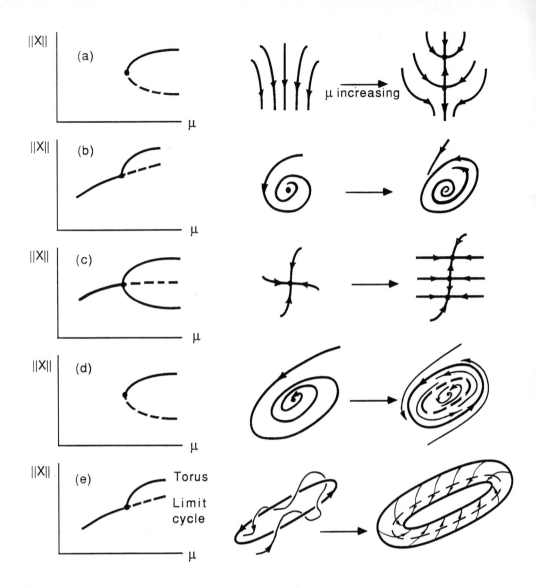

Figure 1. Some codimension one bifurcations, shown via bifurcation diagrams, in which the norm of the solution is plotted versus the parameter, and as phase portraits. (a) Saddle node for fixed points, (b) Hopf for a fixed point, (c) pitchfork for a fixed point, (d) saddle-node for periodic orbits, (e) Hopf for a periodic orbit, in which a 2-torus is created.

grows from a fixed point. Similar results apply to periodic orbits, which can coalesce in saddle-nodes and throw off invariant tori—quasi periodic two-frequency motions—in Hopf or Neimark-Sacker bifurcations; but periodic orbits can also undergo *flip* or period doubling bifurcations in which an orbit having double the period of its parent appears. See Figure 1.

Codimension two bifurcations of fixed points and periodic orbits are also almost completely classified (some technical problems remain); here the behaviors are much richer and the unfoldings reveal *global* bifurcations in embryo (cf. the references above and Golubitsky and Guckenheimer [1986]). These involve the "large scale" rearrangement of solutions such as saddle separatrices; one of the most interesting is the creation of a saddle loop or *homoclinic orbit*: an orbit which is both forward and backward asymptotic (in t) to a hyperbolic fixed point. Figure 2 shows an example in which an attracting periodic orbit grows and forms an "infinite period" homoclinic loop before vanishing as a parameter varies in a planar system. Homoclinic orbits play an important rôle in $n \geq 3$ dimensional systems, often leading to chaotic solutions; the Lorenz attractor is essentially created in a homoclinic bifurcation (cf. Sparrow [1982]).

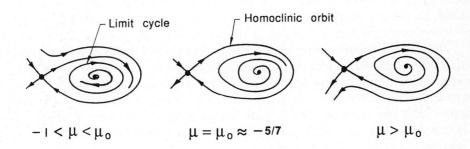

Figure 2.   A global bifurcation. As μ increases for the system $\dot{x}_1 = x_2$, $\dot{x}_2 = 1 - x_1^2 + \mu x_2 + x_1 x_2$ the limit cycle, which appears in a Hopf bifurcation at $\mu = -1$, vanishes in a saddle connection or homoclinic orbit at $\mu \approx -5/7$ (cf. Guckenheimer and Holmes [1983, Section 7.3]).

The words "chaos" and "strange attractor" have been used several times. It is time to explain a little more. For our purposes, a *chaotic invariant set* is an invariant set, $\Lambda \subset M$, composed of solutions of (2.1), having the property that almost all pairs of solutions $x_a(t)$, $x_b(t)$ started arbitrarily $\mathcal{O}(\epsilon)$ close in Λ, diverge by an order one amount after a finite time. If Λ is *hyperbolic*, then divergence proceeds at an exponential rate, locally:

$$|x_a(t) - x_b(t)| \sim c|x_a(0) - x_b(0)|e^{\lambda t} \tag{2.4}$$

for some $c, \lambda > 0$. Of course, unless they are unbounded, exponential divergence cannot continue and the solutions may find themselves close again subsequently. An *attracting set* $\mathcal{A}$ is an invariant set which attracts all solutions starting in some (open) neighborhood of $\mathcal{A}$. An *attractor* is an attracting set containing a dense orbit, so that almost all solutions in it display "typical" behavior (see the references cited earlier on this technical, but important point). The classical attractors are stable fixed points (sinks), limit cycles and $d \geq 2$ dimensional invariant tori carrying quasiperiodic motions with incommensurate frequencies. One of the major contributions of dynamical systems theory was the realization that other, more complex, attractors and attracting sets can exist "stably." (An object is *structurally stable*, if, unlike the homoclinic loop of Figure 2, it cannot be removed by a small perturbation such as a parameter change.) In fact, Ruelle and Takens [1971] argued that quasiperiodic motions having more than three independent frequencies of the type proposed by Landau and Hopf were structurally unstable, while certain strange attractors were stable, and might therefore more likely appear in specific examples. (However, see Grebogi, et al. [1985] on this point.) The attractors considered by Ruelle and Takens were abstract mathematical constructs and it is still not clear how best to define strange attractors in the light of examples occurring in physically relevant models. However, for us, a *strange attractor* will simply be an attractor which contains a chaotic invariant set. Since it also contains a dense orbit, this implies that almost all orbits in a strange attractor, or asymptotic to it, eventually display (exponential) separation and so we have *sensitive dependence on initial conditions*. Alas, while strange *attracting sets* can be shown to exist in specific differential equations, it is much harder to produce true attractors, and many examples, such as that of Lorenz, are in addition structurally unstable in the strict sense (cf. Guckenheimer [1976]).

The above discussion introduced the idea of *structural stability*; a second important notion is that of *generic properties*: properties shared by an open-dense set of systems in some suitably topologized space of systems. In attempting to classify bifurcations, for example, we concentrate on *generic* bifurcations which occur *stably* in parameterized families. But one must be careful; generic properties in one space of systems may be rare in another and structural stability is itself not a generic property. This prompts us to consider the effects of *symmetry*.

Many physical systems are invariant under symmetry groups such as spatial translations, reflections or rotations. This leads to group invariances (or strictly, equivariances; cf. Golubitsky, et al. [1985] [1988]) in the model equations (2.1), and can mean that otherwise unstable or non-generic phenomena *must* appear. For example, if $\mathbf{F}$ is an odd function in $\mathbf{x}$, then nontrivial equilibria always appear in pairs: $+\mathbf{x}_0, -\mathbf{x}_0$. Saddle-node bifurcations do not occur; pitchforks do. This has profound conseqences for bifurcation theory, both local (see Golubitsky, et al. [1985] [1988] and Section 3, below) and global (see below in Section 7).

The general idea of local bifurcation theory, invoking reduction to a (low) dimensional center manifold and subsequent unfolding and analysis of the normal form of the degenerate system restricted to it, has been very successful in many applications.

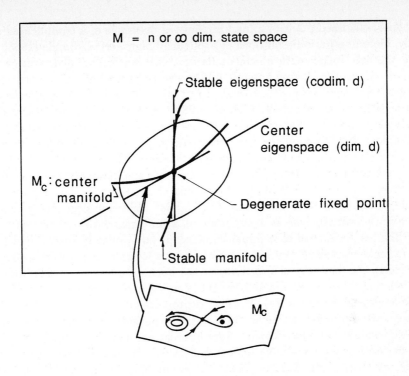

Figure 3. Reduction of dimension via the center manifold theorem. The unfoldings on the center manifold are a parameterized family of invariant subsystems which contain the local attracting sets of the larger system.

Collections such as Campbell and Rose [1983] and Swinney and Gollub [1981] contain examples from fluid mechanics. The process is caricatured in Figure 3.

Although our discussion has thus far proceeded in terms of differential equations, the study of iterated mappings also plays an important rôle in dynamical systems. The Hopf bifurcation from a periodic orbit γ to a two-torus, for example, is best understood via the *Poincaré* or *first return* map induced by the flow of the differential equation near γ on a *cross section* of codimension one in M. Thus the Poincaré map for the Lorenz system is two dimensional (Guckenheimer [1976]). One of the canonical chaotic sets, the Smale horsehoe, was first described for two dimensional diffeomorphisms. Such maps arise naturally in studying periodically forced oscillators and autonomous three dimensional systems such as that of Lorenz, but the ideas extend naturally to $n > 3$ and even infinitely many dimensions (cf. Holmes and Marsden [1981]). As Poincaré [1890, 1899] and Birkhoff [1927] realized, the presence of a transversal homoclinic orbit to a hyperbolic fixed point of a map implies the existence of a complicated set of orbits nearby, including periodic motions of arbitrarily high period. Smale [1963] [1967] completed the description with an uncountable collection of bounded, non-periodic orbits which constitute the chaotic horseshoe. Transverse homoclinic points are relatively easy to find by perturbation procedures such as that of Melnikov [1963] (Guckenheimer and Holmes [1983, Sections 4.5–8], Wiggins [1988]), at least for perturbations of integrable Hamiltonian systems. The method involves a nice interplay among analysis, geometry and topology and is one of the few perturbation procedures which yields global results: one proves the existence of an infinite set of chaotic orbits in a single stroke.

A second phenomenon with a global flavor, is the discovery, by Feigenbaum [1978] and others, that period doubling bifurcations of one dimensional maps typically occur not only in infinite sequences, but that, under mild non-degeneracy conditons, the bifurcation parameter values accumulate at a universal rate independent of the exact map. The results extend to n-dimensional maps (cf. Collet-Eckmann-Koch [1981]) and one can make quantitative predictions of power spectra of the resulting motions. Using these ideas, the *period doubling scenario* has been detected in several fluid mechanical experiments (cf. Swinney and Gollub [1981]) as some stress parameter such as Rayleigh, Taylor or Reynolds number is increased. It provides one example of a "generic" route by which chaotic behavior can develop from simple periodic motions. Other routes for the transition to chaos exist; successive Hopf bifurcations can lead to a structurally unstable quasi-periodic flow on a 3 or 4 dimensional torus, from which a strange attractor might emerge, as Ruelle and Takens [1971] suggested (cf. Newhouse, et al. [1978]), or a *homoclinic explosion* can create a chaotic horsehoe directly, as in the Lorenz example (Sparrow [1982], cf. Kaplan and Yorke [1979]). The mechanism of *intermittency*, a term borrowed from turbulence and applied to rather specific phenomena in low dimensional dynamical systems by Pomeau and Manneville [1980], has also attracted interest (cf. Sreenivasan and Ramshankar [1986]). I will return to it briefly in Section 7.

With the exception of period doubling (cf. Alligood, et al. [1987]), the "routes to turbulence" are really sequences of local bifurcations strung together in a plausible but arbitrary fashion and it is unclear what rôle a list of them can play, other than as indications of what might happen in *any* differential equation or iterated map. The details of successive bifurcations generally change from problem to problem and one can even find several routes in a single experiment in different parameter ranges (cf. Holmes [1984], Holmes and Whitley [1984]). As in the classification of local bifurcations, unfoldings do not reveal what occurs in a specific model: there is no substitute for detailed computation.

In applying the general ideas sketched above to specific models of physical systems, one generally has to use rigorous results to *guide* the application and interpretation of asymptotic and numerical studies. The generic properties of abstract dynamical systems provide *paradigms* of behavior which one seeks in specific ordinary or partial differential equations, both by classical and the newer methods. The numerical discovery of period doubling sequences, say, in an experiment suggests two strategies: in the first, one simply writes down a one dimensional map, fits it to the data, and publishes it. Having no intrinsic connection with the physics of the process, this yields little understanding; it might be compared with Ptolemy's epicycles, although it rarely yields such accurate predictions. In the second, one returns to a model based on rational mechanics and attempts to find such bifurcation sequences in the model and obtain first qualitative and then quantitative comparisons with experiments. Ideally, one hopes to isolate, both mathematically and physically, the important mechanisms at work behind the observed behavior, and hence obtain an understanding which permits predictions. Even partial success in this "Newtonian" approach is worthwhile. It is the approach that I shall take.

So far I have thought in terms of applying dynamical systems methods to the "primitive" equations of fluid mechanics: the Navier-Stokes equation, perhaps coupled to other PDEs, such as the heat equation in convection studies. One can also apply the ideas to low order phenomelogical ODE models of turbulent energy exchange processes, as in the recent work of Bhat, et al. [1989]. Here, however, the modeling process is the most interesting (and problematic) step and the use of dynamical systems methods is relatively direct, so I will not consider such applications further in this paper.

There are many aspects of dynamical systems theory that I have not touched on. A large and useful area is the application of dimension and Liapunov (characteristic) exponent computations to experimental or numerical data. *Embedding theorems* (Takens [1981]) permit one to construct vector (lagged) time series from a single observed variable such as one velocity component; the resulting multidimensional flows are expected to preserve the qualitative features of attracting sets in the "real" system, which is otherwise an inaccessible black box. There are several algorithms available for the computation of Hausdorff, fractal and other dimensions of attracting sets and of Liapunov exponents: roughly speaking, eigenvalues generalized to non-periodic orbits. Chaotic sets have fractional Hausdorff dimension and positive Liapunov exponents and so such computations, together with power spectra and correlation functions, provide evidence of the "signatures" of strange attractors. One can even compute *spectra of dimensions* and seek the distribution of different scales, temporally or spatially. Sreenivasan, et al. [1986] [1988] (cf. Meneveau and Sreenivasan [1987]) have applied such ideas to the spatial structure of turbulent interfaces and mixing layers and, perhaps with less clear results, to the temporal structure of velocities in wakes [1985][1986]. The major problem in applying embedding, dimension and exponent computations to experimental or numerical data is in correctly interpreting the mathematical hypotheses which require "sufficiently large" data sets, etc. The vogue for subjecting data sets to such "tests for chaos" shares much in common with the earlier enthusiasm for frequency spectrum and correlation computations in turbulence studies; the results of these analyses must be interpreted with care since they generally only reflect the behavior at a particular spatial location and the temporal averaging involved can mask much of the significant physics. Also, in using spectral and correlation methods one benefits from over fifty years of experience and rigorous statistical analyses: one has a pretty good idea of the limitations of the methods. I do not believe that one can yet interpret statements such as "the fractal dimension of a shear layer is 4.23" or "the Liapunov dimension of a channel flow is ≈ 350" with the confidence that one places in a power spectrum calculation. Nonetheless, as experience builds up, these newer methods will come into wider use.

Nor have I described the useful rôle that dynamical systems can play in describing the behavior of markers advected in the physical velocity field (a four dimensional—3 space, 1 time—dynamical system). One can seek spatial "signatures" near stagnation points using local phase portraits and bifurcation theory (cf. Perry and Fairlie [1973], Coles [1981], Cantwell [1981]) or use global ideas, including chaotic motions (Aref [1984]

[1986], Ottino, et al. [1988], cf. Khakhar, et al. [1987], Rom-Kedar, et al. [1989a,b]). Such an approach to the problem of "Lagrangian turbulence" appears promising for problems like pollutant dispersion. In view of my limited space, time and energy, I leave it to the other contributors to this section to discuss examples of these applications due to themselves and others.

In spite of the possibility, suggested above, of applying dynamical systems ideas to spatial structures, most of the attempts thus far have concentrated on chaotic time series and ignored spatial features. In closed flow systems, which we briefly address in a moment, such ignorance is perhaps justifiable, since the relatively few active "modes" can often be unambiguously identified with eigenfunctions of a linearized problem (e.g., Rayleigh-Bénard convection rolls or Taylor cells). In open, fully developed turbulence, spatial ignorance seems fatal (Monin [1978]). We will grapple with this in Sections 4 and 5.

3. Bifurcation in a Closed Flow System

Consider the Taylor-Couette experiment in which fluid contained in an annular cylinder is driven by rotating the inner (and possible also the outer) boundary. With a fixed outer boundary, the steady regime in which fluid particles travel on circular paths becomes unstable, as the inner cylinder speed increases, in a (symmetric) pitchfork bifurcation at a critical Rayleigh number and steady cellular motions appear, as first described by Taylor [1923]. Di Prima and Swinney give a good review of the problem in Swinney and Gollub [1981, Ch. 6]. Even in this "simple" bifurcation, the symmetry of the problem plays a rôle. Subsequently the Taylor cells lose their stability and travelling wavy cells and then modulated travelling waves appear. As Rand [1982] realized, the latter could be understood nicely in terms of the SO(2)-equivariance of the vector field (≈ invariance of the equations under elements of the group of planar rotations) governing the dynamics, inherited from the circular symmetry of the experimental apparatus. In particular, he pointed out that "generic" frequency locking of the quasiperiodic modulated motions would not occur if the orbits of the symmetry groups and the dynamical vector field were transverse. Since the latter is the relevant generic property, locking indeed does not occur.

Various attempts have been made to match observations with the routes to chaos outlined in Section 2 (under certain conditions the "quasiperiodic" route from a 2-torus to "weak" temporal chaos is fairly clearly seen) but for our purposes the study of local bifurcations of the Navier-Stokes equations from the circular Couette solution is of greater relevance. In particular, if two parameters (both cylinder speeds, for example) are varied, one can locate codimension two bifurcation points at which Hopf and pitchfork bifurcations coincide. Due to the circular (SO(2)) and axial translational and reflectional (O(2)) symmetries (for infinite cylinders), the resulting reduced system is six dimensional (each eigenvalue has multiplicity two), but the symmetric normal form can nonetheless be analyzed relatively completely. Using these ideas, Golubitsky and Stewart [1986], Golubitsky and Langford [1988] (cf. Golubitsky, et al. [1988, Case Study 6]) and Iooss, et al. [1986] (cf. Laure and Demay [1987]) have classified previously observed and predicted new classes of solutions which should occur in the case of counter rotating cylinders. Some of these have now been found experimentally (cf. Andereck, et al. [1986] and Tagg, et al. [1988]). In addition, in recent experimental and numerical studies, Mullin, et al. [1989a,b] have demonstrated a clear connection between a codimension two bifurcation point at which periodic and steady instabilities interact, and quasiperiodic and irregular flows in a short Taylor-Couette apparatus with rotating end plates.

Here there is a direct and almost rigorous link between experiments and the Navier-Stokes equations, provided by the symmetric normal form on the center manifold: a low (six) dimensional ordinary differential equation which can be shown to exhibit two and three frequency quasiperiodic motions as well as periodic and multiple stationary behaviors. The analysis consists of three main parts: (1) solution of a linear eigenvalue problem to determine the neutral or center eigenspace; (2) computation of the appropriate *general* equivariant normal forms, their unfolding and bifurcation classification; and (3) computation of *specific* coefficients of linear and nonlinear terms of sufficiently high order to determine which of the (finite, but maybe large) list of possible unfoldings actually occurs in a particular system geometry. As hydrodynamic stability theorists know, this latter is a substantial task. This body of work provides a nice model for the appropriate importation of new mathematical methods to fluid mechanics to extend the results of classical hydrodynamic stability studies. Unfortunately, they are only rigorously applicable to weakly nonlinear interactions (the analysis is *local*, so amplitudes must remain small), and any chaotic interactions found are thus of small amplitude and merely represent a weak non-periodic "dancing" of spatially simple structures such as the Taylor cells. (In fact, for technical reasons the parameter sets in which chaos occurs also tend to be exponentially thin near the bifurcation point (cf. Holmes, et al. [1989]).) In this respect, careful (numerical) path following computation of steady and periodic solutions branching from such degenerate bifurcation points would seem well worthwhile. But, in any case, alas! Our business is not with such spatially constrained systems.

4. Coherent Structures in Open Flows

It is now generally recognized that many open flows contain large scale ordered structures, eddies or vortices, which, although not steady in space or time, persistently appear, disappear and reappear. As Roshko [1976] has remarked, discussing shear flows, "Although these flows had been studied for many years and the presence of coherent structures not suspected, once (they) are known to be there it is rather easy to find them." (In this respect, they are a little like chaos and strange attractors; everybody now promotes his own example, when fifteen years ago he would have scorned the very idea.) The reviews of Kline [1967] [1978] (cf. Kline, et al., [1967] and Kim, et al. [1971]) and Willmarth [1975] provide a good background for the boundary layer example which we shall pursue in Sections 6 and 7 and they contain far more detail that is possible or appropriate here. Also, Blackwelder [1989] includes a summary of experimental observations and a discussion of the turbulence generation process in the first part of his recent review. I shall merely summarize some general features.

Flow visualization by injected dye, smoke, hydrogen bubbles or, more recently, Cray XMP $$, has revealed persistent organized structures in many open flow systems. The most striking are those in which shear is dominant, due to the presence of a solid surface or an interface which acts as an anchor and source for the generation of vorticity and thus turbulence. The large eddies in shear layers (Brown and Roshko [1974]) and longitudinal (streamwise) vortices in the wall region (Kline, et al [1967] and Kim, et al. [1971], cf. Kline [1967] [1978]) provide specific examples. Such structures are not stationary, but typically evolve in space and time in a complicated fashion, often exhibiting a repetitive cycle of events, such as the lift, oscillation and ejection of longitudinal boundary layer streaks, followed by sweep and reformation. While easy to see (with the right spectacles), coherent structures are rather hard to pin down, hence, perhaps the great volume of work on conditional sampling techniques, etc. (cf. Van Atta [1973]).

Over thirty years ago Townsend [1956] suggested that a dynamic equilibrium among such large structures might be important in the generation and maintenance of fully developed turbulence (cf. Laufer [1975]). Kline [1967] pointed out the importance of "a nonlinear interaction between the outer and wall layers of the turbulent boundary layer" in which latter the streamwise vortices reside. Roshko [1976], quoting Dryden [1948] with approval, suggested that "...it is necessary to separate the random processes from the non-random processes." Many other researchers have made similar statements. In attempting a Newtonian, rather than a Ptolemaic model, our task is to isolate elements of the "non-random" coherent structures and the "random processes" as *mathematically precise objects* and to derive equations (differential or other) which describe their "nonlinear interaction." (One might also ask where the randomness comes from, if the process is governed by the deterministic Navier-Stokes Equation.)

The strategy I outline in the remainder of this paper represents one start in this direction. The proper orthogonal decomposition, described next, yields a set of basis functions or spatial structures, suitable combinations of which form the coherent structures of the wall region. The flow domain is artificially limited (at $y^+ = 40$ in our example) and the effects of the outer region represented by a quasi-random pressure field which arises naturally from the Navier-Stokes equation in the Galerkin projection process. The nonlinear interaction among the deterministic structures is encoded in the ordinary differential equations, obtained by the Galerkin projection, for the time dependent amplitude coefficients of the basis functions, while the effects of the outer layer appear as a small stochastic forcing term. Thus, as we shall see, each of the rather vague statements quoted above has a precise interpretation in some aspect of the model. It may not be correct, but at least it should be clear where it is wrong, and thus it might be correctable. Of course, I do not intend to suggest that the approach described below is the sole representative, on earth, of the one true faith; merely that it is an *example* of how certain general ideas might be translated into specific models, and of how some recent tools, including those of dynamical systems theory, are essential to the process.

5. The Proper Orthogonal Decomposition

Here we summarize the discussion in Aubry [1988, Sections 2 and 3]. For more details, see that paper or the work of Lumley [1967] [1970] in which the application of the decomposition theorem to turbulent flows was first proposed. We remark that the basic idea has appeared, often independently, in fields as diverse as educational phychology and artificial intelligence and goes by various names, from principal factor analysis to the Karhunen-Loève decomposition theorem (Loève [1955]), not forgetting the collective coordinates, beloved of physicists (Campbell [1987]).

The idea is most easily seen in terms of a scalar field $u(x)$ depending on a single spatial variable $x \in D \subset \mathbb{R}$. Let $\mathcal{U} = \{u^i(x)\}_{i=1}^N$ be a (large) ensemble of realizations of a random field with each u^i belonging to a suitable Hilbert space $\mathcal{H}$ with inner product (f, g) and norm $\|f\| = \sqrt{(f, f)}$ for $f, g \in \mathcal{H}$. Let $\Phi = \{\phi_j\}_{j=1}^\infty$ be a basis for $\mathcal{H}$. Rather than choosing Φ a priori as Fourier modes or Chebyshev polynomials, we ask how Φ can best be chosen to reflect the spatial structures present in $\mathcal{U}$. Specifically, we seek to maximize the projections

$$P_j(u) = \frac{(u, \phi_j)}{\|\phi_j\|} , \qquad j = 1, 2, 3, \ldots \tag{5.1}$$

onto each basis function in turn, of a typical element $u \in \mathcal{U}$, removing the component $\sum_1^j P_k(u)$ already projected at earlier stages. Obviously, *any* function $f \in \mathcal{H}$ can be expressed as

$$f = \sum_{j=1}^\infty a_j \phi_j , \tag{5.2}$$

where the a_j are suitable modal coefficients. The advantage of the basis Φ is that, among *all* basis sets truncated at order k, the approximate reconstruction

$$u_k = \sum_{j=1}^k a_j \phi_j , \qquad \phi_j \in \Phi , \tag{5.3}$$

of a typical member $u \in \mathcal{U}$ is *optimal* in the sense that it maximizes the "kinetic energy" $\|u_k\|^2$. A reconstruction of the same order employing, for example, Fourier modes, would typically capture less energy.

The selection of Φ is a problem in the calculus of variations, and the Euler-Lagrange equation determining the elements ϕ_j turns out to be a Fredholm integral equation of the first kind:

$$\int_D R(x, x') \phi_j(x') dx' = \lambda_j \phi_j(x) , \tag{5.4}$$

where

$$R(x, x') = \langle u(x) u(x') \rangle \tag{5.5}$$

is the autocorrelation function computed by ensemble average, $\langle \ \rangle$. The (normalized) eigenfunctions $\phi_i(x)$ of (5.4) form the desired basis Φ; they are orthogonal and the coefficients a_j resulting from reconstruction of an element $u \in \mathcal{U}$,

$$u = \sum_{j=1}^{\infty} a_j \phi_j , \tag{5.8}$$

are uncorrelated, their mean square values are the eigenvalues,

$$\langle a_j a_k \rangle = \delta_{jk} \lambda_j \tag{5.7}$$

and the latter are ordered $\lambda_1 \geq \lambda_2 \geq \cdots$. In passing, we remark that the eigenfunctions ϕ_i have nothing, a priori, to do with the eigenfunctions of any linearization of the governing equations of the field u, and that they are sometimes called *empirical eigenfuctions*, in view of their derivation from experimental data. If the (flow) conditions under which the ensemble $\mathcal{U}$ is collected are changed, then one expects Φ to change likewise.

In view of (5.7), the ensemble averaged kinetic energy is given by

$$\| \langle u\, u \rangle \|^2 = \sum_{j=1}^{\infty} \lambda_j \tag{5.8}$$

and the rate of decay of λ_j indicates the accuracy, in a mean square sense, of truncations such as (5.3).

To apply these ideas to turbulent velocity fields one assumes stationarity over sufficiently long time intervals and interprets ensemble averages as time averages. The three dimensionality of the velocity field implies that $\mathbf{R}(\mathbf{x}, \mathbf{x}')$ becomes a (symmetric) matrix and the basis functions $\phi_j(\mathbf{x})$, vectors with vector valued arguments. Note that the fact that the autocorrelation matrix derives from a divergence-free flow implies that the basis is likewise divergence-free, a considerable advantage in what follows. Collecting (or computing) the spatial information necessary to generate $\mathbf{R}$ is a considerable undertaking, but depending on the situation, it is not always necessary to compute empirical eigenfunctions $\phi_j(\mathbf{x})$ in *all* directions $\mathbf{x} = (x_1, x_2, x_3)$; for example, the fully developed wall layer is approximately homogeneous in both streamwise (x_1) and spanwise (x_3) directions and a Fourier decomposition suffices and is actually close to optimal in those directions. To make the resulting representation tractible, one chooses reasonable length scales in those directions and isolates a box of size $L_1 \times L_3$ with doubly periodic boundary conditions. The lengths L_i are chosen large enough to capture several "structures," as described below. The velocity field is therefore expanded as

$$\mathbf{u}(\mathbf{x}, t) = \frac{1}{\sqrt{L_1 L_3}} \sum_{k_1, k_3, j} a_{k_1, k_3, j}(t) \exp\left(2\pi i (k_1 x_1 + k_3 x_3)\right) \phi_{k_1, k_3, j}(x_2), \tag{5.9}$$

where the $\phi_{k,j}$ are computed from the vectorial analogue of (5.4) with (5.5) a time averaged spatial autocorrelation matrix (in fact, Aubry et al. [1988] Fourier transform $\mathbf{R}$ in x_1, x_3 to yield a cross power spectrum). Via (5.8) the associated eigenfunctions $\lambda_{k,j}$ reveal the Fourier spectral content of the time averaged energy and permit one to make an intelligent choice of length scales and modal truncations in the 1 and 3 directions.

(Obviously the rate at which eigenfunctions and energy content decay depends on the problem.)

Figure 4 shows slices of the spanwise and streamwise eigenvalue spectra for experimental data obtained in a pipe flow at (centerline) Reynolds number $R_e = 8,750$ by Herzog [1986]. The wavenumbers indicated refer to specific truncations which I discuss in Section 7. Eigenvalues obtained from numerical simulations of Moin [1984] and his colleagues (cf. Kim and Moin [1986]) of a channel flow at $R_e = 13,800$ exhibit similar behavior. In this respect, it is important to remark on the choice of scale in x_2, the "inhomogeneous" direction normal to the wall. Use of Herzog's data forced Aubry, et al. to limit the range to $0 \leq x_2 \leq L_2 = 40^+$ (wall units), but a limit to $x_2 \leq 60 - 80^+$ seems desirable in any case. The more extensive data of Moin [1984] (cf. Aubry, et al. [1988, Figure 3]) indicates that, while convergence of the eigenfunctions $\lambda_{k,j}$ is very rapid in the the wall region (to $x_2 \approx 60^+$, say) and one mode, integrated over the homogeneous directions, captures 60% of the kinetic energy, up to 5 or 10 modes are required if one is to go to $x_2 \approx 400^+$.

At this point one asks how the visually observed coherent structures (Kline, et al.'s [1967] bubble streaks) are related to the empirical eigenfunctions $\phi_{k,j}$. Examining the eigenvalue spectra of Figure 4, it is clear that the peaks, at $k_1 = 0$ and $k_3 = .0035$, reflect the relative streamwise/spanwise scales observed in Kline et al.'s visualizations. If the velocity fields encoded by the eigenfuctions are plotted, one sees pairs of squarish longitudinal vortices of differing spanwise and streamwise scales, depending on the modal wavenumbers, k_3 and k_1. A suitable *combination* of these Fourier modes, chosen with amplitudes reflecting the time averaged energy content of Figure 4, reconstructs pairs of vortices with the appropriate spanwise extent: about 100^+ between roll centers (cf. Aubry [1987] and see Figure 14, below). This is, of course, not surprising; the real test comes when the eigenfunctions are used as a basis for Galerkin projection and one allows the resulting ODE model to dynamically select the modal combination itself.

As I have suggested, to obtain a *reasonably* low dimensional model, it seems wise to attempt to (partially) isolate the inner layer by picking a shallow box and taking $0 \leq x_2 \leq L_2$, with L_2 relatively small. The dimension computations of Keefe, et al. [1987], indicating a Liapunov dimension of 3–400 for turbulent Poiseuille flow, are consistent with this idea; the high dimension is due to inclusion of the outer parts of the boundary layer in their "global" calculation, and one can expect to avoid many of the spatial scales by restricting attention to the inner layer. Here, of course, one makes a *modeling* decision. The coherent structures of interest here are therefore the (small) inner layer "streaks" of Kline, et al. [1967] rather than the larger scale outer eddies.

Here it is perhaps worth mentioning that only now, with tireless computers able to manipulate experimental and numerical databases, has the proper orthogonal decomposition really come of age. The amount of data necessary to assemble correlation matrices and hence produce empirical eigenfunctions in fully three dimensional fields is truly staggering, but examples of simpler model problems are already multiplying rapidly, cf. Chambers, et al. [1988], Sirovich [1987a–c, 1988], Sirovich, et al. [1987] [1989], etc., and references therein. In most cases, homogeneity is assumed in all but one direction (or the problem is a model with only one or two spatial directions, as in Chambers, et al. and most of Sirovich's work), although Glezer and Perlstein [1988] are

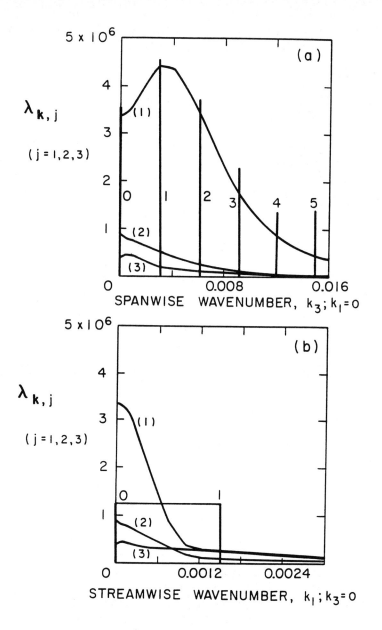

Figure 4. Convergence of the proper orthogonal decomposition in the near-wall region ($x_2^+ = 40$) of a pipe flow according to experimental data. Turbulent kinetic energy in the first three eigenmodes. $\lambda_{\mathbf{k},j}(i = 1, 2, 3)$ as a function of (a) the spanwise wavenumber k_3, (b) the streamwise wavenumber k_1 (from Herzog [1986]).

attempting to derive eigenfunctions for spatially developing structures, which requires nontrivial computation in two (x_1 and x_2) directions.

6. Galerkin Projection, and Energy Loss to Neglected Modes

The Galerkin method is a well established procedure by which one produces a sequence of finite dimensional approximations to a partial differential equation. References relevant to the Navier-Stokes equations include Ladyzhenskaya [1969] and Temam [1984], [1988]. Starting with a partial differential equation written as an evolution equation on a suitable Hilbert space:

$$\frac{du}{dt} = F(u) , \qquad u \in \mathcal{H} , \tag{6.1}$$

possessing an inner product (f, g) for $f, g \in \mathcal{H}$, one chooses a basis $\Psi = \{\psi_j\}_{j=1}^{\infty}$ and expands the dependent variable as:

$$u(x, t) = \sum_{j=1}^{\infty} a_j(t)\phi_j(x) , \tag{6.2}$$

substitutes into (6.1) and projects onto each basis function in turn using the inner product:

$$\left(\left(\frac{d}{dt} - F \right) \left(\sum a_j \phi_j \right) , \phi_l \right) = 0 . \tag{6.3}$$

This process produces a (formally infinite) set of ordinary differential equations in the modal amplitude coefficients a_l:

$$\dot{a}_l = F_l(a_1, \ldots, a_l, \ldots) , \qquad l = 1, 2, 3, \ldots, \tag{6.4}$$

which can be truncated at some finite order, say L, to produce a finite dimensional system like (2.1):

$$\dot{\mathbf{a}} = \mathbf{F}_L(\mathbf{a}) ; \mathbf{a} = (a_1, \ldots, a_L) . \tag{6.5}$$

Each such system lives on a finite dimensional subspace $\Psi_L = \text{span} \left[\{\psi_j\}_{j=1}^{L} \right] \subset \mathcal{H}$ and although its solutions are *not* projections of true solutions in $\mathcal{H}$ onto Ψ_L, but solutions of a projected vector field, the idea is that, as L increases, solutions of (6.5) better and better approximate solutions of (6.1) in the sense that the error

$$\left\| u(t) - \sum_{l=1}^{L} a_l(t)\psi_l(x) \right\| \tag{6.6}$$

approaches zero as $L \to \infty$. Under reasonable conditions such convergence does indeed obtain and many numerical techniques (spectral methods) are based on this. Moreover, existence theorems for Navier-Stokes can be proved by the method; cf. Ladyzhenskaya [1969].

Often one chooses for Ψ a basis of eigenfunctions of the operator $DF(0)$, corresponding to (6.1) linearized at the trivial (= "laminar") solution. While this many be adequate for local bifurcation studies, such as those of the closed Rayleigh-Bènard and Taylor Couette systems of Section 3, it is not likely to yield rapid convergence in strongly nonlinear problems such as fully developed turbulence. Here the empirical basis Φ of Section 5 comes to one's aid. To explain how this works, I must briefly return to dynamical systems theory.

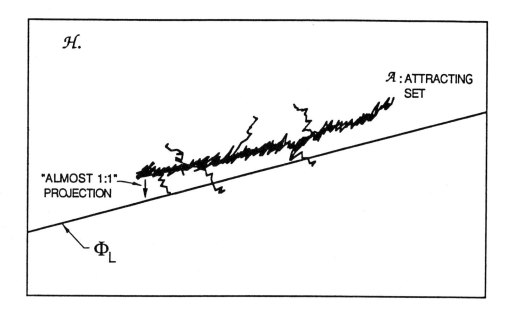

Figure 5.   A geometrical interpretation of the proper orthogonal decomposition. The reader should note that this resembles a "global" version of the local center manifold picture of Figure 3.

Recently, it has been established that, in bounded spatial domains and modulo some technical conditions, the Navier-Stokes equation in 2 and 3 space dimensions, possesses a compact attracting set $\mathcal{A}$ of finite Hausdorff dimension (e.g., Babin and Vishik [1983], Constantin, et al. [1985a] [1985b], Ghidaglia [1986], Temam [1988]). After initial transients decay, *every* solution is close to some orbit in $\mathcal{A}$ and so "typical" long term solutions $u(t)$ define a (collection of) probability measure(s) or invariant density(ies) supported on $\mathcal{A}$. If the resulting measure is uniquely ergodic, then the time averages by which the empirical basis Φ is computed will reflect $\mathcal{A}$ in the sense that solutions represented by the attractors $\mathcal{A}_L$ of the Lth order truncations in Φ_L converge on true solutions in $\mathcal{A}$, as L increases, more rapidly than those represented by any other basis. Now the rigorous dimension estimates are enormously high and I hasten to admit that almost none of the ergodic properties appealed to here have been proved even for reasonable ODEs, let alone the Navier-Stokes equations, but the idea seems at least to provide a rationale for Galerkin projection into the empirical eigenfunctions $\phi_j(x)$. Figure 5 attempts a geometric interpretation of the process. The use of the subspaces $\Phi_L \subset \Phi$ spanned by basis elements obtained from averages over $\mathcal{A}$ guarantee that true solutions $u(t) \in \mathcal{A}$ of (6.1) should project onto solutions $u_L(t) = \sum_{j=1}^{L} a_j(t)\phi_j(t)$ of (6.5) in an "almost one-to-one fashion." Note that the attracting set may be folded and wrinked in a complicated manner and so that typically the dimension of the flat subspace Φ_L will considerably exceed the (Hausdorff) dimension of the attracting set itself. While one cannot hope to capture *all* the behavior of solutions in $\mathcal{A}$ if L is taken too small, the features omitted will correspond to parts of $\mathcal{A}$ on which the invariant density is low: parts which do not contribute significantly to long term time averages. Thus one hopes to capture important dynamical interactions with "fairly small," L-mode truncations.

In the boundary layer problem treated by Aubry, et al. [1988] some "massaging" of the Navier-Stokes equations is necessary prior to the Galerkin projection process. First, since they consider the equations for the perturbation velocity field which rides on the mean profile $\mathbf{u}(\mathbf{x}) = \big(U(x_2), 0, 0\big)$, and the basis Φ is derived for the space $\mathcal{H} = \big(L^2([-L_1, L_1] \times [0, L_2] \times [-L_3, L_3])\big)^3$ of such perturbations, the governing equations involve the temporally and spatially (over x_1, x_3) averaged profile $U(x_2)$ as an energy source. As described in Aubry, et al. [1988, Section 5], it is necessary to account for local modification to U due to the growth of perturbations; this is done by invoking an average involving Reynolds stresses and it leads to global stability in the model equations, since the driving profile tends to collapse when perturbations grow too large. Secondly, since the truncations are fairly severe, some modeling of viscous losses to higher (Fourier and empirical) modes not included seems necessary. This is accomplished by a Heisenberg type model, and introduces a loss parameter, α which should be of $\mathcal{O}(1)$, but which can be varied somewhat to play the rôle of a bifurcation parameter in studies of dynamical behavior (Aubry, et al. [1988, Section 6]). Finally, since the eigenfunctions $\underset{\sim}{\phi}_{\mathbf{k},j}(x_2)$ do not vanish at the outer boundary $x_2 = L_2$, but only as $x_2 \to \infty$, a boundary term survives from the Galerkin projection process and must be dealt with. Specifically, using the inner product $(f, g) = \int_{\mathbf{D}} fg^* dx$, where integration is carried out over the box $\mathbf{D} = [-L_1, L_1] \times [0, L_2] \times [-L_3, L_3]$, the term $\int_{\mathbf{D}} \nabla \underset{\sim}{\phi} \cdot \exp\big(-2\pi i(k_1 x_1 + k_3 x_3)\big) \underset{\sim}{\phi}^*_{\mathbf{k},l}(x_2)\, dx$, after integration by parts and use of the fact that Φ is divergence free, yields a term

$$\hat{p}_{\mathbf{k}}(x_2, t) \underset{\sim}{\phi}^*_{\mathbf{k},l}(x_2) \Big|_0^{L_2} = \hat{p}_{\mathbf{k}}(L_2, t) \phi^{2*}_{\mathbf{k},l}(L_2) \,. \tag{6.7}$$

(^ denotes Fourier transform in the x_1, x_3 directions.) Since it involves a time varying pressure signal, essentially the "footprint" of the outer layer, this term cannot be computed a priori. It appears as a small, quasi-random forcing term in the ordinary differential equations (6.5) and plays a crucial rôle in certain aspects of their behavior, as we shall see.

While the details of modal truncations and modeling will clearly vary from problem to problem, the strategy presented so far seems to have broad applicability. A natural question concerns its domain of validity, since the basis Φ is optimized for a particular ensemble of flows, produced by a single geometry and at a single Reynolds number.

Now the geometry is obviously crucial, but the results of Sirovich and Rodriguez [1987] on the Ginzburg-Landau model problem, suggest that a single basis Φ can be usefully employed over a reasonable parameter range ($\approx \pm 30\%$), although, as I point out in Section 8, this model is rather special. In another model study, the randomly forced Burgers equation, Chambers, et al. [1988] find that the empirical eigenfunctions are largely independent of Reynolds number in the outer part of the flow, but that they strongly depend on Reynolds number in the "wall layer." However, this dependence can be essentially removed by a (linear) scale change. Also see Aubry, et al. [1989] for an example of how a set of basis functions might be "stretched" for use in modified flows. A single set of basis functions therefore may be useful over a range of Reynolds numbers. In this respect it is worth noting that Aubry, et al.'s studies [unpublished] of models derived using the Moin channel flow eigenfunctions ($R_e = 13,800$) share many of the qualitative features of those from the Herzog data ($R_e = 8,750$). See Section 7.4, below.

Having produced the low dimensional model, or models, since one can take several different truncations, there is a choice: one can pick L and integrate a truncated system on Φ_L numerically (it is still, after all, as nasty nonlinear problem), compare its solutions in norm, and via dimension and other computations with the "full" system (or, more likely a projection into few thousand Fourier modes), and publish the hopefully coincident results; or one can attempt to extract an understanding of the interaction processes at work in the full system by studying the truncated projection analytically. I will now outline what Aubry, et al. [1988] were able to do in this latter respect.

7. An Example: Low Dimensional Models of the Wall Region

To bring some order into the welter of analysis in Aubry, et al. [1988] [1989], it seems best to divide my discussion into several subsections.

7.1 Form of the equations, symmetries and nested heirarchies of models

The Reynolds stress mean profile and viscous loss modeling and the Galerkin projection procedure outlined above lead to sets of ordinary differential equations of the form

$$\dot{a} = \mathbf{A}_\alpha a + \mathbf{B}_\alpha(a, a) + \mathbf{C}(a, a, a) + n(t) \tag{7.1}$$

where a is a vector of the complex modal coefficients $a_{k_1, k_2, j}$ (cf. (5.9)), $\mathbf{A}_\alpha$ is a diagonal matrix containing viscous loss and energy production terms resulting from $\nu \nabla^2 u$, $u \cdot \nabla U$ and $U \cdot \nabla u$, $\mathbf{B}_\alpha$ is the quadratic interaction term resulting from $u \cdot \nabla u$ and $\mathbf{C}$ is a cubic "feedback" term due to the Reynolds stress model. The "forcing" term $n(t)$ is the (small) quasi-random outer boundary pressure term of (6.7). The coefficients of $\mathbf{A}_\alpha$ strongly, and those of $\mathbf{B}_\alpha$ weakly, depend on the Heisenberg parameter α (in fact, two parameters could be introduced here). The full equations are given in Aubry, et al. [1988, Appendix A]. The coefficients of $\mathbf{A}_\alpha$, $\mathbf{B}_\alpha$ and $\mathbf{C}$ are computed from integrals of products of pairs, triples and quadruples of the basis functions $\underset{\sim}{\phi}_{k_1,k_3,j}(x_2)$ and their derivatives, which arise as inner products in the projection process.

Symmetries enter (7.1) as follows. The governing equations in the doubly periodic domain $\mathbf{D} = [-L_1, L_1] \times [0, L_2] \times [-L_3, L_3]$ are invariant under spanwise translation $(x_3 \rightarrow x_3 + l_3)$ and reflection $(x_3 \rightarrow -x_3)$ and under streamwise translation $(x_1 \rightarrow x_1 + l_1)$ reflecting the symmetries of a fully developed flow with zero pressure gradient on an infinite flat plate. Via (5.9), these appear in (7.1), *if the forcing term is neglected*, as equivariances with respect to rotations and complex conjugation in the (Fourier) coefficients $a_{k_1,k_3,j}$:

$$a_{k_1,k_3,j} \rightarrow a_{k_1,k_3,j} \exp\left(2\pi i(k_1 l_1 + k_3 l_3)\right), \tag{7.2}$$

$$a_{k_1,k_3,j} \rightarrow a^*_{-k_1,k_3,j}. \tag{7.3}$$

In our initial discussion, we shall omit $\mathbf{n}(t)$, so that (7.2–3) are "exact" invariances. Since the velocity field $\mathbf{u}$ is real, one need only take coefficients whose wavenumbers k_3, say, are nonnegative and appeal to the relation $a_{k_1,k_3,j} = a^*_{-k_1,-k_3,j}$ to construct those with $k_3 < 0$.

Equation (7.3) implies that the subspace given by $a_{k_1,k_3,j} = a^*_{-k_1,k_3,j}$ is invariant for solutions of (7.1), while (7.2) implies that nontrivial equilibrium solutions of (7.1), for which at least one $a_{k_1,k_3,j} \neq 0$, occur in circles or two dimensional tori, and not singly. In particular, all bifurcations from the trivial solution $\mathbf{a} = 0$ are therefore degenerate, with eigenvalues having multiplicity at least two. There are two additional symmetries due to the special structure of the Navier-Stokes equations and the nonlinear interaction terms: the subspace spanned by spanwise Fourier modes having even wavenumber k_3 is invariant and that spanned by modes with no streamwise variation $(k_1 = 0)$ is also invariant.

Before I explore the implications of these symmetries further, I discuss some particular truncations and show how the resulting systems can be nested within one another. Thus far Aubry, et al. have only considered truncations involving a single eigenfunction (the sum on j in (5.9) has one element), but even here choices as to the Fourier (k_1, k_3) modes must be made. The energy spectra of Figure 4 suggest the basic scales. *As an example*, one might start with a model involving six spanwise modes, having wavenumbers $k_3 = 0.003k$, $k = 0, \ldots, 5$ corresponding to $L_3 = 333^+$, and three streamwise modes, $k_1 = 0.0015l$, $l = -1, 0, 1$, corresponding to $L_1 = 666^+$. Within this model a smaller one, containing only the $k_1 = 0$ streamwise mode, lives in an invariant subspace, within that a yet smaller model involving only the $k_3 = 0$, 0.006 and 0.012 spanwise modes occupies a smaller invariant subspace, and the real subspace of that system $(a_{0,k_3,1} = a^*_{0,k_3,1})$ is in turn invariant. Figure 6 shows these modes on the grid of Fourier wavenumbers and indicates some of the nested subsystems and their (complex) dimensions. In giving these dimensions we use the facts that the $k_1 = k_3 = 0$ (purely real) mode decouples and in fact decays to zero (Aubry, et al. [1987]) and so can be excluded, and that reality of $\mathbf{u}$ implies that $a_{-k_1,0,j} = a^*_{k_1,0,j}$, so that $a_{-1,0,1}$ can be excluded. The global analysis of the larger systems relies heavily upon first understanding the smaller invariant subsystems. I emphasize that the nesting requires that one make a choice of "modal grid" (Figure 6) once and for all and that the physical spatial scales assigned to wavenumbers not be changed from model to model.

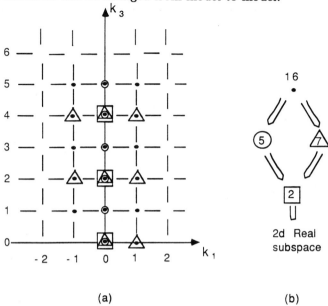

(a)   (b)

Figure 6   (a) Heirarchies of nested systems: $\square$: $k_1 = 0$, $k_3 = 0$, 0.006, 0.012; $\bigcirc$: $k_1 = 0$, $k_3 = 0$, 0.003, 0.006, 0.009, 0.012, 0.015; $\triangle$: $k_1 = -0.0015$, 0, 0.0015; $k_3 = 0$, 0.006, 0.012; $\bullet$: $k_1 = -0.0015$, 0, 0.0015, $k_3 = 0$, 0.003, 0.006, 0.009, 0.012, 0.015. (b) indicates inclusion of subsystems in larger systems and gives complex dimension, excluding $(k_1, k_3) = (0, 0)$ mode.

As Aubry [1987] demonstrates, a reasonable *quantitative* reproduction of the observed "energy budget," even in the first eigenmode alone $(j = 1)$, probably requires inclusion of a further pair of streamwise Fourier modes $k_1 = \pm 2$, leading to a system of complex dimension 27. If three structures were included $(j$ summed from 1 to 3) to bring the time averaged energy up to $\approx 90\%$ of the total, this would yield a dynamical system of (real) dimension over 150. At such a stage it is appropriate to ask if direct Navier-Stokes simulations make more sense. For quantitatively correct models, the dimension problem appears insuperable; one simply cannot include enough relevant spatial scales otherwise, even in the wall layer, but it is still possible that interaction mechanisms might be revealed *qualitatively* by one of the lower dimensional subsystems. Aubry, et al. [1988] found one such mechanism in the six mode model with $k_1 = 0$ and more recently Stone [1989] has found that similar behavior persists in two, three and four mode models, although its observability is rather sensitively dependent upon quadratic and cubic coefficient values and hence on the eigenfunctions, as we shall see.

The fact that lower dimensional systems reside on invariant subspaces of larger systems implies that any invariant set—fixed point, periodic orbit, quasiperiodic torus or chaotic set—of the smaller system is also invariant for the larger one. However, it is important to realize that an attracting set for the smaller system need not be attractive for the larger; each addition of a set of Fourier modes affords additional "directions" of potential instability in the larger phase-space. It is therefore necessary to study the effects of adding modes. Before I do this, however, I introduce some important dynamical phenomena which are common to many of the truncations studied.

7.2 O(2)-equivariance and heteroclinic cycles

In early simulations of the six mode model ($k_1 = 0$, $k_3 = (0)$, 0.003, 0.006, 0.009, 0.012, 0.015), Emily Stone found non-periodic attracting solutions, characterized by quasi-steady phases separated by rapid transitions. In each quasi-steady phase the solution remained close to an element of a circle of fixed points and the transition occurred almost in the real subspace $a_{0,k_3,1} = a_{0,k_3,1}^*$, or in a rotation of it under an element of (7.2). The durations of the quasi-steady phases increased with time, so that the solutions appeared to be approaching a *heteroclinic cycle* connecting a diametrically opposite pair of saddles on the circle of fixed points in the (2.4) ($k_3 = 0.006, 0.012$) subspace.

That the points were saddles was confirmed by direct computation of their eigenvalues, but the question remained of why such solutions occurred for ranges of loss parameter α when heteroclinic cycles are structurally unstable in "generic" dynamical systems (cf. Figure 2, above). Armbruster, et al. [1988] soon realized (as Proctor and Jones [1988], cf. Jones and Proctor [1987], did simultaneously for another fluid problem) that the O(2) equivariance of (7.2–3) stabilized this phenomenon. In particular, they proved the existence of heteroclinic cycles and branches of modulated traveling waves (quasiperiodic motions) for systems involving two complex (Fourier) modes with wavenumbers $k, 2k$. The $k_3 = 0.006, 0.012$ subsystem is precisely such a system, but it is now clear that the behavior can occur in a structurally stable fashion in *any* O(2) equivariant system containing two or more complex modes.

The heteroclinic cycles appear when the unstable manifold of a point A belonging to a circle of nontrivial saddle points intersects the real subspace in a set which lies "in general position" in the stable manifold of another point, B, on the circle of saddles. Restricted to the real subspace, the resulting *saddle-sink* connection γ_{AB} cannot be removed by small perturbations. The cycle is closed by a further connection $\gamma_{BA} =$ from B to A, which is an image of γ_{AB} lying in an image of the real subspace under a suitable rotation by (7.2), so that A is taken to B and B to A. Note that the rotation leaves the circle of equilibria invariant. See Armbruster, et al. [1988] for a specific example. I attempt to caricature the structure in Figure 7, but since it only occurs in phase spaces of real dimension ≥ 4, I doubt that I succeed. In fact, there is a continuous family of cycles forming a heteroclinic manifold, since each infinitesimal rotation of the real subspace contains orbits connecting a diametrically opposite pair of saddle points.

In Aubry, et al. [1988], the effects of the heteroclinic cycles were described in terms of Pomeau and Manneville's [1980] "intermittency of type 2," in which an orbit continually returns to the neighborhood of a saddle point with an unstable manifold containing exponetially growing (spiral) oscillations. Since the return is due to O(2) equivariance, and the unstable manifold need not contain spirals (it can be one dimensional; see below), I now prefer to discuss these cycles in the context of the symmetry groups which

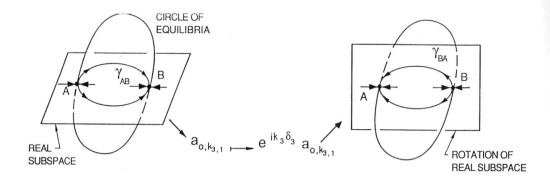

Figure 7. An O(2)-symmetric heteroclinic cycle.

"stabilize" them. In Pomeau and Manneville's discussion it is not clear how or why the continual return and "reinjection" process should in general occur; in the present situation the symmetry makes this return natural.

In addition to circles of fixed points and the heteroclinic cycles, O(2)-equivariant solutions include *travelling* and *modulated travelling waves*. The former occur when the arguments of the (spanwise) phases $\theta_{k_1,k_3,j}$ of the Fourier modes $a_{k_1,k_3,j}$ vary "in step," so that, $\dot{\theta}_{k_1,k_3,j} = k_3 \times$ constant and the resulting velocity field $u(x)$ translates at constant spaced in the spanwise (x_3) direction. (They correspond to circular limit cycles in the phase space on which orbits of the dynamical vector field and group orbits coincide.) Modulated travelling waves occur when a periodic oscillation is superimposed on such a motion, leading to quasiperiodic flow on a two dimensional torus. Both kinds of waves always occur in pairs, left and right going, due to reflection symmetry in x_3. See Figure 5 of Armbruster, et al. [1988] for examples. Travelling waves can also occur (singly) in the streamwise (x_1) direction. Finally, *standing waves* are limit cycles in which the modal phases remain constant and the velocity field oscillates "in place." All of these solutions are found for certain parameter ranges in the model truncations considered below. The analysis of the (general) normal form which revealed them and provided computable conditions for their existence draws heavily upon dynamical systems methods, including both local and global (Melnikov) bifurcation theory (cf. Armbruster, et al. [1988]).

7.3 Global stability and bifurcation diagrams

Henceforth, we restrict our discussion to the model with a single eigenfunction ($j = 1$). In this case, for any Fourier truncation, the coefficients of the cubic terms C of (7.1) are all negative so that, using a Liapunov function of the form

$V = \sum_{k_1,k_2} |a_{k_1,k_3,1}|^2$, one finds that all solutions eventually enter a bounded trapping region in which they remain for all future time. The systems therefore possess attracting sets (Guckenheimer and Holmes [1983]). In fact, if the loss parameter α is sufficiently large, the trivial equilibrium $a = 0$ is the global attractor. It is therefore helpful in understanding the equations to decrease α from this (physically unrealistic) range and examine the branches of solutions emanating from $a = 0$ in local bifurcations. Any such nonzero solution corresponds to an array of steady streamwise vortices, their sizes and relative position depending on the magnitude of the Fourier components $a_{k_1,k_3,1}$. The roots of the branches are easily found analytically from the (multiple) eigenvalues of the diagonal matrix A_α and they can be followed numerically using software such as AUTO (cf. Doedel and Kernévez [1986]) which was used to generate the bifurcation diagrams of Figure 8. In 8a the fixed point and travelling wave solution branches are shown for the $k_1 = 0$, $k_3 = (0)$, 0.006, 0.012 or 2/4 three mode model. The stability descriptions refer to solutions restricted to that subspace. In 8b the diagram for the larger $k_1 = 0$, $k_3 = (0)$, 0.003, 0.006, 0.009, 0.012, 0.015 or 1/2/3/4/5 six mode model in the five dimensional real subspace, omitting $k_3 = 0$, is shown. Note how the smaller subsystem is augmented by the addition of branches bifurcating into the 1, 3 and 5 directions and in particular how the branch of 2/4 mixed modes is destabilized in the 1/3/5 directions in two α-regions. The unstable manifold of these equilibria is one dimensional in region I and two dimensional in region II. In both regions heteroclinic cycles exist; in region I they are essentially those of the pure 1:2 interaction of Armbruster, et al. [1988]; in region II they are more complex, involving oscillatory destabilization of the 2/4 mixed mode in the 1/3/5 directions via a complex conjugate pair of eigenvalues with positive real part. As α decreases these eigenvalues move from left to right in the complex plane, crossing the imaginary axis in a subcritical Hopf bifurcation at $\alpha \approx 1.61$; see Figure 8b. Thus lower values of α correspond to higher growth rates of the instability which develops on the 'steady' streamwise vortices at $\alpha = 1.61$. This instability is evidently related to inflection points in the spanwise variation of streamwise velocity, since there are no streamwise (x_1) modes present in this particular model (cf. Swearingen and Blackwelder [1987]). See Aubry, et al. [1988] and Stone [1989] for details. Figures 9a,b show typical time series of the modal coefficients in region II and just below this region, where the 2/4 branch is additionally destabilized by the bifurcation to 2/4 travelling waves of Figure 8a. In fact, one can indeed see that the heteroclinic cyles are slowly precessing in Figure 9b. (In these Figures I write $a_{k_1=0,k_3=0.003l,j=1}$ as a_l, for brevity.) Aubry, et al. [1988] contains pictures of travelling and modulated travelling waves found for lower α values, but the heteroclinic cycles seem of greatest interest, since they can be related to the bursting phenomenon as I show below. I will discuss the physical implications of the cycles in Section 7.6.

7.4 The effect of subtracting and adding modes

For the purposes of this section, I take the six mode (1/2/3/4/5) model discussed above as a "standard" for comparison. I start by outlining results from the thesis

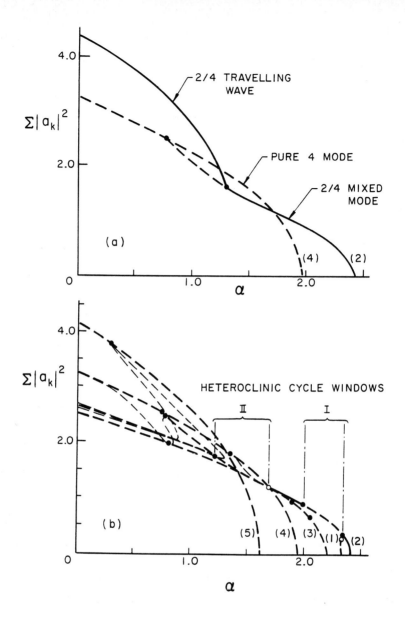

Figure 8. Bifurcation diagrams for (a) the $k_1 = 0$, $k_3 = 0$, 0.006, 0.002 and (b) the $k_1 = 0$, $k_3 = 0$, 0.003, 0.006, 0.009, 0.012, 0.015 models. • denotes a pitchfork bifurcation point, ○ a Hopf bifurcation point to standing waves. In (b) we only show solutions in the real subspace, so that the branch of 2/4 travelling waves of (a) does not appear. — denotes stable solution, - - - denotes unstable solution. From Aubry et. al. [1989].

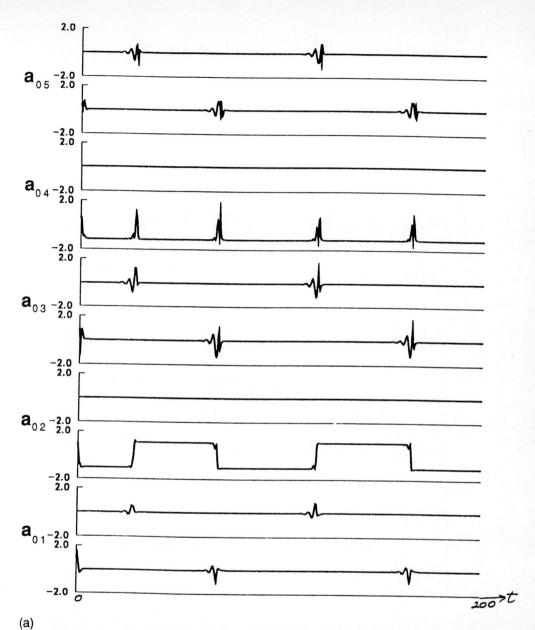

(a)

Figure 9. Time histories of modal coefficients for the $k_1 = 0$, $k_3 = 0$, 0.003, 0.006, 0.009, 0.012, 0.015 model. (a) $\alpha = 1.35$, showing heteroclinic cycles in region II of Figure 8b; (b) $\alpha = 1.30$, showing slowly travelling heteroclinic cycles. From Aubry et. al. [1989].

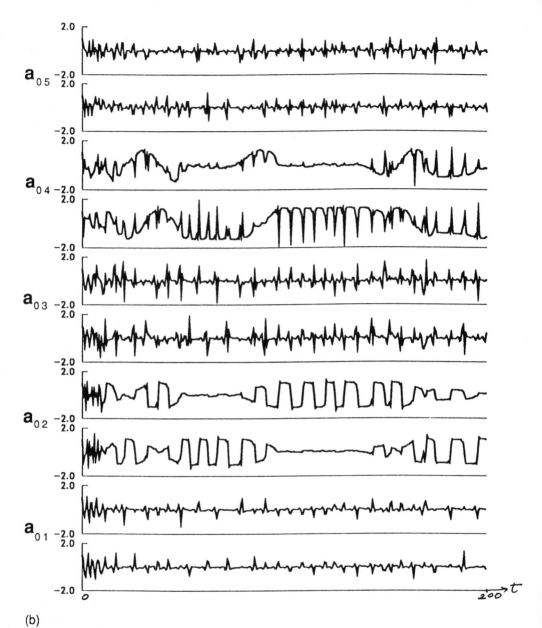

(b)

Figure 9. (continued)

of Stone [1989], who found that heteroclinic cycles persist as spanwise wavenumbers are removed in the absence of streamwise structures ($k_1 = 0$). She studied truncations involving 3, 4, 5 and 6 spanwise modes, the first and last being the $k_1 = 0$, $k_3 = (0)$, 0.006, 0.012 and $k_1 = 0$, $k_3 = (0)$, 0.003, 0.006, 0.009, 0.012, 0.015 nested models of Section 7.3. The four and five mode models have spanwise wavenumbers 0, 0.005, 0.010, 0.015 and 0, 0.004, 0.008, 0.012, 0.016, respectively, chosen to best cover the spanwise spectrum of Figure 4a. These particular models therefore *do not* form invariant subspaces of the larger, six mode model, but they do share its O(2) equivariance and some features of the heteroclinic cycles persist, as Figure 10a indicates. Here I reproduce time series of the three active modes from the four mode model showing heteroclinic transitions between diametrically opposite points on the circle of pure $k_3 = 0.010$ (2) modes. (Recall that the $k_1 = k_3 = 0$ mode is uncoupled and decays to zero.) Here the linear instability is direct, with a single positive real eigenvalue, rather than oscillatory as in the six mode model, and the heteroclinic cycles are "released" via a saddle-node bifurcation in which a stable mixed mode equilibrium vanishes, allowing the unstable manifold of the pure 2-mode to make its connection.

In this model the attracting heteroclinic cycles coexist with attracting travelling waves over part of the parameter range. Figure 10b shows projections of both kinds of solutions onto the $x_1 = Re(a_1)$, $x_2 = Re(a_2)$ plane.

The three and five mode models also exhibit heteroclinic cycles and travelling waves, but for the particular coefficient values in $\mathbf{A}_\alpha$, $\mathbf{B}_\alpha$, $\mathbf{C}$ derived from the eigenfunctions of Herzog and the k_3 wavenumber choices above, the heteroclinic cycles are unstable. Relatively small adjustments to one of these coefficients in $\mathbf{C}$ ($\approx 10\%$) can change repulsion to attraction, however (see Figure 11b, below). For details on this and for the physical implications in terms of acceleration/deceleration of the mean flow profile, see Stone [1989].

The main point is that ranges of α exist in lower dimensional truncations for which heteroclinic cycles persist. Now as I pointed out in Section 7.2, while O(2)-equivariance makes structurally stable heteroclinic cycles *possible*, it does not *guarantee* their existence: for that the coefficients of $\mathbf{A}_\alpha$, $\mathbf{B}_\alpha$, $\mathbf{C}$ must lie in certain ranges (cf. Armbruster, et al. [1988]). The fact that this latter condition obtains (or almost obtains) for several different truncations with different wavenumber choices, is evidence that the cycles reflect true aspects of the physical processes at work.

I now briefly discuss preliminary investigations of eight and seventeen mode models containing the streamwise modes $k_1 = 0$ and ± 0.0015 (the models represented by the symbols by $\triangle$ and $\bullet$ in Figure 6). In deriving these models the eigenfunctions and eigenvalue spectra obtained experimentally present a problem: in particular the k_1 eigenvalue spectrum of Figure 4, peaking at $k_1 = 0$, suggests no obvious length scale L_1. In the $k_1 \neq 0$ computations reported below, the choice $k_1 = 0.0015$, corresponding to $L_1 = 666^+$ was made, while in the $k_1 = 0$ computations described above the effective choice $k_1 = 0.003$ ($L_1 = 333^+$) was made, with a view to including *all* the energetic streamwise wavenumbers available in the experimental data. See Figure 4b: the $k_1 \neq 0$, 0.0015 choice effectively represents the spectrum as a "top hat" as indicated. Thus the $k_1 \neq 0$ models for which computations are available do not include the $k_1 = 0$ models as "exact" invariant subsystems, since the coefficient values in the matrix $\mathbf{A}_\alpha$ change

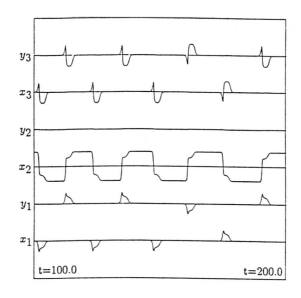

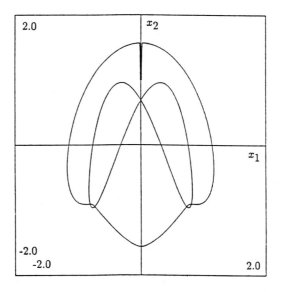

Figure 10. Simulations of the four model model, $\alpha = 0.95$. (a) Time series of hetero-clinic cycles; (b) projection of coexisting heteroclinic cycles and travelling waves onto $Re(a_1)$, $Re(a_2)$ plane. From Stone [1989].

by a multiplicative factor as the k_1 cutoff varies from 0.003 to 0.0015; the coefficients in $\mathbf{B}_\alpha$ and $\mathbf{C}$ also change slightly.

More serious, however, is the choice of different interpolations in the empirical eigenfunctions made by Aubry [1987] in the computation of coefficients of $\mathbf{A}_\alpha$, $\mathbf{B}_\alpha$, $\mathbf{C}$ available to Stone [1989]. While the first model considered, having seven active modes ($k_1 = 0, \pm 0.0015$; $k_3 = 0, 0.006, 0.012$) used eigenfunctions having the same interpolation as the $k_1 = 0$ models discussed earlier, the second, sixteen active mode model ($k_1 = 0, \pm 0.0015$; $k_3 = 0, 0.003, 0.006, 0.009, 0.012, 0.015$) did not. Consequently, the six (five active) mode $k_1 = 0$ subspace of that model is *not* identical to the six mode $k_1 = 0$ model discussed above. Nonetheless, some general indications of the influence of streamwise modes are provided by these computations. For more detail and comment, see Stone [1989].

Figure 11a shows an example of simulations of the eight mode model. Note that all nonzero streamwise wavenumber components decay to zero, indicating that the $k_1 = 0$ subspace is attracting. Solutions started with a range of initial data all converge rapidly on a stable travelling wave in that subspace. As remarked earlier, however, a small change in one of the cubic ($\mathbf{C}$) coefficients within the $k_1 = 0$ model serves to stabilize the heteroclinic cycles, as shown in Figure 11b.

Figure 12 shows time series of all sixteen complex (32 real) active coefficients of the largest model studied so far. Again the nonzero streamwise wavenumbers decay rapidly, indicating that the $k_1 = 0$ subspace is stable, and revealing a stable modulated travelling wave on that subspace (cf. Aubry, et al. [1988, Figure 7d–f.]). For the particular coefficient values obtained from Aubry's [1987] interpolation, the heteroclinic cycles existing for higher α (≈ 1.5, Stone [1989]) are not attracting, although they can be made so by small coefficient changes as in the smaller model of Figure 11.

While more studies are obviously required—L_1 should be varied and a larger number of streamwise modes taken and the eigenvalue discretization problems resolved—the results outlined here do suggest that the addition of nontrivial streamwise modes does not radically affect the dynamical backbone of the $k_1 = 0$ family of systems reported by Aubry, et al. [1988].

I close this section by commenting on studies of Aubry, et al. [1989 and unpublished], in which the $k_1 = 0$, six mode model was modified by applying linear stretching transformations to Herzog's empirical eigenfunctions. (The stretching is an attempt to model scale increases in the wall region due to drag reduction devices such as polymer additives or microbubbles: Lumley and Kubo [1984].) Changes in the eigenfunctions due to a linear stretch by a factor β lead to changes in the coefficients of the ODE's (7.1), certain coeffients being multiplied by β or β^2 and others divided by β, resulting in a *two* parameter family of systems (α, β). The most striking finding is that the major parameter window II (Figure 8b) in α, in which attracting heteroclinic cycles occur, is robust in that it persists for β ranging from 1 (unmodified flow) to 2.5.

In unpublished work the same group found similar heteroclinic cycles, but connecting small standing waves, in a six mode model derived using eigenfunctions obtained by Moin in numerical simulation of channel flows (cf. Moin [1984]). In combination with the evidence presented below, this suggests that the cycles reflect the physics behind the problem rather than being merely quirks of a particular modal truncation. This all

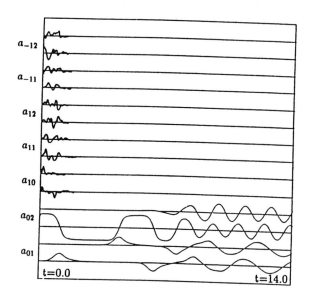

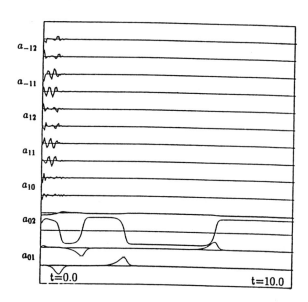

Figure 11. Simulations of the $k_1 = 0$, ± 0.0015, $k_3 = 0$, 0.006, 0.012 model, $\alpha = 0.1$ (scaled coefficients). (a) Travelling wave; (b) attracting heteroclinic cycle observed when one coefficient in C is changed from -2.69 to 3.00. Note that $k_1 \neq 0$ streamwise modes decay to zero. From Stone [1989].

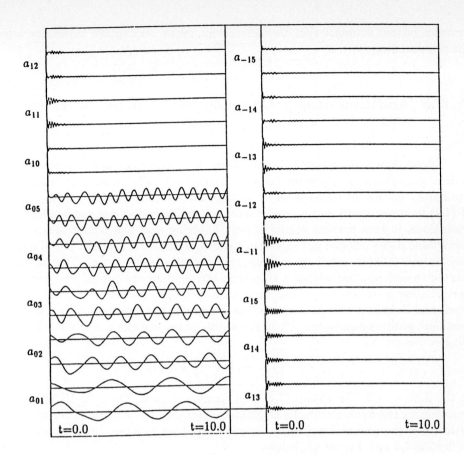

Figure 12. Simulations of the $k_1 = 0, \pm 0.0015; k_3 = 0, 0.003, 0.006, 0.009, 0.012,$ 0.015 model (16 active modes), $\alpha = 1.0$. Note that all streamwise modes $a_{k_1,k_3}, k_1 \neq 0$ decay to zero, leaving a modulated travelling wave in the $k_1 = 0$ subspace. From Stone [1989].

provides further evidence that the heteroclinic cycles do capture important aspects of the physical flow.

7.5 Addition of n(t): Coupling between the inner and outer layer

I pointed out in Section 7.2 that, when the heteroclinic cycles are attracting, solutions asymptotic to them spend periods of continually increasing length near the equilibria. In this situation a perturbation such as the additive "forcing" terms n(t) of (7.1), however small, has a crucial influence; while it does not much affect the rapid transitions, it does prevent solutions from lingering near the saddle points. As Busse and Hiekes [1980] pointed out earlier in a similar symmetric problem (cf. Guckenheimer and Holmes [1988]), small noise leads to "statistical limit cycles"; the motion is attracted to and remains within a thin tube surrounding the heteroclinic cycle, but the perturbation introduces a typical time scale: a mean recurrence or passage time $\langle T \rangle$. Stone and Holmes [1989a,b] compute this in terms of the (real part of the) largest (most unstable) positive eigenvalue, λ^u, of the saddle point and the r.m.s. noise level, ϵ, for a Wiener process, as

$$\langle T \rangle = \frac{1}{\lambda^u} \ln \left(\frac{1}{\epsilon} \right) + \mathcal{O}(1) \tag{7.4}$$

in the limit $\epsilon \to 0$. To obtain (7.4) they analyze a linear Ornstein-Uhlenbeck process representing the dynamics in the neighborhood of the saddle point. They also obtain an estimate of the probability density function $P(T)$ for passage times, to which I return in Section 7.6 and Figure 15, below.

Figure 13 shows an example of computations of mean passage time for the six spanwise mode ($k_1 = 0$) boundary layer model subject to a Wiener process at three different values of α (λ^u increases as α decreases). Comparison with (7.4) in the form

$$T = K_0 + \frac{2}{\lambda^u} \left(\ln |\epsilon| + K_1 \right), \tag{7.5}$$

with the constants K_0, K_1 fitted at a single data point, is excellent. (The factor 2 is introduced because there are two saddle points in the O(2)-symmetric cycle.) Stone and Holmes [1989a,b] contain many further examples.

In the boundary layer application, of course, n(t) is not a vector of Wiener processes, but as Aubry, et al. [1988, Section 11] find, the pressure perturbation (6.7) constructed from numerical data of Moin [1984] yields results with mean in reasonable agreement with (7.5).

To summarize, while the instability (λ^u) which makes the heteroclinic transitions possible, and the resulting "signature" of these events, are present in the deterministic dynamics, the trigger or catalyst which promotes the events and supplies a characteristic time scale is provided by the external perturbation $(\epsilon n(t))$. We comment further below.

7.6 Implications for the wall layer

After locating attracting sets in the phase spaces of various projections, it is necessary to reconstruct the velocity fields via (5.9) and to interpret them as spatio-temporal features in the wall region. I concentrate on the heteroclinic cycles in models with six spanwise modes here, since they appear most interesting.

The unstable equilibria, being composed of 2/4 ($k_1 = 0, k_2 = 0.006, 0.012$) mixed modes, correspond to double pairs of steady, counterrotating streamwise vortices in the basic flow domain $\mathbf{D} = [-L_1, L_1] \times [0, L_2] \times [-L_3, L_3]$. Before and after a homoclinic transition, the velocity field is therefore almost stationary and composed of longitudinal vortices. While such vortices are "encoded" in the model via the empirical eigenfunctions, the dynamical selection of the second mode $k_2 = 0.006$ as the basic scale correctly reflects the spatial scales observed in the wall region (Aubry, et al. [1988] [1989]). The fact that the equilibrium is a saddle point corresponds to a linear instability of these structures. In the rapid homoclinic transition the velocity field "convulses" before the vortices reform, shifted laterally by 1/4 of the domain width (or 1/2 a "structure wavelength"). Figure 14 shows spanwise slices of the velocity field at equally spaced time increments during a heteroclinic transition. Note that, while Figure 14 indicates predominantly reflection symmetric vortices, there are small asymmetric components (otherwise the solution would settle permanently at a symmetric equilibrium in the real subspace or a rotation thereof). One can apparently see an upward lifting and even a coalescence (or breakup?) of the vortices as the Fourier mode $k_1 = 0.003$ momentarily dominates, although the velocity field should be integrated to correctly reflect the visual appearance of dye streaks or bubbles, and the limited extent ($L_2 = 40^+$) of the flow domain precludes one following the migration away from the wall. Nonetheless, there is good qualitative agreement with the "lift, oscillation and breakup" or "burst" events summarized by Kline [1967] [1978], even to the "(streak) pattern often (being) shifted laterally in the neighborhood of the breakup" (Kline [1967, p. 55]). However, as Aubry, et al. [1988] point out, the breakup's effects cannot be followed through by the model because sufficiently high wavenumbers are not included, there are no streamwise variations and the detailed velocity field of the outer part of the layer is entirely excluded.

However, the forcing perturbation $n(t)$ of (7.1) does represent the influence of the outer layer. As we have seen, while it provides a trigger which initiates heteroclinic transitions, the instability mechanism which drives them resides in the inner layer: they are not merely the "footprints" of outer layer distrubances. In this respect the model affords a precise expression of the conclusion of Kline [1978] that an interaction of inner and outer layers is essential to the turbulence production mechanism (cf. Rao, et al. [1971]). While I do not suggest that the mean burst period, for example, should scale with inner (λ^u) and outer (ϵ) variables *exactly* as suggested by the asymptotic limit of equation (7.4), the model does clearly suggest that both inner and outer scales

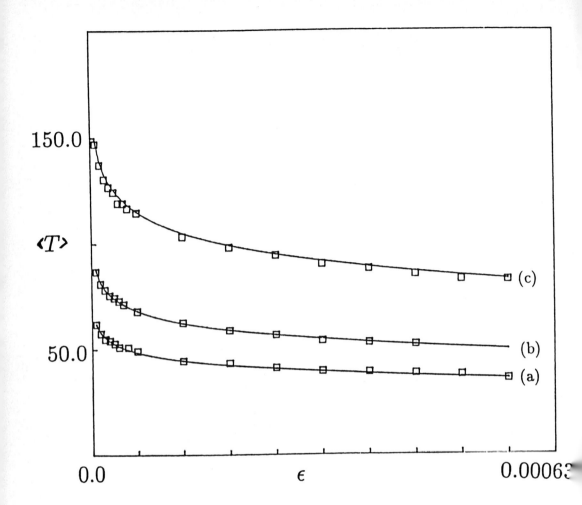

Figure 13. Mean recurrence time versus r.m.s. noise level. Simulations of the O(2) invariant six mode boundary layer model, subject to a Wiener process with components of r.m.s. ϵ. (a) $\alpha = 1.45$, $\lambda_u = 0.3434$; (b) $\alpha = 1.5$, $\lambda_u = 0.2375$; (c) $\alpha = 1.55$, $\lambda_u = 0.1362$. Solid curves show equation (7.5) with K_0 and K_1 fitted to one data point. From Stone and Holmes [1989a, b].

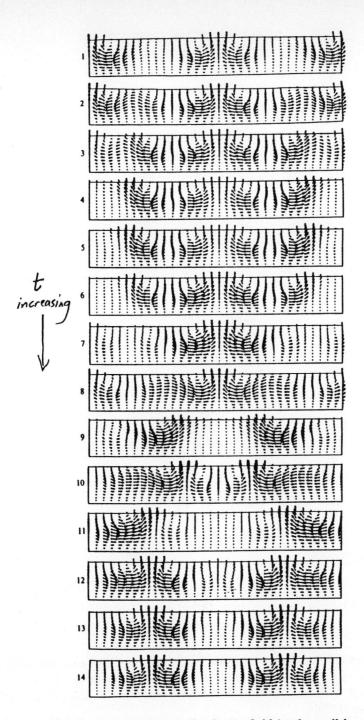

Figure 14. The u_2, u_3 components of velocity field in the wall layer at 14 equally spaced time intervals during a heteroclinic transition. Note lateral shift of quasisteady pairs of streamwise vortices. From Aubry, et al. [1988, Figure 19b].

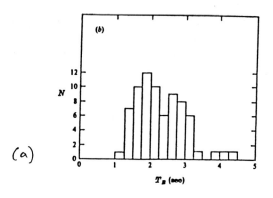

(a)

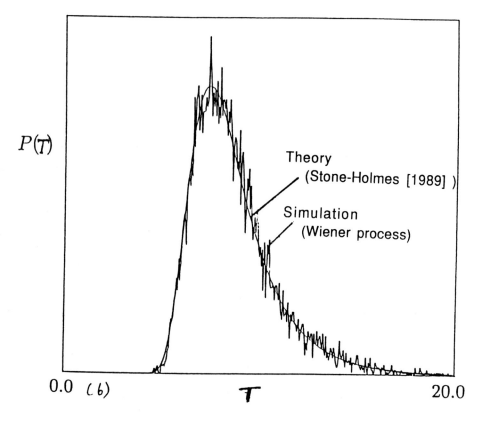

$P(T)$

Theory
(Stone-Holmes [1989])

Simulation
(Wiener process)

0.0 (b)   T   20.0

Figure 15.   (a) Histogram of burst intervals T_B from Kim, Kline and Reynolds [1971, Fig. 21(b)]. (b) Probability distribution $P(T)$ of passage times, T, for heteroclinic cycle subject to additive random (Wiener) noise from Stone and Holmes [1989b].

are important. Moreover, additional evidence (not noted by Aubry, et al. [1988]), appears in a histogram of burst periods, or, in terms of the model, mean heteroclinic passage times. Figure 15 reproduces data of Kim, et al. [1971, Figure 21(b)] and also a probability density function computed by Stone and Holmes [1989b] for an additive Wiener process. The precise flow conditions and other parameter values aside, the two histograms show a striking qualitative similarity; both are strongly asymmetric with substantial tails towards the high period end. Tiederman, et al. [1985] and Bogard and Tiederman [1986] show similar histograms, plotted in the former paper with respect to streamwise distance (x_1). In the latter paper (Figure 7, pp. 404–407) they fit an exponential distribution to the measurments but argue that, while a random model predicts such a distribution, the ejection events are strongly correlated with the 'ordered' low velocity streaks. (Sreenivasan, et al. [1983] also fit exponentials to their zero crossing distribution (Figure 13, pp. 267–269).) I note that the distribution $P(T)$ of Stone and Holmes [1989, equation (2.31)] *also* behaves exponentially:

$$P(T) \sim \frac{\lambda_n^{3/2} e^{-\lambda_n T}}{\epsilon} \tag{7.6}$$

as $T \to \infty$, and so the model presented here may help resolve such issues. Also see Narasimha's paper in this volume.

Aubry, et al. [1988] also show computations of Reynolds stress production during a heteroclinic transition in the $k_1 = 0$, six mode model. While there is a sharp increase and evidence of a double peak (cf. Blackwelder and Kaplan [1976]), it is considerably lower than observed experimentally (eg. Willmarth and Lu [1972]). This is probably due to exclusion of all streamwise wavenumbers and of high wavenumbers in general.

In closing, I remark that the spanwise travelling and modulated travelling waves found for lower α in the modal truncations, when superimposed on the mean stream-wise velocity component $(U(x_2), 0, 0)$, correspond to oblique waves propagating across the boundary layer. Whether or not such structures correspond to observed motions (remnants of transitional instability, for example) is unclear, but it is worth remarking that the "slowly travelling" heteroclinic cycles of Figure 9b correspond to velocity field in which vortices pair and generally interact in a much richer fashion than in the more regular régime of Figure 9a. In particular, far less symmetric vortex structures than those of Figure 14 appear and single (unpaired) vortices are frequently seen, as in the experiments of Smith and Schwarz [1983] and the direct simulations of Kim and Moin [1986]. See Aubry, et al. [1988, Fig. 20] for an example. Further study of this apparently chaotic solution is in progress. However, Aubry, et al. felt that it was necessary to make a reasonably firm connection between the more orderly dynamical events in the model (the heteroclinic cycles) and the physical problem before attempting to interpret the chaotic régime.

To summarize, the orthogonal decomposition, modeling, projection and dynamical systems analysis has effected a translation of the "inherently nonlinear" (Kline [1967, p. 60]) turbulent interactions into mathematically precise and reasonably well understood objects. The coherent structures which lead to streaks are specific combinations of Fourier modes and empirical eigenfunctions. Their internal nonlinear interactions are governed by ODE's obtained by projection of the Navier-Stokes equation into an appropriate "empirical subspace." The external influence of outer layer structures appears in a quasi-random boundary pressure term. The choice of spatial domain makes the inner/outer split precise. The local (linear) instability of these unstable vortices reflects the expected (inflection point) instabilities associated with such structures (Kline, et al [1967], Kim, et al. [1971], Swearingen and Blackwelder [1987]) and global symmetries inherited from the physical problem provide the mechanism for rapid (bursting) transitions and reformation of the unstable vortices. The insights into the range of dynamical behaviors occurring as an (artificial) loss parameter, α, is varied suggests that the models might be used in studies of drag reduction by boundary layer modification, both passive and active. Aubry, et al. [1989] and Bloch and Marsden [1989] represent starts in this direction.

8. Inertial Manifolds and Applicability of the Proper Orthogonal Decomposition

In this section I briefly discuss an important new idea in partial differential equations and infinite dimensional dynamical systems in general—*inertial manifolds*—and indicate how it might be used to put the methods of Sections 5–7 on a firmer basis. I then attempt a critical assessment of the methods discussed above, speculate on where they might be most successful, and indicate some of their limitations.

In a recent series of papers, Constantin, Foias, Nicolaenko, Sell, Temam and others have shown that a fairly wide class of strongly dissipative PDEs not only possess attracting sets of finite Hausdorff dimension, but that these sets lie within smooth inertial manifolds: compact, finite dimensional manifolds M, invariant under the dynamical flow, and attracting *all* solutions at an exponential rate. (Observe that such manifolds are not unique; for example, one can always increase the dimension by adding additional "trivial" modes.) Thus, if a suitable coordinate system can be found for M, solutions are asymptotic to those of a finite (and maybe even computable) set of ordinary differential equations on M and so the long term behavior is rigorously governed by ODEs on a flat finite dimensional model space: $\mathbb{R}^n$. Essentially, inertial manifolds are global, attracting center manifolds. It is important to note that, in this *global* reduction to finite dimensions, one is not ignoring or truncating the high wavenumber modes, one is using the fact that they are "slaved" to the low modes. They are present in the inertial manifold, which is not flat: its curvature describes the coupling ($\approx$ energy flow) to these higher modes. Accounts of the theory can be found in Constantin, et al. [1989], Temam [1988] and references therein.

Inertial manifolds are known to exist for multi-space dimension equations of reaction diffusion type and various one dimensional model equations such as the Kuramoto-Sivashinsky and Ginzburg-Landau equations (Foias, et al. [1988], Doering, et al. [1988]). Unfortunately, the Navier-Stokes equations seem to pose serious technical difficulties (but see Titi [1988]); however, if we assume that for them, or for the "massaged" version of them defined in the wall region, an inertial manifold does exist, then the proper orthogonal decomposition provides a basis for an optimal model space for this manifold.

In all the applications so far, the inertial manifold M is sought as a graph over a subspace spanned by a finite set of eigenfunctions of a (trivially) linearized problem.

This choice of model space is largely ignorant of the nonlinear dynamics and often leads to poor estimates of minimal dimension for M. If the attracting set A within M supports an ergodic measure, or if a suitable ensemble of initial conditions is taken to guarantee that A and M are "well covered," the time or ensemble averages from which the correlation and empirical eigenfunctions of (5.4) are computed ensure that the resulting subspaces $\Phi_L = \text{span} \{\phi_j\}_{j=1}^L$, $L = 1, 2, 3, \ldots$ are well chosen as model spaces. Since M is finite $(n\text{-})$dimensional, it follows that if $L = n$ the correspondence between points in M and points in Φ_n is *exactly* one-to-one and hence that solutions of the PDE in M precisely correspond to solutions of the projected ODE on Φ_n and the scheme of Figure 5 is made rigorous. Recent work of Foias, Manley and Sirovich [C. Foias, personal communication] is relevant here. At the same time, the optimality of the proper orthogonal decomposition might lead to sharper estimates of dimension for inertial manifold in strongly nonlinear systems. It would be a worthwhile project to study a model problem such as the Ginzburg-Landau system from this viewpoint. Actually, the notion of *approximate inertial manifolds* and the associated (reduced) *approximate inertial forms* (Titi [1988], Sell [1989]) is probably more relevant here, since the averaging inherent in the proper orthogonal decomposition process ignores short time transients and events whose average energy is small. As Sell [1989] points out, one may be able to construct such approximate reduced systems of dimension lower than that of the exact inertial forms.

Turning from the prospects for a rigorous underpinning, I now consider the utility of the proper orthogonal decomposion coupled with dynamical systems methods. A number of studies by Sirovich and his colleagues [1987], [1989] have demonstrated that the empirical eigenfunctions allow excellent low order reconstruction of actual solutions and averaged quantities in one and two dimensional "model" problems. However, in most of these cases, the empirical basis functions are very close to the pure sinusoidal eigenfunctions of the original equations linearized at the trivial solution. (See Sirovich [1988] for a model problem, the Ginzburg-Landau equation with "sticky" (zero amplitude) boundary conditions for which this is *not* the case.) This suggests that the spatial constraints of one dimensionality, or, in the case of the Kolmogorov flow (Sirovich, et al. [1989]) spatially sinusoidal forcing, can lead to rather weak nonlinear interactions. While proper orthogonal decomposition works nicely, and could probably be incorporated in a rigorous theory of inertial manifolds and attractors, in these examples it does not tell us much about the spatial structures and their nonlinear interations which is not already available from weakly nonlinear or local bifurcation theory. However, as the work of Keefe, et al. [1987] and, to a lesser extent, Chambers, et al. [1988], suggest, it will be necessary to restrict oneself to problems with some sort of spatial constraints if projections of "dynamically resonable" dimension are to suffice. In the boundary layer application discussed above, this was done by (artificially?) isolating the wall region and selecting the fixed spatial scales L_j of the box $\mathbf{D}$. Such a scale selection process and the concomitant modeling required, such as the Reynolds stress mean velocity model used above, will necessarily be problem dependent.

In view of the importance of symmetries in the boundary layer problem, it is crucial that the isolation of a flow region be performed in a manner which preserves spatial symmetries in the "full" open flow but does not introduce artificial restrictions. For example, if one seeks only spanwise reflection symmetric solutions in the wall layer problem, then one remains in the invariant real (spanwise) subspace and no heteroclinic cycles or travelling waves occur. While the velocity fields of Figure 14 *appear* reflection symmetric, there are in fact small asymmetries present which enable unstable modes to grow and keep the heteroclinic cycles active.

For the proper orthogonal decomposition to work at all, some form of ensemble or time averaging is essential. The flow must therefore possess a reasonable degree of stationarity. The very use of time averages, for example, reflects an a priori decision to neglect brief, possibly violent, events which do not produce much turbulent energy on the average, at the expense of more "regular" processes. This decision may not always be appropriate; to recall the analogy pressed by Kline [1978]: the choice of basis functions determines which part of the elephant we see, or whether, instead, we see a leopard.

In this connection Newell, et al. [1988a,b] have suggested on the basis of asymptotic analysis and numerical simulation that the nonlinear Schrödinger equation modeling Langmuir (plasma) turbulence exhibits randomly occurring sharp coherent events containing large amounts of energy which correspond to "heteroclinic connections to infinity" (HCIs). They envisage an attracting set containing saddle points, or more complex saddle sets such as horseshoes, whose unstable manifolds consist of almost singular solutions whose tendency to focus and blow up at isolated space-time points is mediated by losses at high wavenumber. (It is interesting to note that they also [1988b, Figure 6] find exponential tails in event distributions.) At present, I feel that their coherent structure and heteroclinic transition picture has a primarily metaphorical connection with the (nonsingular, viscosity dominated) work of Aubry, et al. discussed above, but it would be interesting to apply the orthogonal decomposition to the nonlinear Schrödinger equation and attempt to extract a low dimensional model based on the almost singular "collapsing filaments" found by Newell, et al. In fact those authors suggest [1988b] that a local version of the Karhunen-Loève decomposition will be necessary to capture the brief "most significant" events. This would appear to provide a very different test of the decomposition/projection ideas than the spatially smooth, weak interactions of the Ginzburg-Landau model. In any case, the use of time, space or ensemble averages on processes containing violent events is certainly problematic and requires further study.

In general the averaging process in which Φ is derived and the relatively low dimension of the models proposed here necessarily restrict such studies to the dynamics of "larger" spatial scales. In Aubry, et al. [1988] the effect of smaller scales—the unresolved modes—is included only via a simple Heisenberg model. Inclusion of dynamical dissipation mechanisms would require so many additional modes that the system would have dimension in the hundreds rather than the tens. If one regards the "turbulence problem" as addressing the transport of energy in the inertial (Kolmogorov) range and its subsequent dissipation, then these methods do nothing to help solve it. They seem better suited to the study of turbulent energy production.

At this point, it is necessary to repeat that, while the proper orthogonal decomposition provides a relatively small set of basis functions from which an accurate reconstruction of the coherent structures can be made, it does not in itself determine how those basis elements interact to "select" the appropriate combinations; not only is phase information lost, *all* the dynamics is averaged away in producting Φ. Only in combination with a dynamical model for interaction of the (free) modal coefficients, such as that provided by the Galerkin projection process, can one obtain such information.

It is worth comparing this approach with the classical one of hydrodynamic stability theory, in which one seeks the most unstable mode in a linear analysis of the governing equations near a known steady or periodic solution. Weakly nonlinear or local bifurcation theory then describes the nonlinear saturation of this mode and linear theory can again be applied to the resulting (asymptotic or numerical) solution. Orszag and Patera [1983], for example, carry this quite far in their study of three dimensional "secondary" instabilities in shear layers and they apply the results to the transition problem in boundary layers. There one attempts to compute a route to turbulence for a specific problem using essentially "laminar" eigenfunctions. The statistical approach of the proper orthogonal decomposition, in contrast, plunges one directly into a (fully developed) turbulent flow and determines the predominent, strongly nonlinear, spatial structures in an averaged sense. Galerkin projection into the empirical subspace then supplies the (large amplitude) nonlinear interaction mechanisms. From this viewpoint, the methods seem complementary.

Has the application of dynamical systems brought, or does it promise, increased understanding of the boundary layer problem? I answer this as follows. Newton's gravitational theory and the resulting model of planetary motion as sets of ordinary differential equations seem to me to offer greater understanding than Kepler's ellipses or Ptolemy's epicycles. In the application I have described, the rôle of gravitational law is played by the Navier-Stokes equations and the "massage" of them necessary to isolate a projected, truncated model of the wall region. Proper orthogonal decomposition plays a vital part here, but the specific truncation choices and subsequent derivation of a nested heirarchy of models are guided by general ideas from differential equations and dynamical systems theory (invariant manifolds, equivariant vector fields, etc.). In the mathematical analysis of the resulting ODEs, dynamical systems methods play a crucial rôle both in laying out paradigms of anticipated behavior and in the detailed studies of specific systems. For example, in Armbruster, et al. [1988] extensive use is made of center manifolds, normal forms, codimensional two local bifurcation theory and global (Melnikov) perturbation methods in analysis of the heteroclinic cycles. While partial results can be obtained by more classical asymptotic methods (Proctor and Jones [1988]), the new ideas are essential for a complete understanding. To press my analogy: Navier-Stokes and orthogonal decomposition gives me $F = ma$; dynamical systems methods allow me to solve the resulting ODEs.

More significantly, perhaps, the fact that heteroclinic cycles, corresponding to bursting processes, arise naturally in a model describing interactions among the most energetic inner structures under the influence of the outer layer suggests that a crucial part of the basic physics has been captured. Since each component of the mathematical model has its physical counterpart and the mathematical structures can be analyzed fairly completely, I feel that I, at least, have a better understanding of nonlinear interactions in the boundary layer.

9. Conclusion

In this paper I have outlined some ideas and methods from dynamical systems theory and indicated how basically "low dimensional" methods might be applied in concert with the proper orthogonal decomposition to the study of a class of turbulent flows possessing coherent structures. I hasten to say that this is only *one* way in which dynamical systems methods can be applied; much more direct applications to closed flow systems exist, as I have indicated in Section 3, and many other applications show promise, as I remarked in passing in Section 2. However, in describing how three different sets of ideas—coherent structures, proper orthogonal decomposition and dynamical systems methods—can be brought to bear on a problem, I have tried to lay out an *exemplar* of the way in which I anticipate that new ideas might usefully be introduced into turbulence studies. The moral is that new methods must be used with care and sometimes in unexpected combinations; one cannot simply invade another field and score instant success with one's own preconceptions intact. For example, in the identification of randomly perturbed heteroclinic cycles with the bursting process in the boundary layer, strange attractors did not play a rôle (although, as I indicated, they may be important for the model in other parameter régimes). One must let the modeling process and the model itself dictate the pieces of the abstract theory to be used.

My bias has been Newtonian; I find theories which involve a model based on rational mechanics and clear analogies between (mathematical) aspects of the model and (physical) features of the flow much more satisfying than Ptolemaic black boxes. It is easy to iterate a one dimensional map and produce "intermittent" or "turbulent" time series which "compare closely" with hot wire signals from a boundary layer. It is easy to construct arbitrary lattices of coupled quadratic maps which exhibit "spatio-temporal chaos" and "convective instability" (e.g., Kaneko [1988]). It is much harder to provide a rational connection which "explains" the physics involved via the model. In this connection, I regard the original suggestion of Ruelle and Takens [1971] that strange attractors might describe turbulence, or more recent suggestions (e.g., Crutchfield and Kaneko [1988]) that they do not and all is transient, primarily as stimuli to encourage serious studies of specific problems. I hope that the ideas and work outlined here provide not only a stimulus, but suggest some useful directions such work might follow.

Acknowledgments

I would like to thank the Sherman Fairchild Foundation and the California Institute of Technology for the opportunity for development of these reflections, and the Office of Naval Research and the Air Force Office of Scientific Research for providing support under SRO N00014-85-K-0172 and AFOSR-89-0226 for some of the work on which I have reflected. Ron Blackwelder, Donald Coles, Anatol Roshko and George Sell supplied many references and gave me the benefit of their advice. Stephen Wiggins read a draft of this manuscript and provided many helpful comments.

References

K. T. Alligood, E. D. Yorke and J. A. Yorke [1987] *Physica*, **28D**, 197–205. Why period doubling cascades occur: periodic orbit creation follows by stability shedding.

C. D. Andereck, S. S. Liu and H. L. Swinney [1986] *J. Fluid Mech.*, **164**, 155–183. Flow regimes in a circular Couette system with independently rotating cylinders.

H. Aref [1984] *J. Fluid Mech.*, **143**, 1–21. Stirring by chaotic advection.

H. Aref and S. Balachandar [1986] *Phys. Fluids.*, **29**, 3515–3521. Chaotic advection in a Stokes flow.

D. Armbruster, J. Guckenheimer and P. Holmes [1988] *Physica*, **29D**, 257–282. Heteroclinic cycles and modulated travelling waves in systems with O(2) symmetry.

V. I. Arnold [1973] *Ordinary Differential Equations*, M.I.T. Press, Cambridge, MA. (Russian original, Moscow, 1971)

V. I. Arnold [1983] *Geometrical Methods in the Theory of Ordinary Differential Equations*, Springer Verlag.

N. Aubry [1987] *A Dynamical System Coherent Structure Approach to the Fully Developed Turbulent Wall Layer*, Ph.D. thesis, Cornell University.

N. Aubry, P. Holmes, J. L. Lumley and E. Stone [1988] *J. Fluid Mech.*, **192**, 115–173. The dynamics of coherent structures in the wall region of a turbulent boundary layer.

N. Aubry, P. Holmes and J. L. Lumley [1989] (submitted for publication). The effect of drag reduction on the wall region.

A. V. Babin and M. I. Vishik [1983] *Russ. Math. Surveys*, **38**, 151–213. Attractors of partial differential equations and estimates of their dimension.

P. Bergé, Y. Pomean and Ch. Vidal [1987] *Order Within Chaos*, Wiley, Chichester. (French original, 1984)

G. S. Bhat, R. Narasimha and S. Wiggins [1989] (submitted for publication). A dynamical system that mimics shear flow turbulence.

G. D. Birkhoff [1927] *Dynamical Systems*, Amer. Math. for Colloq. Publ., **9**, Providence, RI. (Reissued 1966, 1981)

R. F. Blackwelder [1989]. AIAA paper #89-1009, AIAA 2nd Shear Flow Conference, Some ideas on the control of near wall eddies.

R. F. Blackwelder and R. E. Kaplan [1976] *J. Fluid Mech.*, **76**, 89–112. On the wall structure of the turbulent boundary layer.

A. M. Bloch and J. E. Marsden [1989] (in preparation). Controlling homoclinic orbits.

D. G. Bogard and W. G. Tiedermann [1986] *J. Fluid Mech.*, **162**. 389–413. Burst detection with single point velocity measurements.

G. L. Brown and A. Roshko [1974] *J. Fluid Mech.*, 64, 775–816. On density effects and large structure in turbulent mixing layers.

F. M. Busse and K. E. Heikes [1980] *Science*, **208**, 173–175. Convection in a rotating layer: a simple case of turbulence.

D. K. Campbell [1987] *Los Alamos Science*, Special Issue, **15**, 218–262. Nonlinear Science, from paradigms to practicalities.

D. K. Campbell and H. Rose [1983] (eds.) *Order in Chaos*, North Holland, Amsterdam.

B. J. Cantwell [1987] *Ann. Rev. Fluid Mech.*, **13**, 457–517. Organized motion in turbulent flow.

J. Carr [1981] *Applications of Center Manifold Theory*, Springer Verlag, New York, Heidelberg, Berlin.

D. H. Chambers, R. J. Adrian, P. Moin, D. S. Stewart and H. J. Sung [1988] *Phys. Fluids*, **31**, 2573–2582. Karhunen-Loève expansion of Burger's model of turbulence.

S. Chandrasekhar [1961] *Hydrodynamic and Hydromagnetic Stability*, Oxford University Press, Oxford.

D. Coles [1981] *Proc. Indian Acad. Sci. (Eng. Sci.)*, **4**, 111–127. Prospects for useful research on coherent structure in turbulent shear flow.

P. Collet, J. P. Eckmann and H. Koch [1981] *J. Stat. Phys.*, **25**(1), 1–14. Period doubling bifurcations for families of maps on $\mathbb{R}^n$.

P. Constantin, C. Foias, O. Manley and R. Temam [1985a] *J. Fluid Mech.*, **150**, 427–440. Determining modes and fractal dimension of turbulent flows.

P. Constantin, C. Foias and R. Temam [1985b] *Attractors Respresenting Rubulent Flows*, memoirs of the AMS, **53**, No. 314.

P. Constantin, C. Foias, R. Temam and B. Nicolaenko [1989] *Integral Manifolds and Inertial Manifolds for Dissipative Partial Differential Equations*, Springer Verlag, New York, Heidelberg, Berlin.

J. P. Crutchfield and K. Kaneko [1988] *Phys. Rev. Lett.*, **60**, 2715–2718. Are attractors relevant to turbulence?

E. J. Doedel and J. P. Kernévez [1986] *Applied Math. Rep.*, California Institute of Technology. Auto: software for continuation problems in ordinary differential equations.

C. R. Doering, J. D. Gibbon, D. Holm and B. Nicolaenko [1988] *Nonlinearity*, **1**, 279–309. Low dimensional behavior in the complex Ginzburg-Landau equation.

P. G. Drazin and W. M. Reid [1981] *Hydrodynamic Stability*, Cambridge University Press, Cambridge, U.K.

H. L. Dryden [1948] *Adv. Appl. Mech.*, **1**, 1–40. Recent advances in the mechanics of boundary layer flow.

M. J. Feigenbaum [1978] *J. Stat. Phys.*, **19**, 25–52. Quantitative universality for a class of nonlinear transformations.

C. Foias, B. Nicolaenko, G. R. Sell and R. Temam [1988] *J. Math. Pures. Appl*, **67**, 197–225. Inertial manifolds for the Kuramoto-Sivashinsky equation and an estimate of their lowest dimension.

J. M. Ghidaglia [1986] *SIAM J. on Math. Anal.*, **17**, 1139–1157. On the fractal dimension of attractors for viscous imcompressible fluid flows.

J. Gleick [1987] *Chaos: Making a New Science*, Viking.

A. Glezer and A. Perlstein [1988] Personal communication.

M. Golubitsky and J. Guckenheimer [1986] (eds.) *Multiparameter Bifurcation Theory*, AMS Contemporary Mathematics, **56**, Providence, RI.

M. Golubitsky and W. F. Langford [1988] *Physica*, **32D**, 362–392. Pattern formation and bistability in flow between counterrotating cylinders.

M. Golubitsky and D. G. Schaeffer [1985] *Singularities and Groups in Bifurcation Theory, I*, Springer Verlag, New York.

M. Golubitsky and I. N. Stewart [1986] *SIAM J. Math. Anal.*, **17**, 249–288. Symmetry and stability in Taylor-Couette flow.

M. Golubitsky, I. Stewart and D. G. Schaeffer [1988] *Singularities and Groups in Bifurcation Theory, II*, Springer Verlag, New York.

M. Gorman, P. J. Widmann and K. A. Robbin [1986] *Physica*, **19D**, 255–267. Nonlinear dynamics of a convection loop: a quantitative comparison of experiment with theory.

C. Grebogi, E. Ott and J. A. Yorke [1985] *Physica*, **15D**, 354–374. Attractors on an n-torus: quasiperiodicity versus chaos.

J. Guckenheimer [1976] In *The Hopf Bifurcation and its Applications*, J. E. Marsden and M. McCracken (eds.), 368–381, Springer Verlag, Berlin. A strange strange attractor.

J. Guckenheimer and P. J. Holmes [1983] *Nonlinear Oscillations, Dynamical Systems and Bifurcations of Vector Fields*, Springer Verlag. (Corrected second printing, 1986)

J. Guckenheimer and P. J. Holmes [1988] *Math. Proc. Camb. Phil. Soc.*, **103**, 189–192. Structurally stable heteroclinic cycles.

J. K. Hale [1969] *Ordinary Differential Equations*, Wiley, New York.

S. Herzog [1986] *The Large Scale Structure in the Near-Wall Region of Turbulent Pipe Flow*, Ph.D. thesis, Cornell University.

P. J. Holmes [1984] *Phys. Lett.*, **104A**, 299–302. Bifurcation sequences in horseshoe maps: infinitely many routes to chaos.

P. J. Holmes and J. E. Marsden [1981] *Arch. for Rational Mech. and Anal.*, **76**, 135–166. A partial differential equation with infinitely many periodic orbits: chaotic oscillations of a forced beam.

P. J. Holmes, J. E. Marsden and J. Scheurle [1989] *Cont. Math.*, **81**, 213–244, *Hamiltonian Dynamical Systems*, K. R. Meyer and D. G. Saari (eds.). Exponentially small splittings of separatrices, with applications to KAM theory and degenerate bifurcations.

P. J. Holmes and D. C. Whitley [1984] *Phil. Trans. Roy. Soc. Lond.*, **A311**, 43–102. Bifurcations of one- and two-dimensional maps.

E. Hopf [1948] *Comm. Pure Appl. Math.*, **1**, 303–322. A mathematical example displaying the features of turbulence.

G. Iooss, P. Coullet and Y. Demay [1986] *Preprint #89, Mathématiques, Université le Nice*, Large scale modulations in the Taylor-Coutte problem with counterrotating cylinders.

C. Jones and M. R. Proctor [1987] *Phys. Lett.*, **A**, **121**, 224–227. Strong spatial resonances and travelling waves in Bénard convection.

K. Kaneko [1988] *Europhys. Lett.*, **6** (3), 193–199. Chaotic diffusion of localized turbulent defect and pattern selection in spatio temporal chaos.

J. L. Kaplan and J. A. Yorke [1979] *Comm. Math. Phys.*, **67**, 93–108. Preturbulence, a regime observed in a fluid flow model of Lorenz.

L. Keefe, P. Moin and J. Kim [1987] *Bull. Am. Phys. Soc.*, **32**, 2026. The dimension of an attractor in turbulent Poiseuille flow.

D. V. Khakhar, J. G. Franjione and J. M. Ottino [1987] *Chem. Eng. Sci.*, **42**, 2909–2926. A case study of chaotic mixing in deterministic flows: the partitioned pipe mixer.

H. T. Kim, S. J. Kline and W. C. Reynolds [1971] *J. Fluid Mech.*, **50**, 133–160. The production of turbulence near a smooth wall in a turbulent boundary layer.

J. Kim and P. Moin [1986a] *J. Fluid Mech.*, **162**, 339–363. The structure of the vorticity field in turbulent channel flow, part II: study of ensemble averaged fields.

J. Kim and P. Moin [1986b] *Bull. A.P.S.*, **31** (10), 1716. Flow structures responsible for the bursting process.

S. J. Kline [1967] In *Fluid Mech. of Internal Flow*, G. Sovran (ed.), Elsevier, 27–68. Observed structure features in turbulent and transitional boundary layers.

S. J. Kline [1978] In *Coherent Structure of Turbulent Boundary Layers*, Proc. AFOSR/Lehigh Workshop, C. R. Smith and D. E. Abbott (eds.), 1–26. The role of visualization in the study of the turbulent boundary layer.

S. J. Kline, W. C. Reynolds, F. A. Schraub and P. W. Rundstadler [1967] *J. Fluid Mech.*, **30**, 741–773. The structure of turbulent boundary layers.

O. A. Ladyzhenkaya [1969] *The Mathematical Theory of Viscous Incompressible Flow*, Gordon and Breach, New York.

L. D. Landau and E. M. Lifshitz [1959] *Fluid Mechanics*, Addison Wesley.

J. Laufer [1975] *Ann. Rev. Fluid. Mech.*, **7**, 307–326. New trends in experimental turbulence research.

P. Laure and Y. Demay [1987] *Preprint #120, Mathématiques, Université le Nice*, Symbolic computation and the equation on the center manifold: application to the Taylor-Couette problem.

H. W. Liepmann [1979] *Am. Scientist*, **67**, 221–228. The rise and fall of ideas in turbulence.

M. Loève [1955] *Probability Theory*, Van Nostrand.

E. N. Lorenz [1963] *J. Atmos. Sciences*, **20**, 130–141. Deterministic non-periodic flow.

J. L. Lumley [1967] In *Atmospheric Turbulence and Radio Wave Propagation*, A. M. Yaglom and V. I. Tatarski (eds.), Nauka, Moscow, 166–178. The structure of inhomogeneous turbulent flows.

J. L. Lumley [1970] *Stochastic Tools in Turbulence*, Academic Press.

J. L. Lumley and I. Kubo [1984] In *The Influence of Polymer Additives on Velocity and Temperature Fields*, IUTAM Symposium, Essen, B. Gampert (ed.), Springer Verlag. Turbulent drag reduction by polymer additives: a survey.

J. E. Marsden and M. McCracken [1976] *The Hopf Bifurcation and its Applications*, Springer Verlag, New York.

V. K. Melnikov [1963] *Trans. Moscow Math. Soc.*, **12**, 1–57. On the stability of the center for time periodic perturbations.

C. Meneveau and K. R. Sreenivasan [1987] *Phys. Rev. Lett.*, **59**, 1424–1427. Simple multifractal cascade model for fully developed turbulence.

P. Moin [1984] *AIAA 22nd Aerospace Sciences Meeting*. Probing turbulence via large eddy simulation.

A. S. Monin [1978] *Sov. Phys. Usp.*, **21**, 429–442. On the nature of turbulence.

T. Mullin, S. J. Taverner and K. A. Cliffe [1989a] *Europhys. Lett.*, **8** (3), 251–256. An experimental and numerical study of a codimension-2 bifurcation in a rotating annulus.

T. Mullin and A. G. Darbyshire [1989b] *Europhys. Lett.*, (submitted). Intermittency in a rotating annular flow.

A. C. Newell, D. A. Rand and D. Russell [1988a] *Phys. Lett., A.*, **132**, 112–123. Turbulent dissipation rates and the random occurrence of coherent events.

A. C. Newell, D. A. Rand and D. Russell [1988b]. *Physica*, **33D**, 281–303. Turbulent transport and the random occurrence of coherent events.

S. Newhouse, D. Ruelle and F. Takens [1978], *Comm. Math. Phys.*, **64**, 35–40. Occurrence of strange axiom A attractors near quasi periodic flows on T^m; $m \geq 3$.

D. J. Olinger and K. R. Sreenivasan [1988] *Phys. Rev. Lett.*, **60**, 797–800. Nonlinear dynamics of the wake of an oscillating cylinder.

S. A. Orszag and A. T. Patera [1983] *J. Fluid. Mech.*, **128**, 347–385. Secondary instability of wall bounded shear flows.

J. M. Ottino, C. W. Leong, H. Rising and P. D. Swansen [1988] *Nature*, **333**, 419–425. Morphological structures produced by mixing in chaotic flows.

A. E. Perry and B. D. Fairlie [1973] *Adv. Geophys.*, **18B**, 299–315. Critical points in flow patterns.

V. A. Pliss [1964] *Izv. Akad. Nauk. SSSR. Math. Ser.*, **28**, 1297–1324. A reduction principle in the theory of the stability of motion.

H. Poincaré [1880–1890] *Mémoire sure les cóurbes définies par les équations différentielles I–IV*, (Oeuvres I), Gauthier Villars, Paris.

H. Poincaré [1890] *Acta. Math.*, **13**, 1–270. Sur le problème des trois corps et les équations de la dynamique.

H. Poincaré [1899] *Les Methodes Nouvelles de la Mécanique Celeste*, (3 Vols.), Gauthier-Villars, Paris.

Y. Pomeau and P. Manneville [1980] *Commun. Math. Phys.*, **74**, 189–197. Intermittent transition to turbulence in dissipative dynamical system.

M. R. Proctor and C. Jones [1988] *J. Fluid Mech.*, **188**, 301–335. The interaction of two spatially resonant patterns in thermal convection I: Exact 1:2 resonance.

D. A. Rand [1982] *Arch. Rat. Mech. Anal.*, **79**, 1–37. Dynamics and symmetry: predictions for modulated waves in rotating fluids.

D. A. Rand and L. S. Young [1981] (eds.) *Dynamical Systems and Turbulence*, Springer Lecture Notes in Mathematics, **848**, Springer Verlag, New York, Heidelberg, Berlin.

R. H. Rand [1984] *Computer Algebra in Applied Mathematics: An Introduction to MACSYMA*, Research Notes in Math. 94, Pitman.

R. H. Rand and D. Armbruster [1987] *Perturbation Methods, Bifurcation Theory and Computer Algebra*, Applied Mathematical Sciences 65, Springer Verlag.

K. N. Rao, R. Narasimha and M. A. Badri Narayan [1971] *J. Fluid Mech.*, **48**, 339–352. The "bursting" phenomenon in a turbulent boundary layer.

R. C. Robinson [1989] *Nonlinearity*, (to appear). Bifurcation to a transitive attractor of Lorenz type.

V. Rom Kedar, A. Leonard and S. Wiggins [1989a] (submitted for publication). An analytical sudy of transport, mixing and chaos in an unsteady vortical flow.

V. Rom Kedar and S. Wiggins [1989b] *Arch. Rat. Mech. Anal.*, (to appear). Transport in two dimensional maps.

A. Roshko [1976] *AIAA J.*, **14**, 1349–1357. Structure of turbulent shear flows: a new look.

D. Ruelle and R. Takens [1971] *Comm. Math. Phys.*, **20**, 167–192, and **23**, 343–344. On the nature of turbulence.

G. R. Sell [1989] (submitted for publication). Approximation dynamics: hyperbolic sets and inertial manifolds.

L. Sirovich [1987a] *Q. Appl. Maths*, **45**, 561–571. Turbulence and the dynamics of coherent structures: I.

L. Sirovich [1987b] *Q. Appl. Maths*, **45**, 573–582. Turbulence and the dynamics of coherent structures: II.

L. Sirovich [1987c] *Q. Appl. Maths*, **45**, 583–590. Turbulence and the dynamics of coherent structures: III.

L. Sirovich [1988] *Brown University Center for Fluid Mechanics Report*, #88–92, to appear in *Physica D*. Chaotic dynamics of coherent structures.

L. Sirovich, M. Maxey and H. Tarman [1988] In *Turbulent Shear Flows 6*, F. Durst, et al. (eds.), Springer. Analysis of turbulent thermal convection.

L. Sirovich and J. D. Rodriguez [1987] *Phys. Lett*, **A 120**, 211–214. Coherent structures and chaos: a model problem.

L. Sirovich, N. MacGiolla Mhuiris and N. Platt [1989] (submitted). An investigation of chaotic Kolmogorov flows.

S. Smale [1963] In *Differential and Combinatorial Topology*, S. S. Cairns (ed.), 63–80, Princeton University Press, Princeton, NJ. Diffeomorphisms with many periodic points.

S. Smale [1967] *Bull. Amer. Math. Soc.*, **73**, 747–817. Differentiable dynamical systems.

C. R. Smith and S. P. Schwarz [1983] *Phys. Fluids*, **26**, 641–652. Observation of streamwise rotation in the near-wall region of a turbulent boundary layer.

C. T. Sparrow [1982] *The Lorenz Equations: Bifurcations, Chaos and Strange Attractors*, Springer Verlag.

K. R. Sreenivaran [1985] In *Frontiers of Fluid Mechanics*, S. M. Davies and J. L. Lumley (eds.), Springer Verlag. Transition and turbulence in fluid flows and low dimensional chaos.

K. R. Sreenivasan [1986] In *Dimensions and Entropies*, G. Mayer-Kress (ed.), Springer Verlag. Chaos in open flow systems.

K. R. Sreenivasan and C. Meneveau [1986] *J. Fluid Mech.*, **173**, 357–386. The fractal facets of turbulence.

K. R. Sreenivaran, A. Prabhu and R. Narasimha [1983] *J. Fluid Mech.*, **137**, 251–272. Zero crossings in turbulent signals.

K. R. Sreenivasan and R. Ramshankar [1986] *Physica*, **23D**, 246–258. Transition intermittency in open flows, and intermittency routes to chaos.

K. R Sreenivasan, R. Ramshankar and C. Meneveau [1988] *Proc. R. Soc. Lond.*, **A**, (in press). Mixing, entrainment and fractal dimensions of surfaces in turbulent flows.

E. Stone [1989] *Studies of Low Dimensional Models for the Wall Region of a Turbulent Boundary Layer*, Ph.D. thesis, Cornell University.

E. Stone and P. Holmes [1989a] *Physica*, **D**, (to appear). Noise induced intermittency in a model of a turbulent boundary layer.

E. Stone and P. Holmes [1989b] *SIAM J. on Appl. Math* (to appear). Random perturbations of heteroclinic attractors.

J. D. Swearingen and R. F. Blackwelder [1987] *J. Fluid Mech.*, **182**, 255–290. The growth and breakdown of streamwise vortices in the presence of a wall.

H. Swinney [1983] *Physica*, **7D**, 3–15. Observations of order and chaos in physical systems. (Also in Cambell and Rose [1983].)

H. Swinney and J. P. Gollub [1981] (eds.) *Hydrodynamic Instabilities and the Transition to Turbulence*, Topics in Applied Physics, **45**, Springer Verlag, New York, Heidelberg, Berlin.

R. Tagg, D. Hirst and H. L. Swinney [1988] (preprint, Univ. of Texas, Austin). Critical dynamics near the spiral-Taylor vortex codimension two point.

F. Takens [1981] In *Dynamical Systems and Turbulence*, D. A. Rand and L.-S. Young (eds.), Springer Lecture Notes in Mathematics, **898**, Springer Verlag, New York, Heidelberg, Berlin, 366–381. Detecting strange attractors in turbulence.

G. I. Taylor [1923] *Phil. Trans. R Soc. London.*, **A 223**, 289–343. Stability of a viscous liquid contained between two rotating cylinders.

R. Temam [1984] *Navier Stokes Equations: Theory and Numerical Analysis*, North Holland, Amsterdam.

R. Temam [1988] *Infinite-Dimensional Dynamical Systems in Mechanics and Physics*, Springer Verlag.

W. G. Tiedermann, T. S. Luchik and D. G. Bogard [1985] *J. Fluid Mech.*, **156**, 419–437. Wall layer structure and drag reduction.

E. S. Titi [1988] *MSI Technical Report 86-119*, Cornell University. On approximate inertial manifolds to the 2-d Navier Stokes equations.

A. A. Townsend [1956] *The Structure of Turbulent Shear Flow*, Cambridge University Press.

C. W. Van Atta [1974] *Ann. Rev. Fluid Mech.*, **6**, 75–91. Sampling techniques in turbulence measurements.

S. Wiggins [1988] *Global Bifurcations and Chaos: Analytical Methods*, Applied Mathematical Sciences 73, Springer Verlag.

W. W. Willmarth [1975] *Adv. Appl. Mech.*, **15**, 159–254. Structure of turbulence in boundary layers.

W. W. Willmarth and S. Lu [1972] *J. Fluid Mech.*, **55**, 65–92. Structure of the Reynolds stress near the wall.

Fixed Points of Turbulent Dynamical Systems and Suppression of Nonlinearity

Comment 1.

H.K. Moffatt

Department of Applied Mathematics
  and Theoretical Physics
Silver Street
Cambridge, U.K.

1. Introduction

I see my task as first to comment on the approach to turbulence outlined by Philip Holmes, and secondly to broaden the discussion, to introduce some complementary ideas, and perhaps to be a bit provocative at the same time.

What Holmes has described is a decomposition of the turbulent velocity field in a statistically stationary but inhomogeneous flow in the form

$$\mathbf{u}(\mathbf{x},t) \; = \; \mathbf{U}(\mathbf{x}) \; + \; \sum_{\lambda} a_{\lambda}(t)\mathbf{u}_{\lambda}(\mathbf{x}), \tag{1}$$

where $\mathbf{U}(\mathbf{x})$ is the mean flow and the $\mathbf{u}_{\lambda}(\mathbf{x})$ are a set of structures, the eigenfunctions of the two-point velocity correlation tensor, with the convenient property that the energy of the turbulence is largely concentrated in a small number of leading terms of the series (1). By their construction, the $\mathbf{u}_{\lambda}(\mathbf{x})$ are orthogonal in the sense that

$$\int \mathbf{u}_{\lambda}(\mathbf{x}) \, \cdot \, \mathbf{u}_{\lambda'}(\mathbf{x})d^{3}\mathbf{x} \; = \; \delta_{\lambda\lambda'}, \tag{2}$$

so that substitution of (1) in the Navier-Stokes equations, followed by Galerkin projection onto each $\mathbf{u}_{\lambda}(\mathbf{x})$ in turn, leads to a set of equations for the amplitudes $a_{\lambda}(t)$ of the form

$$\frac{da_{\lambda}}{dt} \; = \; b_{\lambda\mu}\, a_{\mu} \; + \; c_{\lambda\mu\nu}\, a_{\mu}\, a_{\nu} \; + \; d_{\lambda\mu\nu\sigma}\, a_{\mu}\, a_{\nu}\, a_{\sigma}. \tag{3}$$

Here, the linear term $b_{\lambda\mu}\, a_{\mu}$ arises from viscous damping and from a primary part of the interaction of the mean flow with the turbulence; the quadratic term $c_{\lambda\mu\nu}\, a_{\mu}\, a_{\nu}$ arises primarily from the quadratic nonlinearity of the Navier-Stokes equation; and the cubic term $d_{\lambda\mu\nu\sigma}\, a_{\mu}\, a_{\nu}\, a_{\sigma}$ arises from the interaction with the turbulence of the part of the mean flow that is driven by quadratic Reynolds stresses. The set of equations (3) is truncated at a finite level, and the neglected (or 'unresolved') terms are represented by an eddy viscosity $\alpha\nu_{T}$, where α is a dimensionless parameter of order unity. The set of equations (3) then constitutes a dynamical system of order N (the level of truncation) containing a parameter α which can be varied (within reason !). A bridge is thus constructed between the Navier-Stokes equations and the theory of dynamical systems, from which a rich harvest of nonlinear phenomena may be expected, and is indeed found.

The procedure appears, in the abstract, to be extremely attractive and to hold great potential as a tool for investigation of the interaction of characteristic structures, not only for flows which are inhomogeneous with respect to only one space-coordinate, but for more general situations - e.g. turbulent flow in a pipe of varying cross-section. The procedure is a general one, but construction of the $\mathbf{u}_\lambda(\mathbf{x})$ does require detailed input concerning the measured correlation tensor for a given geometry which may take months, if not years, of painstaking experimental effort, to accumulate. Moreover the minimum reasonable order of truncation N is likely to rise rapidly with geometrical complexity, so that a procedure that appears attractive in the abstract may turn out to be prohibitively cumbersome in practice; there are already signs that this is so even for the standard turbulent channel or pipe flow problems.

2.   The quasi-two-dimensional truncation

When the flow is statistically homogeneous in the streamwise (x) and spanwise (z) directions, the structure functions $\mathbf{u}_\lambda(\mathbf{x})$ take the form

$$\mathbf{u}_\lambda(\mathbf{x}) \; = \; e^{i(k_1 x \, + \, k_3 z)} \, \mathbf{u}\left(k_1, \; k_3, \; n, \; y\right), \tag{4}$$

where λ now represents the triple $(k_1, \; k_3, \; n)$, and $n \; = \; 1, \; 2, \; 3, ...$ The particular truncation described by Holmes, whose consequences have been explored by Aubry *et al* (1988), retains only structures for which $k_1 \; = \; 0$, $n \; = \; 1$ and $k_3 \; = \; mk$, $m \; = \; \pm 1, \pm 2$, ..., ± 5. Within the limits of this truncation, the velocity field (1) has the two-dimensional form

$$\mathbf{v}(\mathbf{x}, t) \; = \; \mathbf{U}(y, \; t) \; + \; \mathbf{u}(y, \; z, \; t) \tag{5}$$

where

$$\mathbf{u}(y, \; z, \; t) \; = \; \sum_m a_m(t) e^{imkz} \mathbf{u}^{(m)}(y). \tag{6}$$

Reality of $\mathbf{u}$ for all $(y, \; z, \; t)$ implies that

$$a_{-m}(t) \; = \; a_m^*(t) \; , \; \; \mathbf{u}^{(-m)}(y) \; = \; \mathbf{u}^{(m)*}(y), \tag{7}$$

so that the system (3) is of fifth order in complex amplitudes, i.e. of tenth order in real variables. Each structure function $e^{imkz}\mathbf{u}^{(m)}(y)$ has a non-zero x-component, as well as components in the y- and z-directions. The velocity field (5) is two-dimensional only in the restricted sense that

$$\frac{\partial \mathbf{v}}{\partial x} \; = \; 0. \tag{8}$$

Even so, there are implications that are hard to reconcile with the detailed conclusions of Aubry *et al* (1988).

For, from the incompressibility condition $\nabla \cdot \mathbf{u} = 0$, we may introduce a stream-function $\psi(y, z, t)$ for the flow in the (y, z) plane: writing $\mathbf{u} = (u, v, w)$, this is defined by

$$v = \frac{\partial \psi}{\partial z}, \quad w = -\frac{\partial \psi}{\partial y}, \tag{9}$$

and the vorticity field is given by

$$\boldsymbol{\omega} = \nabla \wedge \mathbf{u} = \left(-\nabla^2 \psi, \frac{\partial u}{\partial z}, -U'(y) - \frac{\partial u}{\partial y} \right). \tag{10}$$

With $D/Dt = \partial/\partial t + v\partial/\partial y + w\partial/\partial z$, the x-component of the Navier-Stokes equation is

$$\frac{D}{Dt}(U + u) = \nu \left(\frac{\partial^2}{\partial y^2} + \frac{\partial^2}{\partial z^2} \right)(U + u), \tag{11}$$

and the x-component of the vorticity equation is

$$\frac{D}{Dt}\omega_x = \nu \left(\frac{\partial^2}{\partial y^2} + \frac{\partial^2}{\partial z^2} \right)\omega_x. \tag{12}$$

Hence the flow in the cross-stream plane is totally decoupled from the mean flow, and evolves as for freely decaying two-dimensional turbulence. In particular, using only

$$v = 0 \ \text{ on } \ y = 0 \ \text{ and } \ pv \longrightarrow 0 \ \text{ as } \ y \longrightarrow \infty, \tag{13}$$

we find that

$$\frac{d}{dt} \int\!\!\int (v^2 + w^2)\, dy\, dz = -2\nu \int\!\!\int \omega_x^2\, dy\, dz \tag{14}$$

Hence the energy associated with the flow in the (y, z) plane necessarily decays to zero, and there is apparently no possibility of a steady state other than that in which $v = w = 0$, and consequently, from (11), $u = 0$ also.

It is hard to reconcile this elementary result with the conclusion of Aubry *et al* (1988) that, for a certain range of values of the eddy viscosity parameter α, non-trivial fixed points of the dynamical system (3) do exist, representing streamwise rolls having no variation whatsoever in the streamwise direction. There is a paradox here that is difficult to track down, because the 'one-way' coupling between the cross-stream flow (v, w) and the streamwise flow $(U + u)$, represented by equations (11) and (12) (i.e. (v, w) obviously affects $U + u$, but $U + u$ does not affect (v, w)), becomes a two-way coupling between each pair $\{u_\lambda(\mathbf{x}), \mathbf{u}_{\lambda'}(\mathbf{x})\}$ when the representation (4) is used, since each $\mathbf{u}_\lambda(\mathbf{x})$ simultaneously involves streamwise and cross-stream components.

It is obviously important to resolve this paradox before attempting to incorporate modes for which $k_1 \neq 0$. The 'invariant subspace' with $k_1 = 0$ provides, as Holmes has said, the backbone on which higher-order systems, incorporating realistic streamwise variation, must be constructed; for the reasons stated above, I am not yet convinced that this backbone is in a sufficiently sound state to support such constructions, but I hope that the paradox to which I have drawn attention here can be swiftly resolved.

3.   The role of fixed points

If the Navier-Stokes equations can be reduced, by proper orthogonal decomposition or otherwise, to a finite-dimensional dynamical system, then a battery of techniques is available, as Holmes has described, for analysis of this system. The natural first step is to locate the fixed points of the system in the N-dimensional space of the variables $a_\lambda(t)$, and to classify these as stable or unstable. If the decomposition is sound, then each such fixed point should correspond to a fixed point of the Navier-Stokes equation regarded as an evolutionary dynamical system in an infinite-dimensional space of, say, square-integrable solenoidal fields; such fixed points correspond to *steady* solutions $\mathbf{u}(\mathbf{x})$ of the Navier-Stokes equations, satisfying

$$(\mathbf{u} \wedge \boldsymbol{\omega})_S \equiv \mathbf{u} \wedge \boldsymbol{\omega} - \nabla h = -\nu \nabla \wedge \boldsymbol{\omega}, \tag{15}$$

where $\boldsymbol{\omega} = \nabla \wedge \mathbf{u}$, for some scalar field $h \left(= p/\rho + \frac{1}{2}\mathbf{u}^2\right)$ satisfying

$$\nabla^2 h = \nabla \cdot (\mathbf{u} \wedge \boldsymbol{\omega}). \tag{16}$$

We use the symbol $(..)_S$ to denote the 'solenoidal projection' of the field, obtained *via* solution of the Poisson equation (16) for h. Equations (15) and (16) must of course be coupled with non-zero boundary conditions on $\mathbf{u}$ and/or p, since otherwise there can be no non-trivial steady state.

In the case of the Euler equations ($\nu = 0$), it has been shown (Moffatt 1985) that steady solutions $\mathbf{u}^E(\mathbf{x})$ exist having arbitrarily prescribed streamline topology, these flows being characterised by subdomains $D_n (n = 1, 2, ...)$ in which streamlines are chaotic and $\boldsymbol{\omega}^E = \alpha_n \mathbf{u}^E$, (i.e. the flow is a Beltrami flow in each D_n), and by the presence of vortex sheets of finite extent, randomly distributed in the spaces between the D_n. The presence of these vortex sheets suggests that these flows will in general be unstable within the framework of the Euler equations.

If, nevertheless, the phase space trajectory representing a turbulent flow spends a large proportion of its time in a neighbourhood of such fixed points (as is not untypical behaviour for low-order dynamical systems in which heteroclinic orbits are known to play a key role) then this behaviour should be recognizable in the statistics of $\mathbf{u}$. This sort of consideration has led Kraichnan & Panda (1988) to examine the evolution of the quantities

$$J \equiv \frac{< (\mathbf{u} \wedge \boldsymbol{\omega})^2 >}{< \mathbf{u}^2 >< \boldsymbol{\omega}^2 >}, \qquad Q \equiv \frac{< (\mathbf{u} \wedge \boldsymbol{\omega})_S^2 >}{< \mathbf{u}^2 >< \boldsymbol{\omega}^2 >}, \tag{17}$$

in a direct numerical simulation of decaying isotropic turbulence, and to compare these with the values J_G, Q_G that pertain to a Gaussian velocity field with (at each t) the same velocity spectrum as the dynamically evolving field. Note that Q would be zero if the flow were a steady Euler flow and that any reduction of Q/Q_G from its initial value of unity may be interpreted in terms of an intrinsic tendency of the system to spend more time near the fixed points (in some sense) than far from them. Similarly, any reduction in J/J_G from unity not only indicates a similar 'hovering' near fixed points, but also provides an estimate of the proportion of the volume that is (typically) occupied by the Beltrami domains D_n in the Euler flows corresponding to the fixed points.

Kraichnan & Panda in fact found the interesting result (corroborated by Shtilman & Polifke 1989) that Q/Q_G decreases to 0.57 and J/J_G decreases to 0.87, the latter decrease being associated with a simultaneous *increase* of the normalised mean-square helicity to 1.20, an effect foreshadowed in previous studies (Pelz *et al* 1985, 1986; Kit *et al* 1987; Rogers & Moin 1987; Levich 1987). These values indeed suggest a significant 'Eulerization' - i.e. suppression of nonlinearity - and a relatively weak 'Beltramization' - i.e. alignment of vorticity with velocity.

Kraichnan & Panda (1988) found a similar suppression of nonlinearity in a system with quadratic nonlinearity but random coupling coefficients, in an ensemble of 1000 realizations, i.e. a system like (3), but with $d_{\lambda\mu\nu\sigma} = 0$, and conjectured that this is "a generic effect associated with broad features of the dynamics". If this is true, then it provides a glimmer of hope that techniques based on weak nonlinearity may yet have some value in systems that are ostensibly strongly nonlinear! We argue this point further in the following section.

4. Turbulence regarded as a sea of weakly interacting vortons

We here use the term 'vorton' in the sense of Moffatt (1986) to represent a vorticity structure of compact support which propagates without change of shape with its intrinsic self-induced velocity $\mathbf{U}$ relative to the ambient fluid. This is in effect a generalised vortex ring which is a steady solution of the Euler equations in a frame of reference translating with the vorton. In a frame of reference fixed relative to the fluid at infinity, the vorticity field has the form

$$\omega(\mathbf{x}, t) = \omega^V(\mathbf{x} - \mathbf{U}t),\qquad(18)$$

and if the associated velocity is $\mathbf{u}(\mathbf{x}, t)$, then

$$(\mathbf{u} \wedge \omega)_S = 0,\qquad(19)$$

where the suffix S again denotes 'solenoidal projection'.

Such vortons provide a wide family of relatively stable structures which are associated, albeit indirectly, with fixed points of the Euler dynamical system, and this suggests that they may provide a natural basis for a description of turbulence which exploits to the full any natural tendency to suppress nonlinearity. To this end, let us suppose that a turbulent velocity field $\mathbf{u}(\mathbf{x}, t)$ can be expressed as a sum of weakly interacting vortons:

$$\mathbf{u}(\mathbf{x}, t) = \sum_n \mathbf{u}^{(n)}(\mathbf{x} - \mathbf{U}_n t) + \mathbf{v}(\mathbf{x}, t)\qquad(20)$$

where each $\mathbf{u}^{(n)}$ satisfies $(\mathbf{u}^n \wedge \omega^{(n)})_S = 0$, and where $\mathbf{v}(\mathbf{x}, t)$ represents the residual velocity field resulting from the interaction of vortons. This interaction process is represented schematically in figure 1, where it is conceived as a Kelvin-Helmholtz type of instability associated with grazing incidence of vortons. Substitution of (20) in the Navier-Stokes equation gives

$$\frac{D\mathbf{v}}{Dt} = \frac{\partial \mathbf{v}}{\partial t} + \mathbf{v} \cdot \nabla \mathbf{v} = -\frac{1}{\rho}\nabla p + \sum_{n \neq m} \left(\mathbf{u}^{(n)} \wedge \boldsymbol{\omega}^{(m)}\right)_S + \nu\nabla^2\mathbf{v}. \tag{21}$$

Note that, although $\mathbf{u}^{(n)} \wedge \boldsymbol{\omega}^{(m)} \equiv 0$ outside the support $D^{(m)}$ of $\boldsymbol{\omega}^{(m)}$, $\left(\mathbf{u}^{(n)} \wedge \boldsymbol{\omega}^{(m)}\right)_S$ includes a pressure contribution which in fact falls off as r^{-4} with distance from $D^{(m)}$.

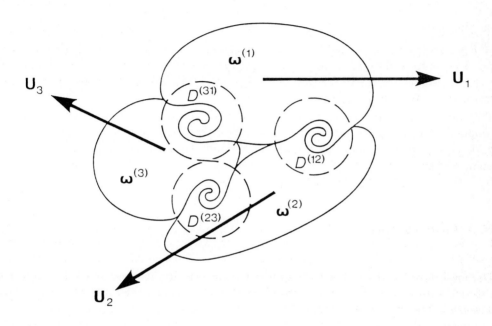

Figure 1. Schematic representation of the interaction of vortons $\omega^{(1)}$, $\omega^{(2)}$, $\omega^{(3)}$ and the production of offspring vortons by Kelvin-Helmholtz instability in the interaction domains $D^{(23)}, D^{(31)}, D^{(12)}$.

As the perturbation $\mathbf{v}$ grows on the smaller scale of effective vorton interaction, energy is of course extracted from the parent vortons which may either adjust in quasi-steady manner, or may after several such collision processes be destroyed. The field $\mathbf{v}(\mathbf{x}, t)$ may be expected to restructure itself as a sum of 'offspring' vortons, convected with the local velocity $\bar{\mathbf{u}}$ associated with larger scales of motion, i.e.

$$\mathbf{v}(\mathbf{x}, t) = \sum_{n} \mathbf{v}^{(n)}\left(\mathbf{x} - (\mathbf{V}_n + \bar{\mathbf{u}})t\right) + \mathbf{w}(\mathbf{x}, t) \tag{22}$$

and the process may now continue, with developing intermittency as the cascade to smaller scale vortons proceeds.

The process envisaged here is similar in spirit to that envisaged by Frisch, Sulem & Nelkin (1978) in their proposed 'β-model' of turbulence, but with the additional dynamical feature that at each length scale, solutions of the Euler equations provide the reference point for the next step (vorton interaction) of the cascade process.

The representation (20) provides an appropriate framework for understanding the suppression of Q (eqn. 17) relative to the Gaussian value Q_G. Neglecting $\mathbf{v}$, we have

$$(\mathbf{u} \wedge \boldsymbol{\omega})_S = \sum_{n \neq m} \left(\mathbf{u}^{(n)} \wedge \boldsymbol{\omega}^{(m)} \right)_S \tag{23}$$

and the (n, m) term in the sum is non-zero only in the domain $D^{(nm)}$ of interaction of the vortons $\boldsymbol{\omega}^{(n)}$ and $\boldsymbol{\omega}^{(m)}$. If we suppose further that these domains do not overlap then

$$< (\mathbf{u} \wedge \boldsymbol{\omega})_s^2 > = < \sum_{n \neq m} \left(\mathbf{u}^{(n)} \wedge \boldsymbol{\omega}^{(m)} \right)_S^2 > . \tag{24}$$

This may be expected to be a factor q less than the Gaussian value, where q is the proportion of the fluid volume in which significant vorton interactions occur. The value $q = Q/Q_G \approx 0.57$ found by Kraichnan & Panda (1988) is not implausible from this point of view.

5. Conclusion

Decompositions such as (1) or (20) of a turbulent velocity field seem to hold promise in capturing the dynamics of long-lived coherent structures and their interactions. In both cases, the fixed points of the underlying dynamical systems play an important role both in understanding the extent to which nonlinearity may be suppressed, and as a starting point for analysis of nonlinear interactions between the basic structures. The study of Aubry *et al* (1988) provides a valuable starting point, and the computations and analysis initiated by these authors now need to be further developed and extended, and reconciled with more primitive considerations concerning two-dimensional turbulence. The concept of interacting vortons, and the associated suppression of $< (\mathbf{u} \wedge \boldsymbol{\omega})_S^2 >$ and (weak) relative enhancement of mean-square helicity $< (\mathbf{u} \cdot \boldsymbol{\omega})^2 >$, have already stimulated a number of experiments, both real-fluid and numerical-simulation. This is an area of intense current interest, in which further rapid developments may be expected.

REFERENCES

1.   Aubry N., Holmes P., Lumley J.L. & Stone E. (1988) The dynamics of coherent structures in the wall region of a turbulent boundary layer. J. Fluid Mech. **192**, 115-173.
2.   Frisch U., Sulem P.L. & Nelkin M. (1978) A simple dynamical model of intermittent fully developed turbulence. J. Fluid Mech. **87**, 719-736.
3.   Holmes P. (1989) Can dynamical systems approach turbulence? This vol.
4.   Kit E., Tsinober A., Balint L., Wallace J.M. & Levich E. (1987)" An experimental study of helicity related properties of a turbulent flow past a grid. Phys. Fluids **30**, 3323-3325.
5.   Kraichnan R.H. & Panda R. (1988) Depression of nonlinearity in decaying isotropic turbulence. Phys. Fluids **31**, 2395-2397.
6.   Levich E. (1987) Certain problems in the theory of developed hydrodynamical turbulence. Phys. Rep. **151**, 129-238.
7.   Moffatt H.K. (1986) On the existence of localized rotational disturbances which propagate without change of structure in an inviscid fluid. J. Fluid Mech. **173**, 289-302.
8.   Pelz R.B., Yakhot V., Orszag S.A., Shtilman L. & Levich E. (1985)" Velocity-vorticity patterns in turbulent flow. Phys. Rev. Lett. **54**, 2505-2508.
9.   Pelz R.B., Shtilman L. & Tsinober A. (1986) The helical nature of unforced turbulent flows. Phys. Fluids **29**, 3506-3600.
10.   Rogers M.M. & Moin P. (1987) Helicity fluctuations in incompressible turbulent flows. Phys. Fluids **30**, 2662-2671.
11.   Shtilman L. & Polifke W. (1989) On the mechanism of the reduction of nonlinearity in the incompressible Navier-Stokes equation. Phys. Fluids **00**, 0000-0000.

Chaotic Fluid Dynamics and Turbulent Flow

Comment 2.

Hassan Aref

Institute of Geophysics and Planetary Physics
University of California, San Diego
LaJolla, CA 92093-0225

and

San Diego Supercomputer Center
P.O. Box 85608
San Diego, CA 92138-5608, USA

Introduction

My task at this meeting, as I understand it, is to "discuss" the paper by Holmes *Can Dynamical Systems Approach Turbulence?* The bulk of Holmes' paper is in the nature of a review of an ongoing investigation of the transition to turbulence in a boundary layer using a combination of "Lumley's orthogonal decomposition" and "chaos theory."[1] This research is being published in premier journals, and has been the object of several lectures at national and international conferences. The remainder of Holmes' paper reviews salient facts about bifurcations and chaos, most of which are hardly controversial. Since I am neither an authority on boundary layers nor on the mathematical theory of chaos, there is really little for me to take issue with in Holmes' paper.

Thus, I shall use my window of space-time at this workshop to make a few general remarks on the epistemological issues of chaos versus turbulence rather than delve into details of one particular application. I shall stress the successes that can legitimately be credited to the "chaos theory" point of view, suggest areas in which this approach will yield further advances, but - and this may really be the essence of my comments - say relatively little about turbulence proper! The basic content of what follows is (1) that chaos is not turbulence; (2) that there are very interesting, sometimes novel areas of fluid mechanics that can profitably use the ideas of chaotic dynamics; and (3) that chaos may have something to say about turbulence, but whatever this is it belongs to a large extent in the future of the subject.

Since chaos from the point of view of turbulence is, in my opinion, still in the "Dark Middle Ages," I shall follow a stylistic favorite of that epoch and cast my message as a series of "theses" each treated in a separate subsection:

Thesis #1: Navier-Stokes turbulence is chaotic

We do not, unfortunately, fully understand what this plausible statement means, nor

---

[1]) I place the first item in quotes because it has a well known composite background and some readers may prefer to call it by another name. I place "chaos theory" in quotes because there hardly is such a thing, as Holmes points out. Nevertheless, I use the term in this paper to designate an emerging body of results.

whether it is a terribly useful characterization. However, most would agree that the three-dimensional Navier-Stokes equation is not integrable in the way the Korteweg-deVries or nonlinear Schrödinger equation is, and although coherent phenomena do appear in the solutions of the 3D Navier-Stokes equation, both computationally and observationally, they are not sufficiently long-lived or stable so that a complete description can be based upon them. It is unclear at present which avenue is the most productive: to focus on (A) the (primarily large-scale) coherent component of a given turbulent flow, to focus on (B) the chaotic "phase jitter" component, or to focus on (C) the fully stochastic (primarily small-scale) component using from the outset a statistical description. It is likely that all three aspects will be important in any ultimate description, but that the relative importance will vary from flow to flow[2]. For example, in a mixing layer it is evident, based on the comprehensive work of Roshko, Browand and others [1], that (A) and (B) go a long way towards a comprehensive description. On the other hand, in grid turbulence (C), as introduced long ago by Taylor, Kolmogorov, and others, seems the prevalent and preferred mode of description.

An important corollary of this thesis is that there is no need to introduce extraneous agents of noise or stochasticity to explain turbulence. The deterministic viewpoint implicit in "chaos theory" says that just as flow instability phenomena can be handled within a deterministic framework so too can turbulence. This is a very important step (if true) in the development of our understanding of turbulence, and I suspect it would have been considered at least a qualitative solution to "the turbulence problem" by many of the grand old men in the field. Indeed, the basic notion of stability theory (as distinct from its many technical details) for a while was hailed as just such a solution, but of course leads to pictures of the onset of turbulence that are today considered incomplete. The fact that it is unnecessary to include extraneous noise in discussing turbulence does not imply that it may not in some cases (for example, in numerical simulation studies) be both expedient and desirable to do so.

There is a long-standing controversy regarding two-dimensional turbulence. Many fluid mechanicians, especially in the engineering communities, consider vortex stretching to be essential for anything they will agree to call turbulence, and, of course, there is no vortex stretching in two-dimensional flow. There was also a suggestion some years ago [2] that the two-dimensional Euler equation is integrable. However, this is apparently not the case in the same sense of the soliton-bearing equations mentioned above, and it is unclear whether the formal statement amounts to anything more than the well known conservation of vorticity in 2D inviscid flow. My view is that 2D flow is chaotic (for which, in general, four interacting, concentrated vortices suffice [3,4]) and that when it encompasses sufficiently many degrees of freedom at high Reynolds number, there is no reason not to call it turbulent.

It has always seemed appropriate to compare the laminar-turbulent dichotomy to the deterministic mechanics-kinetic theory-thermodynamics contrast at the basis of most descriptions of matter at the macroscopic level. Indeed, this comparison yields a basis for introducing the idea of eddy viscosity. From the vantage point of chaos similar issues of recovering stochastic, maybe even ergodic, behavior from deterministic

---

[2]) We may compare the situation to the proverbial one of describing a Swiss cheese. One observer will focus on all the cheese, the other on all the holes!

mechanics plague the turbulence problem as have plagued the foundations of statistical mechanics for years. Although there is no complete solution that I know of, "chaos theory" is the only formulation where the question is at least posed, and the nature of the statistical behavior of turbulence is not just made an *ad hoc* assumption at the outset of the theory.

In equilibrium statistical mechanics we have, in a sense, been extremely fortunate. The theory, more often than not, works far beyond the bounds that a careful theorist concerned about the fundamental assumptions might envision. We know from experience what the role of "islands in a chaotic sea" implies for a system with many degrees of freedom in thermodynamic equilibrium, and the answer seems to be "not very much." It is much less clear what the implications of "chaos theory" are for Navier-Stokes turbulence. When we make statistical assumptions from the start in a turbulence theory, what is being ignored? Answering this kind of question will have a considerable impact on the relevance of "chaos theory" to fully developed turbulence.

Since the evidence for the correctness of this thesis is, at best, circumstantial, hearsay may have a role as well! The second edition of the influential text by Landau and Lifshitz [5] includes several subsections on "chaos theory" written jointly with M.I. Rabinovich, and describing these additions in the preface Lifshitz wrote:

"There have been important changes in our understanding of the mechanism whereby turbulence occurs. Although a consistent theory of turbulence is still a thing of the future, there is reason to suppose that the right path has finally been found."

Thesis #2: Many chaotic flow regimes are not turbulent

Most of the few degree of freedom models that have been studied, from the seminal paper by Lorenz [6] onward, pertain to such regimes. Use of the word turbulent for the flow produced by these is inaccurate and should be avoided. The realization that there are motions in deterministic mechanical systems with only a few degrees of freedom that produce stochastic outputs for some reason took quite a while to be appreciated in the context of fluid mechanical systems. Maybe this was due to the ease with which one excites many degrees of freedom in a fluid and the conviction, based on early theories of the onset of turbulence, that stochastic fluid motion somehow implies many degrees of freedom.

There do exist many-degree-of-freedom flows, such as highly ramified viscous fingering [7], that appear to be "chaotic," in the sense of having very sensitive dependence on initial conditions, and that do produce very complex spatial patterns. It would be unnatural to call these flows turbulent because the dominant dynamics is not inertial. However, such complex, multi-degree-of-freedom flows form a class that can profitably be studied in parallel with "real" turbulence, e.g., many of the statistical methods used in turbulence theory and experiment are suitable here as well. Examples that come to mind in this class include emulsions, sedimentation, and granular flows.

The initial investigations of chaotic flows concentrated on the sequence of transitions leading to turbulence. Hopes were that aspects of the fully turbulent state would be elucidated by understanding the chaotic states leading up to it. I would argue that relatively little has come of these hopes. For example, even with the detailed and unquestionably important understanding that we have of the subharmonic bifurcation route to chaos, we do not understand much more about the resulting turbulent state. The main impact of the application of "chaos theory" to fluid motions has been in the identification of a suitable language of inquiry for a number of situations that previously were immediately addressed by statistical methods. Thus, using the categories given above all turbulence was viewed from the perspective of (C), and this point of view was carried over and applied also to unsteady but essentially laminar flows.

Since most of my own efforts have been spent addressing flow regimes for which a deterministic description with the possibility of chaotic behavior appears to be the appropriate one, I may be allowed to digress here to list a few examples:

Example 1: *Chaotic advection*

The passive advection of an inert fluid particle by any flow is given by the equations

$$\dot{x} = u(x,y,z,t) \ ; \ \dot{y} = v(x,y,z,t) \ ; \ \dot{z} = w(x,y,z,t) \tag{1}$$

This dynamical system, arising from flow *kinematics,* provides a "template" for the Lagrangian motion of individual fluid particles, even in cases where (1) does not describe the motion of the physical particles of interest. Effects associated with particle size, inhomogeneity of physical properties of the fluid, etc., may lead to (1) not being the equation of motion of the fluid particles under consideration. For example, (1) would not be appropriate for following small air bubbles if their buoyancy plays a role.

Clearly Eqs.(1) are rich enough, even for laminar flows, to produce chaotic trajectories. This is the phenomenon of *chaotic advection,* already analyzed in the context of steady three-dimensional flow by Arnold and Hénon [8], on which a considerable and diverse literature has recently been generated[3]. An incomplete list of this work is given in references 14-33.

Important progenitors to this work, that outlined the changes in topology that come about by bifurcations in the streamline patterns of steady flows but did not consider the possibility of chaos, are the papers by Perry & Fairlie [34] and Cantwell [1, 35]. We may in this context also cite the papers by Hama [36] and Williams & Hama [37] on the counter-intuitive nature of pathlines and streaklines in unsteady flows.

It is at once encouraging and depressing that even laminar flow kinematics demands the full machinery of "chaos theory." It is encouraging since many workers in fluid mechanics, in particular chemical engineers working on low Reynolds number problems, know that considerable complexity can ensue in these. Chaotic advection has provided a fresh new component to more traditional descriptions, based among other things on the general theory of continuum mechanics (for a comprehensive exposition

---

[3]) Some writers have used the names "Lagrangian turbulence" [9], "chaotic mixing" [10], "chaotic convection" [11], etc., for this phenomenon. Regarding the first of these it is, as Drazin [12] has so eloquently put it, "felt improper to label 'turbulent' processes that manifestly are not." Regarding the second we prefer, following Eckart [13], to reserve the term *mixing* for processes that involve the molecular diffusivity. The third name causes confusion with common usage in the thermal convection literature, where "chaotic convection" means a flow that is chaotic in the Eulerian sense, e.g., a solution of Lorenz' equations beyond the threshold for chaos.

see the recent monograph [38]). This application of "chaos theory" to fluid flows has generated considerable excitement because the complexity of the system is immediately visible in real space, rather than being an abstract structure in some truncated amplitude space[4]. It is, on the other hand, depressing for someone interested in turbulent flow to realize that the kinematics of laminar advection is in Arnold's words [39] "beyond the capability of modern science."

The appearance of chaotic particle trajectories in the advection problem explains why fractal structure[5] can appear in the signature left by a passive marker. We include here both the observations of fractality in clouds [42] and tracers in shear flows [43], and the observations of individual trajectories [44]. The utility of fractal dimension studies of such signals is still a matter of debate. While the presence of fractals (in one form or another) is presumably beyond question, the ability of fractal dimension concepts to distinguish flows, indeed to distinguish laminar from turbulent agitation, is still unsettled. For example, in [33] the computed agitation of a region of fluid by a few hundred point vortices is compared to experimental images of the cross section of a round jet due to D. Liepmann and M. Gharib. The similarity of the fractal nature of the contours in these two sets of images is considerable, and the dilemma from the point of view of the utility of fractals is that the underlying dynamics differs profoundly. Generalized concepts, such as multifractals, are called for even in a phenomenological description.

This is not the place to embark on a full review. Suffice it to note (1) that recent developments in chaotic advection have expanded the range of concepts from "chaos theory" that are finding a legitimate use in fluid mechanics to include such items as basin boundaries [33] and chaotic scattering [23, 33]; (2) that exciting new avenues have opened up such as chaos in advection-diffusion problems [45, 46], the role of chaos in systems where the particles have inertia [28], and chaos in advection of interacting particles (cf. [47] where the interaction is coagulation); (3) that we are witnessing increased interest in this topic in the context of geophysical fluid dynamics [16, 22, 44, 48, 49], where a number of earlier works [13, 34, 50-52] seem to cry out for interpretation and/or generalization in these terms; (4) that the topic has important consequences for flow visualization, as the papers [1, 34-37, 53] and many others, clearly suggest[6]; and (5) that new modelling tools using global approximations in terms of mappings have been suggested and attempted [21, 55] and may well lead to further important developments.

Example 2: *Vortex dynamics*

Closely related to Example 1 is the subject of vortex dynamics, probably the most "Lagrangian" aspect of classical hydrodynamics. Indeed, the so-called "restricted four-vortex problem" [3-4, 56-57] played a significant role in formulating the notion of chaotic advection. In vortex dynamics the advection is "fed back" to the flow field via the relation that the vorticity is the curl of the velocity field. In what must qualify as the

---

[4]) Indeed, illustrations of chaotic advection appear on the covers of *Nature* and *Scientific American* in conjunction with the articles [29-31], and in *Supercomputing Review*.

[5]) See the monograph by Feder [40] for an exposition of the mathematics and physical science applications of fractals. With particular reference to fluid mechanics we note the brief review [41].

[6]) The connection between chaotic advection and flow visualization was mentioned in [14]; the role of manifold shapes is mentioned in [54].

simplest model, point vortices on the unbounded plane, chaos sets in as the number of vortices is increased from three to four (for review see [3, 4]). For bounded domains fewer vortices are required for chaotic motion, for a general domain as few as two. The one-vortex problem in a domain of any shape is integrable as shown long ago by Routh [58]; the only known boundary for which the two-vortex problem is integrable is the circle. Recent contributions to this subject include special cases for which the four-vortex problem is integrable [59], chaotic scattering of two vortex pairs [60-64], and applications of the Painlevé test to the point vortex equations [65].

Chaotic vortex motions are also present but less completely elucidated in another favorite model of two-dimensional flow: vortex patch dynamics. Here the outer portion of two interacting like-signed vortices can produce intense filamentation [66], that strongly resembles the homoclinic oscillations seen in chaotic advection studies. Filamentation is an important mechanism for vortex patches, and one can, on one hand, speculate about its interpretation as a form of modulated chaotic advection[7], and, on the other hand, think of it in terms of an inertial cascade. Related ideas appear in the recent paper by Pierrehumbert [32].

Example 3: *Pressure fluctuations and sound radiation*

Besides the Lagrangian advection signal Eulerian signals such as far-field pressure and low Mach number sound radiation are sensitive to the integrable versus chaotic nature of a concentrated vortex flow. Again the simplest model situation is the sound radiation from an assembly of point vortices, but the correspondence carries over to interacting vortex filaments in 3D. The pressure fluctuations on a rigid wall bounding a vortex flow calculated using ideal hydrodynamics (Bernoulli's equation) also mirrors the character of the vortex motion [4, 67]. These relations have been apparent for some time and have recently been pursued in more detail [33, 68]. An interesting extension is to consider the Painlevé analysis of the point vortex equations alongside the acoustic signature and the character of the vortex trajectories or Poincaré section [69].

The utility of chaos as a diagnostic in radiated sound or far-field pressure fluctuations has been considered also in the context of cavitation bubble dynamics [70]. A similar Eulerian-Lagrangian dichotomy arises for this problem since intriguing regimes of chaotic behavior have been found in regions of parameter space for the driven Rayleigh-Plesset equation describing cavitation bubble oscillations [71].

Example 4: *Flow-structure interactions*

Vortex shedding phenomena display a variety of interesting applications of "chaos theory." *A priori* one would expect that the two-dimensional motion of interacting vortices in a wake or shear layer would be sufficient for chaotic motions, and certainly model problems will yield such behavior. Nevertheless, precision numerical experiments indicate that purely two-dimensional shedding from a circular cylinder at Reynolds numbers of 40-200 does not lead to chaotic motion [72]. A small amount of periodic forcing will, however, produce spectacular instances of chaotic motion. This forcing can come about by aero-elastic coupling of the flow and vibrations of the

---

[7]) This point of view has been advocated by K. Shariff and A. Leonard in various lectures and private communications.

shedding cylinder [73], by manipulation of boundary layers on a shedding airfoil [74, 75], or by intentionally moving the shedding body [76].

The problem is extremely interesting and promises to become a paradigm for the coupling of temporal chaos and spatial structure both from the Eulerian and Lagrangian points of view. A detailed view of the full three-dimensional structure of the wake of a cylinder set vibrating by an oncoming steady flow shows an intricate interleaving of regions of regular and chaotic fluid motion [77].

Thesis #3: "Chaos theory" has stimulated model-building in fluid mechanics

This I consider to be extremely important, and the relative lack of models in classical fluid mechanics to be a serious weakness. It appears that fluid mechanicians, maybe stimulated by the strong applied mathematics component of the subject, have been much too preoccupied with solving their various equations including every last term. Thus, while perturbation techniques have been honed, physically motivated modelling has languished and has to some extent been frowned upon. The use of maps, truncations, geometrical models, etc., within general "chaos theory" has provoked fluid mechanicians to do the same. Of course, one of the prettiest models in all of physical science, the von Kármán point vortex street model, belongs to fluid mechanics[8] but there should be many more. I believe that stochastic theories are less amenable to intuitive models than deterministic theories, and so I think (and hope) that the trend to determinism fueled by "chaos theory" will pay off handsomely in this area. Indeed, I believe that we are already seeing a good deal of this stimulated in part by "chaos theory" and in part by numerical experimentation, which is often much easier to do with a truncated representation.

Models that come to mind, which have been abetted by the focus on chaotic versus integrable, include point vortices, the localized induction approximation [78, 79] and DLA [80].

Thesis #4: "Chaos theory" has produced new ways of analyzing experimental data

This is obvious from even a cursory examination of the literature, and is again a very important result. A large number of literature references could be given, including essentially every experimental study of a chaotic fluid dynamical system. Suffice it to cite the recent survey [81] where 138 further references may be found. One observes here much the same dichotomy as we have seen on the theoretical side. The signals coming from unsteady flows, including turbulent flows, have presumably not changed all that much over time. However, the attitude that the experimentalist adopts in analyzing them is altogether different depending on whether the signal is expected to come from a deterministic or stochastic process. Thus, we see Lyapunov exponent measurements, extensive use of spectra to seek out "universal" routes to turbulence, embedding of data to determine fractal dimension, etc.

---

[8]) I rank the Kármán vortex street analysis alongside the Ising model, another gem of mathematical physics.

Thesis #5: The turbulent cascade is a chaos phenomenon

This is probably my most controversial thesis. I am drawn by the apparent correspondence between the notion of a cascade in the work of Kolmogorov, Obukhov, von Weiszäcker, and Onsager, and images of island chains surrounding islands as seen in Hénon's 1969 study [82] of the general area-preserving map in the vicinity of an elliptic fixed point. Clearly in both images some kind of hierarchical structure of "eddies" is being produced.

The notion that is surfacing in these images is a relationship between integrability or chaos and spatial structure. Conceptually related studies appear in the work of Moffatt [83], as generalized recently by Gilbert [84], and in the paper by Pierrehumbert [32]. The earlier, purely "topological" introduction of "whorls" and "tendrils" by Berry et al. [85] also belongs in this category. All this applies to 2D flows where the relation between structure, chaos and turbulence is presumably much simplified relative to 3D. I do not believe that we yet understand the real space signature of a cascade, e.g. how one tells whether a constant flow of energy or enstrophy is taking place in wavenumber space, and that until we do, a deterministic understanding of how a turbulent flow works is missing. I should in this context like to call attention to the intriguing kinetic argument put forward by Novikov [86] on why energy migrates to large scales in a system of point vortices.

In 3D we are considerably further from an understanding of how turbulence works, but certain mechanisms that have been stressed in the recent literature, such as strained vortices [87], vortex tube reconnections [88, 89] and "Beltramization," at least have the right "flavor." One vexing problem is that the cascade arguments are, if nothing else, dimensionally correct (!) and so *any* mechanism should indeed lead to the same scaling. The deterministic approach must thus not only produce one or more mechanisms for turbulence, but must find indicators that will tell whether the correct one(s) have been found, and/or whether the appropriate weights have been given to competing mechanisms. The deterministic theory must explain how universal small scale behavior comes about. "Chaos theory" has, of course, yielded some instances of such "universal" behavior. Whether these play any role in fully developed turbulence is, as I see it, an open question.

It is intriguing to speculate that the reason a limit of "inviscid dissipation" occurs in turbulent flows at high Reynolds number [90] is that chaotic motion on small scales "steps in" as a substitute for molecular processes of viscosity of diffusivity.

Conclusion

In conclusion let me say that I believe chaotic behavior in fluids to be one of the most stimulating developments of the subject in our time. While I doubt it will answer all our questions about turbulence, I believe it holds the key to some of them. Maybe even more significantly, I believe that a variety of flow phenomena that hitherto were lacking an appropriate theoretical framework have found one in the mathematics of chaotic behavior, and hence will advance. As far as turbulence is concerned, chaos has taught us that the simple laminar-turbulent dichotomy is an oversimplification. As far as "chaos theory" is concerned, we are gradually leaving an initial sometimes naive era in which flow situations that conformed to some "paradigm" of the theory were sought and found (on occasion accompanied by exaggerated claims of advances in understanding). We are intrigued by pairs of images, one from a situation we know to be a simple numerical model analyzable in terms of "chaos theory," the other an experimental image that shows qualitatively similar features that we, however, do not understand in terms of the basic governing equations. We are, hopefully, entering a more mature and realistic era in which the problems and phenomena of fluid mechanics once again drive the field forward as they have done so successfully in the past.

Preparation of this paper and participation in the workshop was supported by NSF/PYI award MSM84-51107.

References

1. B.J. Cantwell: Organized motion in turbulent flow. *Ann. Rev. Fluid Mech.* **13**, 497-515 (1981).
2. D.G. Ebin: Integrability of perfect fluid motion. *Comm. Pure Appl. Math.* **36**, 37-54 (1983).
3. H. Aref: Integrable, chaotic, and turbulent vortex motion in two-dimensional flows. *Ann. Rev. Fluid Mech.* **15**, 345-389 (1983).
4. H. Aref: Chaos in the dynamics of a few vortices - fundamentals and applications. In *Theoretical and Applied Mechanics*, F. I. Niordson & N. Olhoff eds., Elsevier Science Publ. 1985, 43-68.
5. L.D. Landau & E.M. Lifshitz: *Fluid Mechanics*. Second ed., Pergamon Press, Oxford 1987, 539pp.
6. E. Lorenz: Deterministic non-periodic flow. *J. Atmos. Sci.* **20**, 130-141 (1963).
7. G.M. Homsy: Viscous fingering in porous media. *Ann. Rev. Fluid Mech.* **19**, 271-311 (1987).
8. M. Hénon: Sur la topologie des lignes courant dans un cas particulier. *C.R. Acad. Sci. Paris A* **262**, 312-314 (1966).
9. T. Dombre, U. Frisch, J.M. Greene, M. Hénon, A. Mehr & A.M. Soward: Chaotic streamlines and Lagrangian turbulence: The ABC flows. *J. Fluid Mech.* **167**, 353-391 (1986).
10. W-L. Chien, H. Rising & J.M. Ottino: Laminar mixing and chaotic mixing in several cavity flows. *J. Fluid Mech.* **170**, 355-377 (1986).
11. E. Ott & T.M. Antonsen: Chaotic fluid convection and the fractal nature of passive scalar gradients. *Phys. Rev. Lett.* **25**, 2839-2842 (1988).
12. P.G. Drazin: Lecture at Workshop on *The Lagrangian Picture of Fluid Mechanics*, University of Arizona, Tucson (1988).
13. C. Eckart: An analysis of the stirring and mixing processes in incompressible fluids. *J. Mar. Res.* **7**, 265-275 (1948).
14. H. Aref: Stirring by chaotic advection. *J. Fluid Mech.* **143**, 1-21 (1984).
15. W. Arter: Ergodic stream-lines in steady convection. *Phys. Lett. A* **97**, 171-174 (1983).
16. L.A. Smith & E.A. Spiegel: Pattern formation by particles settling in viscous fluid. *Springer Lect. Notes in Phys.* **230**, 306-318 (1985).
17. H. Aref & S. Balachandar: Chaotic advection in a Stokes flow. *Phys. Fluids* **29**, 3515-3521 (1986).
18. J. Chaiken, R. Chevray, M. Tabor & Q.M. Tan: Experimental study of Lagrangian turbulence in a Stokes flow. *Proc. R. Soc. Lond. A* **408**, 165-174 (1986).
19. D.V. Khakhar & J.M. Ottino: Fluid mixing (stretching) by time periodic sequences of weak flows. *Phys. Fluids* **29**, 3503-3505 (1986).
20. D.V. Khakhar, H. Rising & J.M. Ottino: Analysis of chaotic mixing in two model systems. *J. Fluid Mech.* **172**, 419-451 (1986). ,
21. J. Chaiken, C.K. Chu, M. Tabor & Q.M. Tan: Lagrangian turbulence and spatial complexity in a Stokes flow. *Phys. Fluids* **30**, 687-694 (1987).
22. J.T.F. Zimmerman: The tidal whirlpool: A review of horizontal dispersion by tidal and residual currents. *Netherlands J. Sea Res.* **20**, 133-154 (1986).
23. S.W. Jones, O.M. Thomas & H. Aref: Chaotic advection by laminar flow in a twisted pipe. *J. Fluid Mech.* (In Press, 1989) and *Bull. Amer. Phys. Soc.* **32**, 2026 (1987).
24. D.V. Khakhar, J.G. Franjione, & J.M. Ottino: A case study of chaotic mixing in deterministic flows: the partitioned pipe mixer. *Chem. Eng Sci.* **42**, 2909-2926 (1987).
25. E. Knobloch, & J.B. Weiss: Chaotic advection by modulated travelling waves. *Phys. Rev. A* **36**, 1522-1524 (1987).
26. S. Lichter, A. Dagan, W.B. Underhill & H. Ayanle: Mixing in a closed room by the action of two fans. *J. Appl. Mech.* (submitted, 1987).

27. S.W. Jones & H. Aref: Chaotic advection in pulsed source-sink systems. *Phys. Fluids* **31**, 469-485 (1988).
28. J.B. McLaughlin: Particle size effects on Lagrangian turbulence. *Phys. Fluids* **31**, 2544-2553 (1988).
29. J.M. Ottino, C.W. Leong, H. Rising & P.D. Swanson: Morphological structures produced by mixing in chaotic flows. *Nature* **333**, 419-425 (1988).
30. J.M. Ottino: The mixing of fluids. *Scient. Amer.* **260**, 56-67 (1989).
31. V.V. Beloshapkin, A.A. Chernikov, M.Ya. Natenzon, B.A. Petrovichev R.Z. Sagdeev & G.M. Zaslavsky: Chaotic streamlines in pre-turbulent states. *Nature* **337**, 133-137 (1989).
32. R.T. Pierrehumbert: Large eddy energy accretion by chaotic mixing of small scale vorticity. *Phys. Rev. Lett.* (submitted, 1989).
33. H. Aref, S.W. Jones, S. Mofina & I. Zawadzki: Vortices, kinematics and chaos. *Physica D* (In Press, 1989).
34. A.E. Perry & B.D. Fairlie: Critical points in flow patterns. *Adv. Geophys.* **18**, 299-315 (1974).
35. B.J. Cantwell: Transition in the axisymmetric jet. *J. Fluid Mech.* **104**, 369-386 (1981).
36. F.R. Hama: Streaklines in a perturbed shear flow. *Phys. Fluids* **5**, 644-650 (1962).
37. D.R. Williams & F.R. Hama: Streaklines in a shear layer perturbed by two waves. *Phys. Fluids* **23**, 442-447 (1980).
38. J.M. Ottino: *The Kinematics of Mixing: Stretching, Chaos and Transport.* Cambridge Univ. Press 1989, 375pp.
39. V.I. Arnold: *Mathematical Methods of Classical Mechanics.* Springer-Verlag, New York 1978, 462pp.
40. J. Feder: *Fractals.* Plenum Press, New York 1988, 283pp.
41. D.L. Turcotte: Fractals in fluid mechanics. *Ann. Rev. Fluid Mech.* **20**, 5-16 (1988).
42. S. Lovejoy: Area-perimeter relation for rain and cloud areas. *Science* **216**, 185-187 (1982).
43. K.R. Sreenivasan & C. Meneveau: The fractal facets of turbulence. *J. Fluid Mech.* **173**, 357-386 (1986).
44. A.R. Osborne, A.D. Kirwan, A. Provenzale & L. Bergamasco: A search for chaotic behavior in large and mesoscale motions in the Pacific Ocean. *Physica D* **23**, 75-83 (1986).
45. H. Aref & S.W. Jones: Enhanced separation of diffusing particles by chaotic advection. *Phys. Fluids A* **1**, 470-474 (1989).
46. S.W. Jones: Shear dispersion and anomalous diffusion in a chaotic flow. *J. Fluid Mech.* (submitted, 1989)
47. F.J. Muzzio & J.M. Ottino: Coagulation in chaotic flows. *Phys. Rev. A* **38**, 2516-2524 (1988).
48. M. Falcioni, G. Paladin & A. Vulpiani: Regular and chaotic motion of fluid particles in a two-dimensional fluid. *J. Phys. A: Math. Gen.* **21**, 3451-3462 (1988).
49. R.A. Pasmanter: Anomalous diffusion and anomalous stretching in vortical flows. *Fluid Dyn. Res.* **3**, 320-326 (1988).
50. P. Welander: Studies of the general development of motion in a two-dimensional, ideal fluid. *Tellus* **7**, 141-156 (1955).
51. L. Regier & H. Stommel: Float trajectories in simple kinematical flows. *Proc. Nat. Acad. Sci. (USA)* **76**, 4760-4764 (1979).
52. D.P. McKenzie: The Earth's mantle. *Scient. Amer.* **249**, 66-78 (1983).
53. A.E. Perry & M.S. Chong: A description of eddying motions and flow patterns using critical-point concepts. *Ann. Rev. Fluid Mech.* **19**, 125-155 (1987).
54. K. Shariff, A. Leonard, N.J. Zabusky & J.H. Ferziger: Acoustics and dynamics of coaxial interacting vortex rings. *Fluid Dyn. Res.* **3**, 337-343 (1988).
55. M. Feingold, L.P. Kadanoff & O. Piro: Passive scalars, three-dimensional volume-preserving maps, and chaos. *J. Stat. Phys.* **50**, 529-565 (1988).
56. H. Aref & N. Pomphrey: Integrable and chaotic motions of four vortices. *Phys. Lett. A* **78**, 297-300 (1980).
57. S.L. Ziglin: Nonintegrability of a problem on the motion of four point vortices. *Sov. Math. Dokl.* **21**, 296-299 (1980).
58. C.C. Lin: On the motion of vortices in two dimensions - I. Existence of the Kirchhoff-Routh function. *Proc. Nat. Acad. Sci. (USA)* **27**, 570-575 (1941).
59. B. Eckhardt: Integrable four vortex motion. *Phys. Fluids* **31**, 2796-2801 (1988).
60. S.V. Manakov & L.N. Shchur: Stochastic aspect of two-particle scattering. *Soviet Phys. JETP Lett.* **37**, 54-57 (1983).
61. B. Eckhardt: Irregular scattering of vortex pairs. *Europhys. Lett.* **5**, 107-111 (1988).
62. B. Eckhardt: Irregular scattering. *Physica D* **33**, 89-98 (1988).
63. B. Eckhardt & H. Aref: Integrable and chaotic motions of four vortices II. Collision dynamics of vortex pairs. *Phil. Trans. Roy. Soc. (London) A* **326**, 655-696 (1988).

64. H. Aref, J.B. Kadtke, I. Zawadzki, L.J. Campbell & B. Eckhardt: Point vortex dynamics: recent results and open problems. *Fluid Dyn. Res.* **3**, 63-74 (1988).
65. Y. Kimura: Chaos and collapse of a system of point vortices. *Fluid Dyn. Res.* **3**, 98-104 (1988).
66. D.G. Dritschel: The repeated filamentation of two-dimensional vorticity interfaces. *J. Fluid Mech.* **194**, 511-547 (1988).
67. H. Hasimoto, K. Ishii, Y. Kimura & M. Sakiyama: Chaotic and coherent behaviors of vortex filaments in bounded domains. In *Turbulence and Chaotic Phenomena in Fluids,* T. Tatsumi ed., IUTAM/Elsevier Science Publ. 1984, 231-237.
68. A.T. Conlisk, Y.G. Guezennec & G.S. Elliott: Chaotic motion of an array of vortices above a flat wall. *Phys. Fluids A* **1**, 704-717 (1989).
69. Y. Kimura, I. Zawadzki & H. Aref: Vortex motion, sound radiation and complex time singularities *Phys. Fluids A* (submitted, 1989).
70. W. Lauterborn & E. Cramer: Subharmonic route to chaos observed in acoustics. *Phys. Rev. Lett.* **47**, 1445-1448 (1981).
71. P. Smereka, B. Birnir & S. Banerjee: Regular and chaotic bubble oscillations in periodically driven pressure fields. *Phys. Fluids* **30**, 3342-3350 (1987).
72. G. Em. Karniadakis & G.S. Triantafyllou: Frequency selection and asymptotic states in laminar wakes. *J. Fluid Mech.* **199**, 441-469 (1989).
73. C.W. Van Atta & M. Gharib: Origin of ordered and chaotic vortex shedding from circular cylinders at low Reynolds numbers. *J. Fluid Mech.* **174**, 113-133 (1987).
74. M. Gharib & K. Williams-Stuber: Experiments on the forced wake of an airfoil. *J. Fluid Mech.* (In Press, 1989)
75. K. Williams-Stuber & M. Gharib: Transition from order to chaos in the wake of an airfoil. *J. Fluid Mech.* (In Press, 1989)
76. D.J. Olinger & K.R. Sreenivasan: Nonlinear dynamics of the wake of an oscillating cylinder. *Phys. Rev. Lett.* **60**, 797-800 (1988).
77. C.W. Van Atta, M. Gharib & M. Hammache: Three-dimensional structure of ordered and chaotic vortex streets behind circular cylinders at low Reynolds numbers." *Fluid Dyn. Res.* **3**, 127-132 (1988).
78. L.S. Da Rios: Sul moto d'un liquido indefinito con un filetto vorticoso di forma qualunque. *Rend. Circ. Mat. Palermo* **22**, 117-135 (1906)
79. F.R. Hama: Genesis of the LIA. *Fluid Dyn. Res.* **3**, 149-150 (1988).
80. T.A. Witten & L.M. Sander: Diffusion-limited aggregation. *Phys. Rev. B* **27**, 5686-5697 (1983).
81. J.A. Glazier & A. Libchaber: Quasi-periodicity and dynamical systems: An experimentalist's view. *IEEE Trans. Circ. Syst.* **35**, 790-809 (1988).
82. M. Hénon: Numerical study of quadratic area-preserving mappings. *Q. Appl. Math.* **27**, 291-312 (1969)
83. H.K. Moffatt: Simple topological aspects of turbulent vorticity dynamics. In *Turbulence and Chaotic Phenomena in Fluids,* T.Tatsumi ed., IUTAM/Elsevier 1984, pp.223-230.
84. A.D. Gilbert: Spiral structures and spectra in two-dimensional turbulence. *J. Fluid Mech.* **193**, 475-497 (1988).
85. M.V. Berry, N.L. Balazs, M. Tabor & A. Voros: Quantum maps. *Ann. Phys.* **122**, 26-63 (1979)
86. E.A. Novikov: Dynamics and statistics of a system of vortices. *Sov. Phys. JETP* **41**, 937-943 (1975).
87. T.S. Lundgren: Strained spiral vortex model for turbulent fine structure. *Phys. Fluids* **25**, 2193-2203 (1982).
88. S. Kida & M. Takaoka: Reconnection of vortex tubes. *Fluid Dyn. Res.* **3**, 257-261 (1988).
89. M. Melander & N.J. Zabusky: Interaction and "apparent" reconnection of 3D vortex tubes via direct numerical simulations. *Fluid Dyn. Res.* **3**, 247-250 (1988).
90. L. Onsager: Statistical hydrodynamics. *Nuovo Cimento* **6** (Suppl.) 279-287 (1949).

* * *

The Utility of Dynamical Systems Approaches

Comment 3.

K.R. Sreenivasan

Mason Laboratory
Yale University
New Haven, CT 06520

Abstract

This is a commentary on the utility of the dynamical systems approach to the understanding of transitional and turbulent flows. After a few initial remarks on the position paper by Holmes, I present a summary of three aspects: Universality in transition to chaos in wake flows, the description and dynamics of intermittent fields in fully turbulent flows, and the nature of vorticity and scalar interfaces in turbulent free shear flows. I will show that novel techniques from low-dimensional chaos and fractal geometry yield new and useful information on quantities of central interest in turbulence. The claim is that the dynamical systems appraoch has made definite contributions, not merely enlarged our vocabulary, but the way ahead *vis-a-vis* the turbulence problem has remained hazy.

> *Between the idea*
> *And the reality.....*
> *Falls the shadow*
>
> From T.S. Eliot [1]

> *The nearer we come to the present, of course, the more opinions diverge.*
> *We might, however, reply that this does not invalidate our right to form*
> *an opinion ....*
>
> From J. Burckhardt [2]

1. Introduction

The so-called 'turbulence problem' is not a monolithic entity. Its three essential elements are the origin of turbulence, the dynamics of fully developed turbulence, and the control of turbulence – by which is meant 'making turbulent *flows* behave the way one wants'. The origin of turbulence may have some relation to the onset of complexity in nonlinear systems in general, and hence the currency for notions such as 'universality'. On the other hand, a universal solution to the turbulence control problem is unlikely to exist, because it is specific to a given set of flow and geometric constraints. Fully developed turbulence has a mix of the 'universal' (now in the slightly different sense of being common to a class of turbulent flows) and the particular, both of which are essential to predictive undertaking: The scaling properties of the turbulent energy dissipation and small scale mixing are examples of the former class while the variation with Reynolds number of the drag on a circular cylinder belongs to the latter. The different elements of the turbulence problem are all important in their own right, and the tools of trade are appropriately different. The corollary is that the mastery over no single set of tools will be adequate to address the problem in its full glory.

This section of the meeting was devoted to an assessment of the utility of the so-called 'dynamical systems approach' to the turbulence problem. Just as the turbulence problem is a diffuse one, so is the dynamical systems approach: As Holmes [3] points out, it is a loose but rich mixture of many tools including mathematical theorems, numerical work, experimental studies and model building. As a result of recent developments in dynamical systems, one is now in a position to say essentially everything of interest (at least in a certain parameter range) about the dynamics of some paradigmatic nonlinear systems possessing global universality – the period doubling system being the best known example. The situation is less satisfactory when such global universality is absent. The procedure then is to use the center manifold reduction and classify all possible phase portraits by unfolding the appropriate parameters [4]. In particular, one attempts to study all stable attracting sets. Except when the number of parameters needing unfolding is small (one or two), the possibilities are so huge that one is unlikely to succeed in any generality. One often strings together, by taking recourse to hindsight and symmetry, a number of local bifurcations 'explaining' a particular sequence. In rare circumstances, one has been able to generate dynamical equations by this knowledge.

Holmes [3] summarizes these essential points, and briefly surveys recent developments in 'closed' and 'open' flows; he devotes the remainder of his paper to the elucidation of a qualitative dynamical model [5] for the near-wall structure of the boundary layer at moderate

Reynolds numbers. His basic tool is the proper orthogonal decomposition introduced to turbulence literature by Lumley [6]. I have little to add to the specifics of his model, and so concentrate on a few issues to which he refers in passing or not at all.

2. A broad definition of the approach and the scope of the paper

Nothing will be said here on the control of transitional and turbulent flows. In the context of transition to chaos, there are essentially two points of interest: That chaos (or temporal complexity) does not require many degrees of freedom, and that transition to the chaotic state is universal in character. (This statement needs more precision, and we shall return to it later.) Predictions concerning universality (see [7] for the period doubling route to chaos, and [8] for the quasiperiodic route) have been tested in detail, and there is enough evidence now that low-dimensional temporal chaos has provided new as well as useful ideas and tools for analyzing early stages of transition in (some types of) flow systems which are closed [9] as well as open [10,11]. In Section 3, I present a summary of some recent findings in the wake of a circular cylinder to demonstrate the existence there of universal features.

Experience shows that temporal chaos is of restrictive value once spatial three-dimensionality and classical power-law behaviors set in. Fully developed turbulence is high dimensional [12] and has distinct spatial structures, the gap between what one *can* do with low-dimensional chaos and what one *needs* to do in fully developed turbulence being very wide indeed! One type of advance made in low-dimensional chaos that has found some application in turbulence is the invention of several dynamical measures of stochasticity such as Liapunov exponents [13] and scaling functions [6], or static measures such as the Hausdorff-Besicovich dimension [14], fractal dimension [15], generalized dimensions [16], multifractal spectra [17] and various entropies [18]. Such measures have been made accessible to an experimentalist because of the important notion that an attractor can be constructed by suitably embedding a time series in phase space [19], even though circumstances do exist in which the technique might not be accurate or even useful. These measures cannot be obtained in practice for high-dimensional systems, but progress has been made by treating an instantaneous realization of a flow as a kinematic object consisting of various *objects concentrated on fractal sets embedded in three-dimensional physical space*. I shall summarize this progress in sections 4-6, and remark on the possible utility of such measurements. Brief conclusions are set forth in section 7.

This paper is by choice a summary of results on all three aspects mentioned, rather than a detailed account of any single one. Much of this work has yet to be taken to its logical conclusion, but selective questions can already be asked and partially answered.

3. Universality in transition to chaos in wakes behind cylinders

Briefly, with increasing Reynolds number, the flow behind a stationary circular cylinder first undergoes a Hopf bifurcation [20,21] from the steady state to a periodic state characterized by the vortex shedding mode at a frequency f_0, say. We have shown [21] that the supercritical state (above the critical Reynolds number of 46 when based on the cylinder diameter D and the oncoming velocity) can be modelled by the Landau-Stuart equation, and experimentally determined the relevant constants. These details appear not to depend on the aspect ratio (the length to diameter ratio) if it exceeds about 60. I had shown earlier [10] that a quasiperiodic motion sets in at a somewhat higher Reynolds number, and that the onset of chaos follows. Of interest here are the universal features accompanying this onset of chaos.

A view has been expressed [22] that the quasiperiodicity observed in [10] could be the result of the aeroelastic coupling between the cylinder and the flow, but I must emphasize that no perceptible cylinder vibrations were present in our experiments; see Fig. 1. Recent work [23] has shown that the observed quasiperiodicity is due to the change in the spanwise direction of the vortex shedding pattern at low Reynolds numbers, and *its* onset (as opposed to the onset of vortex shedding) depends on the cylinder aspect ratio, its end conditions and other boundary effects, all of which are not totally under the control of the experimentalist. To observe universality, on the other hand, complete control must be maintained on the sources of quasiperiodicity. It is thus useful to explore quasiperiodic dynamics of the wake by transversely oscillating the cylinder in a controlled manner.

In the work described in [11], the cylinder was placed in a specially designed wind tunnel that allowed more than the usual degree of control on flow parameters, and was oscillated transversely (in the first mode) at various known amplitudes. The Reynolds number was fixed at some value in the supercritical state (55 being the Reynolds number for which the bulk of the data has been obtained). The flow velocity was monitored by a standard hot-wire placed approximately 15D downstream and 0.5D to one side of its rest position. The imposed modulation on the cylinder was at the desired frequency f_e, the amplitude of oscillation A being then a measure of the nonlinear coupling between the two modes. The system thus has two competing frequencies f_0 and f_e, yielding two control parameters f_e/f_0 and the non-dimensional amplitude of oscillation, A/D. Once the external modulation is imposed, we expect f_0 to shift to f_0' (say). By mapping the entire plane of f_e/f_0' and A/D, one can observe many features

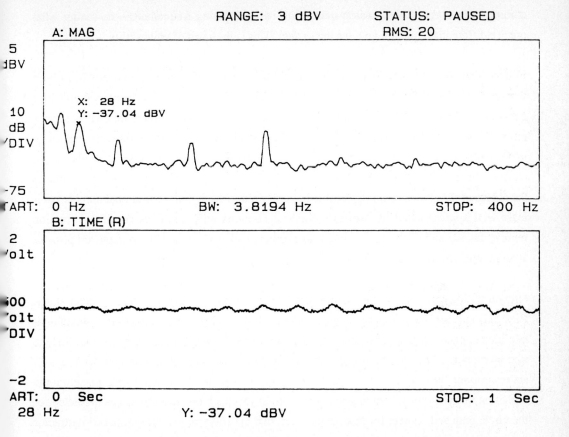

RANGE: 3 dBV   STATUS: PAUSED
A: MAG   RMS: 20

5
dBV

X: 28 Hz
10   Y: −37.04 dBV
dB
/DIV

−75
START: 0 Hz   BW: 3.8194 Hz   STOP: 400 Hz
B: TIME (R)

2
Volt

500
Volt
/DIV

−2
START: 0 Sec   STOP: 1 Sec
28 Hz   Y: −37.04 dBV

Fig. 1. The bottom figure is the time trace from the output of the optical probe (MTI Fotonic sensor) mounted to measure possible transverse displacements of the circular cylinder situated in uniform stream. The cylinder is 'shedding' vortices at a frequency of 287.5 Hz. and, as discussed in [10], the quasi-periodic behavior in temporal dynamics is evidenced by the presence of side-peaks with a difference frequency of 36 Hz. The power spectral density of the time trace is shown in the upper figure. The spectrum shows peaks at 14 Hz, 28 Hz, 60 Hz, 120 Hz and 180 Hz. The last three are related to the response of the optical probe to indicator lights of the electronic instrumentation in an otherwise darkened laboratory. The first two frequencies are related to floor (and thus tunnel) vibrations which have since been damped – the data presented here were obtained in 1985 – with no effect on the observed wake dynamics. More details are forthcoming in [23].

common to a class of nonlinear systems with two competing frequencies – no matter what precise differential equation governs the system. This is the spirit of universality.

In particular, these universal aspects of transition from quasiperiodicity to chaos have been worked out in detail for the sine circle map and are believed to be universal for any map with a cubic inflection point. In the sub-critical state, iterates of the map lock on to rational frequency ratios in the so-called Arnold tongues which increase in width as the nonlinearity parameter increases. At the critical state corresponding to the onset of chaos, universal features occur. To observe them, it is best to proceed without phase locking, this constraint naturally leading to the choice of the golden mean σ_G for the frequency ratios: Note that the irrational number σ_G is least well approximated by rationals (since it contains only 1's in its continued fraction representation), and hence is best suited for avoiding lock-ins – which in principle are possible at all rational frequency ratios.

Olinger & Sreenivasan [11] have demonstrated that the wake of an oscillating cylinder at low Reynolds number is a nonlinear system in which a limit cycle due to natural vortex shedding is modulated, generating in phase space a flow on a torus. They experimentally showed that the system displays Arnold tongues for rational frequency ratios, and approximates the devil's staircase along the critical line. At the critical golden mean point accompanying quasiperiodic transition to chaos, spectral peaks were observed at various Fibonacci sequences predicted for the circle map and, except for low frequencies, had the right magnitudes. A pseudo-attractor was constructed by the usual time delay and embedding methods [19] from the time series of velocity at the critical golden mean point, and Poincaré sections were obtained by sampling data at intervals separated by the period of forcing. The resulting Poincaré section was embedded in three dimensions (in which it was non-intersecting in all three views), and a smoothed attractor was obtained by performing averages locally. The data were then used to compute the generalized dimensions [16] by using standard box-counting methods; in the appropriate log-log plots, the scale similarity regime extended typically over two decades. The multifractal spectrum, or the f(α) curve, was then obtained *via* the Legendre transform discussed in [17]. The multifractal spectrum as well as the spectral peaks showed that the oscillating wake belongs to the same universality class as the sine circle map.

To push this correspondence further, consider the scaling function invented by Feigenbaum [7]. This compact way of describing the dynamics of the system contains complete microscopic description (not merely statistical averages) of the attractor – modulo the information on its embedding dimension. The attractor at the onset of chaos can be regarded as constructed by a process that undergoes successive refinement and leads eventually to the observed scale similar properties. Such a process can be mapped on to a subdividing tree structure whose branches now are the iterates of the dynamical system; the scaling function correctly organizes the intervals to be compared from one level of refinement to another. Extracting Feigenbaum's scaling function experimentally is a nontrivial task, but one can instead use a suitably modified version [24] based on comparing intervals within a single periodic orbit rather than of two different orbits at the same level of stability. Olinger et al. [11] have obtained this modified scaling function at several successive approximations to the critical golden mean point, and shown that it is in good agreement with that calculated for the sine circle map.

These are detailed tests of (metric) universality, comparable to that in some experiments in closed flows. *I find it remarkable that a complex flow such as the wake should conform so well to the predictions for the circle map. This does not mean, however, that transition to chaos and chaos itself can necessarily be defined in every flow equally neatly and in as much detail, or that they all belong to some universality class or another. Our knowledge of what special circumstances or features of the flow render such questions useful is accumulating only slowly, and there is room for further work. But it is clear that the dynamical systems approach is useful even for some class of open flows.* For further remarks on the implications of the results, I refer the reader to the papers cited.

In addition to demonstrating universality in a familiar flow, the work just summarized has the practical value of predicting the width of lock-in regions, and of organizing under a broad umbrella many isolated results on oscillating cylinders. I must remark, however, that *the relevance to fully turbulent state is unclear of these and similar demonstrations of universality accompanying transition to chaos; in fact, the role of chaos in bringing about turbulence is an outstanding question.* In simple dynamical systems – the circle map and the logistic map being two specific examples – one has been able to discover the underlying dynamical structure of the asymptotic state by unraveling the bifurcation sequence as the control parameter evolves. A lesson often emphasized [25] is that the ordering of the asymptotic state may be discerned by understanding the ordering during the evolution of the system to that state. This statement can at best be partially true, if that, for fluid turbulence. If the striking similarity between the coherent motion in a fully turbulent flow and the corresponding motion in its transitional stage is not accidental, it is probably true that the former can be understood in terms of transitional structures. On the other hand, it appears futile to seek the key to the understanding of universal aspects of fully developed turbulence in the transition process because different transition scenarios lead to the same end product. Some thoughts on these issues are taking shape, but I shall now move on directly to the fully turbulent state.

4. Internal intermittency in turbulent flows

As already remarked, an instantaneous realization of a fully turbulent flow such as a jet is an object I wish to study. In this section, I concentrate on aspects of internal structure which, in regions far away from the boundary, are statistically independent of the boundary and initial conditions; the specific quantities examined are the distribution of the turbulent energy dissipation, the 'dissipation' rate of the variance of a passive scalar, and absolute values of turbulent vorticity and the Reynolds shear stress. All these quantities are distributed in some complex way in the three-dimensional real space. Fig. 2 shows the distributions of one component of energy dissipation along a line in the fully turbulent part of a laboratory boundary layer at moderate Reynolds number and in the atmospheric surface layer at a much higher Reynolds number. It is clear that the spatial distribution becomes increasingly intermittent as the Reynolds number increases. The intermittency of the scalar 'dissipation' as measured on a planar cut is shown in Fig. 3.

Two points of interest here are the description of such intermittent distributions and the identification of the dynamics leading to them. Such highly intermittent processes cannot be described efficiently by the conventional moment methods known to be successful for Central Limit type processes. In particular, if a process is Gaussian, its mean and variance describe the process completely; for others close to Gaussian, a few low-order moments contain most of the information. On the other hand, for processes of the type shown in Figs. 2 and 3, the first few moments give little clue to the nature of the process.

Multifractal measures, as they are called in the present parlance of dynamical systems, have built-in intermittency which therefore makes it logical to examine their usefulness in our context; see Mandelbrot [16], Hentschel & Procaccia [16], and [17]. Essentially, multifractals are built up by a procedure (which is often rather simple) that proceeds from one scale (the parent scale) to the next smaller ones (the off-springs) in such a way that the measure (roughly, the amount of a positive quantity such as the rate of energy dissipation) contained in the parent scale is unequally divided among its off-springs. When this procedure repeats many times, the measure on the off-springs of each successively higher generation will become increasingly uneven. If the basic rule determining the unequal division from a parent scale to its off-springs is independent of the generation level, one expects certain scale-similar properties. Because the measure on an arbitrary off-spring at a given generation level is determined by the product of the multipliers (that is to say, numbers characterizing the unequal division of the measure) of all its fore-fathers, a multifractal can be associated with a multiplicative process. The first order of

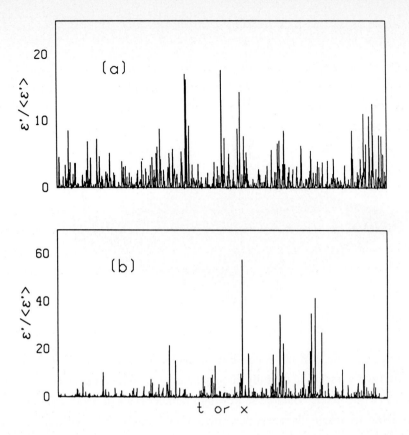

Fig. 2. Typical signals of $\varepsilon' = (du/dt)^2$ normalized by its mean. The upper trace (a) was obtained in a laboratory boundary layer on a smooth flat plate at the moderate Reynolds number R_λ of 150 (based on the Taylor microscale and the root-mean-square fluctuation velocity in the main flow direction). The lower trace (b) was obtained in the atmospheric surface layer a few meters above the roof of a four-storey building. The Reynolds number R_λ is about 1500. It is believed that the statistics of ε' are representative of those of the total energy dissipation. For a description of experimental conditions, see [26,28]. The figure is taken from [28].

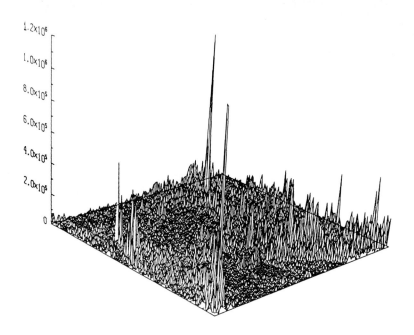

Fig. 3. The dissipation rate χ of the concentration fluctuation c as a function of the two coordinates x (axial) and y (radial) in the fully turbulent part of an axisymmetric jet. The figure covers a grid of 150 x 150 pixels. The nozzle Reynolds number is about 3600, and the center of the picture is about 15 nozzle diameters downstream. $\chi(c)$ was approximated by the sum $(dc/dx)^2 + (dc/dy)^2$, and $<\chi>$ is the average of χ. In [48] it has been shown that the addition of the third component $(dc/dz)^2$ to χ, z being perpendicular to x and y, does not affect the scaling properties. The figure is taken from Prasad et al. [30].

business would be to determine the multifractal properties of the intermittent process interesting to us, and, if possible, identify the associated multiplicative process – if one exists.

Because the measure concentrated on an off-spring at any level is the product of the multipliers of all its forefathers, it is clear that the scaling will only be local; there is therefore the expectation that many, in fact infinitely many, scaling indices will be required for a meaningful description. The purpose of analysis then is to quantify these various scaling indices and unravel their other properties. A possible vehicle for doing this is the $f(\alpha)$ spectrum [17] to which I have made reference already in section 3; the difference, however, is that we are now considering distributions in three-dimensional physical space rather than in a high-dimensional phase space. The $f(\alpha)$ curve has been measured for positive definite quantities characterizing small scale turbulence. (For the energy dissipation see [26-29], for the scalar dissipation rate see [30], for the squared and absolute vorticity see [31, 32], and for the absolute value of the Reynolds stress in the boundary layer see [33]. I invite the reader's attention especially to [28] where there is a detailed discussion of the measurement techniques, signal/noise ratios, the ambiguities in determining the scaling regimes and scaling exponents.) Typical $f(\alpha)$ curve for the energy dissipation is given in Fig. 4. Some of its salient features are the minimum value of α corresponding to the largest singularity in the distribution of ε, the maximum value of α corresponding to the least intense regions of ε, the maximum value of $f(\alpha)$ which is 3 (showing that there is some dissipation everywhere in the flow domain), the point $f = \alpha$ which corresponds to the fractal set on which all the dissipation is concentrated in the limit of infinite Reynolds number [34].

This description of intermittent quantities is more powerful than other descriptions, most of which turn out to be special cases of the present description. For example, Kolmogorov's space-filling dissipation [35] corresponds to the point (3,3) in the f-α plane. Similarly, the β-model [36], in which only the fraction β of the space is occupied by homogeneously distributed dissipative regions, corresponds to another point (D_β, D_β) on the plane depending on the precise value of β. If the $f(\alpha)$ curve can be approximated by a parabola, Kolmogorov's log-normal approximation [37] results. The random β-model [38] is also inadequate because the dimension of the support in that model is less than 3.

These measurements have allowed interesting inferences to be made, and I shall illustrate some of them presently (section 5). I must point out here that *the overwhelming conclusion of this work is that multifractals are a plausible vehicle for describing intermittent fields in turbulent flows.*

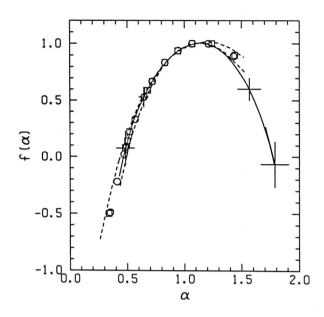

Fig. 4. The multifractal spectrum $f(\alpha)$ obtained from Legendre transforming the generalized dimension data; direct method of measurement [29] yields the same results. Circles correspond to a long data set (10^7 points) in the laboratory boundary layer on a smooth flat plate, and the squares to that in the turbulent wake behind a circular cylinder (5.10^6 points). Both flows have moderate Reynolds numbers (R_λ on the order of 200). The dashed lines represent results from ensemble averaging over many short data segments, each of which is of the order of ten integral scales. The error bars correspond to the standard deviation observed from the analysis of short records. The solid line is the average from [26] in various flows including the atmosphere. To transform $f(\alpha)$ values from one dimension to three dimensions, add 2 to the ordinate. The applicability of this additive law has been discussed in [28] from where this curve has been taken.

In the measurements mentioned above, the spatial resolution was limited to scales of the order of the Kolmogorov scale. If the Schmidt number σ (the ratio of the fluid viscosity to scalar diffusivity) is much larger than unity, the smallest scalar scale, the so-called Batchelor scale [39] $\eta_b = \eta \sigma^{-1/2}$, is much smaller. In such cases, there are two scaling regimes – roughly speaking, that between L and η and another between η and η_b – and it is possible to examine these two scaling regimes separately. In [40], we have explored the multifractal spectrum of the scalar dissipation in the scale range between η and η_b using one-dimensional sections. Finite Schmidt number corrections remain important even when σ of the order 1000, but the primary conclusion is that all generalized dimensions are essentially equal to the fractal dimension of the support in the limit of infinitely large Schmidt numbers or, equivalently, that the multifractal spectrum is simply the point (3,3) in the f-α plane. The natural conclusion is this: *It is not necessary to invoke multifractals – or even ordinary fractals – for describing scalar fluctuations in the range between η and η_b; classical tools should be quite adequate here.*

5. The thermodynamic formalism

Recall that the multifractal description of section 4, general though it is, is only a kinematic description. It is usually not possible to deduce dynamics from such static descriptions, but some headway can be made by noting that the multifractal description of dynamical systems is equivalent to thermodynamic description of statistical mechanical systems [41]. Specifically, then, the question is: Can one deduce something about dynamics given merely thermodynamic information?

Consider the multiplicative process in which each parent interval breaks up into 'a' number of new sub-intervals; we would have at the end of n stages of the cascade a^n pieces. The general procedure is to map such a process on to an a-state n-particle spin system and find, by taking recourse to the measured multifractal properties, the appropriate transfer matrix describing the transition from the n-th stage to the (n+1)-th stage, or from an n-particle system to an (n+1)-particle system. One then has a dynamic process yielding the observed thermodynamics. Usually, one can only get the leading approximations to the transfer matrix [41]. Chhabra et al. [42] have examined the issue in detail with particular reference to turbulence, and shown (not surprisingly) that the procedure yields non-unique solutions; that is, there are many dynamic processes which yield the same thermodynamics. However, a knowledge of the constraints on the system dynamics can render the choice much less ambiguous. Meneveau & Sreenivasan [43] have shown that the measured multifractal properties are in good agreement with a binomial cascade model (designated in [43] as the p-model) in which a parent eddy breaks up (in one dimension) into two equal-sized eddies such that the measure (say, the energy dissipation rate) is redistributed between its two offsprings in the ratio 7/3. (For general multiplicative processes in three dimensions, see [28].) This allows several quantitative predictions (such as the Reynolds number variation of the skewness and flatness factor of the velocity derivative) to be made. In [26,43], it has been shown that they are in good agreement with measurement. As an example of such comparisons, Fig. 5 shows the comparison between the measured and calculated spectral densities of $(du/dt)^2$. It is clear that the agreement is reasonable.

Keep in mind that the inversion is non-unique, which means that I cannot claim that p-models and other multinomial models represent true dynamics of energy cascade – despite impressive agreement with experiment. What I can say is that one can construct simple dynamical models whose outcome is statistically the same, up to some level of approximation, as those of the measured results. *To determine, among the host of existing possibilities, the true dynamic picture requires more information, and is an area of active research.*

Similar binomial and multinomial models have been constructed for other quantities mentioned earlier. Further, the multifractal formalism has been extended [33] to more than one coexisting multifractal measures. The primary motivation for this work is the realization that a high Reynolds number turbulent flow subsumes several intermittent fields simultaneously, and that they display different degrees of correlation among them. The formalism has been applied to simultaneous measurements in several classical turbulent flows of a component of the dissipation rate of the kinetic energy, the dissipation rate of the passive scalar, as well as the square of a component of the turbulent vorticity field. Several joint binomial models, also in agreement with measurement, can be deduced for these joint distributions. One further area of progress [44] has been the recognition that the $f(\alpha)$ curve possesses useful information on spatial correlations, and that it provides a vehicle for quantifying the relative importance to transport of large amplitude but rare events in comparison with small amplitude but ubiquitous events.

6. Interfaces in turbulent flows: An example of the 'outer' dynamics

We now turn to 'outer' dynamics of turbulence, typified by the vorticity interface (that is, the conceptual surface separating domains of intense and zero vorticity fluctuations). It is well-known [45] that an unbounded turbulent flow such as a jet develops at high Reynolds numbers 'fronts' across which vorticity changes are rather sharp on scales larger than the

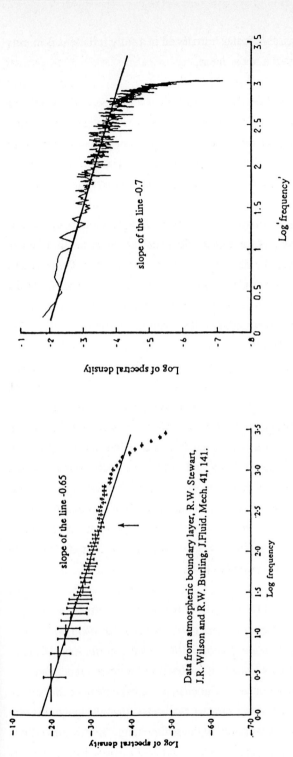

Fig. 5. The left figure is the power spectral density of $(du/dt)^2$ in the atmospheric surface layer. The figure on the right is obtained as follows. Starting with the uniform distribution of a measure on the unit interval, the first step of refinement is initiated by halving the interval and redistributing the measure in the ratio 7/3 (the p-model of [43]). This process is repeated until the smallest scale obtained is in the ratio η/L, η being the Kolmogorov scale in the flow to which the comparison is sought; this ratio is fixed by the Reynolds number. A meaningful comparison between the two requires, from the known Reynolds number of the flow, that 16 iterations be performed in the p-model. The resulting distribution is then Fourier transformed to obtain its spectral density (right figure). It is seen that a reasonable power-law index of -0.7 exists in the right plot, not very different from -0.65 of the left figure; it is possible to fit slightly different lines to the data in the two figures and come up with exactly the same slope.

characteristic thickness of the front. A passive scalar introduced in a fully turbulent flow gets dispersed by turbulence, but itself possesses a scalar interface.

Interfaces are complex objects residing in three-dimensional physical space, and convoluted over a range of scales which, according to conventional wisdom, may be statistically self-similar. It is therefore thought [15] that fractal description of such surfaces is possible. At high enough Reynolds numbers, the scale separation between the largest and smallest interface scales is rather large, and this allows the use of fractals in its characterization. Unlike a mathematical fractal, the scale-similar regime of the interface is bounded on both sides by physical effects: The upper cut-off occurs at around the integral scale of motion, this being comparable to (but distinctly less than) the gross size of the flow such as the jet width, whereas the inner cut-off occurs at a scale where the fluid viscosity is felt directly. This scale is approximately the Kolmogorov scale η (or some multiple of it). The fractal dimension thus characterizes scale similarity in the approximate range between L and η. For the scalar interface corresponding to large Schmidt numbers, there are (as described in section 4) two scaling exponents or fractal dimensions, the second one corresponding to the scale range between η and η_b.

A primary property of a fractal surface being its fractal dimension, much attention has been paid to measuring it. Fractal dimension measurements of vorticity interfaces have been made only in one-dimensional intersections [31,46], but those of the scalar interface in the scale range between L and η have been made using one, two- and full three-dimensional mappings of the scalar field [46-48]. A detailed discussion of the experimental procedures including considerations of the noise effects can be found in [47]. In the scale range between η and η_b, the requirement that the Batchelor scale be resolved allows only trivial extents of the flow to be mapped in two and three dimensions; one therefore has to resort only to one-dimensional intersections. Such measurements have been reported in [40].

Table 1 summarizes the results on the fractal dimensions of both scalar and vorticity interfaces. *One principal result is that the vorticity interface in several of the classical turbulent flows has a fractal dimension of 2.36 ± 0.05, and that the dimension of the scalar interface in the scale range between L and η is also the same.* This last result is consistent with the understanding that the scalar is dynamically passive. The fact that the dimension is independent of the flow is an indication that the scales that produce the fractal character (scales less than the integral scale) possess kinematic similarity at this level of description. In the scaling range between η and η_b, the dimension is close to 3, that is to say, fluctuations in this range are space-filling. This is consistent with the conclusion in section 4 that *one need not invoke fractals to describe these*

TABLE 1. Summary of fractal dimension measurements in classical turbulent flows; from [47].

Flow	scale range between η and L			scale range between[a] η and η_b
	Method of measurement			
	1-D cuts[b]	2-D images	3-D images	
round jet	2.36	2.36	2.36	2.7 (Sc = 1930)[c]
plane wake	2.40	2.36	2.36	2.7 (Sc = 1930)
plane mixing layer	2.39	2.34	--	--
boundary layer	2.40	2.38	--	--

[a]All measurements are from one-dimensional cuts, with Taylor's hypothesis

[b]These one-dimensional measurements for jets and wakes were made both with and without Taylor's hypothesis. Note that one-dimensional measurements often yield slightly higher values for the fractal dimension, but the experimental uncertainties preclude us from attaching much significance to it. The mean value and the statistical error bars, deduced from many measurements, are 2.36 ± 0.05. The slight difference from one flow to another may or may not be significant; the present thinking is that they are not.

[c]The Schmidt number is obtained from [49]. Typical error bars [40] for this estimate are ± 0.03.

sub-Kolmogorov scale fluctuations in scalars. I must remark that this is true only in the limit of infinite Schmidt numbers (if the Reynolds numbers is also proportionately higher), and finite Schmidt number effects reduce this value; for σ of the order of 1000, the measured dimension in this scale range is 2.7 ± 0.03; see [40].

One use to which the fractal dimension has been put is in the explanation of small scale turbulent mixing. The basic idea is that the properties of the scalar interface (say) and the mixing of the scalar with the ambient fluid are related; the area of the surface involves fractal dimension and the ratio of inner to outer cut-off scales [50-51]. The *amount* of mixing is governed by large eddies in the flow, but the small scale mixing is accomplished by diffusion across the surface whose geometry is determined exactly by this requirement. Thus, even though the process is initiated by large scales, one can legitimately try to understand small scale mixing by concentrating on the diffusion end. This approach neither minimizes the role of large eddies nor resorts to gradient transport models usually discredited in turbulence theory. It has been shown [50,51] that this approach yields results of some universality, simply because the small scale features of the flow are, to a first approximation, independent of configurational aspects of the flow. In particular, the arguments presented in [51] yield semi-theoretical estimates for fractal dimensions which agree well with measurements.

Preliminary measurements [52] suggest that the interface in supersonic turbulent flows has a lower dimension than that in incompressible flows. More work is needed before the implication of this result to mixing in compressible flows can be understood.

The outstanding question is to show that the dynamics of the Navier-Stokes equations (or, more precisely, that of the Euler equations, since the effect of viscosity is believed here to be benign in that it merely sets the inner cut-off) somehow imply that the features described here are indeed multifractal in nature – that is, they possess certain type of spatial scale similarity. In spite of much impressive work [52], these issues have remained essentially untouched and invite the attention of a talented reader. In particular, this question has to be reconciled with the existence or otherwise of scale similarity in temporal dynamics.

7. Conclusions

The dynamical systems approach has introduced a variety of new tools and ideas for analysing nonlinear systems. This is an outstanding accomplishment and deserves to become an integral part of the lore of nonlinear phenomena including turbulence. Whether all or some of them are useful for furthering our knowledge and predictive capabilities of transitional and turbulent flows depends to some extent on the ingenuity with which these new tools are applied. Just as initial exaggerations on the importance of the dynamical systems approach to the turbulence problem were unhealthy, so are gloomy thoughts that the approach is but a passing fad to be disparaged. The success one can have with the application of these new tools depends to some measure on how well one understands them, and on how well one can reduce sophisticated mathematical formalisms to realizable measurements. I have presented a summary of some results in transitional and turbulent flows of the open type. These measurements were inspired, in fact made possible, by recent advances in dynamical systems.

It is often believed that the success of the dynamical systems approach in open flows is much less tangible. This view is expressed, among others, by Holmes [3]. Just so, but a part of the reason is the culture of the community which works extensively with such flows. It is true that not all open flows in the early stages of transition can be usefully described by the dynamical systems approach, but so is it true that not all closed flows can be described in this manner! As mentioned already, our understanding of which type of flows can be so described is accumulating only slowly and its detailed discussion should therefore await another occasion; but it seems that the classification of flows as 'open' and 'closed' is probably not specific enough for present purposes. The papers presented in this section of the meeting give samples of the types of questions addressable at present by this approach. These same ideas may be helpful in other contexts such as noise reduction in experimentally obtained signals, an example of the current work being [54].

Mere description of turbulence is no end in itself, and one needs to make predictions by building working models. Some predictions rendered possible from the present work have been summarized at appropriate places; other papers in this section of the meeting present other avenues. Together, they constitute interesting, and possibly useful, contributions to turbulence dynamics. Turbulence has the peculiar status of being at once a classical problem and one at the forefront of physics. There is no complete agreement as to what information is essential and should be acquired, and there is room for fresh air here!

Acknowledgements

This work is the result of a collaborative effort with A.B. Chhabra, M.S. Fan, R.V. Jensen, A. Johnson, P. Kailasnath, C. Meneveau, D.J. Olinger, R. Ramshankar, R.R. Prasad and P.J. Strykowski; to them all, I am grateful. I have benefitted from discussions at various times with M.J. Feigenbaum, D. Gottlieb, B. Knight, B.B. Mandelbrot, M.V. Morkovin, M. Nelkin and L. Sirovich. I am thankful to A.B. Chhabra, B.-T. Chu, C. Meneveau, D.J. Olinger and P.P. Wegener for their comments on a draft and to M.J. Feigenbaum for his hospitality during its writing. Different aspects of this work were supported by grants from AFOSR, DARPA (URI), and NSF.

References

1. T.S. Eliot, *The Hollow Men* (1925); see *Complete Poems and Plays (1909-1950)*, Harcourt, Brace & World, Inc. New York (1934), p. 56.

2. J. Burckhardt, *Reflections on History*, Liberty Classics, Indianapolis (1979), p. 319

3. P. Holmes, this volume (1989)

4. J. Guckenheimer, P. Holmes, *Nonlinear Oscillations, Dynamical Systems and Bifurcation of Vector Fields*, Springer (1983)

5. N. Aubry, P. Holmes, J.L. Lumley, E. Stone, J. Fluid Mech. **192**, 115 (1988)

6. J.L. Lumley, in *Atmospheric Turbulence and Radio Wave Propagation*, edited by A.M. Yaglom and V.I. Tatarsky, Nauka, Moscow (1967), p. 166; J.L. Lumley, *Stochastic Tools in Turbulence*, Academic Press (1970)

7. M.J. Feigenbaum, J. Stat. Phys. **19**, 25 (1979); M.J. Feigenbaum, J. Stat. Phys. Phys. Lett. **74A**, 375 (1979)

8. S.J. Shenker, Physica **5D**, 405 (1982); M.J. Feigenbaum, L.P. Kadanoff, S.J. Shenker, Physica **5D**, 370 (1982); S. Ostlund, D. Rand, J. Sethna, E. Siggia, Physica **8D**, 303 (1983); D. Rand, S. Ostlund, J. Sethna, E. Siggia, Phys. Rev. Lett. **55**, 596 (1985)

9. A. Libchaber, J. Maurer, J. Phys. (Paris) Colloq. **41**, C3-51 (19879); J. Stavans, F. Heslot, A. Libchaber, Phys. Rev. Lett. **55**, 596 (1985); M.H. Jensen, L.P. Kadanoff, A. Libchaber, I. Procaccia, J. Stavans, Phys. Rev. Lett. **55**, 2798 (1985); A.P. Fein, M.S. Heutemaker, J.P. Gollub, Phys. Scr. **T9**, 79 (1985)

10. K.R. Sreenivasan, in *Frontiers of Fluid Mechanics*, edited by S.H. Davis and J.L. Lumley, Springer (1985), p. 41

11. D.J. Olinger, K.R. Sreenivasan, Phys. Rev. Lett. **60**, 797 (1988); D.J. Olinger, A. Chhabra, K.R. Sreenivasan, in *A.S.M.E. Forum on Chaotic Dynamics of Fluid Flows*, edited by K.N. Ghia, Fluids Engg. Div. (1989)

12. L.D. Landau, E.M. Lifshitz, in *Fluid Mechanics (Vol. 6 of the Course of Theoretical Physics)*, Pergamon (1982); P. Constantin, C. Foias, O. Manley, R. Temam, J. Fluid Mech. **150**, 427 (1985)

13. A.M. Liapunov, see *Stability of Motion*, vol. **30** of the series in *Mathematics in Science and Engineering*, Academic Press, New York (1966)

14. F. Hausdorff, Math. Ann. **79**, 157 (1919); A.S. Besicovitch, Math. Ann. **101**, 161 (1929)

15. B.B. Mandelbrot, *The Fractal Geometry of Nature*, Freeman, San Francisco (1982); see also G. Bouligand, Bull. des Sci. Mathematiques, **II-52** (1928), 320, 361.

16. A. Renyi, *Probability Theory*, North Holland (1970); B.B. Mandelbrot, J. Fluid Mech. **62**, 331 (1974); H.G.E. Hentschel, I. Procaccia, Physica **8D**, 435 (1983)

17. U. Frisch, G. Parisi, in *Turbulence and Predictability in Geophysical Fluid Dynamics*, edited by M. Ghil et al., North Holland (1985), p. 84; T.C. Halsey, M.H. Jensen, L.P. Kadanoff, I. Procaccia, B.I. Shraiman, Phys. Rev. A. **33**, 1141 (1986)

18. A.N. Kolmogorov, Dokl. Akad. Nauk. **124**, 754 (1959); see also V.I. Arnold, A. Avez, *Ergodic Problems in Classical Mechanics*, Addison-Wiley, Reading, MA (1968)

19. F. Takens, in *Lecture Notes in Mathematics* 898, edited by D.A. Rand, L.S. Young, Springer (1981), p. 366; J.-P. Eckmann, D. Ruelle, Rev. Mod. Phys. **57**, 617 (1985)

20. C. Mathis, M. Provansal, L. Boyer, J. de Phys. Lett. **45**, L483 (1984); M. Provansal, M. Mathis, L. Boyer, J. Fluid Mech. **182**, 1 (1987)

21. K.R. Sreenivasan, P.J. Strykowski, D.J. Olinger, in *A.S.M.E. Forum on Unsteady Flows*, edited by K.N. Ghia, Fluids Engg. Div. Vol. **52** (1987), p.1.

22. C. Van Atta, M. Ghardib, J. Fluid Mech. **174**, 113 (1987)

23. D.J. Olinger, Ph.D. thesis, Yale University (1989), in preparation

24. A.L. Belmonte, M.J. Vinson, J.A. Glazier, G.H. Gunaratne, B.G. Kenney, Phys. Rev. Lett. **61**, 539 (1988)

25. M.J. Feigenbaum, J. Stat. Phys. **52**, 527 (1988)

26. C. Meneveau, K.R. Sreenivasan, Nucl. Phys. B. (Proc. Suppl.) **2**, 49 (1987)

27. C. Meneveau, K.R. Sreenivasan, Phys. Letts. A. **137**, 103 (1989)

28. C. Meneveau, K.R. Sreenivasan, J. Fluid. Mech. (1989), submitted

29. A. Chhabra, C. Meneveau, R. Jensen, K.R. Sreenivasan, Phys. Rev. A. (1989), to appear

30. R.R. Prasad, C. Meneveau, K.R. Sreenivasan, Phys. Rev. Lett. **61**, 74 (1988); R. Ramshankar, Ph. D. thesis, Yale University (1988)

31. M.S. Fan, *Fluid Mech. Rep. 01,* Yale University (1988)

32. C. Meneveau, K.R. Sreenivasan, P. Kailasnath, M.S. Fan, Phys. Rev. A. (1989), to appear

33. P. Kailasnath, *Fluid Mech. Rep. EAS 753a,* Yale University (1988)

34. K.R. Sreenivasan, C. Meneveau, Phys. Rev. A. **38**, 6287 (1988)

35. A.N. Kolmogorov, C.R. Acad. Sci. U.S.S.R. **30**, 301, 538 (1941)

36. U. Frisch, P.-L. Sulem, M. Nelkin, J. Fluid. Mech. **87**, 719 (1978)

37. A.N. Kolmogorov, J. Fluid Mech. **13**, 82 (1962)

38. R. Benzi, G. Paladin, G. Parisi, A. Vulpiani, J. Phys. A. **17**, 3521 (1984)

39. G.K. Batchelor, J. Fluid Mech. **5**, 113 (1959)

40. K.R. Sreenivasan, R.R. Prasad, Physica **xxD**, 314 (1989)

41. M.J. Feigenbaum, J. Stat. Phys. **46**, 919 (1987); M.J. Feigenbaum, M.H. Jensen, I.

Procaccia, Phys. Rev. Lett. **57**, 1507 (1986)

42. A. Chhabra, R. Jensen, K.R. Sreenivasan, Phys. Rev. A. (1989), to appear

43. C. Meneveau, K.R. Sreenivasan, Phys. Rev. Lett. **59**, 1424 (1987)

44. M.E. Cates, J.M. Deutsch, Phys. Rev. A, **35**, 4907 (1987); C. Meneveau, A.B. Chhabra, Phys. Rev. Lett. (1989), to appear.

45. S. Corrsin, A.L. Kistler, *NACA Tech. Rep.* 3133 (1954)

46. K.R. Sreenivasan, C. Meneveau, J. Fluid Mech. **173**, 357 (1986)

47. R.R. Prasad, K.R. Sreenivasan, Phys. Fluids (1989), to appear

48. R.R. Prasad, K.R. Sreenivasan, J. Fluid Mech. (1989), to appear

49. B.R. Ware, in *Measurement of Suspended Particles by Quasi-Elastic Scattering*, edited by B.E. Dahneke, Wiley, New York (1983), p. 255

50. F. Gouldin, AIAA J. **26**, 1405 (1988)

51. K.R. Sreenivasan, R. Ramshankar, C. Meneveau, Proc. Roy. Soc. **A421**, 79 (1989)

52. A. Johnson, *Fluid Mech. Rep. 5/15,* Yale University (1989)

53. A.J. Chorin, Rep. PAM-425, Dept. Math., Univ. Berkeley (1988); P. Constantin, in Newport Conference on Turbulence (1989), to appear

54. J.D. Farmer, J.J. Sidorowich, Phys. Rev. Lett. **59** (1987), 845; J.D. Farmer, in *Evolution, Learning, Cognition*, edited by Y.C. Lee, World-Scientific Press (1988), p. 277; E.J. Kostelich, J.A. Yorke, Physica D (1989), to appear

Discussion on the Utility of Dynamical Systems Approach

Reporter Nadine Aubry

Levich Institute, Steinman 202
CCNY, Convent Avenue at 140th Street
New York, NY 10031

Tony Perry:

I have a rather provocative question to ask the speakers. What I am wondering is: When you apply the proper orthogonal decomposition - I know this is a procedure you use and it is probably standard - you remove the mean flow and you only look at the perturbations. Why do not you include the mean flow, because if you did you would suppress the cubic terms, and also you would be getting something that would be closer to what people call coherent structures, such as hairpin vortices.

John Lumley:

It is a natural question that has been raised before. It turns out that it does not make any difference and it is not the key to anything as you might expect. If you leave the mean velocity, the first eigenfunction is just the mean velocity and then you subtract it from the orthogonal eigenfunctions and you are right back to where you started.

Philip Holmes:

If I can make a rather general comment. Professor Aref told a charming, delightful story which, in spite of its wit, contains a great deal of truth. He and the other speakers pointed out that dynamical systems methods or what popularly tends to be called chaos theory, provides a tool kit. You still have to use the tool kit with care. If you have a European car and you use an American wrench on it, then you are asking for trouble later on. You have to use the tools with care. There is not a monolithic approach, there is not such a thing as the dynamical systems approach to turbulence. This is why I entitled my paper somewhat tongue-in-cheek: "Can dynamical systems approach turbulence?". Now, that said, when I was asked to speak here, I wanted to go beyond the sort of vague generalities about strange attractors resolving the deterministic/stochastic paradox and so on. I wanted to concentrate on a specific attempt to apply some of these methods, so I had to focus and to exclude a lot, and I am very glad that some of the other discussers pointed out many of the other applications that we can hope for. I concentrated on an approach which attempts to isolate and spatially characterize objects in the flow (objects that some of you may like to call coherent structures) and to derive equations governing their interaction. In deriving these equations, various violences were done, the most serious of which were truncations in physical and wavenumber space. Nonetheless this model reproduces a number of very interesting features. In particular it produces unstable fixed points which are connected by heteroclinic cycles. This seems to me to have actually quite a general message about the evolution and reformation of characteristic structures and it occurs in many equations which possess spatial symmetries. This seems to be one general conclusion coming out of this work. It remains to be seen, of course, how the structures that we have found in our very small truncation would change as we increase the number of modes. There is also the more general question of how many is enough. You may really want to look at 150 or 350 ordinary differential equations. Can you do it in an analytical, qualitative way or do you just want to go back to numerical simulations? These are questions I would like other people to address. However, I would like to ask some of the coherent structures people if the notion of an unstable object evolving and then reforming and the kind of statistics that comes out of it (in particular, the exponential tails that result from perturbed heteroclinic orbits) rings any bells? Does this sound reasonable as a sort of general principle, something that we might be looking for, something suggested to us by ideas of others?

Uriel Frisch:

I would like to make two remarks on the two topics which were touched upon this morning, the first one being dynamical systems, the second one fractals. There is one area where dynamical systems and chaos have been very fruitful, it is in the problem of understanding the origin of magnetic fields. You know that many cosmic objects are surrounded by magnetic fields whose origin is still somewhat of a mystery. In this problem, there is the equivalent of Reynolds number, a magnetic Reynolds number. This problem very often is called the dynamo problem. You have a conducting fluid and a magnetic field. The time it takes for such a magnetic field to grow is controlled of course by some eigenvalue. This eigenvalue may be exceedingly small at very high magnetic Reynolds number, possibly too small for astronomical applications. It would indeed be embarrassing if it took 10^{50} years to develop such a magnetic field because the universe has not been around so long. Therefore, a concept which was introduced by Zel'dovich is the fast dynamo, dynamo where the growth rate of magnetic fields, instead of going to zero when the magnetic Reynolds number goes to infinity, remains finite. It is very hard to prove results about fast dynamos. It has been shown by Arnol'd that a reasonable conjecture is that chaos, in the sense that was discussed by Hassan Aref today, Lagrangian chaos, may be the key for having fast dynamos. There is relatively convincing numerical evidence and significant mathematical progress in this direction and I would not be surprised if, in the next five or ten years, we had some real results for this problem. Also I must point out that Arnol'd considers that studying this problem is a prototype for understanding many aspects of turbulence, in particular vorticity dynamics.

My second remark is on fractals. Could you show the first of the two slides please. Figure 1 from Argoul *et al* . (1989) is shown. This is the well known Cantor set analyzed in a slightly novel way which is called wavelet analysis. You do a lot of convolutions of the cantor set with a large number of filters having a range of widths. Then you represent the whole thing in two dimensions plotting space (or time) horizontally and the width of the filter vertically. Then you see the hierarchical branching that is taking place in the Cantor set. Now can you show picture no. 2 (figure 2 of reference cited). If you do the same analysis on turbulence data obtained from a wind tunnel at very large Reynolds number - for turbulence specialists this is the S1 wind tunnel of ONERA with an R_λ of 2700 and an inertial range of about three decades - and concentrate on scales which are in the inertial range (see the figures b and c), then you see a somewhat similar structure, although a little more complex than in the previous case. Hierarchical branching is taking place, suggestive of a fractal. Of course, this is still a technique which is in its infancy. This is not the final word, but it is an interesting way of looking at turbulence data. If you look at the signal itself with a trained eye, you can find that all the structures that you see of the wavelet decomposition are present in the signal itself. However, having a two dimensional unfolding of the signal makes it somewhat easier for the eye to see this hierarchical branching. Of course, we would like to have even further direct evidence for a fractal process going on in the Richardson cascade. This is part of the grand challenge to experimentalists to give us more than one dimensional data.

Phil Holmes:

Are those velocities function of time?

Uriel Frisch:

This is time indeed but assuming that the turbulence rate is about 7% with Taylor hypothesis, you may consider that this is space. The whole range analyzed corresponds to several tens of integral scales.

Philip Holmes:

Are the point velocity measurements, in pictures b and c, the filtered versions of picture a?

Uriel Frisch:

b and c involve magnification factors 20 and 20^2 larger than a. Figure a shows you the very large scales; b and c show the inertial large and small inertial-range scales.

Ravi Sudan:

In a strongly magnetized plasma, the turbulent fluctuations propagate perpendicular to the magnetic field and are almost isotropic, i.e. almost two dimensional turbulence in directions transverse to the field. The phase changes are slow along the magnetic field. The structure of the turbulence is in the form of long filaments. There is the possibility that small scale modes merge to form larger scale modes. In that case, how would the Galerkin technique help us if we are going to truncate beyond a certain scale size? In other words, the information that is contained in the small modes is not irrelevant and cannot be discarded. In fact, in the problem mentioned by Professor Frisch with respect to the dynamo, the smaller scale turbulence feeds the global large scale magnetic field. Also systems in plasma physics may have more than one invariant.

The second comment is addressed to Professor Moffatt. Do I understand that his conclusion was that the nonlinearity in turbulent flow is local in x-space and that the small scale irregular component of the velocity not contained in the coherent structure is local in the k-space?

Phil Holmes:

If the small scales are important in the dynamical process you have to include them. It seems to me obvious. If you can do some intelligent modelling that somehow incorporates the small scales - if they can in some sense be averaged out, or I should say perhaps averaged in -, then maybe you can get away with looking at the interaction of large space scales with a small scales average force. For example, in our work the energy cascade is completely absent. It was just accounted for by a model. If you want to include the dynamical interactions there, then you are going to need, using this sort of modal basis idea, many modes, hundreds of them. With respect to the second question about building invariance in: In principle, you want to solve your differential equations in a space of functions in which all the invariances and all the symmetries are somehow built in. For example, it seems a good thing to use incompressible velocity fields. You presumably want to build as much as possible into the basis set of the problem. Mathematically, in principle, there is no objection to that. Finding functions of x that do these things for you is of course a difficult matter.

Keith Moffatt:

The question: Do I believe that nonlinearity is localized in x-space? Yes, I do believe that. There are regions where the nonlinear term is small, not identically zero, but small relative to the mean, and other regions where it concentrates. I think this could be a very fruitful line for the numerical simulation experts to actually map out the solenoidal projection of the nonlinear term.

As regards your second question related to k-space, I prefer to focus on what is happening in x-space. I think that the k-space decomposition does actually obscure the physics. I think I would take issue with John Lumley on this point.

Bill Phillips:

I have a comment about the equations Moffatt derived regarding the streamwise rolls that have been observed. He derives evolution equations for the perturbation flow about a laminar mean flow which is homogeneous in the streamwise and spanwise directions. I would submit that his equations are incomplete in the sense that they do not fully simulate the flow in which streamwise rolls form. Such a flow is turbulent and the oscillatory motions of the turbulence itself plays a role. To determine precisely what role we must derive the evolution equations about a turbulent mean flow (with homogeneity as above) formally from the generalized Lagrangian mean equations of Andrews and McIntyre (1978). In doing so two extra terms appear; these in effect couple the u-equation to the vorticity equation and the vorticity equation to the u-equation. Of course, Moffatt's equations are recovered exactly if these terms equal zero. But it can be shown formally (Phillips, 1989) that these terms, which depict the rectified effect of oscillatory turbulence motions and may be expressed in terms of space-time correlations, are not zero. So I would submit that the evolution equations for the perturbation flow about a turbulent mean base are coupled and that they admit non-decaying solutions.

Keith Moffatt:

I was addressing my comments particularly at the contribution of Phil Holmes and that of the paper on which that was based (Aubry *et al.*, 1988). I think the point that Phillips is making now is this: Perhaps the eddy viscosity representation of the neglected modes is not adequate. There are other effects also which have not been incorporated in the dynamical system. I think that is the case, is it not?

Phil Holmes:

In the process of projecting onto these vector valued eigenfunctions, we have essentially forced these structures to exist and interact. Now, the structures as we have exactly put them in do not exist as solutions of the Navier-Stokes equations any more than superpositions of vortons exist as a solution of the Euler equations. One vorton is a solution, two is not. You still think that interactions will give you some insight into how the turbulence production mechanism goes along. In the same way, letting a set of empirically derived coherent structures interact, we feel, gives us some way towards understanding the production mechanism. What remains to be seen is that if we continue to increase the number of modes we will not have exactly the same interactions. At the end, we will not have any fixed points that look exactly like these fixed points for the reasons that you point out. My interpretation of the situation is that we have done the best we can in a low dimensional model to produce an average description of the effects of streamwise interactions. Remember that in our analysis, we built R_{ij}, the various components of the experimental autocorrelation tensors into our eigenfunctions. It is a complicated method, a mixture of modeling, choice of projections and a lot of other things.

Jerry Marsden:

I just wanted to make a comment about Moffatt's comments on the Aubry *et al.* model. This comment is based on an examination of some of the normal forms that are very simplified models, even more simplified than the Aubry *et al.* model that I have been thinking about with Tony Block and Debbie Lewis. And the comment is simply this: If you take some of these normal forms, and you ask: "What is the Hamiltonian part of those normal forms that goes along with it?", one can make this a kind of general question in normal form theory . The philosophy behind it is that it would be very nice if in some kind of dissipationless limit the structure of the homoclinic orbits, and so on, persists right at the limit. This general philosophy, I believe, is a productive one and has been proved rigorously in some cases (cf. Lewis and Marsden, 1989). When one takes one of these normal forms and performs this analysis, one indeed does get a Hamiltonian part that would appear at least in the preliminary discussion, to survive this dissipationless limit. Now the intriguing coincidence that occurs in doing this is that if you look at the full normal form for the 1:1 resonance, some of it is the Aubry *et al.* 4 dimensional model and that the Hamiltonian limit that you end up with has a very interesting Hamiltonian structure. This Hamiltonian structure is one that is closely related to Arnol'd's famous structure for the Euler equations (see for example Marsden and Weinstein, 1983) and the Hamiltonian part of their normal form is indeed an Euler equation with the same Hamiltonian structure that Arnol'd found. Of course the theory has to be developed properly and this has not been done yet. However, it would be very interesting to see if that could recover the kind of structure Professor Moffatt was demonstrating and would provide some extra terms in the normal form which indeed survive this dissipationless limit. So I would just like to say that there is actually a possibility that there is not a conflict here if properly put in a slightly larger context. The two theories may in fact be completely compatible.

Bill George:

I have two questions to Phil Holmes and John Lumley. The first one relates to your symmetry. You end up with these pairs of structures which do not seem to be observed, at least, by Moser and Kim in their simulations. I am wondering if these pairs of structures in fact arise from that symmetry, and if so, is that physically based or is that imposed by some other argument? That is question 1.

The second question relates to Keith Moffatt's comments about the question of how you handle the streamwise modes. I would like to draw your attention to figure 4 of Aubry *et al.* (1988) which is essentially Ziggy Herzog's eigenfunctions (Herzog, 1987). I have always interpreted the behavior at zero k_3 and k_1 as an aliasing brought on by the fact that you have ignored the time dimension and used those measurements of correlations in the two space dimensions without time, so there is no convective effect. I guess I always interpreted the finite value of the eigenspectrum at $k_1=0$ as just representing something which was left out of the problem. In my mind that is consistent with the fact that when you take the $k_3=0$ mode, it in fact dies off and falls out of the problem. What I am asking is: Is it conceivable that perhaps you picked the wrong streamwise mode? There might, in fact, be an off-axis peak in this.

Philip Holmes:

Which question would you like answered first? Let me answer the question about symmetry and then John [Lumley] or Nadine [Aubry] will answer the question about Ziggy Herzog's eigenfunctions. I showed two pictures of flow fields, one was this (figure 14 in paper). It is very symmetrical. It shows an interesting dynamical interaction, but it has more or less a reflection plane of symmetry in the middle. Now, it does not have the exact plane of symmetry, because if it did, we would not get this interaction. In the symmetric subspace, we have a stable fixed point. The fixed point is not stable in the larger subspace. There is a small element of asymmetry. We allow cosines and sines to interact freely. Indeed, in slightly different parameter ranges, we get such free interaction which is what you see in numerical simulations (for example from Stanford), you see single vortices (figure

20b, in Aubry *et al.,* 1988). We allow a relatively free interaction, but the fact that there is an underlying symmetry - this group O(2) - is extremely important and this tells us something about the tendency for the flow often to prefer to be symmetrical. Now there is something, some physical principal going on there which I do not feel I understand well enough to explain, but we do not force symmetries.

Question No. 2. I think someone else who is more familiar with the data, with the eigenfunctions and the derivation of the eigenfunctions should answer. As far as I know there were 3 spatial dimensions which were measured.

John Lumley:

We measured essentially four dimensions. We measured the time also. What we have used is simultaneous data. I really do not know how to answer your question. You are suggesting that we could have moved off the simultaneous time.

Anatol Roshko:

For the last couple of weeks, I have heard excellent exposes of this decomposition technique, today by Philip Holmes and Keith Moffatt, and a couple of weeks ago, I heard John Lumley talk on the same subject. I think that I really now have a good appreciation of what is being done. It is clear from what Phil described today that this technique allows you to introduce structures that are really live structures that we see in the flow, structures that we see for instance in the boundary layer as they are described by Steve Kline. Furthermore, it does not simply give you the structures but tells you something about their dynamics which I think is the thing which is really needed. There remains in my mind one thing that troubled me ever since this technique was first proposed by John. This was alluded to yesterday by showing the picture that Lumley and Payne produced when they first applied this technique to the wake, and they came up with a structure which consists of a pair of vortices normal to the wake. Measurements by Fazle Hussain and Bob Antonia clearly show, I think, that the so called coherent structures, the actual "live" structures, are vortices more or less inclined along the plane of maximum strain. What I am wondering is: Can this be fixed up to take into account conditional averages instead of long time overall averages? Could you come up with something which would be not only a representation of the mean flow field but actually the live structures of the flow? Is it a fair question?

John Lumley:

I think this is a very fair question, but I think that you should remember that the figures that Bob [Antonia] showed were reproduced from Grant's data with an enormous amount of interpolation, extrapolation and so on. The data that existed were very sparse. The data that we took were exceedingly sparse, and it is not clear to me now if those are the right forms. We feel on fairly safe ground on the boundary layer because the structures that we get look like the structures that other people had seen in other circumstances. This is the confirmation that we probably have the right ones. It is not clear to me, that if you find that you are getting the wrong ones in a particular situation, it is conceivable that another piece of information could fix it up. I do not know whether conditional sampling information would necessarily be the right one, but it is a possibility.

Anatol Roshko:

The point is, it seems to me that you get a long time average of structures which in some cases, as in the wake, have various orientations.

Fazle Hussain:

Anatol [Roshko] has raised a very important point, namely the cross-roller structure of Grant, Townsend, Payne and Lumley in the plane wake. This has puzzled me, presumably many others as well for years. We have recently examined this, within limited measurement capability, by recording vorticity maps with one or two (simultaneous) rakes of X-wires (Hayakowa and Hussain, 1985, 1989), and have tried to infer the wake structure topology. The wake structures consist of primarily spanwise structures (rolls) contorted by longitudinal structures (ribs). Structures of one sign or circulation (from one half of the wake) does indeed wander across the centerplane frequently, but there is no evidence of vortex reconnection (like that studied by Melander and Hussain, 1988), even though such reconnection has been proposed by some researchers as an explanation for cross-rollers (see p.119, Townsend, 1976). Because of the wandering of structures across the centerplane, it is not surprising that long-time averaged correlation data would imply cross-rollers. By putting two X-wire rakes across centerplane and examining $\omega_y(x,z)$, no instantaneous cross-rollers were detected. See Hussain and Hayakawa (1987) for details of measurement techniques. Subsequent measurements by Antonia following our technique of X-wire rakes are also consistent with ours.

Regarding the role of small scales addressed by Ravi Sudan, Phil Holmes and Uriel Frisch, I would like to make a point I have done previously. First of all, in spite of what some theoretical physicists would like to believe, there is no clear scale separation in any turbulent flow, at least in the fully turbulent states - there is a continuum of scales covering a wide range (typically many decades). For arguments' sake, if we focus only on large and small scales, their coupling is important, in spite of their disparate sizes. Large scales modify the small scales which then interact among themselves to alter the large-scales. The interactions and coupling are highly localized thus making x-space analysis more useful than k-space approach. I feel this is consistent with Keith's response.

I would also like to answer Phil Holmes' question regarding coherent structure reformation. My answer is likely to be different from other coherent structuralists. Championed by Townsend, many CS researchers believe that CS undergo growth-decay-rebirth cycles. I believe that CS seldom break down and decay. They reform through interactions with themselves and others. This can perhaps be understood better by mutual and self interactions as well as local and non-local interactions - a reason why we have become so heavily involved recently in viscous vortex dynamics (for example, see Melander And Hussain, 1988; Kida *et al.,* 1989).

Bill Reynolds:

In this connection, there may be the question of the degeneracy of the eigenfunctions. Are there two eigenfunctions for the same eigenvalue?

Phil Holmes:

The degeneracy has been taken care of. It is what the symmetries are about, the degeneracy with respect to spanwise and streamwise directions. The eigenfunctions depend on k and n, and multiple in some of the k's because of the streamwise and spanwise invariances. There is a fixed set of spanwise modes; each having a sine and a cosine component. One general comment is that we do not think of the eigenfunctions themselves as coherent structures; they are not. The hope is that by putting in enough eigenfunctions the coherent structures will emerge from the dynamical interaction by the modes which were put in. So I would say: first try to put in a few more modes and then if you do not begin to see anything realistic, then you may need a new kind of averaging. There are all kinds of averagings going on. Yes, we may be losing something. We know that there is a cascade.

Tony Leonard:

I have a comment and a question for Keith Moffatt about his intriguing idea to use a superposition of fixed point solutions to represent the velocity field. Suppose you try to do this in two dimensions. You pick a point vortex which is your fixed point solution. This is an ideal situation. You obtain point vortex dynamics and there are no corrections to deal with. On the other hand you may choose to use elliptical vortex patches which will survive a time dependent strain field. Others have used this notion to generate numerical methods. But the problem comes out when these patches get close together. There is a tremendous generation of boundaries of these contours. Have you thought about this model in two dimensions and worried about these catastrophic results?

Keith Moffatt:

The two-dimensional turbulence problem is a very good testing ground for any idea of this kind. It is interesting that in some two-dimensional experiments, for example, Couder's experiment on the two-dimensional soap film, you do see actual pairs separating away and propagating. These are what I described as generalized two-dimensional vortons. You see these things. There <u>are</u> weak interactions and there are also very strong interactions which occur when two neighbouring vortex patches spin in the same direction. I believe that this sort of process is related, as Hassan Aref touched on this morning, to the cascade to small scales: over most of the space-time domain, nonlinear interactions are weak, but occasionally, very strong interactions occur which rapidly pass the energy to small scales.

Parviz Moin:

I would like to make a few comments. First of all, I am glad that John [Lumley] himself pointed out some deficiencies in the process with which Payne obtained the eddies behind the cylinder. I think that Payne's work has problems; among them is the disturbing presence of some negative eigenvalues (which represent energy). The number of measurement points was only 7, and only diagonal elements of the correlation tensor were measured. I think that the case of flow over a cylinder should be redone with new correlation measurements and decomposition. Second, we keep referring to the eigenfunctions as eddies, but as Phil Holmes pointed out, they are not necessarily coherent structures. The phase information which has been lost in second order moments is important in shaping the eddy.

Now, with regard to the comment on the meaning and utility of conditional averages, we should remember that when one does conditional averaging, one has to provide the condition. If you give me that condition and the two-point correlations which are long time average quantities, I can provide you with the conditional sampled profile (that you would have obtained from conditional sampling) with very good accuracy (Adrian and Moin, 1988). If third order moments are given, I can provide the conditional average with even better accuracy. In other words, it is very important to note that the information that is contained in the conditional average profiles can be reproduced from long-term statistical averages. The way the condition enters in the mathematical expression is in the form of weights for the components of the two-point correlation tensor. I think that is food for thought, especially with regard to the utility of long time averaging. Conditional average profiles have been referred to as coherent structures; there is virtually nothing in the conditionally averaged profile that cannot be extracted from long time averaged statistical correlations (given the prescribed conditions).

References:

Adrian, R. J. and Moin, P., J. Fluid Mech. **190**, 531-559.

Andrews, D. G. and McIntyre, M. E. 1978 J. Fluid Mech.**89**, 609-646.

Argoul *et al.* 1989 Nature **338**, 51

Herzog, S. 1986 Ph.D. thesis. Cornell University.

Hussain, A.K.M.F. and Hayakawa, M. 1987 J. Fluid Mech. **180**, 193-229.

Kida, S., Takaoka, M.and Hussain, F. 1989 Phys. Fluids A **1**, 630-33.

Lewis, D. and Marsden, J. 1989 A Hamiltonian-dissipative decomposition of normal forms of vector fields (preprint, Cornell University).

Marsden, J. and Weinstein, A. 1983 PhysicaD **7**, 305-323.

Melander, M.V. and Hussain, F. 1988 Center for Turbulent Research-88, 254-286.

Phillips, W. R. C. 1989 Cornell University FDA-89-09.

Author's Closure

On Moffatt's Paradox or Can Empirical Projections Approach Turbulence?

Philip Holmes

Keith Moffatt points out that non-trivial solutions to the Navier Stokes equations, having no streamwise variation and driven by a streamwise (mean) velocity dependent only upon distance from the wall, should ultimately decay. This is easily seen from a simplified model with a single cross-stream Fourier mode for each velocity component and a fixed linear mean velocity profile. The streamwise velocity component (u_1) is fed from the mean velocity gradient by the component normal to the wall (u_2). However, neither u_2 nor u_3 has a source of energy. Both u_2 and u_3 decay exponentially from their initial values, with u_1 at first rising, but ultimately decaying exponentially also. The ratio of the Reynolds stress to the energy at first rises, but ultimately decays to zero algebraically.

In our ten-dimensional model, however, the ratio of Reynolds stress to energy cannot decay, but is bounded away from zero, as is easily proved (Berkooz *et al*, 1990). Specifically, in projecting the Navier Stokes equations into the subspace spanned by a single empirical eigenfunction and Fourier modes having various spanwise wavenumbers and zero streamwise wavenumbers, the term $U_{i,j}u_j + u_{i,j}U_j$ yields in the ODE for the $\underline{k}$th mode $(\underline{k} = (k_1 = 0, k_3)$ a linear term of the form

$$- \int F(x_2) \sum_{k=1}^{K} a_k^{(1)}(t)\, \phi_{2_k}^{(1)}(x_2) \phi_{1_k}^{(1)*}(x_2)\, dx_2, \tag{1}$$

where the integration is over the entire region, from the wall to X_2. Here $F(x_2)$ derives from the driving pressure profile (c.f. Aubry *et al*, appendix A and equation 15). Since $\phi^{(1)}{}_{2k}$ and $\phi^{(1)}{}_{2k}$ have opposite sign over the range $(0, X_2]$ for the relevant wavenumbers (Aubry et al. Figure 4), and $F(x_2)$ is of a single sign, this yields a positive contribution to the entries in the diagonal matrix $\underline{A}$ (Moffatt's $b_{\lambda\mu}$), providing the energy source which makes the non-trivial fixed points and heteroclinic cycles possible. The ratio of instantaneous Reynolds stress to energy for the model is

$$\frac{\sum_{k=1}^{K} a_k^{(1)^2}(t)\, \phi_{2_k}^{(1)}(x_2)\phi_{1_k}^{(1)*}(x_2)}{\sum_{k=1}^{K}\sum_{j=1}^{3} a_k^{(1)^2}(t)\, |\phi_{j_k}^{(1)}(x_2)|^2} \;,\tag{2}$$

where the sums are from 1 to 5. This is bounded from below by the minimum, over k, of

$$\frac{\phi_{2_k}^{(1)}(x_2)\phi_{1_k}^{(1)*}(x_2)}{\sum_{j=1}^{3} |\phi_{j_k}^{(1)}(x_2)|^2} \;,\tag{3}$$

which is strictly positive.

Berkooz has also shown (*op cit*) that, since the contributions of some of the higher modes to the Reynolds stress are of opposite sign to that of the first mode, the Reynolds stress for higher-order approximations will not be bounded away from zero. Again in terms of the projection, when additional eigenfunctions are included, terms of the form $\phi^{(n)}{}_{2k}(x_2)\, \phi^{(m)*}{}_{1k}(x_2)$ with m and/or n > 1 and $k_3 = 0$ in the analogue of equation (1) are generally positive over the range $(0, X_2]$ and so yield negative contributions to the $b_{\lambda\mu}$. Thus we expect an "accurate" model, lacking streamwise variations but including many spanwise modes and several eigenfunctions, to exhibit the appropriate decay properties, the trivial solution $\underline{u} = 0$ being a stable fixed point.

The proximal cause for the non-zero Reynolds stress/energy ratio when only the first eigenfunction is included, therefore, is the fact that the vector eigenfunctions have scalar coefficients. Hence, the u_1 and u_2 components in each mode are held in a non-evolving ratio. The eddies which occur in the real boundary layer, of course, have streamwise variation, and temporal variation. They each go through a life cycle, growing to a maximum and decaying. Only in a statistical sense is the ensemble stationary. The stationary behavior of the model reflects the stationary behavior of the ensemble, rather than the non-stationary behavior of the members. The Reynolds stress of the model (relative to the energy) is endowed by the empirical eigenfunctions with the value measured in the real boundary layer. In this way the cross-stream velocity components can extract energy from the mean flow. Hence, the empirical eigenfunctions are, in a sense, a closure approximation that embodies the effects of streamwise structure and unsteadiness in the value of the Reynolds stress represented by the relative sizes of their components. In this sense the model only *appears* to belong to the subspace of fields without streamwise variation.

That a projection should have solutions that are not solutions of the full equations is not surprising or unusual. In fact, it is easy to devise simple model equations for which solutions in an invariant subspace are utterly unlike "typical" solutions for the full equation, yet for which there exist projections having solutions very similar to those of the full equation. The issue is one of choosing an "intelligent" projection which captures key features of typical solutions of the full equations.

In the present context, the vital question is whether the complex and apparently physically significant dynamical behavior of the ten-dimensional model is an artifact of the projection, like the fixed points. Happily we can give strong assurance that this is not so. We have constructed a decoupled model lacking streamwise variations (Berkooz *et al*, 1990) in which the streamwise component and those normal to the streamwise direction have separate coefficients. Solutions of this model decay properly, as described in the first paragraph. The Reynolds stress (relative to the energy) decays to zero. The "fixed points" now drift slowly toward the origin. They are still connected by "ghosts" of heteroclinic cycles, so that the same bursting phenomenon occurs, but the bursts are now modulated by the slow decay. The bursts only occur while the cross-stream components are non-zero. There is a relatively long period after the cross-stream components have decayed during which only the streamwise component remains, no bursting occurs, and the streamwise component decays slowly to zero. I feel that this is probably the explanation for the common observation that the sublayer consists primarily of "streaks" - the streamwise remnants of eddies whose cross-stream components have decayed. The fraction of time during which there is cross stream activity (u_2, u_3 and bursting) is relatively short, and most of the time the scene would be dominated by the streak left behind.

Finally, I may mention some recent computational results of Aubry & Sanghi (1989). They have extended the model to include 1, 2 and 3 streamwise Fourier components, going up to studies of 38 complex (78 real) differential equations. Addition of the streamwise components does not change the basic behavior of the system. The heteroclinic cycles connecting the unstable equilibria remain, and the intermittent bursting process persists over certain parameter ranges. For the most part the streamwise components are relatively quiescent; following a burst, however, they are excited. This, of course, agrees very well with experimental observation.

The fact that the bursting process is robust, persisting in models of many different levels, lends credence to the view, partially supported by mathematical reasoning, that the heteroclinic cycles connecting the saddles are a reflection of the symmetries of the equations, both the full equations and the models of various levels, and thus may be expected to be present in essentially all truncations.

Aubry, N., Holmes, P. and Lumley, J.L. and Stone, E. 198 The Dynamics of coherent structures in the wall region of a turbulent boundary layer *J. Fluid Mech.* 192, 115-173.

Aubry, N. and Sanghi, S. 1989. Streamwise and cross-stream dynamics of the turbulent wall layer. *Proceedings*, July meeting of ASME, New York. ed Ghia.

Berkooz, G., Holmes, P. and Lumley, J. L. 1990. Intermittent Dynamics in simple models of the turbulent wall layer. Report FDA-90-01, Sibley School of Mechanical and Aerospace Engineering, Cornell University, Ithaca, NY.

Session Four

The Potential and Limitations of Direct and Large Eddy Simulations

Discussion Leader: E. A. Novikov, University of
California at San Diego

The Potential and Limitations of Direct and Large Eddy Simulations

W.C. Reynolds

Stanford University and NASA/Ames Research Center
Stanford, California

1.1 Experiments, Theories, Models, and Simulations

Experiments, two-point theoretical closures, one-point turbulence models, and numerical simulations of turbulence *all* have important roles to play in advancing the field of turbulence. The information they provide is complementary, and none will replace any other. However, the evolution of simulations as a viable tool for turbulence research and prediction has changed the way in which the other tools are used.

Experiments can do many things much faster than simulations and can cover ranges not attainable by simulations. Theoretical closures provide a framework for study of simple flows at high Reynolds numbers. Turbulence models are the most important tool for highly repetitive engineering analysis in moderately-demanding flows. Simulations give data that can be obtained in no other way, but only for limited conditions. Because of their general validity, simulations are going to become the primary tool for prediction of complex turbulent flows not amenable to prediction by *simple* turbulence models.

The purpose of this paper is to outline the general approach of direct numerical simulations and large-eddy simulations of turbulent flows, discuss the advantages and problems of these methods, survey some of their important achievements, and forecast their future use and importance. For other recent reviews see Rogallo and Moin (1984) and Schumann and Friedrich (1987). For a recent review of CFD in general see Boris (1989).

2.1 Objective of Direct Simulations

Direct numerical simulations (DNS) are three-dimensional, time-dependent numerical solutions of the Navier-Stokes equations that compute the evolution of all significant scales of motion without using any turbulence models. DNS is used to study turbulence physics, to help in the development of turbulence models, to assess turbulence closure theories, and in some special cases to predict flows of technical interest.

2.2 TECHNIQUE OF DIRECT SIMULATIONS

Because turbulence typically contains a broad range of eddies, the smallest of which change very rapidly, highly accurate numerical methods are required for respectable DNS. This is not an issue to be treated lightly; DNS simply can not be done well by crude numerical methods.

The best work has been done using spectral methods, which provide very accurate spatial differentiation (Orszag and Patterson 1972). Pseudo-spectral evaluation of non-linear interactions, in which the interaction terms are computed in physical space and then differentiated by Fourier transformation, are very effective if adequately dealiased. Spectral methods have been limited to simple domains, but are now being extended to finite-element methods (Orszag and Patera 1984). Methods with near-spectral accuracy based on splines (Lele 1989) are beginning to be used in DNS. Rai and Moin (1989) have evaluated promising higher-order finite difference methods for DNS, and speculate that these methods will be more cost-effective than spectral element methods for complex geometries. Accurate time-differencing is also important. DNS for compressible flows must be done with extreme care; for example, selective smoothing methods, designed to eliminate spurious oscillations near shock waves, must not eliminate dynamically important turbulence.

In homogeneous turbulence, special numerical representations have been used to study the effect of rapid mean deformation. The idea is to use a Lagrangian coordinate system imbedded in the mean motion and a set of dependent variables suggested by rapid distortion theory (Rogallo 1981). This enables a correct solution to be obtained even if a large distortion occurs over one time step. Care must be taken to remove aliasing associated with remeshing.

Resolution requirements

DNS is limited to low Reynolds numbers by the available computational resolution. The simulations must resolve both the largest and smallest eddies that are dynamically significant in the flow. There has been some confusion as to how mesh requirements vary with Reynolds number, and so we shall now put forth our own analysis.

First, we consider the requirements for DNS of *homogeneous isotropic* turbulence. The simulation must capture the viscous dissipation, which peaks at about ten times the Kolmogorov scale $\ell_K = (\nu^3/\epsilon)^{1/4}$ and falls off rapidly for smaller scales (here ϵ is the rate of dissipation of turbulent kinetic energy per unit mass, and ν is the kinematic viscosity). For purposes of estimating the minimum computational resolution, let us say that the smallest scale we need to *resolve* is $\ell = 4\ell_K$. We assume that the largest scale L that we must *resolve* is twice the longitudinal integral scale Λ_f. For a given spectrum type, $\Lambda_f \approx Aq^3/\epsilon$. A varies significantly with the spectral form; we will use $A = 0.3$ as a representative value.

With these assumptions, the ratio of the largest to the smallest scales that must be *resolved* in each direction in a DNS of homogeneous isotropic turbulence is

$$\frac{\mathcal{L}}{\ell} \approx \frac{2\Lambda_f}{4\ell_K} \approx 0.15 Re_T^{3/4} \tag{2.2.1}$$

where $Re_T = q^4/(\epsilon\nu)$ is the Reynolds number of the large-scale turbulence. A parameter more familiar to experimentalists is the Reynolds number based on the longitudinal Taylor microscale λ_f and the rms longitudinal velocity fluctuation. For isotropic turbulence,

$$Re_\lambda = \lambda_f \sqrt{\overline{u'^2}}/\nu = \sqrt{10 Re_T/3} = 1.8\sqrt{Re_T}. \tag{2.2.2}$$

The mesh size Δ must be $\ell/2$ if we are to resolve eddies of scale ℓ. The domain must be larger than $\mathcal{L}$ for two important reasons. First, if we are using periodic boundary conditions, our "turbulence" is fully correlated on opposite sides of the domain, and the domain must be large enough that the correlation is very small in the middle. Second, we must have enough of the important large eddies to get a statistically significant sample in the computational domain for purposes of computing turbulence statistics. Considering both of these factors, let us say that our domain must be at least $L = 4\mathcal{L}$ on each side. The ratio of the domain size L to the mesh width Δ is therefore estimated as

$$\frac{L}{\Delta} \approx \frac{4\mathcal{L}}{\ell/2} \approx 1.2 Re_T^{3/4} \approx 0.5 Re_\lambda^{3/2}. \tag{2.2.3}$$

The number of meshpoints required is therefore

$$N_{xyz} = \left(\frac{L}{\Delta}\right)^3 \approx 1.7 Re_T^{9/4} \approx 0.1 Re_\lambda^{9/2}. \tag{2.2.4}$$

According to this estimate, DNS of homogeneous isotropic turbulence using 128 meshpoints in each direction will be limited to $Re_T \approx 500$ or $Re_\lambda \approx 40$. This is consistent with what has been found in 128^3 simulations that do indeed appear to give good resolution and statistics over both the energy-containing and dissipative ranges (Rogallo 1981, Lee and Reynolds 1985, Rogers and Moin 1987a). The scalings with $Re_T^{9/4}$ or $Re_\lambda^{9/2}$ means that increasing the number of meshpoints by a factor of 10 allows only about a tripling of Re_T or 67% increase in R_λ.

In DNS of isotropic turbulence, the peak in the turbulence spectrum shifts rapidly to lower wavenumbers as the turbulence decays. This leads to a reduction in the number of the most energetic eddies in the computational domain and eventually to loss of sample for the most energetic eddies. But in *homogeneous shear flow* the peak stays in about the same place, and the smallest scale does not change very much. The net effect is that the domain size can be closer to the margin for resolution at both large and small scales, and hence the Reynolds number that can be accommodated is higher ($Re_\lambda \approx 100$).

Next, we will estimate the resolution requirements for the *spatially-developing mixing layer*. We assume that the dissipation is the same order as the production,

$$\epsilon \approx -\overline{u'v'}dU/dy \approx 0.15q^2\frac{(\Delta U)}{\delta} \tag{2.2.5}$$

where (ΔU) is the velocity difference driving the shear layer and δ is its local thickness. Estimating $q^2 \approx 0.2(\Delta U)^2$, the ratio of the shear layer thickness to Kolmogorov scale is estimated as

$$\frac{\delta}{\ell_K} \approx \delta\left(\frac{0.03(\Delta U)^3/\delta}{\nu^3}\right)^{1/4} \approx 0.4Re_\delta^{3/4} \tag{2.2.6}$$

where $Re_\delta = (\Delta U)\delta/\nu$. The mixing layer consists of large-scale two-dimensional vortices at low Re_δ, but undergoes a "mixing transition" in which it contains fine-scale structure important in the mixing. In the spatially-developing mixing layer, this transition begins at about $Re_\delta = 6000$ and extends to about $Re_\delta = 20,000$ (Breidenthal 1981). If we must resolve scales as small as $\ell = 2\ell_K$ and require two meshwidths to do this, the number of meshpoints that must be positioned across the layer is estimated by (2.2.6) as

$$N_y \approx \frac{\delta}{2\ell_K/2} \approx 0.4Re_\delta^{3/4} \tag{2.2.7}$$

At the beginning of mixing transition $N_y \approx 270$ and by the end $N_y \approx 670$. The transition extends over a streamwise distance of about 5 times the final δ, and the large eddies have spanwise scales of 4 δ. So to capture the entire transition process at the resolution needed at the end would require a domain of volume approximately $5\delta \times \delta \times 4\delta = 20\delta^3$, or approximately 6×10^9 meshpoints, 3 orders of magnitude greater than current DNS capabilities but possibly within the capability of the planned TF-1 computer.

Rogers and Moser (1989) have found that strong three-dimensional structures will grow in a *temporally-developing mixing layer* at much lower Re_δ. There the length of the domain can be reduced from 5δ to δ, and the long statistical averaging time needed for the spatially-developing case is avoided by averaging over planes parallel to the mean flow. Thus, it is possible to learn a great deal from DNS on this idealized mixing layer, using current supercomputers.

We will now discuss the resolution requirements for *channel flow*. Here the interesting physics occurs primarily near the wall, where the turbulence Reynolds number is low. Kim, Moin, and Moser (1987) used approximately 2×10^6 meshpoints in a DNS of channel flow at $Re_\delta = U_c\delta/\nu = 3300$, where U_c is the centerline velocity and δ is the channel half-width. The DNS is in good agreement with experiments at the same Re_δ (Niederschulte 1988), so we will use it to estimate the mesh requirements at higher Re_δ. We assume that the thickness of the viscous wall region determines the smallest scales throughout the channel. This thickness scales like ν/u^* where u^* is the friction velocity. Since $u^* \sim U_c\sqrt{C_f}$ and $C_f \sim Re_\delta^{-0.2}$, it follows that the ratio of the channel width to the scale of the smallest eddies in the channel varies like $Re_\delta^{0.9}$. As the Reynolds number increases, mesh refinements will be needed in all three directions. Thus, the number of meshpoints required for a computation comparable with Kim, Moin, and Moser's is

$$N_{xyz} = 2 \times 10^6 (Re_\delta/3300)^{2.7}. \qquad (2.2.8)$$

A simulation at the familiar (and relatively low) laboratory value of $Re_\delta = 13,800$ would require about 10^8 meshpoints, which may be possible on the TF-1.

Finally, consider *boundary layers*, which require a much longer computational domain than channel flow but can be simulated over a much shorter length if nearly in equilibrium. Spalart (1988a) performed a DNS for the flat-plate boundary layer at a momentum-thickness Reynolds number $Re_\theta = 1410$, corresponding to $Re_x = xU_\infty/\nu \approx 5 \times 10^5$, using about 10^7 meshpoints. Assuming the scaling above and noting that $Re_\delta \sim Re_\theta$ and $Re_\delta \sim Re_x^{0.8}$, the number of gridpoints for a computation like Spalart's is estimated as

$$N_{xyz} \approx 10^7 \times (Re_\theta/1410)^{2.7} \qquad (2.2.9a)$$

or alternatively

$$N_{xyz} \approx 10^7 \times (Re_x/500,000)^{2.2}. \qquad (2.2.9b)$$

The wake region of a boundary layer is not fully developed until around $Re_\theta \approx 6000$, for which a calculation similar to Spalart's would require 5×10^8 gridpoints, which may be possible on the TF-1. It must be remembered that these estimates are for a relatively short run and short span of a boundary layer, and that the calculation of the full length and breadth of an airfoil would require many times these amounts.

DNS involving a *scalar field* is limited to low and moderate Prandtl (or Schmidt) numbers. For Prandtl numbers greater than about 0.7, the scalar field requires greater resolution than the velocity field because the pressure fluctuations act to smooth velocity gradients. As a first estimate one might replace R_T in the estimates above by a modified Peclet number $Pe = Re_T Pr/0.7$. Successful DNS has been carried out on 128^3 grids only for $Pr \le 2$ (Rogers, Moin, and Reynolds 1986).

It should be clear from these numbers that DNS is currently limited to relatively low Reynolds numbers, and to Prandtl and Schmidt numbers not much greater than unity. This situation will change some but not a lot. However, it is probably true that the essential physics does not change much once the flow has become turbulent. Therefore, much can be learned about turbulence physics from DNS with current supercomputers.

Time advance

For time-accurate resolution of the smallest eddies, the time step must not carry a fluid particle across more than about one meshwidth. This means that the time step must be decreased as the meshwidth is decreased. If we assume that the operation count scales linearly with the meshpoint count, the time of computation (in megaoperations) would scale as $N_{xyz}^{4/3}$. This is a lower bound on the rate of increase with meshpoints because the operation count per meshpoint also increases with the number of meshpoints, at different rates for different algorithms. If we ignore this complication and use the linear estimate, the $Re_\delta = 13,800$ channel case discussed above would require about 170 times the computational effort as Kim, Moin, and Moser's (1987) $Re_\delta = 3300$ case, which used about 250 Cray X-MP hours. If the operation count per meshpoint does not increase very much, this would appear to be within range of the TF-1.

Boundary conditions

Five types of boundary conditions are required for DNS; *homogeneous, wall, inflow, entrainment*, and *outflow*. Each has its own set of problems, which we will now discuss.

Flows that are *statistically homogeneous* in one or more directions have been handled very well using Fourier representations with *periodic boundary conditions* in the homogeneous directions; care must be taken to be sure that the domain is sufficiently large to contain all unstable Fourier modes.

Boundary conditions at solid *walls* are straightforward, but mesh refinement in the near-wall region is essential in most cases. Uniform blowing or suction can be handled without difficulty, and small-amplitude compliance can be handled with perturbation methods.

Accurate numerical schemes are often very sensitive to *inflow conditions*. Sandham and Reynolds (1987) studied the spatially-developing mixing layer with inlet conditions generated by superposing a fundamental and two subharmonics of the Orr-Sommerfeld modes on the inlet mean profile. They found that vortex merging always occurred at the same location and that the resulting mean velocity and scalar profiles were not self-similar. But by randomly jittering the phases of the Orr-Sommerfeld modes the location of vortex merging was randomized and the mean profiles became self-similar. This suggests that some randomizing of the inflow conditions probably is necessary if simulations representative of natural turbulent flows are to be obtained.

Entrainment boundaries, for example far away from the edges of a boundary layer, jet, or mixing layer, require careful formulation to prevent introduction of spurious disturbances through the pressure field. Mappings to infinity help (Cain, Ferziger and Reynolds 1984. It is necessary to ensure that the potential-flow behavior of the velocity field far from the turbulence is well represented in the basis functions (Spalart 1986, 1988a, Rogers and Moser 1989).

The *outflow condition* presents very subtle problems which have been identified using a combination of theory and DNS for the spatially-developing mixing layer. Lowery and Reynolds (1986) developed a convective outflow condition for simulations of the spatially-developing mixing layer. The idea is to allow all of the turbulence structure to be convected out of the computational domain at a uniform convection velocity. After adopting this condition for his own simulations of the mixing layer using improved numerical methods, Buell found that the mixing layer appeared to be *absolutely unstable*, so that once turbulence had entered the domain the flow would forever be turbulent. It is believed that this flow is *convectively unstable* if both free-stream velocities are downstream, meaning that the turbulence should convect out of the domain if the inlet flow is absolutely steady (Huerre and Monkewitz 1985). Buell and Huerre (1988) examined the stability question by simulations in which a wave-packet disturbance was imposed momentarily at the inlet. They found that the disturbance propagated downstream, and the inlet flow became very quiescent; but when the wave packet reached the outflow boundary it generated a pressure disturbance at the inlet, which caused a perturbation cycle to repeat. Thus, the apparent absolute instability in the DNS was non-physical and arose from the outflow boundary condition. This suggests that one should be skeptical of numerical simulations employing *steady* inflow conditions.

New ideas for handling boundary conditions in DNS (and LES) are needed, and this is an area where people who have stayed away from numerical analysis can contribute to DNS. For example, scientists with backgrounds in chaos theory could help by developing methods for generating chaotic turbulent inflow conditions that match the spectra of typical turbulent boundary layers. Clever ideas, like Spalart's (1986, 1988a) method of using periodic boundary conditions in a spatially-developing boundary layer, can make a big impact.

Initial conditions

Initial conditions are also required. For *homogeneous isotropic turbulence*, where the decay history is to a large degree set by the initial state, the initial conditions are very important. For *homogeneous shear flow*, the developed spectrum and its statistics are less sensitive to the initial field. *Inhomogeneous flows*, such as the channel, establish their own steady-state spectrum and hence the initial conditions are not important.

Initial conditions can be constructed in a variety of ways to match desired spectra. For incompressible flows, the velocity field must be divergence-free unless the algorithm makes this adjustment in the first time step. Established fields from previous DNS can be used to initialize the flow. For compressible flows, initial data must be provided for the density and temperature, and initial species fields must be specified when chemical reactions are involved.

Data management and machine-specific programs

Because of the large memory requirements of DNS, many problems can not be done using just the fast-access core memory of current computers. Therefore, *complicated data management schemes are required*, and these must be optimized for the machine and computational schemes being employed. Parallel processing, if well used, can greatly accelerate the computation. Use of site-specific programming, local languages like VECTORAL, and special computers tends to make DNS programs difficult to exchange between facilities.

Archiving and using the database

The construction and use of databases generated by DNS is still very much an art. At present it is not very easy for someone else to extract information from the database without a lot of help from the originator. The database usually contains three velocity components and perhaps pressure. In order to get other quantities, such as the vorticity or dissipation, operations must be performed on the data base, and these must be done in a manner consistent with the numerical method. Dealiasing may be necessary if non-linear terms are of interest. And some quantities, for example the rms value of eighth derivative of the velocity field, may not be accurately found at all.

Justifiable fear of improper use of the databases has made many simulators very reluctant to release their data to others. Yet the potential for extracting additional information from an old simulation is often so great that all of the time of the scientist-programmer could very well be spent doing service calculations for other scientists. This is a problem with which we grapple daily at the Center for Turbulence Research, and a good solution is sorely needed.

The answer lies in a more global approach to archiving postprocessing the database. A 128^3 DNS will need to save at least three variables per meshpoint (the three velocity components) and perhaps two scalars; at four bytes each this totals to $128^3 \times 5 \times 4 = 40$ Mbytes *per field*. Since each simulation involves many fields, the volume of data is enormous, and the only sensible way to share it at different locations is via optical discs. The current postprocessing workstations do not have enough memory to hold this much data, and hence must process the fields piecemeal. However, single-user minisupercomputers with 64 MBytes of main memory are now available for around $100,000 (*e.g.* Ardent Titan). Coupled to an optical WORM disk drive, these would be ideal for postprocessing of turbulence data.

For remote postprocessing stations of this type to be effective, the database developers must do three things. First, the data must be written on to optical discs so that it can be easily distributed. Second, the data must be documented and its limitations thoroughly discussed in files on the optical disc. Third, post-processing programs in standard languages must be provided with the data so that the user does not have to worry about how to derive data from the database.

A very constructive outcome of this meeting would be the commitment on the part of database originators to move in this direction. This should be coupled with a commitment from the various funding agencies to support the database archival process and the widespread distribution of postprocessing workstations.

2.3 ACHIEVEMENTS OF DIRECT SIMULATIONS

The use of DNS expanded rapidly with the wide availability of supercomputers. Use of the DNS databases developed at NASA/Ames now extends well beyond the local

Ames/Stanford community (Hunt 1988). This section highlights some of the important DNS achievements to date.

Turbulence structure

Because it provides so much information not obtainable from experiments, *DNS has now become the dominant tool for research on turbulence structure*. DNS for channel flows by Kim, Moin, and Moser (1987) showed that the streaky structure in near-wall turbulence does not contain strong streamwise vorticity, as had been conjectured, but is simply an array of high and low-speed regions (jets and wakes). Moreover, the wall-region eddy structures are not nearly as long as the streaks seen by visualization methods. Kim and Moin (1987) studied the effect of the large-scale near-wall structures on scalar transport, and found very strong correlations between the velocity and scalar fields. Current work at NASA/Ames is providing new information on the structure of boundary layers (Robinson, Kline, and Spalart 1988), including flows with pressure gradients, unsteady mean flow (Spalart 1988a), and mean three-dimensionality (Moin, Shih, Driver, and Mansour 1989).

DNS has shown that *homogeneous turbulence has structure*. Hairpin vortices similar to those seen in boundary layers were found by Rogers and Moin (1987a) in homogeneous shear flow. Lee, Kim, and Moin (1986) have shown that at large values of the dimensionless shear rate $S^* = Sq^2/\epsilon$ comparable with the near-wall region of turbulent boundary layers, homogeneous shear flows contain long streaky structures very similar in form and scale to the streaks observed in the near-wall region of boundary layers. Lee and Reynolds (1985) and Rogers and Moin (1987a) examined the family of irrotational homogeneous straining flows, finding very different structures for the two extreme cases of axisymmetric contraction and axisymmetric expansion.

DNS has provided *new understanding of the three-dimensional streamwise vortical structure in mixing layers*. Lowery and Reynolds (1986) showed that these vorticies will arise in a spatially-developing mixing layer from random upstream disturbances. Metcalfe *et al.* (1987) studied the confined time-developing mixing layer, and showed that the initial conditions (relative amplitudes of competing modes) play a big role in determining the history of the streamwise vortices. Rogers and Moser (1989) have developed a new method for accurate representation of the unconfined time-developing mixing layer, and also find that the ultimate strength of the streamwise vortices is strongly related to the strength of initial disturbances.

DNS enables study of the *probabilities for vector alignment* in turbulence. Ashurst *et al.* (1987) studied the structure of the scalar field in homogeneous isotropic turbulence and homogeneous shear flow. They found that the vorticity vector tends to be aligned with the direction of intermediate strain rate, and the scalar gradient vector tends to be aligned with the direction of most compressive strain rate. Ashurst, Chen, and Rogers (1987) examined the DNS for homogeneous shear. They found that the pressure gradient caused by the turbulence (the "slow pressure") has increased probability for alignment with the direction of most compressive strain-rate, but the pressure gradient caused by the mean velocity gradient (the "rapid pressure") was not preferentially orientated.

Experiments

For the *near-wall region* of near-equilibrium boundary layers in mild pressure gradients, *DNS has replaced classic experiments as the standard for comparison of new experimental data.* Spalart's (1988a) boundary layer simulation, Kim, Moin, and Moser's (1987) channel simulation, and Moser and Moin's (1987) curved channel DNS, have all been used by various experimentalists to check the accuracy of near-wall velocity and vorticity measurements (Niederschulte 1988). The near-wall statistics are insensitive to Re_δ, and DNS really has nailed down the nature of the near-wall region.

DNS has been used to *calibrate measurement devices.* For example, Moin and Spalart (1989) used Spalart's (1988a) boundary layer simulation to examine the performance of hot wire probes in the near-wall region, estimating the magnitude of the errors in various statistics and determining probe parameters that would lead to minimal errors.

DNS has been used to *study simple flows that can not be studied in the laboratory.* For example, experimentalists are very interested in two-dimensional simulations of the mixing layer because the simulations tell them what their laboratory flow would look like if it were truly two-dimensional. In order to study the physics of three-dimensional turbulent wall flows, Moin *et al.* (1989) did a DNS of channel flow with a cross-channel pressure gradient. Spalart (1989) simulated an idealized unsteady three-dimensional turbulent boundary layer, and used the simulation to develop a new model.

Turbulence modeling

DNS has *provided new insight and data for development of turbulence models.* For example, Rogers, Moin, and Reynolds (1986) studied scalar transport in homogeneous shear flow with imposed mean scalar gradients in all three directions. They showed that a mean scalar gradient in one direction can drive a turbulent scalar flux in other directions, and identified the mechanism and scaling for this cross-gradient transport. Rogers, Mansour, and Reynolds (1989) developed a model for the scalar flux vector and associated diffusivity tensor. Although based on the DNS for homogeneous turbulence, the model does remarkably well in predicting the scalar flux for simple inhomogeneous flows (Kim and Moin 1987).

DNS has provided *new understanding of the effects of strong rotation on homogeneous turbulence.* The cascade of energy to smaller scales is dramatically inhibited by rotation (Bardina, Ferziger, and Rogallo 1985), an effect was not included in turbulence models until observed in DNS. Speziale, Mansour and Rogallo (1987) showed that a Taylor-Proudman reorganization into two-dimensional turbulence does *not* occur when isotropic turbulence is strongly rotated as some modelers had assumed, but instead the turbulence remains isotropic (in accord with rapid distortion theory). Recently Mansour (1989) studied the rapid rotation of initially anisotropic turbulence. He found that the Reynolds stress anisotropy was quickly changed by the action of the rapid pressure-strain term. Reynolds and Shih (1989) have shown that inviscid rapid distortion theory predicts the same behavior, which therefore can not be attributed to the low Reynolds number of the DNS.

Current Reynolds stress transport models do not predict of this effect, which may explain why they do so poorly on swirling flows.

Mansour, Kim and Moin (1988) used DNS to *examine the terms in the dissipation and Reynolds stress equations* in the near-wall region, and suggested improved near-wall turbulence models. Gerz, Schumann, and Elgobashi (1988) used DNS for improvements in second-order closure models for stratified homogeneous turbulent shear flow. Further improvements were suggested recently by Holt, Koseff, and Ferziger (1989). Coleman, Ferziger, and Spalart (1989) used a DNS of the turbulent Ekman layer to study surface stresses, and developed a new wind surface stress model that will be of interest in large-scale meteorology.

DNS has provided *data on the one-point pdf of velocity and scalar* needed for turbulence models based on the pdf. For examples see Mortazavi, Kollman, and Squires (1987) and Eswaran and Pope (1988)

DNS has been used to *evaluate turbulence models for LES*. Clark, Ferziger, and Reynolds (1979) used DNS to test models for LES in isotropic turbulence. Bardina, Ferziger and Reynolds (1980) used DNS to evaluate several models and developed an important new model (see below). Most recently Piomelli, Moin, and Ferziger (1988) used this approach to improve near-wall modeling for LES. They also used DNS to sort out the influence of the residual-scale model from the influence of the filter used to define the resolved scales.

Compressibility effects

DNS for compressible homogeneous turbulent shear flow by Feiereisen, Reynolds and Ferziger (1981) showed that *most of the turbulent shear stress is carried by the "incompressible" part of the velocity field.* Lele (1989) has developed improved numerical methods for compressible DNS, and is using this to study the physics of noise generation. Sandham and Reynolds (1989) used this method in DNS of the time-developing compressible mixing layer. This DNS *explains why the supersonic mixing layer spread less rapidly* than the subsonic layer: the baroclinic term generates vorticity that cancels the vorticity of the mean shear. Dilitational terms also play an important role. Together with stability analysis, this work suggests that above a convective Mach number of about 0.6 the mixing layer will tend to be three-dimensional rather than two-dimensional.

Eddy shocklets, found in two-dimensional simulations, have not yet been found in three-dimensional simulations up to a convective Mach number of 2. Therefore, *insight into supersonic turbulent shear flows can not be obtained from two-dimensional DNS.*

Chemically reacting flows

DNS is beginning to provide information on *turbulence with chemical reaction.* Most of this work has been done for incompressible flows, so it is not really characteristic of combustion. Riley, Metcalfe, and Orszag (1986) used DNS to study reactions in the time-developing mixing layer between two chemically reacting streams. El-Tahry *et al.* (1987) used DNS to study the speed of the reaction front (turbulent "flame" speed) in a premixed mixture, and Leonard *et al.* (1988) used DNS to study a "diffusion flame", both in homogeneous incompressible isotropic turbulence. Boris and Oran (1988) studied chemically-reacting compressible mixing layers, but only with two-dimensional turbulence. The main impact of all of this work has been to lend credence to the idea that DNS can be a valuable tool for study of chemically reacting turbulent flows.

Practical applications

DNS is beginning to be used for *engineering flows at low Reynolds number*. For example, Spalart (1988b) used DNS to obtain information pertinent to transition on a swept wing. By looking at contamination of a laminar boundary layer with finite-amplitude disturbances originating at the wing root of a swept wing, he arrived at design criteria for the suction that is required to maintain laminar flow on the wing.

Turbulence control

One of the most imaginative uses of DNS is in the *exploration of new ways to control turbulence*. Kim and Moin (1989) performed a numerical control experiment in which they applied suction or injection on the wall of a channel exactly opposite to the wall-normal component of velocity at about $y^+ = 10$. They found that this produced about a 20% reduction in the average wall shear stress. Calculations like this, designed to guide experiments, are very important in the developing field of flow control.

Two-point closures

DNS for isotropic turbulence has been used to *assess two-point closure theories*. In a recent study of decaying isotropic turbulence, Domaradski and Rogallo (1988) found that most of the transfer was *local* (from one wavenumber shell to another nearby), and that most of that transfer was due to non-uniform advection by the largest scales rather than a local (in wave space) break-up of eddies. What little non-local transfer was present was a transfer *to* the large scale (low wavenumber in the triad), which would correspond to a small negative eddy viscosity. This was not too surprising since the Reynolds number of this simulation was very low ($Re_\lambda \approx 15$) and the spectrum did not have the $k^{-5/3}$ character of high Re_λ turbulence. However, the same results were recently found by Chasnov (1989) in a forced simulation that did have a $k^{-5/3}$ spectrum. There again the transfer was still primarily local, but the small non-local transfer was now from the large to the small scales (positive eddy viscosity). These results call into question theories that assume that the effect of the unresolved (small) scales on the resolved (large) ones can be treated through an eddy viscosity *even when the resolved and unresolved scales are not separated*. This issue is far from resolved, but clearly the DNS is heating up the controversy and probably will be called upon to settle it.

Evaluation of new concepts

DNS has been used to *evaluate new concepts* in turbulence. For example, Lumley (1967) proposed the *proper orthogonal decomposition* as a more efficient means for representing turbulent fields, but the idea proved very difficult to evaluate because one needs the full two-point statistics of the flow in order to calculate the eigenfunctions used in the representation. Moin and Moser (1989) were able to use their DNS for channel flow to determine these eigenfunctions. They indeed confirmed that a few modes in the inhomogeneous directions carry most of the important statistics, but found that a broad spectrum of modes is needed in the directions of homogeneity. If one desires to capture something like 80% of the turbulence energy, the eigenfunctions provide a very efficient basis set. However, of one wants to capture more nearly all of the energy, then Chebyshev expansions are only slightly less efficient. The Chebyshev functions have the operational advantage that they are known functions whereas the eigenfunctions can be found only after the two-point correlation is known. In related work, Moser (1988) and Adrian and Moin (1988) used DNS to study a stochastic estimation approach that is an alternative for revealing deterministic structures in turbulent fields.

DNS has *settled a debate about helicity*. Prompted by suggestions that coherent structures are regions of high helicity and low dissipation, Rogers and Moin (1987b) carefully examined the helicity in homogeneous shear flow, irrotationally strained homogeneous turbulence, and channel flow, and did not find evidence to support the helicity conjecture. They showed that much of the evidence that had been put forth for this concept by others arose because of choices of initial conditions and poor numerical resolution.

One of the most interesting fundamental studies involving DNS is study of the *cut and connect mechanism* by Melander and Hussain 1988. Their detailed calculations examine the merging of perpendicular vortex filaments by viscous diffusion, and identifies bridging between filaments as the cut-connect mechanism. The DNS enabled them to put forth a very comprehensive description of these complicated processes that seems likely to cause some rethinking about vortex methods for turbulence simulation.

Keefe and Moin (1989) used DNS to *explore dynamical systems theory* as it applies to turbulence. They carried out a set of 450 simultaneous DNS calculations at a very low Re using 16 streamwise and 8 spanwise Fourier modes, and 33 cross-stream basis functions, starting each simulation with a slightly different initial condition. The idea is to determine the Lyapunov exponent spectrum and from that estimate the dimension of the attractor. The significant finding is that the dimension is about 350, far less than the number of basis functions. This suggests that the concept that turbulence is confined to a strange attractor is correct, but the attractor dimension may be too high to allow useful decomposition in terms of a lower-dimensional set of basis functions.

Challenges to old concepts

Finally, DNS has given some hints that perhaps some of what we have always believed about turbulence may not be quite correct. There is some evidence that the anisotropy in the dissipation tensor $\overline{(\partial u'_i/\partial x_j)^2}$ is proportional to the anisotropy in the Reynolds stress, with the proportionality factor vanishing as $Re_\lambda \to \infty$ (Schumann and Patterson 1978). It is generally believed that the spectral anisotropy of the small scales decreases with increasing wavenumber (*local isotropy*). However, in DNS of homogeneous turbulence subjected to persistent mean strain, the (radial) spectra of the velocity components E_{11}, E_{22}, and E_{33}, often appear to be *parallel* (rather than converging at high wavenumbers) when plotted against the wavenumber magnitude k on log-log paper. Similar evidence can be seen in the channel flow simulations.

This suggests that there may be a *range of turbulence scales for which the anisotropy is a constant fraction of the energy at the local scale* (a fractal concept), which would cause one to rethink the concept of local isotropy. This question is clouded by the low Re_λ of the simulations, but the hint is nonetheless there and it should be explored with an open mind. The answer may be derivable from a DNS in which the large scales are forced, if one can find a suitable forcing that does not assume the answer.

2.4 OUTLOOK AND LIMITATIONS OF DIRECT SIMULATIONS

The value of DNS in turbulence research should be clear to anyone who compares the list of achievements discussed above with the advances in turbulence by other means over the same time period. There is no doubt that *DNS will be of growing importance in providing new insights in basic turbulence physics and guidance in turbulence modeling.* As computational capabilities expand, DNS will be used more and more in scientific studies of complex effects in simple geometries, for example shear flows in various rotating passages.

DNS will remain *limited to relatively low Reynolds numbers and to Prandtl numbers not much greater than unity* for the foreseeable future and perhaps forever. Increasing computational capabilities will be useful for attaining improvements in statistical sampling, for handling more complex geometries, and for rather modest extensions in Reynolds number.

On the practical side, application in engineering problems is likely in the *study of special effects in complex flows* (curvature, rotation, transpiration). Early application in turbomachinery is expected because these flows are very complex and the Reynolds numbers are relatively low. DNS will find increasing use in the *study of new concepts for turbulent flow control*, and in the *study of compressible turbulent flows*, especially very low density transitional flows at very high Mach numbers.

Peterson *et al.* (1989) have projected the impact of advances in computers on DNS for turbulence physics (Fig. 1). They believe that the new Cray-3 should be able to handle a DNS for a full boundary layer at $Re_x = 10^5$, and they foresee the possibility of DNS for the boundary layer at $Re_x = 10^6$ sometime during the next decade. According to their estimate, doing a full DNS on an aircraft would require machines 10^9 as fast with 10^6 times the memory of today's largest computers. Therefore, DNS on this scale is highly unlikely to occur for many decades, if ever. That does not diminish the important uses of DNS for study of special effects in complex flows. While DNS as a general engineering tool will be very limited, the payoff from DNS in better turbulence physics, modeling, and control would seem more than ample to justify a vigorous program in DNS.

3.1 OBJECTIVES OF LARGE EDDY SIMULATIONS

The objective of large eddy simulations (LES) is to compute the three-dimensional time-dependent details of the largest scales of motion (those responsible for the primary transport) using a simple model for the smaller scales. LES is intended to be useful in the study of turbulence physics at high Reynolds numbers, in the development of turbulence models, and for predicting flows of technical interest in demanding complex situations where simpler modeling approaches (*e.g.* Reynolds stress transport) are inadequate.

LES came to the attention of the engineering research community about a decade after it had been pioneered by scientists in weather prediction (Lilly 1967). Some of the most important contributions and achievements of LES have come from the atmospheric sciences community. Here we will concentrate on LES methods and achievements in fundamental science and engineering, mentioning where appropriate key carry-over from the work in meteorology.

Very large eddy simulation (VLES) combines a more sophisticated turbulence model with coarser resolution than LES. The impetus for VLES came originally from the atmospheric sciences community, but engineers are now interested in developing VLES for predicting flows of technical interest.

3.2 FORMULATION OF THE LARGE EDDY EQUATIONS

We like to separate the formulation of the LES problem from the numerical method used for its solution. Therefore, the large-eddy field and the small-scale turbulence model are first defined without reference to the solution grid. The large-eddy field is called the *resolved* field and the small-eddy field is referred to as the *residual field* (also called "sub-grid field").

Filtering

A precise definition of filtering is needed to define the resolved field. It is important that the filter commute with differentiation so that equations for the filtered field can be developed from the Navier-Stokes equations. For homogeneous turbulence, one can define the filtered field $\overline{f}$ by

$$\overline{f}(\mathbf{x}, t) = \int G(\mathbf{x} - \mathbf{x}', t) f(\mathbf{x}', t) d^3 \mathbf{x}' \tag{3.2.1}$$

where $G(\mathbf{x} - \mathbf{x}', t)$ is a filter function. The *Gaussian filter*

$$G(\mathbf{x} - \mathbf{x}', t) = \left(\frac{6}{\pi \Delta_f}\right)^{3/2} \exp\left[-6\frac{(x_i - x_i')^2}{\Delta_f^2}\right], \tag{3.2.2}$$

where Δ_f is the filter width (Δ_f sets the scale of the smallest resolved eddies). The coefficient is adjusted so that the filtered value of a constant is that constant, and the factor of 6 in the exponent (chosen for historical reasons) makes the width representative of the scale of smallest resolved eddies. The Gaussian filter is widely used for homogeneous turbulence and for inhomogeneous turbulence in directions of homogeneity, often with separate filter widths in the three directions.

Another common filter is the *sharp cut-off filter* (Leonard 1974), in which all Fourier modes having wavenumbers greater than a specified cutoff are put into the residual field, and all modes with smaller wavenumbers retained in the resolved field. The physical-space representation of this filter is, for each direction x_α,

$$G(x_\alpha - x_\alpha') = \frac{2\sin[\pi(x_\alpha - x_\alpha')/\Delta_\alpha]}{\pi(x_\alpha - x_\alpha')}. \tag{3.2.3}$$

This filter removes modes with wavenumbers $|k_\alpha| > \pi/\Delta_\alpha$.

Another filter, representing uniformly-weighted averaging over a finite width in the y direction in a mesh where Δ_- is the spacing below the node point and Δ_+ is the spacing above, is

$$G(y, y') = \begin{cases} 2/[\Delta_- + \Delta_+] & \text{for} \quad y - \Delta_-/2 < y' < y + \Delta_+/2 \\ 0 & \text{otherwise} \end{cases}. \tag{3.2.4}$$

This is the filtering implicit in a low-level finite difference representation.

Filtering in LES is different than the conventional averaging used in turbulence theory. Unlike conventional averaging, in LES filtering in general

$$\overline{f'} \neq 0 \quad \text{and} \quad \overline{\overline{f}} \neq \overline{f} \tag{3.2.5a, b}$$

since a second smoothing removes additional structure from the resolved field. However, the equalities do hold for sharp cut-off filters.

Some workers reserve the term LES for the case where the filter scale Δ_f corresponds to a wavenumber in the $k^{-5/3}$ inertial range, and use the term VLES when the residual field begins before the inertial range.

For channel flow, the homogeneous filtering in the streamwise and spanwise directions is used, but no explicit filtering is used in the direction normal to the walls. This is not a very satisfactory situation because one does not have a clear definition of what it is that has been computed. This lack of precision is one of the unsettling aspect of current LES practice.

Germano (1986a,b) proposed differential filters which have the interesting property that the full field can always be recovered by simple processes. These have not yet been carefully tested in LES, but they may offer new ways to improve LES calculations and statistics estimation.

The filtered equations

Decomposing the velocity into filtered and residual fields by $u_i = \bar{u}_i + u'_i$, then filtering the Navier-Stokes equations, one obtains the equations for the resolved field. For incompressible flow with constant viscosity, these are

$$\frac{\partial \bar{u}_i}{\partial x_i} = 0 \tag{3.2.6a}$$

$$\frac{\partial \bar{u}_i}{\partial t} + \frac{\partial \overline{u_i u_j}}{\partial x_j} = -\frac{1}{\rho}\frac{\partial \bar{P}}{\partial x_i} - \frac{\partial \tau_{ij}}{\partial x_j} + \nu \frac{\partial^2 \bar{u}_i}{\partial x_j \partial x_j}. \tag{3.2.6b}$$

These equations hold only for filters that commute with differentiation. The effect of the residual field appears as

$$\tau_{ij} = \overline{\bar{u}_i u'_j} + \overline{\bar{u}_j u'_i} + \overline{u'_i u'_j} \tag{3.2.6c}$$

and must be modeled. The second term on the left in (3.2.6b) can be obtained by the indicated filtering of the resolved field, and hence need not be modeled (we previously introduced the name *Leonard stress* for the difference between this term and its filtered value and modeled the Leonard stress, but we no longer follow this approach).

Given a model for τ_{ij}, the filtered equations can be solved numerically by any appropriate means. If a spectral method is used, the usual approach is to pick the computational mesh width Δ_m so that $\Delta_m = \Delta_f/2$.

Extracting statistical information

Once the computation is made, the desired statistics can be extracted from the numerical solution. The mean of the filtered field is the mean of the total field. However, the residual turbulent stresses τ_{ij} are often very significant, especially if the filtering is very coarse, and so it is necessary to add these stresses to the average of the filtered stresses to obtain an estimate for the total turbulent stress. Denoting the appropriate average by $<>$ (for example volume averages for homogeneous turbulence),

$$< u_i u_j > = < \bar{u}_i \bar{u}_j > + < \tau_{ij} >. \tag{3.2.7}$$

In doing LES for comparison with experiments, one must be sure to compare the *filtered* experimental statistics with the statistics of the filtered field; alternatively, one can compare the total-field statistics from the simulation with the experiment.

3.3 RESIDUAL SCALE MODELS FOR LES

The residual turbulence model is a key element in accurate LES. Most commonly used is the Smagorinski (1963) model,

$$\tau_{ij}^S = q_R^2 \delta_{ij}/3 - 2\nu_R S_{ij} \tag{3.3.1a}$$

where q_R^2 is the energy of the residual turbulence,

$$S_{ij} = \frac{1}{2}\left(\frac{\partial \bar{u}_i}{\partial x_j} + \frac{\partial \bar{u}_j}{\partial x_i}\right) \tag{3.3.1b}$$

is the strain-rate of the resolved field, and ν_R is the effective viscosity of the residual field. The Smagorinski model for ν_R, for the special case of equal-width filtering in all three directions, can be cast as

$$\nu_R = \ell_S^2 \sqrt{2 S_{mn} S_{nm}} \tag{3.3.1c}$$

where ℓ_S is a length scale proportional to the filter width Δ_f. A common model is

$$\ell_S = C_S \Delta_f; \quad C_S = 0.23. \tag{3.3.1d}$$

If the smallest resolved scale is in the inertial range, the residual stress may be shown to be of $O(\Delta^{2/3})$ (Leith 1969). q_R^2 is usually unknown, but can be taken in with the pressure ($p^* = p + q_R^2/3$) for purposes of calculation.

Few LES calculations of interest are done with the same filter width in all three directions. If the Gaussian filter is used with Δ_1, Δ_2, and Δ_3 as the filter widths in the three orthogonal directions, then a common approach has been to replace Δ_f in (3.3.1c) by (Deardorff 1970)

$$\Delta_f = (\Delta_1 \Delta_2 \Delta_3)^{1/3}. \tag{3.3.2}$$

Recently Lilly (1989) showed that this approach works very well up to aspect ratios of at least 20:1.

Bardina, Ferziger, and Reynolds (1980) suggested an important improvement to the Smagorinsky model. The idea is to add an extra term representing the contribution from the range of scales in which the (Gaussian-filtered) resolved and residual fields overlap. This term contains a constant factor that Bardina found empirically should be very near unity. It may be shown that this coefficient must be exactly unity if the terms are to have the same Galilean invariance properties as those that they replace (Speziale 1985). The resulting *mixed* model is

$$\tau_{ij}^M = \tau_{ij}^S + (\overline{\bar{u}_i \bar{u}_j} - \overline{\bar{u}}_i \overline{\bar{u}}_j). \tag{3.3.3}$$

Speziale *et al.* (1988) have proposed extensions of models of this type for use in LES of compressible flows.

Piomelli, Ferziger, and Moin (1988) clarified the importance of the relationship between the filter and the model. Since the Gaussian filter produces a filtered field that overlaps the residual field in wavenumber space, the mixed model *must* be used with the Gaussian filter. This overlap is absent when using the sharp cut-off filter, which therefore should *not* be used with the mixed model.

In VLES the pertinent length scale of the residual field may be different than the filter width. A good turbulence model equation for this scale is needed, and this is an ideal research problem for experienced turbulence modelers who want to contribute to LES development.

Model evaluation

Clark, Ferziger, and Reynolds (1979) developed an approach for using DNS to test LES models (now called *a priori* testing). The idea is to filter a velocity field obtained by DNS and compare the residual stresses with the residual stress model. The statistical correlation coefficient between the modeled and exact stress provides a convenient measure of success. Using the Gaussian filter in this approach, Bardina, Ferziger, and Reynolds (1983) found that the correlations between the DNS stress and the modeled stress is considerably better for the mixed model than for the Smagorinski model when the turbulence is anisotropic.

However, the LES of Piomelli, Ferziger, and Moin (1988) shows that *a priori* testing gives an overly pessimistic view. Rarely is the correlation coefficient between the residual stress and that of its model greater than 0.5, and the average correlation (for channel flow) might be as low as 0.2. This does not seem to matter very much, as the LES using the model produces statistical predictions in rather good agreement with those of DNS. Somehow the LES is very resilient to the residual turbulence model, and it is this fact that gives one great hope for LES as a tool for solution of tough practical problems.

Near-wall modifications

In near-wall regions additional modeling is required. In their early work, Moin and Kim (1982) found that it was necessary to use a different turbulent viscosity for the mean flow and resolved turbulence (Schumann 1975) in order to obtain self-sustained turbulence. This rather unsatisfactory situation was improved by the recent work of Piomelli, Ferziger, and Moin (1988). Following a suggestion of Horiuti (1985), they used the Arakawa form of the non-linear terms rather than the rotational form; the rotational form leads to large truncation errors when second-order differences are used. With this change in the numerical method *it is not necessary to use the two-viscosity model*. However, it is necessary to diminish the model length scale in the near-wall region, although the form of that reduction is not particularly important. Piomelli, Ferziger, and Moin used a form that gives the proper limiting behavior at the wall,

$$\ell_{PFM} = C_s[1 - \exp(-y^{+3}/A^{+3})]^{1/2}(\Delta_1\Delta_2\Delta_3)^{1/3} \quad C_5 = 0.065 \quad A^+ = 25. \quad (3.3.4)$$

Very good LES predictions were obtained with this model.

LES with no wall-region resolution

Piomelli, Ferziger, and Moin (1988) also examined coarser LES in which the wall region is not resolved. This is the type of LES that one would expect to use for high Reynolds number engineering calculations. Their tests against DNS included an adaptation of a model proposed by Schumann (1975) and extended by Grötzbach (1983). In this model the instantaneous stress at the wall is set proportional to the tangential velocity at the first meshpoint away from the wall, with the constant of proportionality such that the average over the plane matches the logarithmic law of the wall. The large-eddy velocity normal to the wall is set to zero at the wall. Mason and Callen (1986) assumed that the law of the wall was satisfied instantaneously, an unlikely assumption.

Piomelli, Ferziger, and Moin improved on Schumann's condition by setting the instantaneous wall shear stress proportional to the tangential component of velocity at a point away from the wall an appropriate distance *downstream* from the wall point. The idea is that there are structures inclined away from the wall in the downstream direction, so the tangential velocity at a point away from the wall is most strongly correlated with an upstream wall shear stress. They also proposed an *ejection model* in which the wall shear stress fluctuation is proportional to the wall-normal component of velocity at the first meshpoint away from the wall at an appropriate downstream distance. In both cases the optimal shifts correspond to an inclination away from the wall of about 8°. They found that the two shifted models yield more accurate mean predictions than either Schumann's or Mason and Callen's condition, especially when coarse meshes were used.

Piomelli, Ferziger, and Moin found that calculations in which the wall region was modeled *used less than 15% of the computational effort required for an LES in which the wall region was resolved.* The advantage will be even greater at higher Reynolds numbers. They also found that the turbulence intensities were relatively insensitive to which wall model was employed, indicating that they are controlled primarily by outer layer dynamics.

Residual turbulence models based on RNG

Renormalization group methods (RNG) are currently getting a great deal of attention (Yakhot and Orszag 1986). The procedure produces a modified form of the Smagorinski model with a correction for low local Reynolds numbers. Models of similar form can be developed in other ways, so the magical content attributed to RNG may be somewhat overstated. If the model removes the need for wall damping (3.3.4), then it would certainly have merit. At this writing we have not seen any definitive comparisons with the best recent work (*e.g.* Piomelli, Ferziger, and Moin 1988). We are preparing to do a direct test of the RNG-based model against the models of Piomelli, Ferziger, and Moin, using the same basic numerics, so that a proper evaluation of the different models can be made.

Dynamic residual turbulence models

The Smagorinski model in effect assumes that the residual turbulence is always in equilibrium with the local resolved field. This certainly is not the case in flows undergoing transition to turbulence, relaminarization, or rapid change from one type of flow to another. Therefore, *a dynamic residual turbulence model may be necessary for LES in rapidly adjusting flows.*

Dynamic residual turbulence models, in which the energy of the residual turbulence is tracked by an evolution equation, have been used by Schumann (1975), Horiuti (1985), Horiuti and Yoshizawa (1985), and by Schmidt and Schumann (1988). They use algebraic equations to relate the residual turbulence energy and large-eddy deformation rate to the residual stresses, very similar in structure to k-ϵ second-order closure models that use algebraic stress equations obtained assuming structural equilibrium. However, no ϵ equation is needed because the length scale of the residual turbulence is set by the filtering (*i.e.* by the mesh resolution). An important benefit of a dynamic residual stress model is that it provides a value for the residual turbulence energy $q_R^2/2$, which is needed if one wants to predict the *total* turbulent stress.

In problems with *combustion*, the residual turbulence models are crucial if the chemical reactions occur on residual scales faster than the time scale of the resolved eddies. Schumann (1988) has used a fairly sophisticated second-order dynamical closure model for concentration covariances and concentration-velocities in LES modeling of the atmospheric boundary layer.

LES with no models?

One sees remarkable pictures of turbulent flows that look very realistic, obtained from simulations with no explicit residual turbulence model at all. Typically these use some sort of higher-order upwind difference scheme (Tsuboi, Tamura, and Kuwahara 1989), and are put forth as a "high Reynolds number" result. In real turbulence the dissipation is set by the non-linear cascade in the inertial range, and the only role of viscosity is to set the smallest scales of motion. Is it conceivable that in LES one needs only to calculate the inertial eddies properly, allowing whatever dissipative mechanism is present (*i.e.* upwind differencing) to set the smallest scale?

This approach might actually work if the dissipative process is confined to small scales. In this case the simulation could produce useful results in flows where Reynolds number effects are not important and separation points are set by sharp edges rather than by viscous fluid dynamics. But as there is no relationship between numerical viscosity and fluid viscosity, *this approach simply can not predict important Reynolds number effects* and consequently should be used with caution.

3.4 NUMERICAL TECHNIQUES FOR LARGE EDDY SIMULATIONS

What has been said above about the numerical methods for DNS applies to LES. A new element is that the numerical error must be kept small compared to the residual stress terms. Spectral methods and other methods with near-spectral accuracy, such as those developed by (Lele 1989) and Rai and Moin (1989), are among the best techniques for this purpose.

The same sort of inflow-outflow problems that arise in DNS occur in LES. Some sort of chaotic inflow conditions are required. If the LES model requires the residual scale energy, then q_R^2 must be prescribed chaotically at inlets.

Since LES is intended primarily for use in engineering computations and meteorology, the ability to handle complex domains is important. The large-eddy field is governed by elliptic equations (*e.g.* the Poisson equation for the pressure), and so LES inherently involves elliptic solvers. There is a great need for fast and accurate elliptic solvers for complex domains.

3.5 ACHIEVEMENTS OF LARGE EDDY SIMULATIONS

The widest application of LES is currently in atmospheric sciences, and this work is expanding at a great rate. VLES is now a *routine element in weather forecasting* as well as an important tool for study of special environmental problems (Leith 1978). For recent examples see Moeng 1986, Moeng and Wyngaard 1986, Nieuwstadt and de Valk 1987, Mason 1987, Mason and Thomson 1987, Schumann *et al.* 1987, and Schmidt and Schumann 1988.

LES *paved the way for use of direct simulations in turbulence structure research* (Deardorff 1970; Moin and Kim 1982, Kim 1983, Moin and Kim 1985). A great deal of physical insight was obtained about turbulence structure in channel flow from this early LES work, which *demonstrated that simulation is a viable tool for turbulence research.*

The achievements of LES have not been as spectacular as those of DNS, primarily because of the intense effort now focused on DNS and the backseat that has been taken by LES. Nevertheless, LES has made some definite contributions to turbulence physics and modeling. For example, *anomalous effects in experiments on rotation of homogeneous turbulence were explained* and turbulence models were improved for rotation, using LES (Bardina, Ferziger, and Reynolds 1983). LES has also *aided the development of meteorological models* (Schumann *et al.* 1988).

There have not been many applications of LES for solution of engineering problems as yet. There is considerable effort going into developing LES technology for engineering in France (see for example Laurence 1985). Work in the U.S. has concentrated on fundamentals; *the U.S. appears to be spending much less effort than Europe on the development of practical LES.*

3.6 OUTLOOK AND LIMITATIONS OF LARGE EDDY SIMULATIONS

Peterson *et al.* (1989) have projected the impact of advances in computers on LES for aerothermodynamics (Figs. 1). They foresee the onset of *extensive use of LES for engineering research* in the immediate future, with simple geometries being handled with current machines and LES of a wing being possible with machines expected by the mid 1990s. LES for *a complete aircraft configuration seems beyond reach in this century*, although the first attempts at such ambitious calculations may be possible with only three orders of magnitude increase in speed and memory over current machines.

The great bulk of routine engineering calculations of turbulent flows always will be made with the most economical representations of the turbulence that provide adequate predictions. Given that situation, *it is very important to continue the development of simpler turbulence models*, which will carry the burden of routine engineering analysis for the lifetimes of all participants at this meeting.

However, for complex flows that are three dimensional in the mean, three-dimensional calculations must be made with any approach. For these flows *LES may actually prove to be faster and cheaper than turbulence models having the sophistication necessary to capture the complex effects.* This is particularly true in the meteorology area, where *LES is likely to be better than turbulence models in dealing with stable and unstable stratification.*

LES is *not* a technology limited to massive supercomputers. A 32^3 LES can be done today in reasonable time on a 25 MHz 80386 machine with a Weitek floating point co-processor or an Inmos Transputer (either of which will give about 1 MFLOP) and 4 MBytes of memory (total cost about $10,000). The next generation of personal work stations will do even better, at less cost. *It is not our computer arsenal that limits LES deployment, but LES technology itself.* Better turbulent inflow conditions, wall-region models with heat transfer and chemical reaction, simple dynamic residual turbulence models, and fast elliptic solvers are among the immediate needs.

Creative ideas for meeting these needs are most likely to come from collaboration between people who understand simulation problems very well and people who understand turbulence. One possible constructive outcome of this meeting could be a strategic plan for these developments.

ACKNOWLEDGEMENTS

This paper draws very heavily on the work of the author's colleagues, particularly those at NASA/Ames and Stanford University. Discussions with P. Moin were particularly helpful in shaping this paper. Much information and many useful comments on the draft were provided by him and by M.M. Rogers, R.S. Rogallo, R.D. Moser, P.R. Spalart, J. Kim, N.N. Mansour, J.H. Ferziger, and U. Schumann (DFVLR), who have each made very important contributions to this field.

REFERENCES

ADRIAN, R.J. & MOIN, P. 1988 Organized turbulent structure; homogeneous shear flow. *J. Fluid Mech.* **190** 531-559.

ASHURST, W.T., CHEN, J.Y. & ROGERS, M.M. 1987 Pressure gradient alignment with strain rate and scalar gradient in simulated Navier-Stokes turbulence. *Phys. Fluids* **30**, 3293-3294.

ASHURST, W.T., KERSTEIN, A.R., KERR, R.M. & GIBSON, C.H. 1987 Alignment of vorticity and scalar gradient with strain rate in simulated Navier-Stokes turbulence. *Phys. Fluids* **30**, 2343-2353.

BARDINA, J., FERZIGER, J.H. & REYNOLDS, W.C. 1980 Improved subgrid scale models for large eddy simulation. *AIAA paper 80-1357.*

BARDINA, J., FERZIGER, J.H. & REYNOLDS, W.C. 1983 Improved turbulence models based on large eddy simulation of homogeneous, incompressible, turbulent flows. *Dept. of Mech. Engrg. Rept. TF-19*, Stanford U., Stanford, California.

BARDINA, J., FERZIGER, J.H. & ROGALLO, R.S. 1985 Effect of rotation on isotropic turbulence; computation and modelling. *J. Fluid Mech.* **154**, 321-336.

BORIS, J.P. 1989 New directions in computational fluid dynamics. *Ann. Rev. Fluid Mech.* **21**, 345-385.

BORIS, J.P. & ORAN, E. 1988 The numerical simulation of compressive and reactive turbulence structures. *Proc. Joint U.S./Fr. Workshop Turb. React. Flows* , Rouen, Ed. B.G. Murthy & R. Borghi, Springer-Verlag, in press.

BREIDENTHAL, R. 1981 Structure in turbulent mixing layers and wakes using a chemical reaction. *J. Fluid Mech.* **109**, 1-24.

BUELL, J.C. & HUERRE, P. 1988 Inflow/outflow boundary conditions and global dynamics of spatial mixing layers. In *Studying Turbulence Using Numerical Simulation Databases - II*. Proc. Summer Program 1988, Center for Turbulence Research, Stanford U., 19-27.

CAIN, A.B., FERZIGER, J.H. & REYNOLDS, W.C. 1984 Discrete orthogonal function expansion for nonuniform grids using the fast Fourier transform. *J. Comp. Phys.* **56**, 272-286.

CHASNOV, J. 1989 (private communication), Columbia U.

CLARK, R.A., FERZIGER, J.H. & REYNOLDS, W.C. 1979 Evaluation of sub-grid scale models using an accurately simulated turbulent flow. *J. Fluid Mech.* **91**, 1-16.

COLEMAN, G.N., FERZIGER, J.H. & SPALART, P.R. 1989 A numerical study of the turbulent Ekman Layer. *J. Fluid Mech.*, to appear.

DEARDORFF, J.W. 1970 A numerical study of three-dimensional turbulent channel flow at large Reynolds numbers. *J. Fluid Mech.* **41**, 453-480.

DOMARADZKI, J.A. & ROGALLO, R.S. 1988 Energy transfer in isotropic turbulence at low Reynolds numbers. In *Studying Turbulence Using Numerical Simulation Databases - II*. Proc. Summer Program 1988, Center for Turbulence Research, Stanford U., Report CTR-S88, 169-174.

EL-TAHRY, S., RUTLAND, C.J., FERZIGER, J.H. & ROGERS, M.M. 1987 Premixed turbulent flame calculation. In *Studying Turbulence Using Numerical Simulation Databases*. Proc. Summer Program 1987, Center for Turbulence Research, Stanford U., Report CTR-S87, 121-132.

ESWARAN, V. & POPE, S.B. 1988 Direct numerical simulations of the turbulent mixing of a passive scalar. *Phys. Fluids* **31**, 506-520.

FEIEREISEN W.J., REYNOLDS, W.C. & FERZIGER, J.H. 1981 Simulation of compressible turbulence. *Dept. of Mech. Engrg. Rept. TF-13*, Stanford U., Stanford, California.

GERMANO, M. 1986a Differential filters for the large eddy simulation of turbulent flows. *Phys. Fluids* **29**, 1755-1757.

GERMANO, M. 1986b A proposal for a redefinition of the turbulent stresses in the filtered Navier-Stokes equations.*Phys. Fluids* **29**, 2323-2324.

GERZ, T., SCHUMANN, U. & ELGOBASHI, S.E. 1988 Direct numerical simulation of stratified homogeneous turbulent shear flows. *J. Fluid Mech.*, to appear.

GRÖTZBACH, G. 1983 Direct and large eddy simulation of turbulent channel flows. In *Encyclopedia of Fluid Mechanics*, **6** (Ed. N.P. Cheremisinoff), Gulf Publishing, West Orange.

HOLT, S.E., KOSEFF, J.R. & FERZIGER, J.H. 1989 The evolution of turbulence in the presence of mean shear and stable stratification. *Seventh Symp. Turb. Shear Flows*, Stanford (to be presented).

HORIUTI, K. 1985 Large eddy simulation of turbulent channel flow by 1-equation model. In *Proc. Int. Symp. Computational Fluid Dynamics*, Tokyo.

HORIUTI, K. & YOSHIZAWA, A. 1985 Large eddy simulation of turbulent channel flow by 1-equation model. In *Proc. EUROMECH Colloquium No. 199* (eds. U. Schumann and R. Friedrich), Vieweg and Sohn, Braunschweig.

HUERRE, P. & MONKEWITZ, P.A. 1985 Absolute and convective instabilities in free shear layers. *J. Fluid Mech.* **159**, 151-168.

HUNT, J.C.R. 1988 Studying turbulence using direct numerical simulation: Center for Turbulence Research NASA Ames/Stanford Summer Programme. *J. Fluid Mech.* **190**, 375-392.

KEEFE, L. & MOIN, P. 1989 Applications of chaos theory to shear flow turbulence. Abstracts for AAAS meeting Jan 14-19 1989, San Francisco.

KIM, J. 1983 On the structure of wall-bounded turbulent flows. *Phys. Fluids* **26**, 2088-2097.

KIM, J. & MOIN, P. 1987 Transport of passive scalars in a turbulent channel flow. *Proc. Sixth Symp. Turb. Shear Flows*, Toulouse.

KIM, J. & MOIN, P. 1989 Active turbulence control in a wall bounded flow using direct numerical simulation. Submitted to *IUTAM Symposium on Structure of Turbulence and Drag reduction*, July 1989, Zurich, Switzerland.

KIM, J., MOIN, P. & MOSER, R.D. 1987 Turbulence statistics in fully-developed channel flow at low Reynolds number. *J. Fluid Mech.* **177**, 133-166.

LAURENCE, G. 1985 Advective formulation of large eddy simulation for engineering flow. In *Proc. EUROMECH Colloquium No. 199* (eds. U. Schumann and R. Friedrich), Vieweg and Sohn, Braunschweig.

LEE, M.J., KIM, J. & MOIN, P. 1986 Turbulence structure at high shear rate. *Proc. Sixth Symp. Turb. Shear Flows*, Toulouse.

LEE, M.J. & REYNOLDS, W.C. 1985 Numerical experiments on the structure of homogeneous turbulence. *Dept. of Mech. Engrg. Rept. TF-24*, Stanford U., Stanford, California.

LELE, S.K. 1989 Direct Numerical Simulation of Compressible Free Shear Flows. AIAA-89-0374, Reno Meeting, January 9-12.

LEONARD, A. 1974 Energy cascade in large-eddy simulations of turbulent fluid flows. *Adv. Geophys.* **18A**, 237-248.

LEONARD, A.D., HILL, J.C., MAHALINGHAM, S. & FERZIGER, J.H. 1988 Analysis of homogeneous turbulent reacting flows. In *Studying Turbulence Using Numerical Simulation Databases - II*. Proc. Summer Program 1988, Center for Turbulence Research, Stanford U., Report CTR-S88, 243-255.

LEITH, C.E. 1969 Numerical simulation of turbulent flows. In *Properties of Matter Under Unusual Conditions*, Interscience, New York, 267-271.

LEITH, C.E. 1978 Objective methods for weather prediction. *Ann. Rev. Fluid Mech.* **10**, 107-128.

LILLY, D.K. 1967 The representation of small-scale turbulence in numerical experiments. In *Proc. IBM Scientific Computing Symposium on Environmental Sciences*, IBM, White Plaines, NY.

LILLY, D.K. 1989 The length scale for sub-grid-scale parameterization with anisotropic resolution. Preprint, Center for Turbulence Research, Stanford.

LOWERY, P.S. & REYNOLDS, W.C. 1986. Numerical simulation of a spatially-developing, forced, plane mixing layer. *Dept. of Mech. Engrg. Rept. TF-26*, Stanford U., Stanford, California.

LUMLEY, J.L. 1967 The structure of inhomogeneous turbulent flows. In *Atmospheric Turbulence and Radio Wave Propagation*, ed. A.M. Yaglom & V.I. Tatarsky, NAUKA, Moscow, 166-178.

MANSOUR, N.N. 1989 Private communication, NASA/Ames Research center.

MANSOUR, N.N., KIM, J. & MOIN, P. 1988 Reynolds-stress and dissipation rate budgets in a turbulent channel flow. *J. Fluid Mech.* **194**, 15-44.

MASON, P.J. 1987 Large eddy simulation of a convective atmospheric boundary layer. *Proc. Sixth. Symp. Turb. Shear Flows*, Toulouse; to appear, Springer-Verlag.

MASON, P.J. & CALLEN, N.S. 1986 On the magnitude of the subgrid-scale eddy coefficient in large eddy simulation of turbulent channel flow. *J. Fluid Mech.* **162**, 439-462.

MASON, P.J. & THOMSON, D.J. 1987 Large-eddy simulations of the neutral-static stability planetary boundary layer. *Q. J. R. Meteorol. Soc.* **113**, 413-443.

MELANDER, M.V. & HUSSAIN, F. 1988 Cut-and-connect of two antiparallel vortex tubes. In *Studying Turbulence Using Numerical Simulation Databases - II*. Proc. Summer Program 1988, Center for Turbulence Research, Stanford U., Report CTR-S88, 257-286.

METCALFE, R.W., ORSZAG, S.A., BRACHET, M., MENON, S. & RILEY, J.J. 1987 Secondary instability of a temporally growing mixing layer. *J. Fluid Mech.* **184**, 207-244.

MOENG, C.H. 1986 Large eddy simulation of stratus-topped boundary layer: Part I, Structure and budgets. *J. Atmos. Sci.* **43**, 2886-2900.

MOENG, C.H. & WYNGAARD, J.C. 1986 An analysis of closures for pressure-scalar covariances in the convective boundary layer. *J. Atmos. Sci.* **43**, 2499-2531.

MOIN, P. & KIM, J. 1982 Numerical investigation of turbulent channel flow. *J. Fluid Mech.* **118**, 341-378.

MOIN, P. & KIM, J. 1985 The structure of the velocity field in turbulent channel flow. part I. Analysis of instantaneous fields and statistical correlations. *J. Fluid Mech.* **155**, 441-464.

MOIN, P. & MOSER, R.D. 1989 Characteristic eddy decomposition of turbulence in a channel. *J. Fluid Mech.* **200**, to appear.

MOIN, P., SHIH, T.S., DRIVER, D. & MANSOUR, N.N 1989 Numerical simulation of a three-dimensional turbulent boundary layer. *AIAA-89-0373*, Reno, Nevada.

MOIN, P. & SPALART, P.R. 1989 Contributions of numerical simulation data bases to the physics, modeling, and measurement of turbulence. In *Advances in Turbulence*, Ed. George and Arndt, Hemisphere, 11- 38 (also *NASA TM 100022*).

MORTAZAVI, M., KOLLMAN, W. & SQUIRES, K. 1987 A statistical investigation of the single-point pdf of velocity and vorticity based on direct numerical simulations. In

Studying Turbulence Using Numerical Simulation Databases. Proc. Summer Program 1987, Center for Turbulence Research, Stanford U., Report CTR-S87, 121-132.

MOSER, R.D. 1988 Statistical analysis of near-wall structures in turbulent channel flow. *NASA TM 100092.*

MOSER, R.D. & MOIN, P. 1987 Direct numerical simulation of curved channel flow. *J. Fluid Mech.* **175**, 479-510.

NIEDERSCHULTE, M.A. 1988 Turbulent flow through a rectangular channel. *P.D. Thesis*, Department of Chemical Engineering, Univ. Illinois, Urbana.

NIEUWSTADT, F.T.M. & DE VALK, P.J.P.M.M. 1987 A large eddy simulation of buoyant and non-buoyant plume dispersion in the atmospheric boundary layer. *Atmos. Environm.* **21**, 2573-2587.

ORSZAG, S.A. & PATERA, A.T. 1984 A spectral element method for fluid dynamics; laminar flow in a channel expansion. *J. Comp. Phys.* **54**, 468.

ORSZAG, S.A. & PATTERSON, G.S. 1972 Numerical simulation of three-dimensional homogeneous isotropic turbulence. *Phys. Rev. Lett.* **28**, 76-79.

PETERSON, V.L., KIM, J., HOLST, T., DEIWERT, G.S., COOPER, D.M., WATSON, A.B. & BAILEY, F.R. 1989 Supercomputer requirements for selected disciplines important to aerospace. *J. IEEE*, to appear.

PIOMELLI, U., FERZIGER, J.H. & MOIN, P.1988 Models for large eddy simulation of turbulent channel flows including transpiration. *Dept. of Mech. Engrg. Rept. TF-32*, Stanford U., Stanford, California.

PIOMELLI, U., MOIN, P. & FERZIGER, J.H. 1988 Model consistency in the large eddy simulation of turbulent channel flow. *Phys. Fluids* **31**, 1884-1891.

RAI, M.M. & MOIN, P. 1989 Direct simulation of turbulent flows using finite-difference schemes. *AIAA-89-0369*, Reno, Nevada.

REYNOLDS, W.C. & SHIH, T.S. 1989 Rapid distortion of anisotropic homogeneous turbulence. *Unpublished work notes*, Center for Turbulence Research, Stanford.

RILEY, J.J., METCALFE, R.W. & ORSZAG, S.A. 1986 Direct numerical simulation of chemically reacting turbulent mixing layers. *Phys. Fluids* **29**, 406-422.

ROBINSON, S.K., KLINE, S.J. & SPALART, P.R. 1988 Quasi-coherent structures in the turbulent boundary layer: Part II. Verification and new information from a numerically simulated flat-plate layer. *Zoran P. Zaric Memorial International Seminar on near-Wall Turbulence*, Dubrovnik, Yugoslavia, May 16-20.

ROGALLO, R.S. 1981 Numerical experiments in homogeneous turbulence. *NASA TM-81315.*

ROGALLO, R.S. & MOIN, P. 1984 Numerical simulations of turbulent flows. *Ann. Rev. Fluid Mech.* **16**, 99-138.

ROGERS, M.M, MANSOUR, N.N. & REYNOLDS, W.C. 1989 An algebraic model for the turbulent flux of a passive scalar. *J. Fluid Mech.*, to appear.

ROGERS, M.M & MOIN, P. 1987a The structure of the voriticity field in homogeneous turbulent flow. *J. Fluid Mech.* **176**, 33-66.

ROGERS, M.M. & MOIN, P. 1987b Helicity fluctuations in incompressible turbulent flows, *Phys. Fluids* **30**, 2662-2670.

ROGERS, M.M., MOIN, P. & REYNOLDS, W.C. 1986 The structure and modeling of the hydrodynamic and passive scalar fields in homogeneous turbulent shear flow. *Dept. of Mech. Engrg. Rept. TF-25*, Stanford U., Stanford, California.

ROGERS, M.M. & MOSER, R.D. 1989 *Private communication*, NASA/Ames Research Center.

SANDHAM, N.D. & REYNOLDS, W.C. 1987 Some inlet plane effects on the numerically simulated spatially-developing mixing layer. *Proc. Sixth Symp. Turb. Shear Flows*, Toulouse, France, Springer-Verlag, in press.

SANDHAM, N.D. & REYNOLDS, W.C. 1989 The compressible mixing layer; linear theory and direct simulation. *AIAA 89-0371*, Reno, Nevada.

SCHMIDT, H. & SCHUMANN, U. 1988 Coherent structure of the convective boundary layer derived from large-eddy simulations. *DFVLR report IB-553 2/88*, to appear in *J. Fluid Mech.*.

SCHUMANN, U. 1975 Subgrid scale model for finite difference simulation of turbulent flows in plane channels and annuli. *J. Comp. Phys.* **18**, 376-404.

SCHUMANN, U. 1988 Large-eddy simulation of turbulent diffusion with chemical reactions in the convective boundary layer. *DFVLR report IB-553 23/88*, submitted to *Atmospheric Environment*.

SCHUMANN, U. & FRIEDRICH, R. 1987 On direct and large eddy simulation of turbulence. In *Advances in Turbulence*, Springer-Verlag, 88-104.

SCHUMANN, U., HAUF, T. HÖLLER, H., SCHMIDT, H. & VOLKERT, H. 1987 A mesoscale model for the simulation of turbulence, clouds, and flow over mountains; formulation and validation examples. *Beitr. Phys. Atmos.* **60**, 413-446.

SCHUMANN, U. & PATTERSON, G.S. 1978 Numerical study of pressure and velocity fluctuations in nearly isotropic turbulence. *J. Fluid Mech.* **88**, 711-735.

SMAGORINSKI, J. 1963 General circulation experiments with the primitive equations. I. The basic experiment. *Monthly Weather Review* **91**, 99-164.

SPALART, P.R. 1986 Numerical simulation of boundary layers; part 1; weak formulation and numerical method. *NASA TM 88222*.

SPALART, P.R. 1988a Direct numerical simulation of a turbulent boundary layer up to $R_\theta = 1410$. *J. Fluid Mech.* **187**, 61.

SPALART, P.R. 1988b Direct numerical study of leading-edge contamination. *AGARD Symposium on Fluid Dynamics of Three-dimensional Turbulent Shear Flow and Transition*, Oct 3-6, Cesme, Turkey.

SPALART, P.R. 1989 Theoretical and numerical study of a three-dimensional turbulent boundary layer. *J. Fluid Mech.*, to appear.

SPEZIALE, C.G. 1985 Galilean Invariance of subgrid-scale stress models. *J. Fluid Mech.* **156**, 55-62.

SPEZIALE, C.G., ERLEBACHER, G., ZANG, T.A. & HUSSAINI, M.Y. 1988 The subgrid-scale modeling of compressible turbulence. *Phys. Fluids* **31**, 940-942.

SPEZIALE, C.G., MANSOUR, N.N. & ROGALLO, R.S. 1987 Decay of turbulence in a rapidly rotating frame. In *Studying Turbulence Using Numerical Simulation Databases*. Proc. 1987 Summer Program, Center for Turbulence Research, Stanford U., Report CTR-S87, 205-212.

TSUBOI, K., TAMURA, T. & KUWAHARA, K. 1989 Numerical study of vortex induced vibration of a circular cylinder in high Reynolds number flow. *AIAA 89-0294*, Reno, Nevada.

YAKHOT, V. & ORSZAG, S.A. 1986 Renormalization group analysis of turbulence. I. Basic Theory. *J. Scientific Computing* **1**, 3-51.

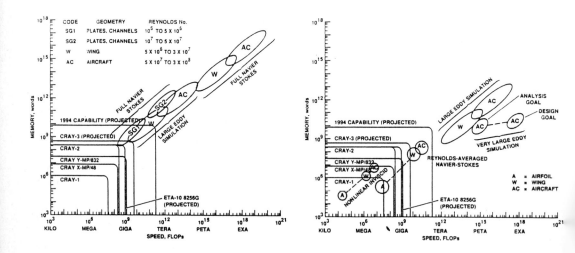

Figure 1. Projected use of simulations in turbulence physics (left) and aerothermodynamics (right). From Peterson *et al.* (1989).

SPALART, P.R. 1988b Direct numerical study of leading-edge contamination. *AGARD Symposium on Fluid Dynamics of Three-dimensional Turbulent Shear Flow and Transition*, Oct 3-6, Cesme, Turkey.

SPALART, P.R. 1989 Theoretical and numerical study of a three-dimensional turbulent boundary layer. *J. Fluid Mech.*, to appear.

SPEZIALE, C.G. 1985 Galilean Invariance of subgrid-scale stress models. *J. Fluid Mech.* **156**, 55-62.

SPEZIALE, C.G., ERLEBACHER, G., ZANG, T.A. & HUSSAINI, M.Y. 1988 The subgrid-scale modeling of compressible turbulence. *Phys. Fluids* **31**, 940-942.

SPEZIALE, C.G., MANSOUR, N.N. & ROGALLO, R.S. 1987 Decay of turbulence in a rapidly rotating frame. In *Studying Turbulence Using Numerical Simulation Databases*. Proc. 1987 Summer Program, Center for Turbulence Research, Stanford U., Report CTR-S87, 205-212.

TSUBOI, K., TAMURA, T. & KUWAHARA, K. 1989 Numerical study of vortex induced vibration of a circular cylinder in high Reynolds number flow. *AIAA 89-0294*, Reno, Nevada.

YAKHOT, V. & ORSZAG, S.A. 1986 Renormalization group analysis of turbulence. I. Basic Theory. *J. Scientific Computing* **1**, 3-51.

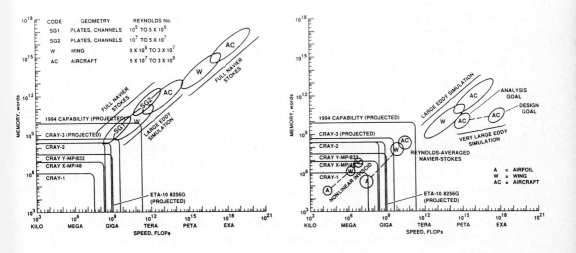

Figure 1. Projected use of simulations in turbulence physics (left) and aerothermodynamics (right). From Peterson *et al.* (1989).

On Large Eddy Simulation Using Subgrid Turbulence Models

Comment 1.

Jay P. Boris

Laboratory for Computational Physics and Fluid Dynamics
Naval Research Laboratory, Code 4400
Washington, D.C. 20375

1. Introduction

This short paper contains a preliminary and abbreviated statement of a perspective on Large Eddy Simulation using subgrid turbulence models which differs somewhat from the "classical" approach developed and referenced in Professor Reynolds position paper at this conference. The goals are the same, to study and predict complex fluid dynamic flows and turbulence in situations of engineering concern and scientific interest. Further much of the terminology is the same. The differences, where they exist, arise from alternate views of how certain necessary tradeoffs should be made and how best to optimize the overall simulation models. No proofs or derivations will be attempted for this perspective here. Most aspects of this perspective, however, are based on two decades of experience in CFD fields related to turbulence, thousands of completed simulations, and on common sense in regard to making the necessary numerical tradeoffs.

Some aspects of this evolving viewpoint have been published earlier, e.g., [1-5]. Recently my colleagues and I have been trying to understand what we have really learned about modeling complex fluid dynamics (turbulence) and about the intrinsically imprecise notion of Large Eddy Simulation (LES). We are working to organize, quantify and substantiate the strong evidence suggesting that monotone convection algorithms, which were designed to satisfy the physical requirements of positivity and causality, in effect have a minimal LES filter and matching subgrid model already built in. The positivity and causality properties, not built into other commonly used convection algorithms, seem to be sufficient to ensure efficient transfer of the residual subgrid motions, as they are generated by resolved field mechanisms, off the resolved grid with minimal contamination of the well-resolved scales by the numerical filter. Some of these conclusions should at least become plausible through the discussion to follow. We begin with four hopefully self evident statements:

1) If the fluid dynamic interactions at any particular scale are not resolved, no numerical model cannot be expected to give more than an approximation to the true behavior at that scale.

2) No model is perfect. Therefore, any simulation model contains imperfections arising from the necessary tradeoffs that have to be made to construct the model.

Topical Discussion relating to the position paper entitled *The Potential and Limitations of Direct and Large Eddy Simulations* by Professor W.C.Reynolds at the conference: **Whither Turbulence? or Turbulence at the Crossroads**, Cornell University, Ithaca, NY, March 21-24, 1989.

3) The effect on numerically resolved macroscopic scales in the flow of each individual unresolved small structure is itself small though the composite effects can be appreciable.

4) The largest unresolved scales are generally most important; smaller unresolved scales contribute less to the resolvable behavior.

Truly definitive numerical experiments are still very difficult to define because the range of scales available to direct simulation, while growing, is not very large. Furthermore, theoretically analyzing exactly how a given Large Eddy Simulation (LES) model and a given subgrid turbulence model will actually fit together is not practical and, as pointed out by Uriel Frisch at this conference, probably isn't even possible.

2. Discussion of CFD Algorithms and Requirements

Professor Reynolds has described the generic view of large eddy simulations (LES), derivation of the large-scale filtered equations, and the subgrid phenomenology added to the resolved field equations to approximate the large-scale effects of the unresolved small scale motions. He stresses the crucial importance of accuracy in the numerical methods and I agree completely. However, everything in large simulations involves tradeoffs. Certainly Rai and Moin [6] recently have employed finite differences in Direct Numerical Simulation (DNS) of the Navier-Stokes equations quite effectively because of their relative efficiency even though DNS problems have generally been treated by spectral algorithms in the past because of their greater accuracy in the well-resolved wavelength regimes.

The CFD requirements for DNS and LES are in fact different. In direct simulations, the smallest resolved scales are continuously being smoothed by viscous diffusion. The motions at the smallest dynamical scales are quite slow and the highest harmonics of the corresponding field variables are quite small, so local numerical errors have little effect. Since spectral methods shine at intermediate and long wavelength where physical viscosity gives relatively little smoothing, they generally have been a good match for DNS problems. In LES, the Reynolds number of the flow is so large that viscosity isn't effective in removing numerically-difficult steep gradients even on the smallest resolved scales. The spectral and energy content of gradients on these scales is thus correspondingly larger in LES problems than in DNS problems.

The global view of LES expressed here differs in an important way from Professor Reynolds position paper. I do not believe it is practical "to separate the formulation of the LES problem from the numerical method used for its solution." This separation is valuable in principle as it enables some numerical analysis of the resulting methods but this analysis is unsatisfying and incomplete. Furthermore, inverting the filtered-equation solutions to obtain the physically meaningful fields is a nuisance and the required subgrid models still require empirical calibration via experiments and simulations. In practice, the "subgrid fields" have to be matched to the "resolved fields" at the smallest resolved scales – just where the distinctions between various methods and algorithms are the very largest. Since this matching should be done with some representation of the fluid dynamics at all scales included once but not twice, the short-wavelength errors in the specific algorithms chosen cannot be ignored in practice.

By filtering the mathematical model in the usual way to obtain LES equations at the resolved scales, sufficient smoothing is added that the otherwise underresolved Navier-Stokes equations will be well behaved at the grid scale – even for algorithms not designed to control gradients at this scale. The price is a rather substantial influence of the filtering at larger scales where most algorithms would be accurate, even on the unfiltered equations. The "subgrid" model, in turn, must make up for the effects of this filtering on the well resolved scales, effectively extending the phenomenological modeling far into regimes of scale length where it shouldn't be needed.

As monotone nonlinear convection algorithms were designed to limit errors in the shortest resolved scales in a physically meaningful way (where sensible connection to a subgrid model is also required), they seem a better choice for use in LES models than linear convection. The cost of satisfying positivity and causality and the enhanced accuracy at short wavelength of monotone methods is somewhat larger errors at long wavelengths than found in spectral methods. Since these errors in long wavelengths are small in any case, the comparative advantage shifts to the monotone methods when accurate treatment of the smallest resolved scales is of paramount importance.

It is my experience that nonlinear monotone CFD algorithms really have a built-in filter and a corresponding built-in subgrid model. These monotone "integrated" LES algorithms are derived from the fundamental physical laws of causality and positivity in convection and do minimal damage to the longer wavelengths while still incorporating, at least qualitatively, most of the local and global effects of the unresolved turbulence expected of a large eddy simulation. These convection algorithms, when properly formulated, accept and transform the unresolved variability in the fluid field variables that is pushed to short wavelengths by nonlinear effects and instabilities. This variability is locally converted to the correct macroscopic variables, e.g., viscous dissipation of the unresolved scales appears as heat. Diffusion of the eddy transport type is automatically left in the flow as required but the fluctuating driving effects of random phase, unresolved eddies on the large scales is missing unless specifically included as a subgrid phenomenology. This deficiency, however, is common to all the subgrid models in current use.

Reynolds position paper discusses the notion of LES with no subgrid models, but doesn't address the possibility that these LES systems may already have crude built-in subgrid models, particularly if they are effectively monotone. In these LES models, as in fact in all others, Reynolds number effects cannot be simulated without viscosity appearing explicitly somewhere and boundary layer phenomenologies are always needed in any case for underresolved wall regions. Reynolds does observe that "LES is very resilient to the residual turbulence model". I agree fully and would carry this further, a factor of two increase in the spatial resolution of such monotone integrated LES models will bring more improvement in the accuracy of the resolved scales than all the work in the world on the subgrid model, particularly when a linear filter is being applied in the usual LES procedure.

Most of my experience comes from a particular class of monotone methods, the Flux-Corrected Transport (FCT) algorithms (e.g., [7–9]). Below the acronym FCT will occasionally be used when it should be understood that the comments could apply equally for a number of other suitably formulated monotone methods as well. Indeed, many comparable monotone methods now exist; see [10–12] and the extensive references in [3] for examples. Significant three-dimensional LES simulations using these methods have been performed for a number of problems but the concentration has been on highly transient systems. Originally these models were applied to problems with strong shocks and blast waves [5, 11, 13, 14], to detonations [5, 19], and to chemically reactive flows; see [3, 4, 20, 21] and the references therein. Recently our applications of monotone convection algorithms have migrated into the compressible shear flow and turbulence arenas [5, 15–18] so detailed comparisons in the usual DNS and LES contexts are beginning to become available and more can be expected in the next few years.

3. Properties a Good Subgrid Turbulence Model Should Have

One of my strongly held opinions is that as wide a range of scales as possible should be directly simulated; the subgrid models in LES should be restricted to the unresolved scales to the greatest extent possible with minimal effect on the resolved scales. The four "statements" in the introduction seem to point in this direction. The first two statements suggest that accurately calculating as much as possible is a good idea. The second two statements suggest that the fluid does not strongly conspire against the necessary LES partitioning into resolved and unresolved subgrid fields. Indeed, Reynolds mentioned, as an example of the results obtained from Direct Numerical Simulation (DNS), the fact that nearby scales seem to interact most strongly. The main effect of the large scales on widely separated small scales seems to be vortex stretching arising in the mean strain fields.

Next consider the properties that an ideal subgrid model should have.

1. It should apply without restriction to the fluid dynamic model being solved macroscopically, e.g., include compressibility, high Mach number, multispecies effects, etc., as appropriate to the problem at hand.

2. It should satisfy the global conservation laws of the system as integrated over the unresolved scales.

3. It should minimize the contamination of macroscopic scales by the inaccurately resolved flow structures on the grid scale and by the numerical filtering. This allows the resolvable linear and nonlinear processes which physically drive the subgrid dynamics can be calculated as accurately as possible.

4. It should accomplish the physical mixing and averaging expected of the complex but unresolved flows on the correct macroscopic space and timescales.

5. It should match smoothly onto the resolved macroscale solutions at each point in space, even for variable grid size. The effects of all scale lengths, whether modeled or resolved, should be included exactly once.

Several assumptions have to be made about the high Reynolds number fluid dynamic system being modeled and the set of equations being used for the LES–subgrid model approachs to make sense. These assumptions are based in part on the self-evident statements in the introduction and partly state distinctions between LES and Very Large Eddy Simulation (VLES) as introduced by Professor Reynolds.

1. The problem being solved is such that the macroscopic LES model can resolve the dynamics of the energy containing, turbulence-driving scales.

2. The macroscopic convection velocities are sufficiently larger than the unresolved turbulence velocities that small-scale turbulent motion of material, mass, momentum, and energy accounts for a small portion of the global transport in the problem.

3. Unresolved "turbulent" diffusion dominates molecular transport or else the molecular effects are explicitly included in the LES model equations.

Without these three conditions being satisfied, the expectation of any subgrid turbulence model working is small. Fortunately, any system being tackled by large eddy simulation satisfies these conditions essentially by definition. The first and second assumptions above guarantee that the component of the fluid dynamics which is done accurately contains most of the information of interest. Were this not the case, most of the solution would depend on the subgrid model, casting the predictive capability into the realm of phenomenology. The third assumption says that any transport phenomenon which is not resolved convection, and that is at least as important as the unresolved convection, is also included in the macroscopic model.

Thus, it remains to discuss the extent to which monotone, flux-limiting convection algorithms such as FCT contain an adequate built-in filter and subgrid model.

4. Monotone Convection Algorithms as Built-In Subgrid Models

Previous use on high-speed flow problems has given some indications that FCT and other monotone CFD methods are in fact LES models with an integrated subgrid algorithm for compressive (shock) phenomena in the gas dynamic equations. However, the FCT techniques actually were developed for linear and nonlinear convection with no distinction between compressional, rotational, and potential aspects of the flow. Thus, it should come as no surprise that FCT is an equally good LES algorithm for rotational flows. The nonlinear flux-correction procedure takes the role of a minimal subgrid model for rotational and compressible turbulent flows. Extensive time-dependent simulations of shear layers, jets, detailed acoustic-vortex interaction studies, reactive shocks, detonation cell structures, and turbulent diffusion flames show good agreement with experiments and known analytic solutions, where available, without addition of an external turbulence model [3].

The impression that some of the monotone methods were specifically developed for shock problems is true. For the simulation of nearly incompressible, low Mach-number flow, the repeated solution of the Rankine-Hugoniot problem to find fluxes seems unnecessarily wasteful. This current bias toward high-speed applications coupled with the origin of many of the monontone methods naturally has led to the impression that all monotone methods are limited to high Mach number. This impression is incorrect for FCT and many of the other monotone algorithms.

The Navier-Stokes equations with adequate spatial resolution have been solved using FCT, giving excellent results for vortex shedding from cylinders at intermediate Reynolds number. When the physical viscosity is small, or alternately the cells large, solutions of the Navier-Stokes equations and the Euler equations are essentially identical using FCT, both showing the effects of the flux-correction procedure as a residual, nonlinear filtering of short wavelengths. This filtering influences long wavelengths negligibly and yet is strong enough at short wavelength to prevent aliasing of high frequencies into the long wavelengths. This was shown for Berger's equation in detailed comparisons with a spectral model. It has subsequently been checked repeatedly for fluid dynamics through spatial convergence tests in every major configuration where FCT has been applied to jets, shear layers, and reacting flows [15–18]. These tests have been done in three dimensions as well as two, and generally show a converged long wavelength behavior when the system size is large enough to support at least a modest ratio between the energy containing long wavelengths and the eddies of a few cells wavelength which dissipate quickly. Coherent vortex structures more than 15–20 cells across change negligibly when the resolution is doubled or quadrupled to allow resolution of much smaller scales.

Monotone algorithms, like other algorithms that use knowledge of the grid relative to variations of the evolving solution, cannot be Galilean invariant. Adding a constant velocity to the flow everywhere moves real structures in the computed solution to different locations relative to the grid at the end of each timestep. However, the Gibbs phenomenon error, which arises from finite resolution associated with convection across the grid, is present regardless of the solution algorithm. This Gibbs error is also not Galilean invariant. In fact, the non-Galilean feature of monotone algorithms is designed to cancel this non-Galilean aspect of the solution arising from the representation. The composite interaction of a monotone algorithm on the representation gives a solution which is essentially Galilean invariant.

Any algorithm which itself is Galilean invariant will be unable to cancel the Gibbs error without extensive diffusion. Thus the solution will either be highly diffusive at all wavelengths or it will not be even approximately Galilean invariant. In DNS applications the real viscosity provides adequate diffusion for the resolved velocity field. Here adequate diffusion is defined to be at least as much as occurs in first order upwind algorithms. The price of this approximate Galilean invariance is a severe limit on the Reynolds number which can be reached. In multimaterial flows or flows with contact surfaces, viscosity is not generally adequate to ensure Galilean invariance for physical variables which are absent in constant density, incompressible flow.

Monotone convection algorithms such as FCT implicitly provide a minimal subgrid model which enforces, at least in qualitatively correct form, the five desired properties of a subgrid turbulence model which were identified earlier. Properties 1, 2 and 3 are built into the formulation of the basic flux-corrected convection algorithm. Existing explicit subgrid models are generally limited to constant density, incompressible, non-reacting flows on uniformly spaced meshes, effectively violating property 1.

Property 4 is enforced by FCT through the residual local diffusion left to enforce property 3. This feedback clearly loses phase information about the unresolved small scales but other subgrid models also lose this information. A random local excitation of long wavelengths is required and could easily be added, since the exact amount of FCT averaging is known, but this has not been done to date. Other subgrid models generally lack these random feedback effects.

Property 5, that the subgrid model, match smoothly onto the LES model, is the most attractive aspect of montone convection methods. A consistent and integrated viewpoint is used to convert unresolvable fluid dynamics (and grid resolution limitations) into subgrid field. Even with spatially and temporally varying grids, the residual long wavelength transport from the flux limiters acting on the subgrid variability is included causally and consistently while the (Property 3).

The intrinsic filter in monotone algorithms is solution and grid dependent but the solution converges to the solution of the underlying partial differential equations being solved. This means that the well resolved field solutions differ at most slightly from the exact (laminar) solutions of the equation set being modeled. Inverting this built-in filter is not possible but this inversion also should not be necessary for at least the well resolved scales of motion. The filter can be applied to experimental data or to theoretical models if more detailed comparison with the resolved scales is desired.

Using linear filters in the standard LES approach is simplified appreciably by selecting filters that commute with the derivative operators. This commutation usually fails with variably spaced grids so standard LES also generally assumes uniformly spaced grids in any direction where the formalism is applied, clearly a serious liability for engineering-level simulations. This restriction appears not to be necessary in principle, however, as the monotone algorithm filters clearly do not commute with derivatives and yet the numerical solutions converge to the solutions of the desired equations.

Since these monotone methods have not long been thought of as LES models in this light, there are clearly areas where work and extensive verification are needed. It has to be demonstrated that the residual average transport at long wavelength from unresolved subgrid turbulence is large enough. Almost certainly additional eddy viscosity must be added to the minimal amount provided by the algoritms. The essentially random and fluctuating components of the subgrid fields are also missing from these integrated LES models as well as from other LES models, and ought eventually to be modeled. The cell-averaged source terms which drive these fluctuations are available, however, as they are contained in the components of the fluxes removed in the nonlinear limiting process. Physical assumptions about the short timescale temporal behavior (cascade) and spatial characteristics of the unresolved motions have to be made. All that is known about the subgrid fields during the simulation has to be inferred from the resolved fields. RNG makes one set of assumptions about these scales. Other models make different assumptions.

Source terms in the LES equations can be included for these subgrid fluctuations once the community has decided what is scientifically appropriate. Clearly this will be another phenomenology. It appears that such terms should have a local and random aspect on the macroscale so that resolvable scale flow instabilities and hence turbulent structure can be triggered by dynamics on the unresolved scales. This, also, is easy to do. Furthermore, these fluctuating subgrid source terms automatically will lead to additional macroscopic transport because the monotone flux-limiters will work on these subgrid-determined effects as well as on the macroscopic effects.

5. Summary

Above I have argued that nonlinear monotone methods really have at least a minimal built-in LES filter and a matched subgrid model which do minimal damage to the longer wavelengths while still incorporating, at least qualitatively, most of the local and global effects of the unresolved turbulence expected of a large eddy simulation. When properly formulated, a wide variety of these monotone convection algorithms transform unresolvable structure in the fluid field variables, as it is pushed to short wavelengths by nonlinear convective effects and instabilities into the appropriate resolved fields. This grid scale variability is locally converted to the correct macroscopic variable averages, e.g., viscous dissipation of the unresolved scales appears as heat. Furthermore, these methods are quite capable of capturing quantitatively how much unresolved structure from the long wavelengths is actually present. Diffusion of the eddy transport type is automatically left in the flow as required but the fluctuating driving effects of random phase, unresolved eddies on the large scales is missing unless specifically included as a subgrid phenomenology. A factor of two increase in the spatial resolution of such LES models will bring more improvement in the accuracy of the well resolved scales than all the work in the world on the subgrid model of a more coarsely resolved LES model with the usual filtering procedure contaminating the long wavelengths unnecessarily. Satisfying proofs of these statements have not been provided here, but work is underway to do exactly this.

Acknowledgements

Research on numerous systems using this large eddy simulation, or under-resolved Navier-Stokes approach, has been supported by a number of sponsors including the Office of Naval Research, the Naval Research Laboratory, DNA, DARPA and DOE. The author also wishes to acknowledge the varied and creative contributions of his co-authors at NRL and colleagues around the world to understanding complex and turbulent fluid phenomena through numerical simulation using monotone methods.

References

1. J.P. Boris and E.S. Oran, Modelling Turbulence: Physics or Curve Fitting?, Presentation & Proceedings, 7th International Colloquium on Gasdynamics of Explosions and Reactive Systems (ICOGER), Gottingen, W. Germany, 20–24 August 1979, in **Combustion in Reactive Systems**, J.R. Bowen, et al. (eds), AIAA Progress in Astronautics and Aeronautics, Vol 76, pp 187–210, 1981.

2. E.S. Oran and J.P. Boris, Detailed Modelling of Combustion Systems, *Progress In Energy and Combustion Sciences*, Vol 7, pp 1–72, 1981.

3. E.S. Oran and J.P. Boris, **Numerical Simulation of Reactive Flow**, Elsevier Science Publishing Company, New York, 1987.

4. J.P. Boris, E.S. Oran, and K. Kailasanath, The Numerical Simulation of Compressible and Reactive Turbulent Structures, Invited Presentation & Proceedings, Joint US/France Workshop on Turbulent Reactive Flows, Rouen, France, 7–10 July 1987, Springer-Verlag, Eds. B.G. Murthy and R. Borghi, 1988.

5. J.M. Picone and J.P. Boris, Shock Generated Turbulence, also E.S. Oran, J.P. Boris, and K. Kailasanath, Computations of Supersonic Flows: Shear Flows, Shocks, and Detonations, to appear in Proceedings, International Workshop on the Physics of Compressible Turbulent Mixing, Princeton University, Princeton NJ, 24–27 October 1988, Springer, 1989.

6. M.M. Rai and P. Moin, Direct Simulation of Turbulent Flows Using Finite-Difference Schemes, AIAA Paper 89–0369, 27th Aerospace Sciences Meeting, Reno NV, AIAA, Washington DC, January 1989.

7. J.P. Boris, New Directions in Computational Fluid Dynamics, **Annual Reviews of Fluid Mechanics**, Vol 21, pp 345–385,1989.

8. J.P. Boris and D.L. Book, Flux-Corrected Transport I: SHASTA - A Fluid Transport Algorithm that Works, *Journal of Computational Physics*, Vol 11, pp 38–69, 1973.

9. J.P. Boris and D.L. Book, Solution of the Continuity Equation by the Method of Flux-Corrected Transport, Chapter 11 in **Methods in Computational Physics**, Vol 16, pp 85–129, Academic Press, New York, 1976.

10. B. van Leer, Towards the Ultimate Conservative Difference Scheme. V. A Second-Order Sequel to Godunov's Method, *Journal of Computational Physics*, Vol 32, pp 101–136, 1979.

11. P.R. Woodward and P. Colella, The Numerical Simulation of Two-Dimensional Fluid Flow with Strong Shocks, *Journal of Computational Physics*, Vol 54, pp 115–173, 1984.

12. P. Colella and P.R. Woodward, The Piecewise Parabolic Method (PPM) for Gas-Dynamical Simulations, *Journal of Computational Physics*, Vol 54, pp 174–201, 1984.

13. J.D. Baum, S. Eidelman, and D.L. Book, Numerical Modeling of Shock Wave Interaction with Conical Blast Shields, Proceedings of the 9th International Symposium on Military Applications of Blast Simulation, Oxford, U.K., June 1985.

14. J.D. Baum and S. Eidelman, Time-Dependent Three-Dimensional Simulation of Shock Wave Interaction with a Cylindrical Cavity, in **Computational Fluid Dynamics**, G. de Vahl and C. Fletcher (editors), Elsevier Science Publisher, 1988, pp 267-281.

15. R.H. Guirguis, F.F. Grinstein, T.R. Young, E.S. Oran, K. Kailasanath and J.P. Boris, Mixing Enhancement in Supersonic Shear Layers, AIAA Paper 87-0373, 25th Aerospace Sciences Meeting, Reno NV, AIAA, Washington DC, 1987.

16. K. Kailasanath, J. Gardner, J. Boris and E. Oran, Numerical Simulations of Acoustic-Vortex Interactions in a Central-Dump Ramjet Combustor, *J. Prop. Power*, 3, 525–533, 1987.

17. F.F. Grinstein, R.H. Guirguis, J.P. Dahlburg and E.S. Oran, Three-Dimensional Numerical Simulation of Compressible, Spatially Evolving Shear Flows, Proceedings, 11th International Conference on Numerical Methods in Fluid Mechanics, Williamsburg, Springer Verlag, to appear, 1988.

18. F.F. Grinstein, F. Hussain and E.S. Oran, Three-Dimensional Numerical Simulation of a Compressible, Spatially Evolving Mixing Layer, AIAA Paper 88-0042, Reno NV, 11–14 January 1988, see also A Numerical Study of Moving Control in Spatially Evolving Shear Flows, AIAA Paper 89-0977, AIAA 2nd Shear Flow Conference, Tempe AZ, AIAA, Washington DC, March 1989.

19. K. Kailasanath, E.S. Oran, J.P. Boris, and T.R. Young, Determination of Detonation Cell Size and the Role of Transverse Waves in Two-Dimensional Detonations, *Comb. Flame*, 61, 199–209, 1985.

20. K. Kailasanath, J.H. Gardner, E.S. Oran, and J.P. Boris, Effects of Energy Release on High-Speed Flows in an Axisymmetric Combustor, AIAA Paper 89-0385, AIAA 27th Aerospace Sciences Meeting, Reno NV, AIAA, Washington DC, January 1989.

21. G. Patnaik, K. Kailasanath, and E.S. Oran, Effect of Gravity on Flame Instabilities in Premixed Gases, AIAA Paper 89-0502, AIAA 27th Aerospace Sciences Meeting, Reno NV, AIAA, Washington DC, January 1989.

The Potential and Limitations of Direct and Large-Eddy Simulations

Comment 2.

M. Yousuff Hussaini and Charles G. Speziale

Institute for Computer Applications in Science and Engineering
NASA Langley Research Center
Hampton, VA 23665

Thomas A. Zang

Computational Methods Branch
NASA Langley Research Center
Hampton, VA 23665

1. Introductory Remarks

There is no question that both direct and large-eddy simulations will play an important role in turbulence research for many years to come. The position paper of Professor Reynolds makes two major points in this regard: (a) direct numerical simulations (DNS) will be used primarily to gain a better understanding of basic turbulence physics and to assess the performance of models and theories, and (b) large-eddy simulations (LES) will be used as a prediction and design tool for complex turbulent flows of scientific and engineering interest. Professor Reynolds also alluded to the point that numerical simulations will not supplant the need for physical experiments and turbulent closure models but rather will complement them. Physical experiments will be necessary to gain a better understanding of the physics of turbulent flows whose high Reynolds numbers and/or geometrical complexities make them inaccessible to direct numerical simulations. Likewise, turbulent closure models will probably need to be used for engineering design purposes in the forseeable future.

We agree with most of the points made in the Reynolds position paper, but feel that it would be useful to clarify and elaborate on some issues concerning the accomplishments and limitations of direct and large-eddy simulations. In particular, some clarifications are needed in regard to the current status of numerics for turbulence simulations, and Renormalization Group Methods. In addition, we shall present our views concerning some critical problems that must be overcome before large-eddy simulations can be used on a routine basis for the prediction and control of engineering turbulent flows. These points will be discussed in detail along with our views of the prospects for future research.

2. Numerics and Mathematical Theory

The choice of a numerical method for incompressible DNS and LES remains a compromise between the accuracy produced by a high-order method and the speed achieved by a low-order scheme. By now there is overwhelming numerical evidence that the spatial accuracy in direct simulations ought to be at least fourth-order, while there is substantial sentiment that large-eddy simulations should be high-order as well. On this last point, however, considerable philosophical disagreement remains. One camp argues that subgrid-scale modeling inaccuracies vitiate any practical advantages of high-order methods over simple second-order schemes (Schumann, Grotzbach, and Kleiser [1]). The other camp contends that high-order schemes permit clear distinctions to be made between numerical and modeling errors. In special cases, such as problems amenable to tensor-product grids, many (including our own), but not all, groups have preferred fully spectral schemes (see Canuto, Hussaini, Quarteroni, and Zang [2]) to high-order finite-difference or finite-element methods (Browning and Kreiss [3]). For more general geometries the choice is between fourth- or perhaps sixth-order finite-difference methods (see Krist [4] and Rai and Moin [5]) and spectral domain decomposition methods (reviewed in Chapter 4 of Zang, Streett, and Hussaini [6]). The paper by Rai and Moin [5], referred to by Reynolds, describes a (mostly) sixth-order finite-difference scheme that shows promise, although it has been tested thus far only on a tensor-product grid for which fast elliptic solvers are available and for which the deterioration in accuracy near the boundary seems not to be significant. We must note, however, that this paper contains no evidence to support Reynolds' speculation that high-order finite-difference methods may be more efficient than spectral domain decomposition approaches. It will take several years for sufficient experience with both types of methods to accumulate before even tentative conclusions about their relative efficiency can be drawn. We suspect that in the end neither approach will prove uniformly preferable.

We concur with Professor Reynolds' statement that the key to the efficiency of these methods is rapid elliptic solvers, and observe that the multigrid community has already produced fairly general techniques that are applicable to complex three-dimensional geometries. What remains is for these to be implemented in DNS and LES codes (see Krist and Zang [7] and Krist [4] for some DNS applications in tensor-product environments). Moreover, the prospects are quite high that the computers that achieve the 100 Gigaflops to 1 Teraflop speeds projected by Reynolds and many others will be highly parallel machines with programming considerations quite unlike those for conventional supercomputers (Cray, Cyber-205, Fujitsu, etc.) on which most large-scale turbulence simulations have been performed. Thus, a major effort is required on highly parallel algorithms for turbulence simulations in general geometries.

Professor Reynolds notes that Horiuti [8] demonstrated that the so-called Arakawa form (more properly called the skew-symmetric form) of the nonlinear terms in the momentum equation was superior to the rotation form for LES simulations. However, as documented for a variety of fully spectral transition and turbulence simulations by Zang [9], the problem with the rotation form is not one of greater truncation error but rather one of greater aliasing errors. Indeed, Zang's results indicate that the difference between aliased and de-aliased spectral calculations is very much less when the rotation form is replaced by the skew-symmetric form. Moreover, there is now substantial additional numerical evidence (Ronquist [10]) that even for spectral domain decomposition methods, the skew-symmetric form yields much more accurate results than the rotation form.

Numerical simulations of turbulence have traditionally used time-discretizations with an accuracy ranging between second- and fourth-order, with an implicit treatment of the viscous terms and an explicit treatment of the advection terms. The codes are typically operated close to the stability limit. This approach is certainly adequate for the lower-order statistics but appears to be questionable for the higher-order ones since the smaller scales are treated less accurately by the time-discretization.

The major question facing simulations of compressible flow is what to do about the shock waves: resolve them, fit them, or capture them. At sufficiently low Reynolds number they can, of course, be resolved. This has been the approach adopted, for example, by Passot and Pouquet [11], Lele [12], and Erlebacher, Hussaini, and Kreiss [14]. But there are pressing technological issues involving strong shocks whose internal structure is impractical to resolve numerically. Some basic issues can be addressed numerically by shock-fitting techniques (Zang, Hussaini, and Bushnell [13]), but this approach is not practical for most flows. The dilemma facing shock-capturing methods is that the numerical viscosity required for capturing the shock may well seriously distort small-scale turbulent features of interest near the shock. The model problem studied by Zang, Hussaini, and Bushnell [13] is a good test bed for shock-capturing methods and has been used recently to check some modern upwind methods (Shu and Osher [16], Zang, Drummond, and Kumar [17]). The recent studies of free shear layers by Soetrisno, Eberhard, Riley, and McMurtry [18] are one example of the use of shock-capturing in this context.

In recent years mathematicians have made significant progress in the analysis of the relevant range of scales in turbulent flow that has important implications for the numerical simulation of turbulence. Foias, Manley, and Temam [19] estimate the cut-off wavenumber at which exponential decay of the spectrum ensues. Henshaw, Kreiss, and Reyna [20], [21] show that the minimum scale of turbulence is inversely proportional to the square root of the Reynolds number based on the viscosity and the maximum of the vorticity. This information can be used to predict accurately the resolution requirements for DNS. The latter work also shows that in two-dimensional incompressible turbulence, the flow passes from a stage with a k^{-3} spectrum to a stage with a k^{-4} spectrum. In the first stage the exact dissipation mechanism is extremely important. For example, a hyperviscosity subgrid-scale dissipation produces a physically incorrect result. In the second stage the enstrophy decays very slowly and the Euler equations are a good approximation. Results for three-dimensional turbulence appear to require an assumption on the maximum vorticity in the flow. Henshaw, Kreiss, and Reyna [20] show rigorously that if the maximum norm of the vorticity scales as $\nu^{-1/2}$, then the Kolmogorov power law follows.

3. Direct Numerical Simulations

Incompressible Flows

Existing data bases on homogeneous turbulence have been quite useful in studying turbulence structure as discussed in the Reynolds position paper. The most useful data bases have been those on plane shear, plane strain, and the axisymmetric contraction/expansion. While the homogeneous data bases at NASA Ames/Stanford based on the pioneering work of Rogallo [22] are quite extensive, it should be mentioned that there are some notable deficiencies. For instance, no simulations have been conducted on plane shear or plane strain with rotations (only coarsely resolved large-eddy simulations were conducted for rotating shear flow by Bardina, Ferziger, and Reynolds [23]). Such flows can be extremely useful in calibrating one-point turbulence closure models as demonstrated by Speziale and MacGiolla Mhuiris [24]. The large-eddy and direct simulations on rotating isotropic turbulence conducted by Bardina, Ferziger, and Rogallo [25] and Speziale, Mansour, and Rogallo [26] are interesting but possibly inappropriate for calibrating one-point turbulence closure models; the suppression of the energy cascade by a system rotation (which is the primary physical effect manifested in rotating isotropic turbulence), is decidedly a two-point phenomenon that is not likely to be described properly within the framework of the usual Reynolds stress closures.

The DNS of inhomogeneous turbulent flows have primarily concentrated on boundary layer and plane and curved channel flow. As pointed out in the Reynolds position paper, this work has shed some new light on the turbulence structure near solid boundaries. However, a word of caution is but proper. The Reynolds number of the channel flow simulations are so low that serious questions can be raised about the applicability of the results to high Reynolds number turbulent channel flow with fully developed turbulence statistics. A minimization of Reynolds number effects (i.e., for the turbulence statistics outside the wall layer to be independent of the Reynolds number) would require at least an eight-fold increase in Reynolds number (i.e., from $Re \approx 4000$ to $Re \approx 32,000$). Such computations would require approximately a three orders of magnitude increase in computational power – a figure that will be achievable in the not too distant future. Such an increase in computational capacity would also allow for homogeneous turbulence simulations that are within the Reynolds number range of many of the physical experiments that have been conducted. Hence, while we feel that some significant strides have been made in studying turbulence physics based on DNS data bases, we believe that results of a lasting impact are more likely to be achieved by the higher Reynolds number data bases that are likely to be generated after the next decade (i.e., by the turn of the century).

The work of Gilbert and Kleiser [27] has shown that the constant flow rate formulation of turbulent channel flow is more economical than the constant pressure gradient formulation that has been customary. The recent landmark calculation by Gilbert [28], in collaboration with Kleiser, of the complete transition to turbulence process in channel flow bears mention. For the first time they have reliably computed both the transitional region and the fully turbulent regime. As recent calculations by Zang and Krist [29] suggest, this achievement was due in part to a fortunate choice of parameters, for at higher Reynolds numbers or with different initial conditions the transitional regime may actually require substantially more resolution than the turbulent regime. Nevertheless, Gilbert's data does offer the first opportunity to examine the final stages of breakdown into turbulence. We should also note that the results of Zang and Krist for transitional flow clearly indicate the mechanisms responsible for inverted vortices.

Compressible Flows

Work done at Langley Research Center (Erlebacher, Hussiani, and Kreiss [14] and Speziale, Erlebacher, Zang, and Hussaini [15]) has also shown that in compressible DNS of homogeneous turbulent flow, the flow statistics and the kinetic energy remain mostly incompressible when starting from essentially incompressible initial conditions. We have also shown that this is primarily due to the initial conditions for 2-D isotropic turbulence. Passot and Pouquet [11] state that it is the relative importance of the fluctuating Mach number, Ma, relative to the rms density fluctuations, ρ_{rms} which determines the final outcome of the turbulent structures, and these results have been confirmed by Erlebacher, et al.[14]. If $Ma^2 << 1$ and $\rho_{rms} = O(1)$, then weak shocks prevail and most of the kinetic energy is compressible; on the other hand, if $Ma^2 << 1$ and $\rho_{rms} = O(Ma)$, no shocks occur and most of the kinetic energy is solenoidal. The energy spectra and contours of the divergence of velocity associated with these two regimes which are taken from the direct simulations of Erlebacher et al. [14] on 2-D isotropic turbulence are illustrated in Figures 1-4. Of course, strong shocks appear if $Ma = O(1)$. The curvature of these shocks then generates localized regions of intense vorticity. Passot and Pouquet as well as Erlebacher, Hussaini and Kreiss [14] have observed shocks/shocklets in 3-D isotropic, homogeneous turbulence as well. The dynamics seem to agree with the 2-D case, but at a higher Ma in 3-D.

The fact that shocks have not yet been found in three-dimensional shear flow simulations at NASA Ames Research Center does not prove that they do not exist. Erlebacher, et al have already found eddy shocklets in three-dimensional isotropic turbulence, and it seems reasonable to expect that a proper choice of initial conditions will induce them in shear layers as well. There is a large parameter regime to study. Although not all of the important mechanisms of supersonic compressible turbulence can be extracted from two-dimensional simulations, the uniquely compressible effects arising from the interactions of large-scale shocks with turbulence are present in two-dimensions and can provide some useful information.

4. Large-Eddy Simulations

It has been speculated since the mid 1970's that large-eddy simulations would some day become a routine tool for the analysis of complex turbulent flows that cannot be predicted by existing turbulence models. To date, the accomplishments of LES have been somewhat disappointing. As mentioned in the position paper of Reynolds, LES has "paved the way for the use of direct numerical simulations in turbulence". However, there have been few practical uses of LES outside of the geophysical fluid dynamics community (e.g., in weather prediction) and there the details of the turbulence structure do not seem to be so important. There is no doubt that the lack of wide availability of supercomputers and the redirection of research efforts toward direct simulations have stunted the growth of LES as pointed out in Reynolds paper. However, it is our opinion that there are some major difficulties with LES that need to be overcome before it can yield reliable and economically feasible predictions for the complex turbulent flows of scientific and engineering interest. These problems are as follows:

(i) the implementation of LES in spectral domain decomposition or high-order finite-difference codes so that complex geometries can be treated;

(ii) the development of improved subgrid scale models for strongly inhomogeneous turbulent flows (e.g., flows with localized regions of relaminarization or large mean velocity gradients);

(iii) the development of reliable a priori tests for the screening of new subgrid scale models;

(iv) the problem of defiltering;

(v) the problem of modifying subgrid scale models to accomodate integrations to solid boundaries.

In so far as point (i) is concerned, it must be emphasized that most of the successes of LES have been for either homogeneous turbulent flows or simple inhomogeneous turbulent flows (e.g., channel flow). To accommodate complex geometries, either spectral domain decomposition techniques or finite-difference or finite-element methods must be used. Proper filtering techniques which are appropriate for such methods must be developed. Furthermore, to compute turbulence statistics, either ensemble or time averages must be implemented in lieu of the spatial averages used for homogeneous turbulent flows or the simple planar averages used in turbulent channel flow. These complicating features can increase by up to two orders of magnitude the computational time needed in comparison to the LES of more simple homogeneous turbulent flows. In fact, the latter problem of constructing ensemble or time averages instead of spatial averages makes it, in our opinion, highly unlikely that the LES of complex three-dimensional flows will be less expensive than computations with more sophisticated Reynolds stress models such as second-order closures.

In order to obtain reliable predictions for strongly inhomogeneous turbulent flows with large mean field gradients there is the likely prospect that simple eddy viscosity models will not suffice as alluded to in point (ii). Subgrid scale transport models that account for non-local and history-dependent effects may well be needed (cf., Deardorff [30]). Such models have the potential of increasing the computational efforts required by as much as a factor of five. Furthermore, there is the need for a substantial research effort to develop transport models that can yield reliable predictions and are well-behaved.

The more efficient development of improved subgrid scale models would be aided considerably by the development of reliable a priori tests as mentioned in point (iii). The commonly used a priori tests (see Clark, Ferziger, and Reynolds [31] and Bardina, Ferziger, and Reynolds [23]) are rather unreliable in predicting the performance of LES as alluded to in the position paper of Reynolds. To elaborate further on this point, it has long been speculated that the success of LES hinges on the use of a subgrid scale model that dissipates the right amount of energy in order to compensate for the energy cascade to scales that are not resolved by the computational mesh. However, the Smagorinsky model has been shown to yield a correlation coefficient of roughly 0.5 based on dissipation. To gain an appreciation as to how poor this correlation is, it should be noted that the correlation between the functions $y = x$ and $y = \exp(-x)$ (two functions with entirely different qualitative behavior) is 0.696 on the interval [0,1]! Yet, despite this extremely poor correlation of 0.5, the Smagorinsky model has performed well in many large-eddy simulations. Consequently, it is clear that alternative means of correlating subgrid scale models with the results of DNS are needed. One possible alternative might be to base the correlations on dissipation subject to the condition that the subgrid scale dissipation is some specified, substantial amount larger than the viscous dissipation. This issue needs to be explored in more detail in the near future.

Anisotropy Tensor	Large-Eddy Simulations	Experiments
b_{11}	0.305	0.201
b_{22}	-0.265	-0.148
b_{12}	-0.15	-0.142

Table 1. Comparison of the anisotropy tensor obtained from the large-eddy simulation of Bardina, et al. [23] for homogeneous turbulent shear flow with the physical experiments of Tavoularis and Corrsin [32].

If higher-order turbulence statistics are needed beyond the mean velocity, the problem of defiltering arises (point (iv)). This problem becomes most critical when more than 10-20% of the turbulent kinetic energy is in the subgrid scale motion. Typically as much as 50% of the turbulent kinetic energy is in the SGS fields in practical LES. (Certainly if the value were only 10%, little computational savings would be gained for a given Reynolds number!). With so much energy in the subgrid scale motions, considerable errors can be introduced in the computation of second and higher-order moments (see Table 1). This points to the need for defiltering wherein an estimate must be made for the contributions from the subgrid scale fields (see Bardina et al. [23]). It must be said at the outset that there is no general, reliable method for defiltering. The problem of defiltering is tantamount to solving a Fredholm integral equation of the first kind

$$\overline{\Phi} = \int G(\mathbf{x} - \mathbf{x}') \Phi(\mathbf{x}') \, d^3 \mathbf{x}', \tag{1}$$

where Φ is any given flow variable, $\overline{\Phi}$ is its filtered form, and G is the filter function. Since a Fredholm integral equation is ill-posed in the sense of Hadamard (i.e., any numerical solution for Φ yields complete uncertainty in regard to the high frequency Fourier components of Φ), the higher-order statistics associated with $\overline{\Phi}$ can only be crudely estimated. Hence, LES can only generally give accurate information on first-order moments. More formal probabilistic methods for estimating the SGS contributions to second-order moments should be studied in the future.

Finally, we would like to make a few remarks on point (v) concerning the use of subgrid scale models near a solid boundary. Most of the LES conducted have used ad hoc modifications such as Van Driest damping near solid boundaries (e.g., Schumann [33], and Moin and Kim [34]). While such an approach is satisfactory for attached flows in simple geometries, it cannot be reliably applied to turbulent flows with separation or complex geometrical singularities. Here, we believe the Renormalization Group (RNG) approach of Yakhot and Orszag [35] holds more promise than other existing methods in bridging the eddy viscosity to the molecular viscosity as a solid boundary is approached. The LES of turbulent channel flow based on RNG (Orszag 1987, private communication) shows considerable promise in predicting the turbulence structure near solid boundaries. In our opinion, the RNG approach to LES deserves more future research.

5. Concluding Remarks

As outlined in the Reynolds position paper, significant strides have been made in the analysis of turbulence physics based on direct numerical simulations of the Navier-Stokes equations. The low Reynolds numbers of these simulations do somewhat minimize the long-range impact that the specific results obtained will have on our understanding of basic turbulence physics. Nevertheless, an important methodology has been developed which, by the turn of the century (with a projected three orders of magnitude increase in computational speed), will allow for the direct simulation of basic turbulent flows at Reynolds numbers that are high enough to be representative of the physical mechanisms observed in the turbulent flows of scientific and engineering interest.

In so far as large-eddy simulations are concerned, we believe that there are major operational problems that must be overcome (as outlined in this discussion) before they can become a more commonly used tool for the analysis of complex engineering flows. Nonetheless, we feel that there is the potential for making a major impact with LES in scientific and engineering computations so that research along these lines should be vigorously pursued.

Acknowledgements

The authors are indebted to Dr. Gordon Erlebacher for his assistance with the preparation of this paper. This research was supported by the National Aeronautics and Space Administration under NASA Contract NAS1-18605 while the authors were in residence at ICASE, NASA Langley Research Center, Hampton, Virginia 23665.

References

[1] Schumann, U.; Grotzbach, G.; and Kleiser, L.: Direct numerical simulation of turbulence, in Prediction Methods for Turbulent Flows, W. Kollmann (ed.), Hemisphere Pub. Corp., Washington, pp. 123-258 (1980).

[2] Canuto, C.; Hussaini, M. Y.; Quarteroni, A.; and Zang, T. A.: Spectral Methods in Fluid Dynamics, Springer-Verlag, Berlin (1988).

[3] Browning, G.; and Kreiss, H. O.: Comparison of numerical methods for the calculation of two-dimensional turbulence, Math. Comput., Vol. 52 (in press).

[4] Krist, S. E.: Direct Solution of the Navier-Stokes Equations Using Finite-Difference Methods, Ph.D. Dissertation, George Washington University (1989).

[5] Rai, M. M.; and Moin, P.: Direct Simulations of Turbulent Flow using Finite-Difference Schemes, AIAA Paper No. 89-0369 (1989).

[6] Zang, T. A.; Streett, C. L.; and Hussaini, M. Y.: Spectral methods for CFD, ICASE Report No. 89-13 (1989).

[7] Krist, S. E.; and Zang, T. A.: Simulations of Transition and Turbulence on the Navier-Stokes Computer, AIAA Paper No. 87-1110 (1987).

[8] Horiuti, K.: Comparison of conservative and rotational forms in large eddy simulation of turbulent channel flow, J. Comput. Phys., Vol. 71, pp. 343-370 (1987).

[9] Zang, T. A.: On the rotation and skew-symmetric forms for incompressible flow simulations, Appl. Numer. Math. (in press).

[10] Ronquist, E. M.: Optimal spectral element methods for the unsteady three-dimensional incompressible Navier-Stokes equations, Ph.D. Thesis, Massachusetts Institute of Technology (1988).

[11] Passot, T.; and Pouquet, A.: Numerical simulation of compressible homogeneous flows in the turbulent regime, J. Fluid Mech, Vol. 181, pp. 441-466 (1987).

[12] Lele, S. K.: Direct Numerical Simulation of Compressible Free Shear Flows, AIAA Paper No. 89-0374 (1989).

[13] Zang, T. A.; Hussaini, M. Y.; and Bushnell, D. M.: Numerical Computations of Turbulence Amplification in Shock Wave Interaction. AIAA J., Vol. 22, pp. 13-21, (1984).

[14] Erlebacher, G.; Hussaini, M. Y.; and Kreiss, H. O.: Direct Simulation of Compressible Turbulence, (in preparation).

[15] Speziale, C. G.; Erlebacher, G.; Zang, T. A.; and Hussaini, M. Y.: The Subgrid-scale Modeling of Compressible Turbulence, Phys. Fluids, Vol. 31, pp. 940-942 (1988).

[16] Shu, C. W.; and Osher, S.: Efficient implementation of essentially non-oscillatory shock capturing schemes II, ICASE Report No. 88-24 (1988).

[17] Zang, T. A.; Drummond, P.; and Kumar, A.: Numerical Simulation of Enhanced Mixing and Turbulence in Supersonic Flows, in The Physics of Compressible Turbulent Mixing, (W. P. Dannevik, A. C. Buckingham, and C. E. Leith, eds.) Lecture Notes in Engineering, Springer-Verlag, in press.

[18] Soetrisno, M.; Eberhardt, S.; Riley, J. J.; and McMurtry, P.: A Study of Inviscid, Supersonic Mixing Layers Using a Second-Order TVD Scheme, AIAA Paper No. 88-3676 (1988).

[19] Foias, C.; Manley, O.; and Temam, R.: On the interaction of small and large eddies in two dimensional turbulent flows, ICASE Report No. 87-60 (1987).

[20] Henshaw W. D.; Kreiss, H. O.; and Reyna, L. G.: On the smallest scale for the incompressible Navier-Stokes equations, Theor. Comp. Fluid Dyn., Vol. 1, pp. 65-95 (1989).

[21] Henshaw W. D.; Kreiss, H. O.; and Reyna, L. G.: On the exponential decay of the energy spectrum, Arch. Rational Mech., (submitted for publication).

[22] Rogallo, R. S.: Numerical Experiments in Homogeneous Turbulence, NASA TM-81315 (1981).

[23] Bardina, J.; Ferziger, J. H.; and Reynolds, W. C.: Improved turbulence models based on large eddy simulation of homogeneous, incompressible, turbulent flows. Report TF-19, Dept. of Mech. Eng., Stanford Univ. (1983).

[24] Speziale, C. G.; and MacGiolla Mhuiris, N.: On the prediction of equilibrium states in homogeneous turbulence, ICASE Report No. 88-27 (1988) (to appear in J. Fluid Mech.).

[25] Bardina, J.; Ferziger, J. H.; and Rogallo, R. S.: Effect of rotation on isotropic turbulence: computation and modelling, J. Fluid Mech., Vol. 154, pp. 321-336 (1985).

[26] Speziale, C. G.; Mansour, N. N.; and Rogallo, R. S.: The decay of isotropic turbulence in a rapidly rotating frame, in Studying Turbulence Using Numerical Simulation Databases, Report CTR-S87, pp. 205-211, Stanford Univ (1987).

[27] Gilbert, N.; and Kleiser, L.: Near-wall phenomena in transition to turbulence, in Proceedings of the International Seminar on Near-Wall Turbulence, Hemisphcrc, Washington (1988).

[28] Gilbert, N.: Numerische Simulation der Transition von der laminaren in die turbulente Kanalstromung, DFVLR-FB 99-55, Gottingen, W. Germany (1988).

[29] Zang, T. A.; and Krist, S. E.: Numerical experiments on stability and transition in plane channel flow, Th. Comp. Fluid Dyn., Vol. 1, pp. 41-64 (1989).

[30] Deardorff, J. W.: The use of subgrid transport equations in a three-dimensional model of atmospheric turbulence, J. Fluids Eng., Vol. 95, pp. 429-438 (1973).

[31] Clark, R. A.; Ferziger, J. H. and Reynolds, W. C.: Evaluation of subgrid scale models using an accurately simulated turbulent flow, J. Fluid Mech., Vol. 91, pp. 1-16 (1979).

[32] Tavoularis, S.; and Corrsin, S.: Experiments in nearly homogeneous turbulent shear flows with a uniform mean temperature gradient. Part I, J. Fluid Mech., Vol. 104, pp. 311-347 (1981).

[33] Schumann, U.: Subgrid scale model for finite difference simulations of turbulent flows in plane channels and annuli, J. Comput. Phys., Vol. 18, pp. 376-404 (1975).

[34] Moin, P.; and Kim, J.: Numerical investigation of turbulent channel flow, J. Fluid Mech., Vol. 118, pp. 341-378 (1982).

[35] Yakhot, V.; and Orszag, S. A.: Renormalization group analysis of turbulence. I. Basic theory, J. Sci. Computing, Vol. 1, pp. 3-51 (1986).

2-D ENERGY SPECTRA

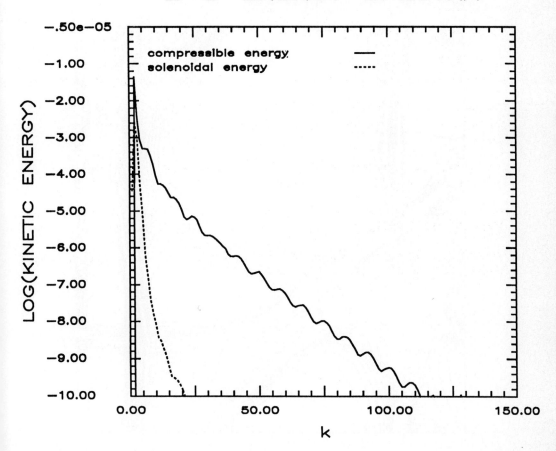

Figure 1. Energy spectra for 2-D isotropic decay of compressible turbulence taken from the direct simulations of Erlebacher, et al. [14]: the eddy shocklet regime; $Re_\lambda \doteq 10$, Ma = 0.028, $\rho_{rms} = 0.10$.

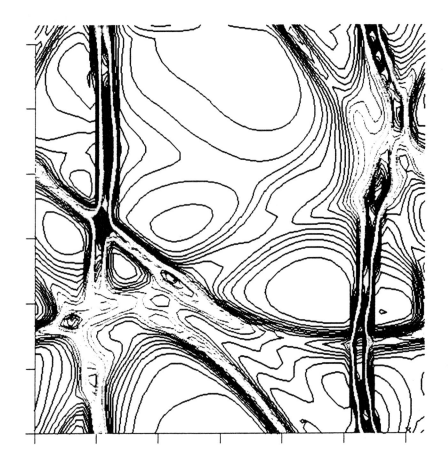

Figure 2.   Contours of the divergence of velocity for 2-D isotropic decay of compressible turbulence taken from the direct simulations of Erlebacher, et al. [14]: the eddy shocklet regime; $Re_\lambda \doteq 10$, $Ma = 0.028$, $\rho_{rms} = 0.10$.

2-D ENERGY SPECTRA

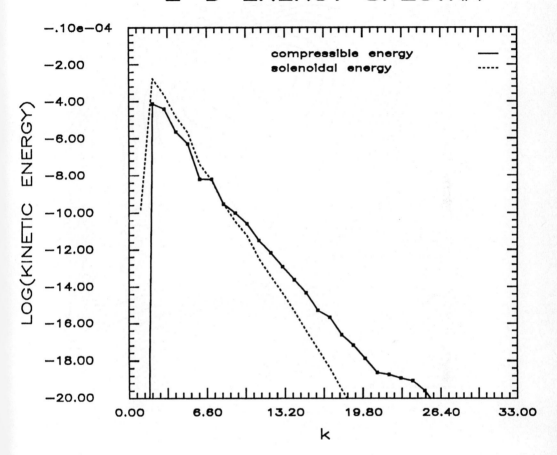

Figure 3.   Energy spectra for 2-D isotropic decay of compressible turbulence taken from the direct simulations of Erlebacher, et al. [14]: the quasi-incompressible regime; $Re_\lambda \doteq 10$, $Ma = 0.028$, $\rho_{rms} \doteq 0$.

Figure 4. Contours of the divergence of velocity for 2-D isotropic decay of compressible turbulence taken from the direct simulations of Erlebacher, et al. [14]: the quasi-incompressible regime; $Re_\lambda \doteq 10$, $Ma = 0.028$, $\rho_{rms} \doteq 0$.

The Potential and Limitations of Direct and Large-Eddy Simulations

Comment 3.

J. Wyngaard

Geophysical Turbulence Program
National Center for Atmospheric Research
P.O. Box 3000
Boulder, CO 80307

1. Introduction

I commend Prof. Reynolds for his splendid survey of turbulence simulation. He has left little unsaid, but I will expand on his remark that the widest application of LES (large-eddy simulation) is currently in atmospheric sciences.

2. Atmospheric and Laboratory Turbulence: Contrasts

First, I should emphasize the evolutionary differences of the engineering and atmospheric turbulence communities. I suspect that a researcher from one side, on first exposure to the other, knows something of the feeling Darwin experienced on his visit to the Galapagos Islands.

These differences were much more pronounced 30 years ago, however. The publication in 1964 of the Lumley-Panofsky monograph *The Structure of Atmospheric Turbulence* seemed to stimulate a new era of cultural exchange. Today our communities speak very similar languages; I think most of our remaining separation is maintained by real differences in our research problems and in our turbulence.

How does turbulence in the atmospheric boundary layer (ABL) differ from that in the laboratory and in engineering devices? First, it is much more difficult to measure its full spatial structure. The ABL can be 2–3 km deep during the day; to date, we have done little more than probe its vertical structure. Second, it is often complicated by effects that in the laboratory are controllable or nonexistent— e.g., nonstationarity of boundary conditions and the mean flow, radiation, phase change, thermal stratification, and gravity waves. Third, its stochastic variability—the difference between time and ensemble averages—is usually much larger. Scattered data are part of the facts of life outdoors.

Even so, atmospheric turbulence researchers have done very well. The first complete measurements of the budgets of Reynolds stress and heat flux were made not in a laboratory flow, but in *the atmospheric surface layer*. The experimental section in the Lumley-Panofsky monograph is long out of date. Nonetheless, atmospheric turbulence researchers today seem to rely more heavily on numerical approaches than on experiment.

There are two other salient differences in our flows. One, we rarely see unstratified, shear-driven turbulence in the atmosphere; eddies are large enough and slow enough that very small temperature differences cause dynamically controlling density changes. The lower atmosphere is more apt to resemble free convection or stably stratified shear flow than a neutral boundary layer. Two, Reynolds numbers are much larger in the atmosphere, giving a huge range of scales (order-of-magnitude values for the daytime boundary layer are η, the Kolomogorov microscale, 1 mm; λ, the Taylor microscale, 10 cm; and ℓ, the integral scale, 1000 m—so R_λ is 10^4.) Thus, there typically are 3–4 decades of residual (subgrid-scale) turbulence in atmospheric LES, even with the grid scale in the inertial range! Perhaps these explain the atmospheric community's attraction to second-order closure, which in principle handles stratification effects directly, and its preference of LES over direct numerical simulation (DNS).

3. LES and VLES in the Atmospheric Sciences

LES is indeed widely used today in the atmospheric sciences, but its practitioners tend to call it "three-dimensional nonhydrostatic numerical modeling." As Reynolds suggests, much of it is *very* large eddy simulation, with the grid size in the energy-containing rather than inertial range.

Reynolds' "dynamic residual turbulence models" are standard in atmospheric LES codes; Deardorff (1974) even tried a full set of second-moment equations! With a wave-cutoff filter and closure constants determined as outlined by Lilly, LES velocity and temperature spectra show the beginning of an apparent inertial range (Moeng and Wyngaard, 1988), although Mason (1989) questions this interpretation. In the

ABL with a 96^3 grid and away from the lower surface, virtually all of the energy and flux are carried by the resolvable eddies.

I believe the achievements of LES in the atmospheric sciences are as spectacular as those of DNS in turbulence research. LES has provided a new approach to the study of severe storms, allowing them to be generated under controlled conditions and yielding complete kinematic, dynamic, and thermodynamic data both in and near the storm (e.g., Klemp, 1987). It has given deep insight into topographically forced clouds and turbulence (Smolarkiewicz and Clark, 1985), and downslope winds and mountain-wave turbulence (Clark and Farley, 1984). It has recently been used (in VLES form) to calculate the eddy circulation over 30 years in the Atlantic ocean basin (60 N to 15 S), an effort that consumed 1400 hours of CRAY X-MP time! (F. Bryan, NCAR, personal communication). Some of the earliest LES studies (Deardorff, 1972) established the basic scaling parameters for the convective boundary layer. Over the past 15 years LES has revolutionized research in small-scale meteorology.

An example of a problem very difficult to investigate experimentally in the atmosphere but well suited to LES is one-dimensional (i.e., area-source) diffusion of passive, conservative species through the boundary layer. LES allowed us to expolit the linearity of scalar diffusion by decomposing this into "bottom-up" and "top-down" problems and studying each separately in the same simulated ABL. The well-known singularity in the eddy diffusivity occurs only in the bottom-up problem; we traced it to the strong, positive skewness of the vertical velocity field (Wyngaard and Brost, 1984; Moeng and Wyngaard, 1984; Wyngaard, 1987).

LES can allow progress where the traditional route—experiment—has been blocked. Not long after the first attempts to use second-order closure models for the PBL it was found they needed a good deal of "tuning" in order to perform well over the required range of stratification. Experiments with third-order closure were encouraging, but the data necessary to develop and test them fully were not available. Laboratory experiments have provided valuable data, but some flow properties (e.g. the pressure field) cannot be measured. LES "data bases" are now helping in both areas (Moeng and Wyngaard, 1989).

Another example is the cloud-topped boundary layer off the west coast of continents. It has a sizeable impact on the earth's energy budget, since when present it reflects most of the incoming solar radiation that otherwise would be absorbed. This geophysical flow, strongly influenced by the thermodynamics of phase change and radiative transfer, cannot be simulated in the laboratory. There are zero-order problems in measuring its flow fields (e.g., in-cloud temperatures are notoriously difficult to obtain). LES has given insight into its structure and dynamics that has been impossible to achieve any other way (e.g., Deardorff, 1980; Moeng, 1987).

LES has also been used to study the late-afternoon decay of the ABL after the cutoff of surface heating (Nieuwstadt and Brost, 1986); the response of the ABL to spatially nonuniform surface heating (Hadfield, 1988); and the structure of the stably stratified ABL (Derbyshire, 1989).

4. Needed Development Work

To my knowledge, all atmospheric LES codes have what Reynolds calls no wall-region resolution; the first grid point is typically 5–50 m from the surface. Thus, for the first few grid points virtually all of the fluxes are residual scale. Mason (1989) has modified the usual closure there in an attempt to improve its poor physics, but found some disturbing evidence that it changed the resolvable-scale structure aloft. Much remains to be done here.

VLES does not share LES' insensitivity to residual-scale parameterizations, since its grid can fail to resolve most of the flux, for example. We may see renewed interest in the use of second-order closures for the residual scales in VLES, particularly if they are more economical and fundamentally based than in the past. LES can help here.

Existing residual-scale closures for LES use a deterministic relation between resolvable and residual-scale properties, but in principle it should be random. Furthermore, to my knowledge the Reynolds number enters none of the LES codes presently used (except in near-wall adjustments) so that for a given geometry and boundary conditions—those of an ABL, for example—laboratory and atmospheric simulations are indistinguishable. Therefore the LES codes currently available cannot reveal whether the energy-containing structure varies with Reynolds number (again, except for near-wall effects). One expects Reynolds number dependence of the residual scale turbulence; for example, the intermittency of its fine structure depends on the length of the inertial subrange and therefore on turbulence Reynolds number (Kolmogorov, 1962). Thus, the effects of residual-scale turbulence could depend on the turbulence Reynolds number. One can explore this with turbulence closures or possibly with DNS forced at the large scales, but I suggest it is best pursued the old-fashioned way: experimentally.

Acknowledgments

In preparing this I had useful discussions with T. Clark, P. Gallacher, J. Herring, B. Kerr, P. Mason, J. McWilliams, C.-H. Moeng, and P. Smolarkiewicz.

References

Clark, T.L., and R.D. Farley, 1984: Severe downslope windstorm calculations in two and three spatial dimensions using anelastic interactive grid nesting: A possible mechanism for gustiness. *J. Atmos. Sci.*, **41**, 329–350.

Deardorff, J.W., 1973: Three-dimensional numerical modeling of the planetary boundary layer. In *Workshop on Micrometeorology*, D.A. Haugen, Ed., Amer. Meteor. Society, Boston, MA 271–311.

Deardorff, J.W., 1974: Three-dimensional numerical study of turbulence in an entraining mixed layer. *Bound.-Layer Meteorol.*, **7**, 199–226.

Deardorff, J.W., 1980: Stratocumulus-capped mixed layers derived from a three-dimensional model. *Bound.-Layer Meteorol.*, **18**, 495–527.

Derbyshire, S.H., 1989: Ph. D. thesis, Univ. of Cambridge, in preparation.

Hadfield, M.G., 1988: The response of the atmospheric convective boundary layer to surface inhomogeneities. Ph.D. thesis, Dept. of Atmos. Sci., Colorado State Univ., Fort Collins, CO, 401 pp. Available as Atmos. Sci. Paper No. 433.

Klemp, J.B., 1987: Dynamics of tornadic thunderstorms. *Ann. Rev. Fluid Mech.*, **19**, 369–402.

Kolmogorov, A.N., 1962: A refinement of previous hypotheses concerning the the local structure of turbulence in a viscous incompressible fluid at high Reynolds number. *J. Fluid Mech.*, **13**, 82–85.

Lumley, J.L., and H.A. Panofsky, 1964: *The Structure of Atmospheric Turbulence.* Interscience, Wiley, 239 pp.

Mason, P.J., 1889: Large eddy simulation of the convective atmospheric boundary layer. To appear, *J. Atmos. Sci.,*

Moeng, C.-H., 1987: Large-eddy simulation of a stratus-topped boundary layer. Part II: Implications for mixed layer modeling. *J. Atmos. Sci.*, **44**, 1605—1614.

Moeng, C.-H., and J.C. Wyngaard, 1984: Statistics of conservative scalars in the convective boundary layer. *J. Atmos. Sci.*, **41**, 3161–3169.

Moeng, C.-H., and J.C. Wyngaard, 1988: Spectral analysis of large-eddy simulations of the convective boundary layer. *J. Atmos. Sci.*, **45**, 3573–3587.

Moeng, C.-H., and J.C. Wyngaard, 1989: Evaluation of turbulent transport and dissipation closures in second-order modeling. To appear, *J. Atmos. Sci.*

Nieuwstadt, F.T.M., and R.A. Brost, 1986: The decay of convective turbulence. *J. Atmos. Sci.*, **43**, 532–546.

Smolarkiewicz, P.K., and T.L. Clark, 1985: Numerical simulation of a three-dimensional field of cumulus clouds. Part I: Model description, comparison with observations and sensitivity studies. *J. Atmos. Sci.*, **42**, 502–522.

Wyngaard, J.C., 1987: A physical mechanism for the asymmetry in top-down and bottom-up diffusion. *J. Atmos. Sci.*, **44**, 1083–1087.

Wyngaard, J.C., and R.A. Brost, 1984: Top-down and bottom-up diffusion of a scalar in the convective boundary layer. *J. Atmos. Sci.*, **41**, 102–112.

Discussion of "The Potential and Limitations of Direct and Large Eddy Simulations

Reporter Laurence Keefe

Nielson Engineering & Research
510 Clyde Avenue
Mountain View, CA

Jack Herring:

If you take the point of view of two point closures, and go through an anisotropic turbulence calculation, you get an estimate, in a crude way, for the rate of return to isotropy in spectral form. It seems to me, using those estimates, that the result you showed is expected for low Reynolds number flows. If you estimate the return to isotropy that way, there are two terms: One is the interaction of the locally isotropic part of the turbulence with the large scale mean strain, and this produces small-scale anisotropy. The other is the decay of anisotropy at the small-scale wave numbers k, which is roughly one eddy circulation time at this scale. However, if you look at the latter more carefully, it turns out to be something like the rms strain of scales larger than k, and so if the spectrum falls off faster than k^{-3} this is a constant, and you get a constant anisotropy at all scales. Could that be a possibility?

Bill Reynolds:

The concept of constant anisotropy at all scales is very appealing. Sreenivasan suggests one look for a process by which a structure could fractally repeat itself on down in scale, which would produce this behavior. This would not be a low Reynolds number effect.

Jack Herring:

Well in this case I was trying to say that spectra steeper than k^{-3} may not return to isotropy. I assume that the slope for your case is steeper than k^{-3}. If it's more like $k^{-5/3}$ I would expect a return to isotropy.

Marcel Lesieur:

I also have a comment regarding Bill Reynold's paper. Concerning LES I would like to mention the possibility of working with a spectral code using Kraichnan's spectral eddy-viscosities. We applied them in Grenoble with Chollet and Metais, and at CTR with Bob Rogallo, to isotropic turbulence[1] [2] [3]. In particular they were applied to decaying isotropic turbulence, but it works quite well for stably stratified turbulence. There is very nice work done by Olivier Metais at Grenoble on that,[4] and recently a student of Grenoble, Pierre Comte,[5] has calculated a 3-D temporal mixing layer at high Reynolds number using these techniques, of which I showed some slides yesterday. There are possibilities, though they have not so far completely succeeded, of extending this technique to wall flows, and also compressible flows, at least when the cut-off wavenumber is in a range where you may assume it is not too far from isotropy. I prefer to use these spectral quantities, rather than those which can be derived from RNG techniques because, in particular, they do not apply to decaying situations.

Ulrich Schumann:

I have to make some comments on the paper of Bill Reynolds and I would also like to restate questions we have been discussing for fifteen years. One point, the general point, is that some of the technical complications which were deduced fifteen years ago have hindered progress rather than supported it. One example of this is the concept of filter. I think if you look at the theoretical description of a large-eddy-simulation you find there are schools. There is the school of Stanford, and there are other schools, and the school of Stanford says that first you have to filter your

---

1Chollet, J. P. and Lesieur, M. 1981, J. Atmos. Sci., 38, p. 2747.

2Lesieur, M. and Rogallo, R. 1989, Phys. Fluids A, 1, p. 718.

3Lesieur, M., Metais, O., and Rogallo, R. 1989, C. R. Acad. Sci., 308, Ser II, p.1395.

4Metais, O. and Chollet, J. P., 1988, in "Turbulent Shear Flows VI, Springer-Verlag, p. 398.

5Comte, P., 1989, PhD Thesis, Grenoble.

equations and then solve them by a finite difference or spectral scheme which no longer experiences strong approximations. In principle this is a good approach. But in practice it is totally unfeasible. To do a real large eddy simulation you must resolve all scales up to the inertial range, otherwise it's not a large eddy simulation. You have to have a quite large number of modes in the interior of the flow, typically 50 or more independent grid points in one direction, and that's already close to what one realistically can do. One cannot filter away part of this simulation because the resolution isn't fine enough. The grid scale must be smaller than the filter and that must still be smaller than the scale at which you reach the Kolmogorov inertial spectrum. And that's not feasible to satisfy.

The second point is a reiteration of a statement by Boris which I want to support: A factor of two increase of spatial resolution will bring more improvement than better sub-grid scale models. What is really crucial in LES is resolution, you have to see the scales. No matter how good the approximation is, you have to see the eddies. That's the first condition. Then the second condition is to approximate them well. But the first condition is you have to see them, otherwise if you don't model them at all you make a greater error than if you approximate them crudely. In fact the filtering totally fails in models close to the wall because you are never able to satisfy the separation between these scales close to the wall. Schemes that use variable grid spacing in the direction normal to the wall have not used the filtering method because there it's infeasible.

The other point which has made more complication than has supported progress is the concept of higher order schemes. One must be aware, and Bill Reynolds pointed this out, that the sub-grid scale viscosity is of the order grid scale to the 4/3 power, and the sub-grid scale kinetic energy is grid scale to the 2/3. So the sub-grid scale contributions are of the order of scale which is large in comparison to δ^2. Thus any numerical scheme which has numerical approximation errors of $O(\delta^2)$ is fine already compared to the errors that come from the sub-grid scale model. There are a few exceptions. For example, if you have strong advection, a large mean flow, then it's important to describe this mean advection problem accurately, and for that particular purpose, pseudo-spectral methods or fourth order schemes might even be necessary, but not in general. Don't say in general you need fourth order schemes, this generalization hinders more than it supports progress.

The third point is one concerning the built in filter described by Boris and proposed and applied by several Japanese colleagues using a monotonic finite difference scheme. They apparently seem to produce results which look like turbulence, but I haven't really seen any paper in which someone has applied these methods and has produced statistics, say just of rms values, or spectra or length scales, and compared them to experimental data. You still have to do this! At least nothing on that seems to be reported in the open literature. And of course I have serious doubts that such simulations can be successful. And the reason, as you said, is that the numerical scheme is not Galilean invariant. So the filter at one position in the shear flow is different from that at another position. Your filter varies as a function of distance. I have serious doubts that you can ever get results to compare with experimental data to high accuracy with such numerical procedures.

Some people argue for applying sub-grid scale kinetic energy equations in sub-grid scale models. In a paper in 1975[6] I stated that in the channel flow simulation, the inclusion of a transport equation for sub-grid scale kinetic energy does not change the results very much in comparison to the simple Smagorinsky model. That statement is true for the channel flow; it is true for a shear flow; it's probably not true for cases where you have other body forces, particularly buoyancy forces or rotational forces, in addition to shear forces. If at the sub-grid scale level buoyancy forces are comparable to shear forces, then its very essential to have these additional forcings in the model, and a simple way to do it is by using a sub-grid scale kinetic energy equation.

---

[6] U. Schumann: Subgrid Scale Model for Finite Difference Simulations of Turbulent Flows in Plane Channels and Annuli, J. Comput. Phys., 18 (1975) 376-404.

A general point is that in the past, direct simulations and large-eddy simulations have identified many problems with existing second-order closure models. I myself often found that many simple models proposed, say, for the pressure-strain correlation or dissipation rates, are insufficient[7] . One can clearly identify the reasons why they are insufficient, but it is very difficult to propose a better one. There are few new proposals, but the number of proposals to make the models better are certainly much less than the number of papers that identify problems. I think this process will not close, will not converge. In the long trend, the large eddy simulations have to take over the responsibility for the results alone, they cannot trust the simpler models. They have to produce the results themselves.

Jay Boris:

I agree with much of what Dr. Schumann said. I believe there are a couple of references in my paper to Grinstein, who certainly shows comparisons with experiment. He got good comparison with Bowman's experiments at Stanford. These are 3-dimensional calculations. We've also done a number of 2-D calculations where we get good agreement with things which you would hope and expect 2-D calculations to work for, for example, with respect to experiments that Fazle Hussain and others have done. It's not yet the complete body of literature you'd like to see, but I expect that will be coming along in time.

I'd like to correct an earlier comment regarding Galilean invariance. The errors which come from the existence of the grid, are not Galilean invariant because the grid itself is not Galilean invariant. Thus, the fact that a particular numerical method may be defined in a way which is not Galilean invariant, shouldn't bother you. If it were otherwise, the method would not be able to remove the numerical error that comes from the non-Galilean representation! This isn't to say there isn't some residual non-Galilean error. We have done shear flow problems where the whole problem has been moving and the answers are substantially the same as in the calculations performed in

---

7T. Gertz, U. Schumann, S. E. E. Elghobashi, Direction Numerical Simulation of Stratified Homogenous Turbulent Shear Flows. J. Fluid Mech, 200 (1989) 563-594.

a different frame of reference. I think that kind of test, documented carefully, is what we all want. Parenthetically, I have not seen this kind of test from other methodologies either!

Let me change the subject for a moment. I would like to ask a question of Bill, Parviz, yourself, and anyone else who has done direct simulations. You are solving the Navier-Stokes equations and resolving them over a reasonable range of scales, though not as large a range as you would like. Do you see evidence, with these infinite numbers of vortices and all sorts of fluid contortions, of singularities forming in the Navier-Stokes equations, and if so can you describe the geometry and configuration? Or does the fact that viscosity enters the model always resolve the singularity?

Bill Reynolds:
I'll answer as an observer of what the most careful DNS people do. They are extremely careful to resolve things, so when shocks are encountered there are enough points put in the shock to get a really nice resolution. If you ask if numerical things are preventing true singularities from occurring, I think the numerical work is done so carefully that this is not the case.

Jay Boris:
I am asking if anyone sees a signature, of the sort of singularity mathematicians claim may invalidate the Navier-Stokes equations, toward short wavelengths where you know a code is not accurate because you're getting such large gradients. Are you always resolving this situation? Is the lack of an existence (uniqueness?) therorem for Navier-Stokes solutions a real problem?

Bill Reynolds:
You see computational limitations by spectral pile-up at high wavenumbers and should be cautious in using such fields.

Anatol Roshko:
I wanted to ask you about the experiments Bill described this morning in which you

observed vortical structures in highly sheared flows. I have two pictures of those vortical structures. The old picture, that came from the experiments in boundary layers, left me with the idea these vortices occurred in pairs, as parts of hairpin vortices, and were somehow connected with instability of the flow, possibly near the wall. In the direct simulations of the channel flow at the Center for Turbulence Research, they don't really occur in pairs, they can occur singly and randomly. Now the experiments on strained homogeneous flow show that at high strain rates you have these events also occurring. One interpretation is that random vorticity is picked up in the high strain-rate regions and intensified, and that's what you see. That seems to be different from a model in which stability somehow plays a part, which is really the picture you get from the things Phil Holmes talked about this morning. So it's not clear they're talking about the same thing in both cases or two different things. I'd like to hear your views on that.

Bill Reynolds:

These are not vortices, these are streaks, like jets and wakes. The vorticity is not in the streamwise direction, its wound around the streaks, in a spiral. Would you like to add to this Parviz?

Parviz Moin:

This is correct. Anatol Roshko pointed to the early experiments. Some of the problems stemmed from making only measurements of two-point correlation of streamwise velocity fluctuations that showed there are elongations. From these single component measurements it was deduced that everything is elongated near the wall. Streamwise vortices are elongated; you name it, it is elongated. Furthermore, it was extrapolated that there were pairs of these things. Part of it has been lack of data, part of it has been statistical artifact. Any statistical approach like the ones used for calculating the two point correlations of velocity put in some symmetry in the calculation. In other words, from two-point correlation data, without any additional conditions, you always get pairs of vortices, you will never get a single vortex. That part has been a statistical artifact.

Uriel Frisch:

I think that was a very interesting question of Jay Boris concerning the possible errors coming from singularities. We are constrained by what we know from mathematics. Now there is an illusion sometimes that we know nothing about the mathematics of flows. That's wrong! There are some hard and proven mathematical results which can be used to answer, at least in part, the kinds of questions that have come up. For example, one can show that in a flow, with or without viscosity, which has no real singularities, the Fourier transforms fall off exponentially. Its a non-trivial result, it has to do with analyticity, it has a rigorous proof. Now that implies that you will have exponentially small errors, as long as the mesh is significantly smaller than the distance to the nearest complex space singularities, provided you are using spectral (but not finite difference) methods and have only periodic boundary layer conditions. For Reynolds numbers up to a few thousand, the numerical evidence is that there is a very clean exponential fall-off ruling out real singularities. An open question is: if you have no viscosity at all, will there be a singularity after a finite time? There are only conjectures. People don't even agree on what the right conjecture is. Everybody agrees that these singularities, whether or not they are formed after finite time at zero viscosity, will certainly approach arbitrarily close to the real domain if you wait long enough. Lesieur believes singularities are formed in a finite time; I do not want at this time to commit myself one way or the other.

E. Novikov:

There is a difference between singularities in space and singularities in time.

Uriel Frisch:

I was speaking about singularities in the real space-time domain.

E. Novikov:

But in time then there may still be a singularity.

Uriel Frisch:

The problem is: in the presence of viscosity are the singularities confined to the complex domain? Singularities are away from the real domain by roughly one Kolmogorov scale. That is the conjecture. It may be wrong. But I must say the very clean numerical evidence at Reynolds number up to a few thousand is that it is approximately right. The scaling law may not be exactly the one predicted by Kolmogorov (1941, Dokl. Acad. Nauk, 30, 299), but still it is believed to be close.

Jay Boris:

This seems to mean that well resolved Navier-Stokes equations do not show singularities.

Uriel Frisch:

Yes. However, I promised I would start a controversy. Take the best channel flow simulation at Reynolds number around 3,000 from NASA-Stanford, where they use Chebyshev polynomials, and collocation points are very closely packed in the boundary layer and not so in the inside of the flow. It is absolutely clear that the boundary layer is fully resolved. It may be questionable if its fully resolved inside the flow. There may be a few small eddies inside which are not fully resolved. I don't think that matters very much, but such questions should be investigated. But do you realize, dear participants, that if you take, not a simple geometry like a channel flow, but a flow around a body, like a car, what the highest speed is at which you can reliably, fully reliably, simulate it with a Navier-Stokes code using the best existing computer and the best code? It is something less than a centimeter per second!

Julian Hunt:

I wanted to make a remark connecting this afternoon's session with this morning's session. Let me put up a slide. One of the exciting and interesting notions of this morning's session was the importance of chaos theory and dynamical systems theory, in analyzing flows and experiments. Stimulated by the work of Perry and a conversation with Frisch, we did some work at Ames in the summer with Alan

Wray[8] and Parvis Moin[9] (Wray & Hunt) taking direct numerical simulations of homogeneous turbulence and trying to characterize regions, which we call 'eddy' regions and 'convergence' or straining regions, using the notions of the second invariant Perry[10] was talking about yesterday. I think the point has not been made that you can use these dynamical systems ideas first to analyze these flows, and then, to provide criterion for deciding what are the essential features of the flow if you want to do 'cheaper' simulations. Our 'cheaper' simulations (at Cambridge[10]), follow the ideas of Kraichnan and involve summing many independent random Fourier modes so as to simulate a turbulent velocity field. A test of whether it looks like a direct numerical simulation is to compare the distributions of these regions of eddy motions or convergence motions (defined using the objective criterion based on invariants). The point was raised by John Wyngaard this afternoon that at the moment we can not directly (or even through L.E.S.) analyze the properties of turbulence when there is an inertial subrange of four or five decades; but it may be possible to represent some features of the flow in this 'cheap' way (modifying Kraichnan's ideas a bit with the large modes randomly advecting the small modes) and then look at properties such as Lagrangian spectra, over several decades of turbulence. For example, we find the correct form of the Lagrangian spectra going like ω^{-2} in the inertial range, with a coefficient quite close to atmospheric turbulence data. This technique of cheap simulation has been calibrated at moderate Reynolds number by comparison with D.N.S., using dynamical systems concepts. This may be one of the ways of simulating turbulent flow fields up to very high Reynolds numbers and looking at some properties of them!

Steve Kline:

I want to go back to Anatol Roshko's question because it provides an opportunity to

8Perry A. E. and Chong, M. S. 1987 Ann. Rev. Fluid Mech., 125-155.

9Hunt, J. C. R., Wray, A. A. and Moin, P. 1988 Proc. C. T. R. Summer Program 1988 pp. 193-208.

10Fung, J. & Perkins, R. J. 1989 Proc. 2nd European Turb. Conf., Springer-Verlag, Berlin.

clarify several points about wall-streaks and vortices which have been the subject of considerable confusion. What I will say in part is clarification of old work, but is primarily new clarification from recent study of P. R. Spalart's DNS data base for the boundary layer. First the low-speed/ high-speed regions one observes as the dominant structure feature in the zone $0 < y+ < 10$, are not vortices. They were never vortices. The data have been clear on this point from the earliest observations. The low-speed/high-speed streaks are simply regions of high and low streamwise velocity. The variation is large; in the flat plate layer the ratio of high to low speed is about 3 to 1. The ratio is higher in adverse pressure gradients. The mean transverse wave length is 100 wall units, and the mode 80. These numbers appear stable for a wide range of Reynolds numbers and pressure gradients; they alter only slowly as one approaches quite extreme conditions of additives, relaminarization, etc. Many laboratories have confirmed these results.

Second, recent study of DNS data for the canonical flat plate at R-theta = 670 created by P. R. Spalart confirms that the low-speed regions are long streaks, often more than 1000 wall units in length, although they wander a bit in z as one traces them in x. Wall temperature observations by R. J. Moffat and his students show the same results. However, the high-speed regions are much shorter, usually not more than a 100 or so wall units in streamwise extent. Hence, I will refer to low-speed "streaks" and high-speed "regions".

Third, although neither the low-speed streaks or high-speed regions are themselves vortices, there is a relation between the high-speed/low-speed pattern and the vortical elements near the wall. To be clear about this relation, I will use the nomenclature from a paper to be given by S. K. Robinson at the IUTAM meeting in Zurich in June 1989[1] . The word "vortex" will denote a structure in which there is a circular, or near circular, pattern of velocity vectors in a plane normal to the core when moving with the vortex. A "vortex element" denotes a straight, or nearly

11 Robinson, S. K. "A Review of Vortex Structures and Associated Coherent Motions in Turbulent Boundary Layers," IUTAM meeting on Near Wall Turbulence and Control, ETH, Zurich, June 1989, procs to be published by Springer Verlag.

straight, vortex. A "vortical structure" denotes a compound shape of two or more vortical elements with a common core threading through the elements. In this notation, the primary type of vortex element observed near the wall is a <u>nearly streamwise vortex of short streamwise extent</u>. The word "nearly" implies the average vortex element is tilted upward from the wall 5-10 degrees in the downstream direction along the vortex. These tilted near-streamwise vortex elements frequently exist alone; however, they also are frequently observed as part of a hook like structure consisting of a foot, leg, neck, and head. These hook-like structures appear in both left and right handed realizations, as they must in a two-dimensional mean-flow. Distributions in y+ space of location, diameter, and circulation for both transverse and near streamwise vortex elements will be given by Robinson.

Given this description, we see three observations are central to answering Anatol's question. <u>First, vortex elements and the part of vortical structures observed near the wall are nearly always a foot, that is a slightly-tilted streamwise vortical element of short x-extent whether or not it is attached to a leg, neck, and/or head. Second, when a foot vortex is observed, it lies at the interface between a low-speed streak and high-speed region. Third, the circulation (and the circulation per unit area) of the typical foot vortex is significantly higher than the mean background vorticity of the boundary layer.</u> The location of these near-wall vortices has been independently observed by Lian at Beijing about two years ago using hydrogen bubble visualization for a flow with high adverse pressure gradient (nearing separation). Hence, thus far, the spatial relationship between vortices and the low-speed/high-speed pattern appears uniform. We must also note that not every high-speed/low-speed pair at the wall has a vortex at its interface, but when the vortex is there, as it often is, then it lies at the interface in z-location.

As Robinson shows, one must distinguish between these short tilted streamwise (foot) vortex elements and horseshoe or arch shaped patterns or vorticity lines. (Here "vorticity line" has the standard definition: a locus of points parallel to the vorticity vector). On careful study, this distinction between vortex elements and vorticity lines is seen to be important for many reasons. However, it is sufficient for this discussion to note only one; it is a kinematic necessity arising from presence of a strong foot vortex. Such a vortex will induce tilted, head-up horseshoe-shapes of vorticity lines on the low-speed streak (ejection) side and inverted (head down) horseshoe shapes of vorticity lines on the high-speed (sweep) side. Thus we again see that the vortices are not the low or high speed regions, and more important must be distinguished from them to understand the flow pattern.

Fourth, these spatial relations suggest the vortices may "create" the low-speed streak and the high-speed regions which then trail behind, as John Lumley suggested elsewhere in the discussion. However, one must not take this "creation" in a linear causal sense, because the DNS studies also show that a large fraction of the lifted, low-speed streaks on the ejection side of foot vortex elements roll up, thus forming vortex head elements. In addition, the shear layers along the sides of the lifted streaks sometimes roll-up into leg vortex elements as suggested by Blackwelder some time ago, and this occurs both separately and connected to roll-up of a head vortex element. Moreover, one also observes leg vortices growing back from large heads which initially rolled up on lifted-low-speed-streaks. Thus more than one kind of closed repetitive sequence of events occurs <u>frequently</u>. In such sequences one ought not assign a "cause" because what is seen as a "cause" <u>depends only on where one start in the closed sequence</u>. Argument about cause thus recreates the old conundrum about the chicken and the egg.

Fifth, in regard to Anatol's question about symmetry of structure, the DNS data do show some symmetric "horse-shoe" like vortical structures; these can be visualized as a head, two necks and two extensive legs. However, <u>such horseshoe-like vortical structures are observed only rarely;</u> only three percent in the total sample of several hundred vortical structures have a horseshoe-like shape. Arches (that is heads with one or two necks) are common, and so are four other forms: (i) the one-sided "hook-like" shapes, (ii) legs alone, (iii) heads alone, and (iv) complex "pile-ups" of vortex elements.

In our 1967 paper we drew a symmetric shape of vortical structure above a lifted low-speed streak; this occurs, but it represents at most an average structure; it is more symmetric than most observed structures, and it is not the only structure which seems to have statistical relevance. Unfortunately, we drew it symmetrically without emphasizing sufficiently that it was an average structure, because we could not see vortices clearly with the available techniques, and did not realize either the rarity of complete symmetry or the full variability of the structure from realization to realization. That the structures are typically asymmetric was verified statistically and independently by Guzennec in the 1987 Summer Institute at the NASA-Ames Stanford Center for Turbulence Research; see that volume.

Sixth, the vortical structures are not static over time. The heads move at approximately $0.8U_e$, but the foot vortex elements being much closer to the wall move far slower. Thus, when the two structures are connected, rapid stretching of the leg vortex occurs. This creates the region of highest rate of dissipation of TKE in the boundary layer. Hence the difference between a purely streamwise and slightly tilted vortex element is important dynamically. Other time dependent motions of vortical structures are also observed, but will be described elsewhere. When this time-dependence of structure is taken with the remarks about types of vortices above, it suggests strongly that the search for "a single dominant vortical structure which moves through the flow" is an illusion; more than one vortical shape occurs commonly, and the shapes evolve in time. As Robinson's DNS data show, the types of statistically-dominant vortical structures near the wall differ from those in the outer flow. To obtain a good "statistical structural" model, we may need to account for the frequency of these different vortical forms as a function of y+.

Seventh, we ask, "Is all this detail important in the physics of boundary layers?" We can answer the question by looking at the location of $u'v'_2$ and $u'v'_4$, the controlling Reynolds stresses in the second and fourth quadrant of the perturbation hodograph. The $u'v'_2$ lies largely in two places: (i) underneath and behind vortex head elements; (ii) on the ejection side of leg vortex elements on top of lifted low speed streaks. The $u'v'_4$ lies primarily in two places: (i) alongside the vortex leg elements on the sweep side; (ii) alongside vortex neck elements on the sweep side. It is also significant that no significant $u'v'_2$ is observed along near-wall-shear-layers formed by a lifting-low-speed streak. These near-wall-shear-layers are ubiquitous in the zone $12 < y+ < 80$. However, as soon as one of these near-wall-shear-layers rolls up into a vortex head element, then strong $u'v'_2$ is observed beneath and behind the vortex element. Such roll ups occur on a large fraction of the near-wall-shear-layers rolls up into a vortex head element, then strong $u'v'_2$ is observed beneath and behind the vortex element. Such roll ups occur on a large fraction of the near-wall-shear-layers. Moreover, large amounts of $u'v'_2$ continue to be observed so long as the head vortex element lasts, including heads lying at large values of y+ in the outer flow. This strongly suggests that these details are important in the production of Reynolds stress and TKE in the boundary layer. These implications will be more fully documented elsewhere.

In sum, the sublayer streaks are not vortices but appear to have a complex relationship with short, tilted, near streamwise, vortices; these vortices occur both alone and as part of the more complex usually asymmetric vortical structures. This relationship appears to play an important role in the physics of turbulence in the boundary layer.

R. Narasimha:

I have two questions concerning streaks addressed to the people who are doing the direct simulations as well as to those who are doing dynamic systems. First of all we have been told this afternoon by Bill Reynolds and Parviz Moin that these streaks are found not only near a wall, but wherever the shear is sufficiently high, including e.g. in homogeneous shear flow. The first question concerns the relation between those streaks and the rolls that went into the dynamical system that Phil Holmes described this morning - are they the same or are they different? The second question stems from the following remarkable fact about the turbulent boundary layer, that the structure of a large part of it, except very near the wall, is the same whether the wall is smooth or rough. Clauser (1956) showed that this was so for the mean velocity profile. Grass showed in 1971 that the bursting cycle is the same on smooth and rough walls. Over most of the atmospheric boundary layer (which always corresponds to a rough surface), the signatures obtained in conditions of neutral stability are remarkably similar to those found in the laboratory on smooth walls (Narasimha & Kailas 1987, 1988). All this suggests that most of the boundary really doesn't know whether the wall is smooth or not, and doesn't care (at least as long as the roughness is not extremely high). So if streaks are not really specially characteristic of the wall layer, and in any case what happens near a wall is not so terribly important to the rest of the boundary layer, should we not really be doing a theory where the modes are from the outer flow and not from near the wall?

John Lumley:

A number of people have done calculations over the years. Several years ago we did an energy method stability analysis of simple shear flow,[1,2] both laminar and turbulent, and we found that the most unstable mode is in the form of streamwise rolls. People have different feelings about energy method stability analysis and there are all kinds of questions about their connections to reality, but I think it suggests that if you have a soup of motions, and you apply a shear, out of that soup certain motions are going to be amplified more than others, and the ones that are going to be amplified are rolls[13] in the streamwise direction.

The streaks are produced by the rolls. The streaks are produced from the shear by the rolls. If you have a shear and rolls, it produces streaks because it lifts up or pushes down fluid from each side. That produces a low speed region. In fact, when you do the energy method stability analysis and get the circulation in the y-z plane, you also get a defect, or excess, in the x direction to go with it.

Joel Ferziger:

If the vortices keep flipping around (in the dynamical model) is there enough time to develop the streaks?

---

[12]Lumley, J. L. 1971. Some comments on the energy method, In <u>Developments in Mechanics</u> 6, eds. L. H. N. Lee and A. H. Szewczyk, Notre Dame Press,pp. 63-88.

[13]Note that the vorticity is not in the streamwise direction but tilted in the direction of the mean positive strain rate.

John Lumley:

Flipping around like what?

Joel Ferziger:

Well, as Phil Holmes showed this morning, the vortices keep changing positions.

John Lumley:

Oh yes, in between the jumps there is plenty of time to develop the streaks.

Marcel Lesieur:

I would like to come back to the question of singularities asked by Boris. I do not completely agree with what Uriel Frisch said - he does not agree that there might be a singularity in the Euler equations. I would like to be more precise on this point. If you consider unforced turbulence, and work in the framework of the Euler equations, there are a lot of mathematical problems that are unclear, if you start initially with a sharp peak in the kinetic energy spectrum in the large scales and you accept the fact that at some time, perhaps infinite, you are going to form an infinite Kolmogorov energy spectrum (it will be infinite because the Kolmogorov scale will have tended to infinity). I am quite sure, in this case, you will form the spectrum at a finite time and there are lots of phenomenological or more sophisticated models which make you believe this. The second point is this: if this spectrum forms at a finite time, that means the entropy will blow up at this finite time, since it is the integral of k^2 times $k^{-5/3}$. What is the corresponding singularity for the velocity field? I would like to mention some work done by Alan Wray at NASA-Ames. When running Euler computations with these kinds of initial conditions, even though the system is truncated, as long as the excitation has not reached the cutoff wavenumber you have an accurate calculation describing the Euler equations. Alan showed that very quickly you form very intense (he called them "hot spots") regions of vorticity. I can show a slide of these things. This is not precisely what Alan Wray found, but comes from the large-eddy simulations described before. You pick up the point where the vorticity is maximum, you make 3 cross sections, and you see very well these hot sports of vorticity. I think in homogeneous isotropic decaying turbulence, starting initially with energy at large scales, this is the kind of singularity you might form. I believe much further work needs to be done on that.

Joel Ferziger:

I am replying to both Jay Boris and Schumann. Although the issues have appeared in the literature I think they need to get into the record. I think the reason we have always favored filtering is that it very carefully defines what it is you're trying to calculate and what it is you have to model. This does two things for you. It makes comparisons with experimental data much easier because you know exactly what you need to compare with. The other thing is it makes the development of sub-grid scale models, on the basis of direct numerical simulation, much easier because again you know exactly what you need to calculate, and you know what it is you're comparing. From a practical point of view I think the difference between you and us is really not very much; when you get down to what you are programming in the computer, we're doing almost the same thing. From a philosophical point of view it is important to think about filtering the equations before finite differencing.

With regard to Jay's point, I am very uncomfortable with methods which rely on numerical error to be the sub-grid scale model, because for one thing you don't know what to compare with, and because you don't know what filter you used. The filter may be space dependent, different everywhere. Secondly, if the results don't compare well with experiment, you don't know who to blame, is it the numerics, or is it the sub-grid scale model? You never quite know. I think at low Reynolds numbers it probably doesn't make much difference, the energy in the sub-grid scale is relatively small and almost anything you do is going to give reasonable results. I think if you try to do very large eddy simulations with these methods, you could be headed for trouble. A few more things about high Reynolds numbers, or very large eddy simulations: when the filter width gets very large all the turbulence will eventually wind up in the sub-grid scale and eventually you'll be back to a Reynolds average. At that point there should be no difference between a sub-grid scale model and a Reynolds averaged turbulence model. However, you're going to stop somewhat short of that, and what that means is that you are going to need something like the models used in Reynolds averaging, but the coefficients will represent something different, so the kinetic energy equation will not be the same as the one used in Reynolds averaged models. Furthermore, when the cut-off becomes smaller than the integral scale of the turbulence, it is even necessary to introduce a length scale into the sub-grid scale model because the filter width or grid size is no longer sufficient. For very large eddy simulations we don't know what the appropriate models are and we need some work in that area.

Hassan Nagib:

I want to go back to the comments made by Professors Narasimha, Kline, and Roshko. I think we should realize that the streaks are not the rolls, as was already pointed out. Rolls and shear will create streaks, but the existence of streaks does not necessarily imply the existence of rolls or longitudinal vorticity. We should be very careful about this. Another aspect of my comments is that experiments for a range of Reynolds numbers in the boundary layer show that the scales represented by the modes in Aubry's work, discussed this morning, are actually a hierarchy of scales. In Aubry's work, they picked the first one, i.e., the closest one to the wall, which may scale with wall variables. In reality, there is a whole range of them. If you take that range and evaluate an integral scale measure of it, it is very intriguing (and this is work presented at the APS meeting in 1988 by Candace Wark) that this integral scale does not scale with the inner variables, but rather with the outer variables. Let's go back also to the comment about roughness and the statements made that in rough boundary layers the outer part of the flow doesn't care about the roughness. It is not quite correct that the outer part doesn't care for all rough wall conditions. This you can see for example, in a review paper in the AIAA Journal by Frank Dvorak, about roughness. Past a certain range of roughness, the outer flow is influenced by the roughness, and the coupling may be different. In many ways we have to be very careful in just taking one condition and generalizing. More recently, Tani showed that mild roughness may act to reduce the drag, similar to riblets. But the fact that mild roughness, for some range, does not interact with the outer flow, doesn't mean that roughness in general will do that.

Finally, I wanted to couple this with something else. This hierarchy of scales is a very complex hierarchy, with a broadening range with Reynolds number, but I think to start capturing one piece of it in a model, as Phil Holmes pointed out this morning, and then start filling up that hierarchy, is a very useful exercise for us. At the 1988 APS meeting I suggested that based on our experiments at high Reynolds numbers, a more symmetric hierarchy exists as compared with lower Reynolds numbers. The reason for this is that the smallest scales are axisymmetric and the evolution tends to distort them in a number of ways. The importance of what's going on near the wall is that it may be the mechanism leading to what exists in the outer flow.

Sid Leibovich:

Professor Roshko asked a question of Parviz Moin, and I'd like to ask it again. Parviz, do you see any kind of organized streamwise vorticity near the wall?

Parviz Moin:

Yes, of course we see streamwise vortices. The point of confusion and disagreement is whether they appear in pairs and the extent of their streamwise elongation. By elongated, to be precise, we mean of the order of a thousand wall units . It has been proposed by several investigators that the wall-layer streaks are created by the pumping action of pairs of streamwise vortices that are as long as the low-speed streaks and surround them. We see streamwise vortices, some short, some long, sometimes singles, sometimes in pairs that are sprinkled randomly in space. Yes, we see them.

E. Novikov:

The discussion was very productive.

Session Five

What Can We Hope For From Cellular Automata?

Discussion Leader: P. Moin, Stanford University

What Can We Hope For From Cellular Automata?

Gary Doolen

Center for Nonlinear Studies
Los Alamos National Laboratory

ABSTRACT

Although the idea of using discrete methods for modeling partial differential equations occured very early, the actual statement that cellular automata techniques can approximate the solutions of hydrodynamic partial differential equations was first discovered by Frisch, Hasslacher, and Pomeau. Their description of the derivation, which assumes the validity of the Boltzmann equation, appeared in the Physical Review Letters in April 1986. It is the intent of this article to provide a description of the simplest lattice gas model and to examine the successes and inadequacies of a lattice gas calculation of flow in a two-dimensional channel. Some comments will summarize a recent result of a lattice gas simulation of flow through porous media, a problem which is ideal for the lattice gas method. Finally, some remarks will be focused on the impressive speeds which could be obtained from a dedicated lattice gas computer.

MOTIVATION

There are several reasons for the recent rapid growth in lattice gas research. The method provides very high resolution because it is very memory efficient. In the simplest algorithm, over 10 cells are stored in each CRAY word. Problems with 5,000,000,000 cells can now be run on a CRAY X/MP. The algorithm is quite fast. 300,000,000 cells can be updated each second on a CRAY X/MP using four heads and about an order of magnitude higher speed can be achieved on a Connection Machine 2.

Also, the algorithm is totally parallel. This parallel feature is easily exploited on existing computers. In addition, an enormous gain can be made by constructing dedicated hardware. Already, inexpensive dedicated boards are available which allow small PCs to run lattice gas problems near CRAY speeds. Dedicated boards are now planned for delivery in 1990 which are expected to be a thousand times more powerful. It is possible to build with existing technology a dedicated machine which has the complexity of existing CRAYs but which would execute lattice gas algorithms many millions of times faster. (One should interpret this impressive gain in computer speed cautiously. For periodic problems on existing machines , lattice gas methods are slower than spectral methods at least by an order of magnitude. But for complicated boundary conditions, lattice gas methods can solve problems which are not solvable by other methods. An example discussed below is flow through porous media.)

Other advantages of the lattice gas algorithm include their ability to conserve energy and momentum exactly, their inherent stability with no mesh tangling or time step crashing, and their capability of implementing complex boundary conditions quickly and easily.

THE SIMPLEST MODEL

To illustrate the lattice gas technique, it is worthwhile devoting a few paragraphs to the simplest two-dimensional model. This model is often called the FHP model, using the last names of the three authors who showed that hydrodynamics can be simulated by the model.

This model uses a triangular lattice to fill the two-dimensional plane. One can equivalently consider the plane to be tiled with hexagons with a lattice point at the center of each hexagon. Each lattice point has six nearest neighbors. At each lattice site, at most one particle is allowed for each of the six directions. Each particle has a velocity vector which points toward one of the six neighboring lattice sites. All particles have the same speed, so that every time step each particle moves to the neighboring lattice site in the direction of its velocity vector. After the move, all particles at each site are allowed to collide and change directions, if and only if the collision conserves energy and momentum.

There are many sets of collision rules which can be chosen. The simplest model allows only two-particle collisions and three-particle collisions. These rules are illustrated in Figure 1. If only two- particle collisions are permitted, then momentum is conserved along each row of lattice sites in a periodic lattice. Thus too many conserved quantities exist and unphysical results are obtained. The total momentum along each row always remains at its initial value.

The three-particle collision rule allows the momentum along each row to vary with time, in such a way that the total momentum for the whole system is conserved. (A few extra constants of the motion are generated in many single-speed models, but these are usually harmless. An example of an unphysically conserved quantity in this simplest model is the fact that the momentum perpendicular to a row summed over even rows on even time steps equals the same quantity summed over odd rows on odd time steps.)

In order to fully appreciate the speed and memory efficiency of this model, it is useful to understand a few details of an actual lattice gas computer program. At each lattice site, we only need to know whether or not a particle exists in each of the six directions. One bit of memory for each direction is all that is required to completely specify the particles at this site. The information for all sites can be packed into memory so that every single bit of memory is used equally effectively. Coined "bit democracy" by von Neumann, this efficient use of memory should be contrasted with standard floating-point calculations in which each number requires a full word.

When the two-dimensional hydrodynamic lattice gas algorithm is programmed on a computer with a word length of 64 bits, then each word contains the full information required to specify 10 2/3 sites. To appreciate the significance of this efficient use of memory, consider how many sites can be specified in the solid-state storage device presently used with the CRAY X-MP/416 at Los Alamos. This device stores 512,000,000 words, with each word containing 64 bits. With 10 2/3 sites per 64 bit word, we see that this device can store information for over 5,000,000,000 sites. This corresponds to a two-dimensional lattice with 100,000 sites along one axis and 50,000 sites along the other. This number of sites is a few orders of magnitude greater than the number of sites normally treated when other methods are used. (Although this high resolution may appear to be a significant advantage of the lattice gas method, one should keep in mind the fact that some averaging over space or time is required to obtain smooth results for physical quantities such as velocity and density.)

There are two commonly used methods to store the bits. One method groups the bits into words according to direction; the other method groups bits according to sites. The direction-oriented method is most natural for logic-based codes, in which logical commands are used to implement the collision rules. The site-oriented method is best for table look-up implementation of the collision rules. We will describe the direction-oriented method.

If we label the six directions with the letters A,B,C,D,E, and F as shown in Figure 2, then we can store the information for the bits in the A-direction in one set of CRAY words in the following way. Starting in the lower left corner of the two-dimensional grid, we set the left-most bit of the word labeled A(1) to be 1 if there is a particle in the A-direction at the site in the lower left corner. The bit will be set to be 0 if there is no particle in the A-direction. Moving one site to the right, we can set the second bit from the left in the word A(1) to be 1 if there is a particle at this site in the A-direction. This procedure is followed for 64 sites and the corresponding 64 bits in the A(1) word. The information for the A-direction for site 65 in this row will be stored in the left-most bit of the A(2) word. If there are 1024 sites in the first row, then words A(1) through A(16) will contain all of the information for the A-direction for the first row. The left-most bit of A(17) will contain the information for the A-direction for the left-most site of row two. If there are 1024 rows in the simulation, then all of the A-direction information is contained in words A(1) through A(16384). An equivalent strategy is used for the remaining five directions.

Using this storage scheme, the FORTRAN coding for implementing the lattice scattering algorithm for the three particle collision rule is as follows;

```
      do 1  i=1,16384
      T=AND(AND[AND(B(i),D(i)), F(i)], NOT(OR[OR(A(i),C(i)),E(i)]))
      A(i)=XOR(A(i),T)
      B(i)=XOR(B(i),T)
      C(i)=XOR(C(i),T)
      D(i)=XOR(D(i),T)
      E(i)=XOR(E(i),T)
    1 F(i)=XOR(F(i),T)
```

This simple do-loop checks to see that particles exist in the B-, D-, and F-direction and that no particles exist in the A-, C-, and E-direction at each site. If this condition is met for a site, then the bit corresponding to the site location is set in the T word and the T word is used to reverse the directions of the B, D, and F particles at this site. This coding updates each site at the rate of about two sites per nanosecond on a single head of a CRAY X-MP.

FLOW THROUGH POROUS MEDIA

Several features of the lattice gas algorithm are exploited in flow through a porous medium. These features include the ability to describe arbitrarily complicated boundary geometry with little effort and the ability to simulate accurately Navier-Stokes for low flow speeds. In this section we briefly describe a simulation which indicates that the lattice gas method has much to contribute in this field of research.

The simulation describes a Navier-Stokes solution for the microscopic flow through an experimentally measured geometry. The sample simulated was obtained from a thin- section of oil-bearing rock. This thin-section geometry was etched into a glass plate sandwich which was then used to experimentally measure enhanced oil recovery processes, including permeability. The etched plate geometry was digitized to produce the complex pattern representing the rock geometry shown in dark blue in Figure 3. A grid size of 1024 by 572 lattice sites was chosen. The flow pressure is color coded with the high pressure at the top represented by red and the low pressure at the bottom represented by yellow/green. No fluid is allowed to exit through the sides. A no-slip boundary condition is assumed at the interface between the rock and the fluid. The lattice gas model used was the six-bit FHP model with all possible collisions allowed. The Reynolds number was approximately 1.0 and the fluid speed in the simulation was less than 0.1 times the sound speed.

The simulation was allowed to equilibrate for 30,000 cycles and then an average was taken over 20,000 cycles to obtain the equilibrium velocity and pressure distributions. The permeability of this geometry was calculated and found to agree with the laboratory-measured permeability within 5 percent.

This technique may provide us with a method for calculating many pore scale phenomena as functions of the geometry and dimensionless flow parameters. In general, many of the Darcian properties that have only been empirically available as average properties may now be calculated directly from the pore geometry. An understanding of the detailed flow structure underlying these averages and the relationship of these properties to the geometry of the pore space will help to improve our ability to define the constitutive relationships for use in many practical problems.

THE PROMISE OF DEDICATED HARDWARE

In 1986, the JASON committee was asked to evaluate the desirability of constructing hardware which was dedicated to the lattice gas algorithm. After several meetings, a brief summary of the conclusions was prepared in June 1988. A copy of this summary appears in the appendix. One conclusion was that, with existing technology, a CRAY-sized computer could be constructed, which could execute the lattice gas algorithm several millions of times faster than the fastest existing computer.

Since the report was issued, techniques have been developed which allow calculations to execute about a thousand times faster. It appears that the algorithm is still developing rapidly at the present time and that it is too early to commit to a particular implementation in dedicated hardware. Some of the features which are being explored now include implementation schemes for several other partial differential equations and minor modifications which would allow implementation of neural net algorithms.

FUTURE EXPECTATIONS AND CHALLENGES

At least four separate communities are closely following lattice gas developments with quite different expectations. The molecular dynamics community consider the lattice gas method to be a minimal bit strategy for solving Newton's equations of motion for orders of magnitude more particles than are usually simulated. Finite difference theorists consider the method to be an over-restricted set of finite difference equations, rightly expecting all finite difference disease to be amplified. Statistical mechanicians hope to use the method to gain new insight into the relation between microdynamics and macrodynamics. Parallel computer hardware scientists see the method as the simplest and fastest totally parallel algorithm with broad applications.

One can think of the lattice gas method as filling a niche between molecular dynamic methods and continuum methods. A present challenge is to determine the boundaries of parameter space where lattice gas methods are most appropriate. It is possible to add complexity to these methods until the results become indistinguishable from both molecular dynamics and continuum methods. However, the method becomes slower as the complexity increases. At the present time, the lattice gas method appears to be ideal for describing flow through porous media.

The lattice algorithm can be shown to approximately solve the Navier-Stokes equations in the long wavelength limit. But the algorithm can go far below this limit and possibly give considerable insight into the correct macroscopic treatment in situations where gradients are important and also in situations where the mean-free-path is not negligible. The understanding of how to go from the microscopic rules to the correct macroscopic equations remains a challenge, and the lattice gas has much to contribute here.

Another challenge is to determine the optimal lattice gas algorithm. At present, table look-up methods are very fast. For example, 300,000,000 sites can now be updated each second on a CRAY X/MP 416. The table size grows, however, as 2 to the Nth power, where N is the number of bits required at each site. For 24-bit models, 16 million word tables are required. It appears that reduced-size tables are possible if the symmetry of the table is fully exploited, but the restrictions which these symmetries place on the collision rules may significantly limit the range of allowed viscosities. Several studies are in progress to determine the class of algorithms which ought to be implemented in the type of dedicated lattice gas computer described in the appendix.

The most significant challenge is the implementation of the lattice gas algorithm on large-scale dedicated hardware. The gain in speed over existing computers is a factor of the order of many millions. This opportunity does not belong exclusively to lattice gas techniques but applies to all algorithms which have a parallel implementation.

ACKNOWLEDGMENTS

The research reported in this paper includes contributions from work at Los Alamos by the following scientists: Hudong Chen, Shiyi Chen, Jill Dahlburg, Karen Diemer, Jerome Dangmann, Ken Eggert, Castor Fu, Simeon Gutman, Brosl Hasslacher, Robert Kraichnan, Y. C. Lee, Lishi Luo, David Montgomery, Harvey Rose, Tsutomu Shimomura, Bryan Travis, and Greg Valentine.

APPENDIX. PROSPECTS FOR A LATTICE GAS COMPUTER

Prospects for a Lattice Gas Computer

Report of a Workshop held in La Jolla, California

June 17-18, 1988

A. Despain and C. E. Max
JASON, Mitre Corporation

G. Doolen and B. Hasslacher
Los Alamos National Laboratory

Abstract

A two-day workshop was held in June of 1988, to discuss the feasibility of designing and building a large computer dedicated to lattice gas cellular automata. The primary emphasis was on applications for modeling Navier-Stokes hydrodynamics. The meeting had two goals: 1) To identify those theoretical issues which would have to be addressed before the hardware implementation of a lattice-gas machine would be possible; and 2) To begin to evaluate alternative architectures for a dedicated lattice-gas computer. This brief paper contains a summary of the main issues and conclusions discussed at the workshop.

I. INTRODUCTION

This note summarizes a two-day workshop held in La Jolla California in June of 1988. The workshop was co-sponsored by the Center for Nonlinear Studies at the Los Alamos National Laboratory, and by the JASON group, Mitre Corporation. The purpose of the workshop was to identify, define, and begin to resolve substantive issues which must be addressed before a lattice gas computer can be implemented in hardware.

The workshop attendees were:

George Adams. Purdue University
Gary Doolen. Los Alamos National Laboratory
Paul Frederickson. NASA Ames. RIAC project
Castor Fu. Stanford University
Brosl Hasslacher. Los Alamos National Laboratory
Fung F. Lee, Stanford University
Norman Margolus, MIT Laboratory for Computer Science
Tsutomu Shimomura, Los Alamos National Laboratory
Tom Toffoli. MIT Laboratory for Computer Science

and the following members of the JASON group:

Kenneth Case. University of California at San Diego
Alvin Despain, University of California at Berkeley
Freeman Dyson. Institute for Advanced Study
Michael Freedman, University of California at San Diego
Claire Max. Lawrence Livermore National Laboratory
Oscar Rothaus, Cornell University.

The primary emphasis of the workshop was on the use of cellular automata for simulations of three-dimensional incompressible Navier-Stokes hydrodynamics. Within this context, there are two types of applications for which a special-purpose computer might offer important potential advantages over conventional numerical hydrodynamics techniques implemented on general-purpose supercomputers:

1) Studies of flows with complex boundary conditions. For example, one might look at a boundary-layer and study various techniques that have been suggested for drag-reduction and boundary-layer modification.

2) Studies of three-dimensional incompressible flows at high Reynolds numbers. These could include studies of the onset of fluid turbulence, free-boundary problems (such as ship wakes and drag), or the combination of hydrodynamics and simple chemical reaction systems.

The issues discussed at the workshop fall into three general categories: theory, computer simulation, and hardware.

II. THEORETICAL ISSUES

The most prevalent use of cellular automata for modeling hydrodynamics has been the so-called lattice gas. In this approach, one follows the motions of many individual particles which interact via given collision laws at fixed lattice sites. The individual particles are allowed to have only one (or at most a few) discrete speeds relative to the grid of lattice sites. The hydrodynamic limit is regained by averaging over a large number of these discrete particles, to obtain the first few moments of their distribution function.

In two spatial dimensions, the properties of possible sets of collision rules for the particles and lattice geometries for the sites are now reasonably well understood. There are two practical ways to represent a given rule set: via a look-up table which enumerates all the possible incoming and outgoing configurations, or via an algorithm or computation which generates the rules anew at each timestep and each collision site. However in three spatial dimensions the possible rule sets are far more complicated, and there are many unsolved questions regarding appropriate collision rules and their efficient execution. For maximum efficiency, a special-purpose lattice-gas computer should probably contain a hard-wired implementation of a particular rule set. However the general consensus at the workshop was that there is not yet a sufficient understanding of rule sets that have been proposed for three spatial dimensions to settle upon an optimum one for hardware implementation.

Important issues that remain to be solved concerning collision rules for three-dimensional hydrodynamics are the following:

1) What is the "best" rule set to use for modeling three-dimensional hydrodynamics?

    a) How does the choice of this "best" rule set change with the type of application one wants to solve? For example, are some rule sets better for studies of boundary-layer effects or free-boundary problems, while others are optimum for studying the onset of turbulence at high Reynolds number?

    b) How can rules be "tuned" to get optimum results for given problem parameters? For example, how can one optimize for high Reynolds number, or for specific types of boundary conditions?

2) Rules for lattice gases representing three-dimensional hydrodynamics tend to be very complicated. One way to implement them computationally is to use look-up tables, but these become very large. If there are n bits at each lattice site, then there are 2^n table entries. For example, the 24 bit model requires 16 million entries. How can this large number of rules be reduced by "factoring" or "grouping" them, to reduce the size of the rule representation in the look-up table? What is the fundamental dimension of the rule set?

3) In several proposed rule sets, one has to choose whether the same collision will always have the same outcome, or whether one will implement a randomization process within the rule set to "mix up" the collision outcomes. The addition of an explicit randomization procedure is expensive computationally. Under what circumstances can one rely on the inherently high frequency of particle collisions to achieve randomization, so that it does not have to be explicitly included in the rule engine?

4) A related question concerns the desirability of adding a "collision bit" to the algorithm. This is an additional bit determining whether a particle will or will not undergo a collision at the next lattice site that it reaches, if all the other conditions for a collision at that site are satisfied. If all particles undergo collisions whenever they can (no collision bit), one obtains a more "collisional" rule set, leading to the potential for attaining higher Reynolds numbers. Are there circumstances in which a less collisional rule set would be desirable?

5) What advantages are there to using nonperiodic tiling or quasilattices for modeling three-dimensional hydrodynamics, as compared with the so-called four-dimensional schemes or other periodic tiling schemes?

6) Is there a lattice gas analogue for adaptive-mesh hydrodynamic techniques, so that greater spatial resolution can be achieved in regions where it is needed? Can sub-grid scaling rules be derived to extend the spatial resolution of the lattice gas method?

7) What physical laws or partial differential equations do the various rule sets represent? Can the differences between the Navier-Stokes equations and the lattice gas implementations be systematically understood?

a) Under what conditions (limits on the Mach number, particle density, Reynolds number) does the lattice gas model with a given rule set reduce to three-dimensional Navier-Stokes hydrodynamics?

b) Given a set of physical constraints, can an algorithm be developed that will systematically generate a corresponding lattice gas rule set?

c) Each given rule set implies a particular functional form for the viscosity as a function of density. Given that the density is nearly constant in space for incompressible flows with Mach numbers small compared to unity, does it matter whether or not the density- dependence of the viscosity law is physical?

d) It has been suggested that the nonphysical function g(rho) appearing in front of the $u \cdot \nabla u$ term in the momentum equation can be eliminated by using rules which include two or more discrete (nonzero) velocities. Is this generally valid? Under what conditions would it be desirable to use more than one particle velocity? What is the gain in accessible Reynolds number when additional speeds are allowed? Are there advantages to these schemes that would allow lattice gases to satisfy statistics other than Fermi statistics? (The latter prevail for most currently used rules.)

In addition to the above questions concerning rules for lattice-gas representations of hydrodynamics, there are a set of issues involving extensions of the cellular automata methodology to other physical models:

1) Can hydrodynamics be modeled by using cellular automata particles to represent vorticity, in analogy with finite-difference vorticity-tracking algorithms? What range of Reynolds numbers could such a technique model?

2) How practical would it be to add some simple extensions to three-dimensional lattice-gas hydrodynamics? Some extensions that would be useful include gravity or other body-forces, two or more different fluid types, or simple chemistry. It was generally agreed at the workshop that extensions which involve action-at-a-distance, such as Maxwell's equations, would require a very different algorithmic approach.

III. ISSUES CONCERNING NUMERICAL SIMULATIONS

The workshop participants felt that there was a need to develop "benchmark" simulation problems. These would consist of a few canonical two- and three-dimensional hydrodynamics problems, for which the numerical results of various lattice-gas models could be compared with each other, with conventional hydrodynamics simulations, and with experimental results. This is particularly important because of the fact that different lattice gas rule sets may represent different approximations to the Navier-Stokes equations (i.e., they may approach the Navier-Stokes equations in different asymptotic limits).

A parallel effort should be made to compare lattice-gas simulation results with standard analytic solutions to the Navier-Stokes equations, in cases where these are known. Possible examples are channel flow, pipe flow, Pouseille flow, Couette flow, and so forth. This has been done to a limited extent for two-dimensional lattice gas models, but three-dimensional applications have not yet been well studied.

A different type of test of lattice gas algorithms was thought to be important as well. One should perform the standard numerical test of increasing the grid resolution, while holding fixed all of the "physical" parameters describing the problem. The goal would be to check that the higher-resolution result is identical to that obtained with lower numerical resolution.

A final numerical simulation issue thought to be important by the workshop participants concerns how to generate adequate graphical visualizations of the results of a three-dimensional lattice-gas simulation. It was pointed out that the amount of data storage needed for a three-dimensional simulation at high Reynolds number will be very high. Therefore, thought must be given to how to integrate input-output and graphical display within the process of the numerical computation itself. For the types of physical problems which one wants to address using lattice gases, it may not be adequate to obtain graphical displays of the results based entirely upon post-processing.

IV. ISSUES CONCERNING THE HARDWARE PERFORMANCE OF A LATTICE-GAS COMPUTER

In order to focus on the issue of hardware design for a lattice-gas machine. a set of performance measures was chosen. The idea was to outline hypothetical specifications for hardware components, so that when different candidate architectures were compared with each other. they would all be making the same assumptions about the capabilities of commonly used hardware components.

The following table gives a rough overview of the capabilities of VLSI technology today and in five years.

VLSI TECHNOLOGY (CMOS)

TODAY	IN FIVE YEARS
1 cm^2 active area	1 cm^2
200 pins	400 pins
1 Mbit DRAM	4 Mbit
50K "random" transistors	200K
10 nsec internal clock (on-chip communications)	1 nsec
80 nsec external drive (off-chip communications)	8 nsec

Using these characteristics, which are of course only approximate. one can outline the characteristics and performance of various architectures for a lattice gas supercomputer.

There appears to be a practical limit on the total number of chips it is plausible to include in a supercomputer. Existing supercomuters have about a third of a million chips. Workshop participants hypothesized that in the future one might build supercomputers with up to a million chips. Since the total number of lattice points required for a lattice-gas computation of high Reynolds number three-dimensional hydrodynamics is much larger than a million, one is led to a design in which many cellular automata lattice points are placed on each chip.

The next hardware issue is how to implement the set of collision rules. Since only a few rule sets for three spatial dimensions have been studied to date, the workshop participants felt that it was premature to choose a specific rule set for implementation in hardware. It was suggested that even after more three-dimensional rules have been studied, it would be desirable to leave flexibility in the choice of rules for the lattice gas supercomputer. There are two reasons for this choice. First, a new and better set of collision rules for hydrodynamics might be invented at any time; and second, one may at some later point want to use the lattice gas computer to study other physical models such as the mixing of two different gases, or hydrodynamics with simple chemistry.

Flexibility in the choice of rule set would have the most straightforward implementation if collision rules were executed via look-up tables. In that case one could feed an alternative look-up table into the computer when one wanted to change rules. The difficulty with this approach is that the rules suggested to date for hydrodynamics in three spatial dimensions would require very large look-up tables, with the disadvantages that the tables would use up large amounts of memory and would be slow to compute collisions.

Thus there is a lot to be gained by understanding the symmetries underlying each proposed rule set, so that the look-up table can be collapsed into a considerably smaller amount of memory space.

The second method of implementing a rule set is to design a computational engine in hardware that would re-calculate the rules "on the fly" for each collision. The advantage of this technique relative to a look-up table will depend on the details of the chosen rule set. At present for three spatial dimensions it seems that the look-up table approach is preferable from the point of view of speed and feasibility. In addition, the hardware rule-engine is less flexible than a table look-up approach, unless a software layer can be added to customize the rule engine for a choice of several different rule sets.

Since many lattice sites reside on each chip. and since off-chip communications are slower than those that remain on-chip, it seems desirable to locate on each chip the look-up tables or hardware rule-engines which calculate the collision outcomes. This avoids the time delays which would occur if on had to go off-chip to calculate collision outcomes. If there are many lattice points on a chip, one may want to have many "computational nodes" on each chip. (Here a "computational node" is defined to be a look-up table or a rule-engine for calculating collision outcomes.) This would avoid the time delays inherent in updating all of the lattice sites on a given chip sequentially.

Thus one must choose how to trade-off the number of lattice sites which can be stored on a chip with the number of "computational nodes" that will fit on a chip. The results of this trade-off will probably vary with the specific type of rule set chosen. since the size and complexity of the "computational node" and the number of bits required for a lattice site will in general vary. In the example shown in the following table, it was decided to allocate half of the chip space to lattice sites and half to "computational "nodes".

With the above discussion as background. the workshop arrived at the following the target performance characteristics of a hypothetical lattice gas computer:

Target Performance Parameters

> **Problem definition:**
>> three-dimensional incompressible hydrodynamics flexible boundary conditions
>> some flexibility in the rule set, if possible high input-output rate

> **Hardware aspects:**
>> 512 lattice sites per "computational node"
>> 64 "computational nodes" per chip
>> 32.000 lattice sites per chip
>> about a million chips total
>> 3×10^{10} lattice sites total ($> 10^{11}$ within 5 years)
>> 10 nsec update rate (on-chip)
>> about 64×10^8 site updates per chip per second
>> about 6×10^{15} site updates/sec total
>> ($> 6 \times 10^{16}$ site updates/sec within 5 yrs)

V. CONCLUSIONS

The workshop in La Jolla produced a considerable amount of enthusiasm about the potential of a dedicated lattice gas computer. Preliminary estimates based on the above performance numbers suggest that such a machine could surpass the present performance of a general-purpose supercomputer such as the CRAY II (3×10^7 site updates/sec) by a factor of about 10^8 and possibly considerably more. Of course the target machine could be expensive; first-of-a-kind supercomputers can cost from tens to a hundred million dollars. The cost of this machine is proportional to the number of chips, so a change in speed by a factor of 10 would result in a factor of 10 change in cost.

In view of the combination of large cost and high scientific potential for such a machine, it will be imperative to proceed along two parallel paths: 1) refinement of the theoretical understanding of cellular automata rules and lattices in three spatial dimensions, and 2) building of intermediate-scale hardware implementations of dedicated cellular automata computational engines, so as to gain expertise in the practical areas of architecture trade-offs and implementations. A promising example of such an intermediate engine is the CAM-8 machine recently proposed by Margolus and Toffoli, which would deliver 2×10^{10} site updates per second with 16 bits per site. Progress along both of these paths will be necessary in order to learn how to best exploit the potential of a dedicated lattice-gas supercomputer.

Work performed in part under the auspices of the U.S. Department of Energy, Office of Energy Research, Basic Energy Sciences.

Cellular Automata and Massively Parallel Physics*

Comment 2. [+]

C. E. Leith

Lawrence Livermore National Laboratory
Livermore, CA 94550 USA

1. Cellular Automata

A cellular automaton (CA) has been proposed[1] as an effective computer for fluid flows and thus for turbulence simulations. One deduces macroscopic hydrodynamic motion by simulating crudely but efficiently on a CA the microscopic dynamics of molecules in a gas and then averaging the results. Unfortunately, bad molecular dynamics gives poor hydrodynamics[2,3]. The best known difficulties are the lack of Galilean invariance, the large and ill-defined viscosity, the velocity dependent equation of state, statistical problems of low Mach number, and the need for a wastefully low density.

In addition to these more obvious problems there are subtle difficulties coming to light. For example, McNamara and Zanetti[4] have discovered false momentum integrals in the hexagonal lattice gas which are more obscure than those in the square lattice gas. Colvin, Ladd, and Alder[5] find for the lattice gas an unrealistic approach to the $1/t$ long-time tail of two-dimensional molecular dynamics, but this is improved in their maximally discretized molecular dynamics (MDMD) of colliding hard hexagons on a CA at no great difference in cost.

Much progress has been reported at this meeting in reducing these problems and finding the moderate Reynolds number domain where a CA may be useful.

Even if the CA hydrodynamics were good, it is noisy by the statistical nature of the underlying process. It is necessary to average over the molecular motions to get hydrodynamic variables; the averaging has inevitable sampling errors; for these to be moderate the sample must be large; and it turns out that the bit manipulations in a CA space-time averaging block may be put to better use in floating point operations at the corresponding single space-time grid point in conventional Navier-Stokes solvers[2].

A simple way to get rid of the noise is to carry a probability rather than a bit for each molecular state on the lattice. The bit collision rules are now

---

*Work performed under the auspices of the U.S. Department of Energy by the Lawrence Livermore National Laboratory under Contract No. W-7405-Eng-48.

[+]Note that the manuscript for Comment 1, by S. A. Orszag, is not available for publication.

replaced by floating point evaluation of a Boltzmann collision term. The resulting lattice Boltzmann model[6] is an intermediate step between the CA lattice gas and conventional numerical models.

Although a CA appears to be useless as a device for high Reynolds number turbulence simulation, it may make an important although indirect contribution to turbulence modeling. Experiments with lattice gas molecular dynamics or perhaps better with MDMD can and have shed considerable light on the deduction of continuum mean equations from random primitive processes. In particular, the many approximations made in such deductions can be explicitly checked. The crude analogy between turbulent eddies and molecular motions seems to extend to the deduction of eddy viscosity in turbulence models and of molecular viscosity in continuum hydrodynamics. Each requires closure assumptions, each truncates a divergent process at first order, and yet each obtains sensible results. Why?

The analogy is particularly close in the deduction of models of inhomogeneous turbulence in terms of evolving single-point probability distributions following the work of Lundgren[7] and Pope[8]. As the underlying random process, one might try to formulate a discrete hydrodynamics, as suggested by Karweit[9], executed on a CA from which one could extract corresponding turbulence statistics. Bad hydrodynamics probably would give poor turbulence but here the standards are lower.

The clear benefit that CA lattice gas dynamics has over its present competition is that it scales perfectly to ever larger CA's and thus represents a perfect massively parallel paradigm for hydrodynamics. The future competition is likely to be other such paradigms with a more accurate representation of the physics. One possibility appropriate for compressible hydrodynamics is discussed next.

2. Massively Parallel Physics

In many continuum physical systems, whose evolution we desire to simulate numerically, the velocity of propagation of influence is finite. Space-time causal cones are thus defined with characteristic speeds such as that of light or of sound. In such systems the values of fields at a particular point at an advanced time level depend only on field values within a local spatial domain at an earlier time level. The evolution equations to advance to the new time can thus be written as integrals over a local patch in space at the old time. For a small enough time step between the old and new time, the local problem may be treated as linear even for nonlinear systems.

Such a causal constraint in a physical system permits a natural mapping onto a massively parallel computer with a multiple instruction stream, multiple data (MIMD), distributed memory, message passing architecture such as that of a hypercube. A cluster of representation points scattered over a spatial domain can be assigned to each processor node. An interpolation scheme over each such cluster defines fields and any integrals of them at the old time, and thus the processor can compute field values at the new time at represention points in a somewhat smaller interior domain. Clearly there must be enough overlap in the spatial domains in each processor so that these interior domains cover the universe of interest. Message passing is required to transfer boundary information between processor nodes. Load balancing also may require the transfer of representation points between nodes. In any case, since only local information is required, the universe may be mapped onto an indefinitely large number of processors working simultaneously to advance a single time step. In this way massively parallel scaling is achieved in which the speed of execution is proportional to the number of processor nodes.

There are three general problems to be solved in connection with this general programming paradigm.

First, a choice must be made of the number and location of representation points at the new time level. The general guiding principle is to do this in such a way that truncation error is relatively uniform across the universe. The representation point density should clearly be related to the size of the causal patch. The usual explicit stability condition is achieved in this paradigm by representing the causal patch adequately. In a sense this means, for the specified δt, making δx big enough, rather than the more conventional approach of making δt small enough for the given values of δx. Clearly some estimate of truncation error is needed for guidance, and this should come from the interpolation scheme.

Second, an interpolation scheme is needed which can provide an accurate estimate of the required integrals over the old causal patches associated with each new representation point. In this regard the multiquadric scheme developed by Hardy[10], worked on by Kansa[11], and given mathematical foundation by Buhmann[12] looks to be the most promising. It may be necessary to utilize some sort of double representation point coverage in order to extract the necessary error information. Questions must be answered about the optimal number of fitting points and whether their spatial distribution can and should be controlled in the new point selection process in order to optimize condition number for the coefficient linear solve.

Third, communication between processor nodes must be worked out. The boundary domain for which information must be exchanged between processors is about as wide as the causal patch width, and this is supposed to be small compared to the cluster size. Message passing time should be covered by cluster arithmetic time, and this may determine the minimum desirable cluster size. There is also here a synchronization problem. Data should have a time value attached and be transferred into a cleared buffer of a neighboring processor. The buffer can then at the appropriate stage be cleared with its contents moved into the old time array. Although it is made simpler by the use of a specified time step independent of space and time, synchronization locks must be set and released carefully owing to the inherently asynchronous nature of the MIMD processing. It is here that some dynamic load balancing scheme must be used in order to avoid processor idle time.

Some work has been done on the formulation of three-dimensional compressible hydrodynamics within this paradigm using the Poisson integral with a singular kernel over a spherical surface in space at the old time. This is a generalization of the method of characteristics to three dimensions. Diffusion is similarly characterized by a Gaussian kernel for the integration, and this must be cut off to remain strictly local. In this paradigm, the details of the physics being done only influences the choice of kernel for the integrals and does not change the overall logical structure.

References

1.   U. Frisch, B. Hasslacher, and Y. Pomeau: Phys. Rev. Lett. **56** 1505 (1986)
2.   S. Orszag and V. Yakhot: Phys. Rev. Lett. **56** 1691 (1986)
3.   J. Dahlburg, D. Montgomery, and G. Doolen: Phys. Rev. A **36** 2471 (1987)
4.   G. McNamara and G. Zanetti: "The Hydrodynamics of Lattice Gas Automata", preprint, University of Chicago (1989)
5.   M. Colvin, A. Ladd, and B. Alder: Phys. Rev. Lett. **61** 381 (1988)
6.   F. J. Higuera, S. Succi, and R. Benzi: "Lattice Gas Dynamics with Enhanced Collisions", preprint (1989)
7.   T. Lundgren: Phys. Fluids **12** 485 (1969)
8.   S. Pope: in Theoretical Approaches to Turbulence, Dwoyer, Hussaini, and Voigt, eds., (Springer-Verlag, Berlin and New York, 1985)
9.   M. Karweit: in Frontiers in Fluid Dynamics, Davis and Lumley, eds., (Springer-Verlag, Berlin and New York, 1985)
10.   R. L. Hardy: J. Geophys. Res. **76** 1905 (1971)
11.   E. J. Kansa: Proc. Multiconference on Computer Simulation, San Diego **4**, 111 (1986)
12.   M. D. Buhmann: IMA J. Numer. Anal. **8** 365 (1988)

A New Strategy for Hydrodynamics - Lattice Gases

Comment 3.

U. Frisch

CNRS, Observatoire de Nice
B.P. 139, F-06003
Nice Cedex

To appear in Proceed. 2nd European Turbulence Conference, Berlin, Sept. 1988,
Springer.

1. Introduction

Very interesting perspectives of massively parallel and interactive simulations for phenomena in fluid mechanics have appeared in 1985, thanks to a technique of a drastically new conception called lattice gas hydrodynamics. It has been developed by scientists from the observatory of Nice (M.Henon, J.P. Rivet and the author) together with scientists and engineers from Ecole Normale Superieure in Paris (A. Clouqueur, D. d' Humieres, P. Lallemand and Y. Pomeau) and of Los Alamos (B. Hasslacher, G. Doolen, et T. Shimomura). About ten other research centers in France, Belgium, Italy, the Soviet Union and particularly in the United States also work in this field.

In the present paper we shall not deal with theoretical questions at a technical level. For this see Frisch, d'Humieres, Hasslacher, Lallemand, Pomeau et Rivet (1987) and Kadanoff, McNamara and Zanetti (1988). We also shall limit ourselves to only a few references. Let us mention that the August 1987 issue of *Complex Systems* is entirely devoted to this subject.

The purpose of the present paper is to show the richness of the approach by lattice gases mixing theory, algorithms, simulations on super computers and construction of special purpose machines. The technological mutations of computer sciences forces us to take the path of parallelism (see for example the turn taken by the company IBM in 1988) and thus also probably to think of a new way to deal with scientific computations. Lattice gas hydrodynamics is a perfect example of this new way of thinking. At the end of the paper we shall also indicate some other possible applications of massive parallelism.

2. Floating point methods in hydrodynamics

Traditional methods of simulations in hydrodynamics which we shall call floating point methods are based on the description of a fluid by field (velocity, temperature etc.) satisfying partial differential equations. After suitable discretisation these fields are represented in the machine by floating points. The details of these methods depends mostly on the type of approximation chosen: finite difference, finite elements, spectral methods, vortex methods, etc. In going from one method to another one what changes is the base of function on which the fields are projected (for example on trigonometric - or Chebychev functions in spectral methods). The implementation of boundary conditions is the main difficulty of those methods; the complexity of the algorithms can become rather cumbersome and so does the computer time. However these methods have benefitted from about forty years of experience, their massive use going back to the Manhattan project of the A-bomb. The super computers are highly optimized for the solution of partial differential equations by floating point methods. Aeronautics, astrophysics, meteorology and many other fields have benefitted very much from these developments. Increase of the power of the machines allows us now to deal with the modeling of three-dimensional compressible flows in turbulent regimes (such as for example the collapse of a self-gravitating cloud).

3. From the true microscopic world to the fictitious microscopic world: Cellular automata

Any fluid is made out of molecules. The equations of fluid mechanics such as for example the Navier-Stokes equations are reflecting at the macroscopic level the microscopic laws of conservation in the collision between molecules of mass, momentum and energy. Conservation laws as well as symmetries, that is invariance groups of Newtonian physics (translations, rotations and Galileian transformation) determine completely the macroscopic equations.

One way of simulating hydrodynamics is to introduce explicitly molecules with their interactions. This approach of molecular dynamics requires many more resources than floating point methods. It can nevertheless be of interest when one wants to determine the transport coefficients or to take into account complex chemical-physical phenomena near a boundary, for example during re-entry of a space shuttle.

If one is only interested in macroscopic aspects of hydrodynamics one can completely ignore the microscopic level. One may also, and this is the central idea of the new method, introduce a fictitious microscopic world, simpler than the real world, better adapted to the structure of electronic components. If such a fictitious world presents conservation laws and invariance laws of suitable character (we shall come back to this), its macroscopic properties will be indistinguishable from that of a real fluid. As one would have guessed, the encoding of such a fictitious microscopic world will have to be Boolean, that is to use bits.

One is then led to explore the possibility to use a cellular automaton. This term, due to von Neumann and Ulam, designates boolean computation structures which are highly decentralised and repetitive. One of the best known examples is the game of live of J. Conway: one has a kind of checker-board where the squares represent two states, say, black and white. One updates the automaton by counting the number n of white squares among the eight adjacent squares of the given square. If n equals 2 the colour of the square is left unchanged; if n equals 3 is becomes white, otherwise it becomes black. This automaton rule is applied simultaneously to all squares. One is given arbitrary initial states and then the game is continued indefinitely. More generally, a cellular automaton is constituted of a regular lattice (one - or multidimensional) with an elementary machine (automaton) at each node. This elementary machine possess a finite number of states (coded in general with bits). A rule, mostly deterministic and identical for all nodes, allows to update the state of a node as a function of its preceding state and of a finite number of neighbours. It is clear that this is amenable to massive parallelism in its electronic implementation. One can imagine (which is not necessarily the most efficient way), that an elementary processor is associated to each node. An important point is that each processor needs to be physically connected only to a finite number of other processors, independently of the total number of processors. Thus one does not have to fear a bottleneck of communication.

Is it possible to find a cellular automaton, the large scale properties of which are that of a real fluid? It is useful for that to have particles associated to the state bits; one is then speaking of a "lattice gas". Consider for example the square lattice of unit mesh of figure la. Let us imagine that particles of unit mass and speed are moving in one of the four directions of the lattice (east, north, west, south) with an exclusion principle (not more than one particle per node and per direction). It is clear, that the state of each node is codable on four bits, each of which represents the presence (1) or the absence (0) of a particle. But is it possible to find a rule for that automaton, leading to the equations of fluid mechanics? Rules of this nature should be interpretable as interactions between particles. These interactions should possess firstly the laws of conservation and secondly suitable symmetries. The first point does not lead to serious difficulties, the second, however, does: the macroscopic and the microscopic properties of the real fluid are invariant under arbitrary rotations; our automata, however, will have at best a descrete invariance group (multiples of 90° rotations for the square lattice). This is where a major conceptual and central difficulty of our approach lies. Let us now see how it has been solved.

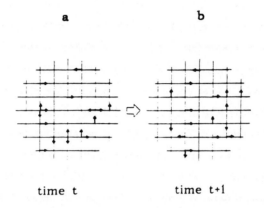

Fig. 1: The HPP model. The black arrows indicate the occupied states. The automaton is shown at two consecutive times in (a) and (b).

4. Looking for a symmetry group: a detour through the fourth dimension

A lattice gas automaton of a particular simple type now called HPP has been introduced about fifteen years ago by Hardy, Pomeau and de Pazzis (1973). This automaton resides on the square lattice previously introduced (fig. 1). The updating rule can be decomposed into two steps. First step: simultaneously on each node, collisions are performed between the particles present. The only non-trivial possible interaction which conserves mass and momentum is the head-on-collision: two particles arriving from opposite directions, say, East and West, leave at right angles in the directions of North and South, assuming of course that these directions were not previously occupied. Second step: particles are propagated by performing a shift on the four boolean fields in such a way that the particles associated to East bits are moving one mesh to the East (right) and similarly for the three other directions. This two-step rule is illustrated in figs. 1a and 1b which shown two consecutive states of the HPP automaton.

In spite of its particle interpretation, the macroscopic properties of the HPP model are not those of a real fluid (except for sound wave propagation). The symmetry group of the square lattice is indeed too small. This point is not obvious and we are tempted to spare the reader some of the details. Unfortunately the reader would then also be deprived of the pleasure of the promenade through the forth dimension. Therefore let us step in.

In a Newtonian fluid at rest when a shear is applied, viscous stresses appear, which depend linearly on the rate of shear (the symmetrical part of the second order tensor of velocity gradients). The viscous stresses themselves are forming a second order tensor. To go over from one tensor to the other one we have to contract with a fourth order tensor. In a real fluid these tensors are isotropic and can be expressed entirely with the help of the two coefficients of shear and bulk viscosities. In our lattice gases the tensors of fourth order, invariant in the symmetry group of the lattice are not necessarily isotropic. This problem is well known in elasticity theory: the elastic properties of a crystal (that is to say the relation between strain and elastic stresses) are not in general isotropic. Fortunately there is at least one exception: a two-dimensional crystal forming a *triangular lattice*, invariant under 60° rotation, is isotropic. This is a result in group theory, possessing an elementary proof to which we shall spare the reader.

These considerations have led Frisch, Hasslacher and Pomeau (1986) to modify the HPP model and to propose the FHP model for which they proved that the macroscopic behaviour is that of a real fluid satisfying the two-dimensional Navier-Stokes equations. The FHP model uses a triangular lattice, where each node has six neighbours. The model has thus six bits per node, but there exists a more efficient seven bit variant. We shall not dwell on the collision laws of this model here.

Even in the presence of boundaries, one or two dimensional, a flow becomes generally three-dimensional as soon as the non-linearity (measured by the Reynolds number) becomes sufficiently large. Can the above method be extended to three dimensions? At first sight, no. Indeed in three dimensions there exists no regular lattice, the symmetry group of which is sufficiently large to ensure isotropy of fourth order tensors. Still, there is a solution. It goes through the fourth dimension (d'Humiere, Lallemand and Frisch, 1986). In four dimensions the face centered hyper-cubic lattice (FCHC) possesses all the required symmetries (this is connected to the existence in four dimensions of six regular polytypes). The FCHC model uses 24 bits per node, because there are 20 neighbours in such a lattice. From there one can go back to three dimensions. One then obtains the pseudo-4D-model which resides on a cubic lattice and possesses again 24 bits per node. Henon (1987a) has proposed collision rules adapted for this model and has then perfected them (see below). The validation of this model was done by Rivet (1987), by comparison with a reference three-dimensional flow.

5. Optimization of collisions

For application of lattice gas methods it is of interest to have a viscosity as small as possible. Lowering the viscosity by a factor of 2 doubles the accessible Reynolds number. The same result can be achieved by doubling the size of the lattice, but the number of needed operations is then multiplied by 8 in two-dimensions and by 16 in three-dimensions. The viscosity depends only on the choice of collisions. With the 24 bit pseudo-4D model (which can be used both in two and three dimensions), the number of output states after collision for a given input state can be extremely high. (Henon 1987b) has shown that the minimization of the viscosity can be reduced to a problem of optimal pairing of input and output states. The solution of this problem produces a collision table giving for each 24 bit input state a unique output state. Such a table has therefore $24*2^{24}$ bits. This optimization problem has been completely solved. Between the simplest early two-dimensional FHP model and the pseudo-4D-model with the optimal collision rules obtained by Henon, the viscosity has been lowered by more than a factor of 16. This is one of the elements which has allowed the method of lattice gases to become a tool for simulations which is interesting and possibly even competitive. Let us now take a look at the use of this method.

6. Implementation

The use of lattice gases for the simulation of hydrodynamics is in its principle very simple. Macroscopic quantities such as density or the hydrodynamic speed are obtained from the boolean microscopic quantities by taking spatial means over cells of suitable size (for example 8*8*8 in three dimensions). The initial conditions are obtained by randomly selecting the boolean variable with appropriate distributions. All the evolution is then determined by the automaton rule except for those nodes located at boundaries. Rigid boundaries are obtained by bouncing back particles from them. The overhead due to the introduction of boundaries is usually negligible. Although randomness plays only a limited role in the method, which is mostly deterministic, still, the lattice gas method is closely related to Monte-Carlo methods, in so far as the signal to noise ratio varies like the square root of the size of samples (as does here the number of particles per cell). The practical implementation of the method can be done on special purpose machines or by simulating the automaton on a general purpose machine.

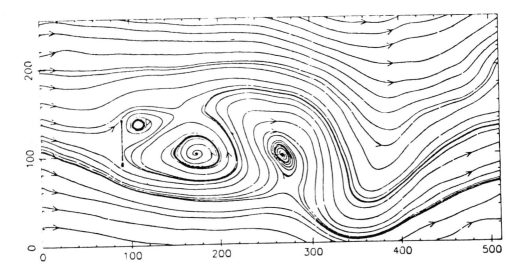

Fig. 2:   Simulation of a two-dimensional v.Karman street on the special purpose machine RAP-1 using the FHP model.

7. Special purpose machines

Fig. 2 gives an example of a simulation of a v. Karman street behind a flat plate perpendicular to the flow in a kind of "numerical windtunnel". The simulation was done on the special purpose machine RAP-1 (Reseau d'Automates Programmable). This machine has been built by Clouquer et d'Humieres (1987) at Ecole Normale Superieure in Paris. It comprises 256 by 256 noes of 16 bits. The propagation phase of the updating of the automaton is done by shifting the bit planes. The collision phase is done by a look-up table of $16*2^{16}$ bits operating sequentially on all nodes. The boundaries are taken into account in the collision phase and can be modified interactively during the simulation.

Using exclusively off-the-shelf electronic components RAP-1 operates at the video frequency of 50 complete updates per second. The memory of the automaton is also a video memory, this allows the direct visualization of flows simulated on a colour monitor (figure 2 has required some post-processing to reveal the streamlines). RAP-1 is limited to two-dimensional flows with Reynolds number of the order of a few hundred. The typical duration of a simulation is a few minutes. The cost of the components is about 1500 dollars. Getting the prototype ready has taken approximately one man-year.

Two points are worth stressing. First RAP is a *programmable* machine which can deal with a large variety of problems which can be formulated as cellular automata. This involves for example image processing (for which there exist already quite a few special purpose machines), the formation of fractal aggregates in the presence of diffusion, the propagation and the diffraction of acoustical waves in a heterogeneous ground and so on. Second, the architecture of RAP is such that it allows the interconnection of an arbitrary number of machines in parallel to achieve large scale simulations. RAP-2, which is now being built, will allow two-dimensional simulations of several thousand nodes in each direction. RAP-3, intended to be a three-dimensional machine, is under study. Let us mention that the RAP operation is supported by the European community (stimulation action).

Other teams of scientists in the world are working on special purpose machines for cellular automata. The pioneer in this matter is Toffoli who, with his collaborators at MIT, has built in the early eighties of CAM (Cellular Automaton Machine). It was on the CAM that the first simulation on the HPP model, showing wave propagation phenomena has been realized in 1984. The MIT team is presently preparing a 512 * 512 * 512 machine.

8. Simulation on general purpose machines

Figure 3 shows a detail of a three-dimensional simulation of flow around a circular plate normal to the flow. The Reynolds number is about 200. The flow is strongly unsteady. Figure 3a corresponds to approximately two circulation times. the axial symmetry is only weakly broken at that time. After 3.5 circulation times (figure 3b) the flow has become fully three-dimensional and detached eddies are observed.

This simulation has been done by Rivet on the CRAY-2 of the "centre de calcul vectorial pour la recherche" in Palaiseau near Paris. It makes use of the pseudo-4D lattice gas method with the collision rules discussed above. The updating uses a 16 million word table calculated once for all. The code is highly optimized for the CRAY-2 (vectorisation, multitasking on four processors, reduction of memory conflicts) and allows thus updating of approximately 30 $*10^6$ nodes of 24 bits per second. The simulation shown on figure 3 had 128*128*256 nodes and took approximately 2 minutes of CRAY-2 per circulation time. See reference: Rivet, Henon, Frisch and d'Humieres (1988).

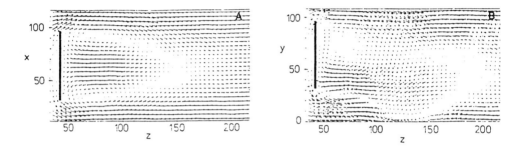

Fig. 3:   Three-dimensional simulation of the flow behind a circular plate with the pseudo-4D-model. (a) Detail of an axial cut at about 2000 steps of the automaton. (B) Detail after 3500 steps.

This result is surprising in at least two ways. First because it shows that supercomputers, like the CRAY-2, used intelligently, are well adapted to simulation of lattice gases (and more generally of cellular automaton), while in principle they have been optimized for floating point calculations. Also because floating point methods can only with great difficulty reach these fully three-dimensional regimes.

Lattice gas simulations were actually started already in the summer of 1985, after the first two-dimensional FHP model was obtained. The goal at the beginning was to validate the model on known examples of fluid mechanics. The first simulations done at Ecole Normale Superieur (on FPS 164) and at Los Alamos (on SUN and CRAY-XMP) did show a pair of recirculation eddies for a fairly low Reynolds number flow behind a plate. Such eddies are easily observed by slowly moving a spoon in a cup of coffee. The simulation of an unsteady two-dimensional v. Karman street by d'Humieres, Pommeau et Lallemand (1985) was already an nontrivial example of the possibilities of the new method.

Fig. 4: Instability of a flame front. The fresh gases are arriving from below. We observe the Darrieus-Landau instability. (Figure communicated by P. Lallemand and G. Searby)

Rather spectacular simulations were also done at Cambridge/Massachusetts by Salem and Wolfram (1986) on the hyperparallel Connection Machine (CM-1) which, on this kind of application, was running several hundred times faster than the FPS 164. We will have the opportunity to come back to the Connection Machine below. At this stage the principal interest of the method seems to be its simplicity, particularly for the implementation of boundaries.

As one could have expected, lattice gas methods were then subject to criticism. According to Yakhot and Orszag (1986), floating point methods would be much more efficient at high Reynolds numbers. This (friendly) argument is far from being settled. At the moment floating point methods (particularly spectral methods) have, at large Reynolds numbers, an important advantage for trivial geometries, such as 2π - periodic- or simple ones (flow behind two parallel plates). However, lattice gases seem now to have an advantage for three-dimensional external flow around obstacles.

The *quantitive* comparison of lattice gases with theoretical and experimental results was done by d'Humieres and Lallemand. The results were very encouraging, including cases considered delicate by numerical specialists, such as recirculation behind a step.

Rather rapidly the question arose of how adaptive the method is: is it limited only to hardcore incompressible Navier-Stokes? It doesn't seem so. Works started by the scientists of Ecole Normale Superieure in collaboration with the Laboratoire de Recherche en Combustion (P. Clavin and G. Searby) have led to simulations involving thermal phenomena, multi-phase phenomena, reactions in combustion - for example instability of flame fronts (see fig.4)j. Convective phenomena have been simulated in particular by S. Zaleski from MIT. d. Rothman and J. Keller, also from MIT have recently shown how to simulate the phenomena of surface tension between two non-miscible fluids.

9. Perspective for massive parallelism

The lattice gas method has been conceived for allowing easy parallel processing. Even before the construction of special purpose machines with large parallelism, simulations of lattice gas methods on general purpose machines (FPS, CRAY etc.) have revealed an interesting potential (three-dimensional flows, multiphasic problems, interactions between combustion and flow, etc.). Practical applications are in sight, for example the simulation of the cooling of chips, For further validation of the method it will be of interest to compare three-dimensional lattice gas simulations with very well controlled laboratory experiments for flows around simple objects such as spheres or cylinders; the corresponding three-dimensional regimes are indeed rather purely understood.

One difficulty of the lattice gas method, but also perhaps one of its most fascinating aspects, is that each time that we want to simulate an equation we have to imagine a microscopic world that is suitable. When the phenomenon which we are trying to incorporate has a molecular origin (such as the interfacial tension between two non-miscible fluids), this can lead to very efficient algorithms. Incorporating electromagnetic phenomena is considerably more difficult (except possibly for MHD effects).

Lattice gas methods and more generally cellular automaton methods are not meant as substitutes for floating point methods, but rather are meant to supplement them for specific applications (for example numerical windtunnels or meteorological forecast). Still, we expect that increasingly frequent use of parallelism will have a strong effect on our ways of approaching scientific computations. Most traditional algorithms have indeed been devised for sequential computers; running them in parallel can be a source of great problems. It is then often better to invent new algorithms, using floating points or cellular automata.

This becomes particular urgent with the advent of hyper-parallel machines such as the Cosmic Cube or the Connection Machine. The later can have up to 2^{16} boolean processors and 2^{11} floating point processors. The communication network of these machines is of the hypercube type, that is, there is a physical communication between processors the addresses of which differ by one bit. Thus the number of wires out of one processor goes like the logarithm of the total number of processors. This type of machine is intermediate between general purpose machines and special purpose machines. For suitable applications, powers of several gigaflops are accessible at a fraction of the cost of a supercomputer. This is for example the case for the gravitational N-body problem (with approximately uniform distribution of the bodies). In the traditional approaches the computation time grows like N^2. On a hyper-parallel machine a processor can be devoted to each body and thus the computer time will grow only linearly with N. Another application, well adapted for hyper-parallelism, is the processing of very large astronomical images (up to maybe 30.000 * 30.000), in order to improve the understanding of the large scale structure of the universe. These structures may themselves reflect the properties of primordial turbulence.

Of course we should not underestimate the effort that will be needed to get used to these new high- parallelism architectures. Fortunately, Nature which may be viewed as a sort of grand parallel computer, is mostly indicating us the right direction.

References

1. Clouqueur, A., d'Humieres, D., 1987, Complex Systems, 1, 584.
2. d'Humieres, D., Lallemand, P. and Frisch, U., 1986, Europhys. Lett., 2, 291.
3. d'Humieres, D., Pomeau, Y. and Lallemand, P., 1985, C.R. Acad. Sci., II 301, 1391.
4. Frisch, U., d'Humieres, D., Hasslacher, B., Lallemand, P., Pomeau, Y. and Rivet, J.P., 1987, Complex Systems, 1, 648.
5. Frisch, U., Hasslacher, B. et Pomeau, Y., 1986, Phys. Rev. Lett., 56, 1505.
6. Henon, M., 1987a, Complex Systems, 1, 475.
7. Henon, M., 1987b, Complex Systems 1, 762.
8. Hardy, J., Pomeau, Y. et Pazzis, O., 1973, J. Math. Phys., 14, 1746.
9. Kadanoff, L., McNamara, G. and Zanetti, G., 1988, Phys. Rev. A., in press.
10. Rivet, J.P., 1987, C.R. Acad. Sci., 305, 751.
11. Rivet, J.P., Henon, M., Frisch, U. and d'Humieres, D., 1988, Europhys. Lett., 7, 231.
12. Salem, J. and Wolfram, S., 1986 in "Theory and Applications of Cellular Automata", S. Wolfram ed., p. 362 (World Scientific).
13. Yakhot, V. and Orszag, S., 1986, Phys. Rev. Lett., 56, 169.

Discussion of "What can we hope for from Cellular Automata"

Reporter M.J. Karweit

Dept. of Chemical Engineering
The Johns Hopkins University
Baltimore, MD 21218

Brian Cantwell:

I was curious to know how energy dissipation is actually accomplished. Is the method exactly energy conserving? It s got an isothermal equation of state. How is energy dissipated and where does it go?

Uriel Frisch:

Energy is conserved but in a trivial way. In the simplest lattice gas schemes, energy is conserved because all the particles have the same speed, say unity. Kinetic energy per particle is the same as its mass; therefore mass conservation and energy conservation are not independent. The kinetic energy is just a very tiny perturbation on top of the microscopic world, and the kinetic energy just randomizes into equilibrium motion. You have the same apparent paradox in the real world. If you look at kinetic energy, in the incompressible limit you have to go to very low Mach number. Therefore the kinetic energy is a negligible part of the total energy.

Brian Cantwell:

But that is my question. What is a measure of the temperature?

Uriel Frisch:

It is more interesting to point out that when one does combustion or compressible flow-there are lots of models which were not presented here which have several speeds and which have a genuine energy equation which is not identical with the mass equation-one can have burned and unburned gases and have expansion just the same as in the real world-with the mass and energy and momentum relations being treated in the equations. (See Clavin, Lallemand, Pomeau, and Searby, **JFM**, 1988, vol 188, pp. 437-464.)

R. Narasimha:

This topic of particle propagation that you mention. Is it the sort of thing you get from the nonlinear steepening mechanisms that is presented in the N-S equations?

Uriel Frisch:

No. No. It is a microscopic phenomenon and has nothing to do with the nonlinear terms. It is a purely kinematical effect of how particles move along the lattice without any collisions between nodes.

Hassan Nagib:

Nobody made comments about flows with body forces, MHD. Is there a special advantage with this approach?

Gary Doolen:

There are three different algorithms I've implemented for solving problems with MHD. The hydrodynamic aspects are evolving too rapidly now for one to seriously contemplate calculations.

Uriel Frisch:

Body forces are very easy. Just flip particles.

Hassan Nagib:

That's what I wanted to ask. How's it done? Are the collision rules different?

Uriel Frisch:

In order to have uniform body forces, you have to occasionally flip particles, say, from up to down motion. Similarly for a coriolis force, you occasionally have them turn 90 degrees.

Gary Doolen:

For MHD there's an addition quantity, the vector potential, where you just take derivatives of the potential to get $\vec{J}$ and $\vec{B}$ to implement $\vec{J} \times \vec{B}$); it's essentially N-S.

Phil Holmes:

This question has sort of come up implicitly. 300 years ago Newton and (depending on whether or not one lived in England) Leibnitz invented calculus and we got differential equations. And ever since then, most of our physics has been based on continuum modeling. In particular, this community of engineers and applied mathematicians have loved continuum models. So a lot of the objection to cellular automata, or other objects people study, such as coupled map lattices, is that they don't look like the equations of continuum mechanics. You try very hard to make them look like the N-S equations, but you've said implicitly: Maybe we shouldn't do that. Maybe we should be rethinking the way we model the universe.

Uriel Frisch:

Some physicists believe that the basic laws of physics at the lowest level are better described on a lattice. This is a philosophical issue, of course, which has not been settled among physicists.

Phil Holmes:

It's hard to believe that modeling 10^{23} particles is really the right way to understand the construction of things.

Uriel Frisch:

Not 10^{23}. That is a misconception. There's a problem of universality classes. There are different models which, provided they have the right conservation laws and symmetries, will have their large scale behavior governed by the same equations-Navier-Stokes. And we can keep in that class any member which happens to be convenient for a certain stage of our hardware or software development. That can change in time. At the microscopic level, lattice gases completely misrepresent the true world. The same is true also of one of the most popular models of statistical mechanics: the Ising model for ferro-magnetics which has been so popular for understanding critical phenomena. If you look at its microscopics-the way atoms are interacting in a ferromagnet-it has little to do with the Ising model. It is much more complicated. But this is irrelevant for critical phenomena. The same thing is happening here for fluid dynamics: our artificial Boolean microworld is macroscopically indistinguishable from the real world.

Javier Jimenez:

If you really want to solve the problem why not use a good finite difference or finite element approach? Finite elements are very good for complex geometries.

Uriel Frisch:

It depends on which problem you want to do. It is a bit illusory, for example, to completely suppress the work of half a century of development in CFD starting mostly with the Manhattan Project. And say: "Oh, let's forget about that. Funding agencies! Stop supporting traditional CFD! Lattice-gas modelling is the way." Actually this is highly unlikely. Based on the history of all the approaches introduced within CFD, it appears that each method has its niche. Vortex methods have certain applications. And we're beginning to understand where the most likely niches are for lattice gases. And they're in complex flows and multiphase flows and such things. And I would not, for example, in any foreseeable future try to compete with the scheme of Parviz Moin for channel flow with a lattice-gas approach. Maybe for an external flow, but certainly not for a channel flow.

Steve Traugott:

There were some examples of forced flow shown-with two-phase flow, I gather-and there were some interesting cases where some of the little spaces were left behind by the fluid, while it went through others. There was an implication that somehow surface tension would be included and that one could do those kinds of problems. It would seem to me there are other than surface-tension-driven phenomena going on there. Is it claimed that flow through a single capillary chasing an interface through a single capillary could be done this way?

Uriel Frisch:

Yes, it can be done. I've seen it done on a Sun station. Rather spectacular.

Steve Traugott:

It contains all the features of that flow which has interfacial motion and so forth?

Uriel Frisch:

I'm not familiar with the capillary flow, but what I saw was rather convincing. Not having D. Rothman (Dept. Geophysics, MIT) here, maybe Gary Doolen has something to say.

Gary Doolen:

It appears to be a question of detail. How well can you mock up the pairwise potential between an oil molecule and sand or silicon? I believe the answer is you can mock it up as closely as you want, but you pay a price.

Jay Boris:

I'd like to ask anyone here who's looked at the lattice Boltzmann approach. There we go from a discrete form to continuum mechanics. Can you use the same lattice, or do you use a coarser lattice?

Uriel Frisch:

In the lattice Boltzmann equation one does not have to restrict oneself to the original lattice. It has not been done, but it could be done. I'm pretty sure that the six- or seven-speed model, the MHD kind, could be solved on a square grid, for example, or a non-structured grid or a non- uniform grid, because the quantities that are varying smoothly in space and time could be interpolated. You may lose stability if you do that, but that's another issue.

Javier Jimenez:

I might add to that. In lattice gasses, using the classical approach, you need to use a much finer lattice and then average. With the Boltzmann approach each site is already averaged.

Jay Boris:

So you have to solve the continuity equation?

Javier Jimenez:

No. That is automatically taken care of through the transition rules.

Jay Boris:

Steve came up with continuity equations for each of the different populations. They could move a fraction of a site, perhaps.

Javier Jimenez:

No. Move just one site per step.

Jay Boris:

OK, each direction. So you might have a lattice which is 10 or 100 times coarser than a lattice gas, but it's still one site per step.

Javier Jimenez:

Yes.

Robert Brodkey:

What is the prognosis of being able to incorporate in a hard sphere model or a plate model of chemical kinetics from the standpoint of having a distribution of sources and sinks of different identities and activation energies? Potentially you might have something very useful in the kinetic regime.

Gary Doolen;

This has been implemented, and the idea is to have two reactive products come in at one speed, and the different products go out at different speeds. Simulations like that have been done and you can get very unstable flame fronts that look very realistic.

Uriel Frisch:

This was done in **JFM** (1988, vol 188, pp. 437-464) by Clavin, Lallemand, Pomeau, and Searby. There you see the wavefront initially flat and then it develops well-documented Darrieus-Landau instability. This was done in order to see that the wavelength is what is predicted by theory. The same question was asked for capillaries. There are non-dimensional parameters in combustion. And for simple problems in combustion you can get them all essentially right. Of course, if you want to model something that has dozens of chemical reactions, that's a bad idea. You can only do combustion models which are themselves derived from microscopic models after smoothing out a number of reactions.

Simon Goren:

You're calling these things lattice gases. But they really are rarefied gas calculations unless you're looking at small objects or narrow channels, because you're using a relatively small number of objects possessing only mass and linear momentum and subject to Newton's laws. The objects which are being moved about are not molecules but large groups of molecules, and you cannot use the standard molecular potential energies of interaction. You must dream up something different. If that's the case, why don't you imbue these objects with additional properties other than linear momentum? Give them angular momentum, give them the ability to dissipate energy, and perhaps you'll be able to obtain more interesting results.

Uriel Frisch:

Absolutely. There have been many speculations in these directions. We have complete freedom to modify the rules of the game. For example, I've been trying, in a way I don't consider successful from the practical point of view, to give them vorticity-a mixture of lattice gas and vortex methods. In the lattice model, in the simplest ones, what you are discretizing is momentum. In vortex methods you're discretizing vorticity. Why not mix them? Depending on where you are in the flow, it may be better to discretize momentum or vorticity. In principle, it's doable, but I haven't found, so far, and efficient scheme. You have to solve the Biot-Savart problem for the velocity, and you have to propagate with a finite speed. The cost you pay for that seems to be really too high. Let me also observe that the way of modifying models is not the usual one: by continuity. You need to be inventive. The same remark can be made about mathematics. P. Holmes pointed out that since Newton, there's been continuum mathematics. The strongest momentum in mathematics today may be discrete mathematics. It's very hard. Here, you can't reason by continuity. You need to make quantum jumps to get new ideas.

Parviz Moin:

As a reference point, with a single Cray processor, how long does it take to calculate a Blasius boundary layer with cellular automata?

Uriel Frisch:

I think that's the last thing you want to do with lattice gas dynamics. Perhaps you want to do it by Boltzmann methods. Lattice gas methods with the Boolean strategy are intrinsically noisy. So typically you get errors of a few percent to ten percent, depending on the application. So you want to limit yourself to applications where you are quite happy to have such accuracy. And this is very easily the case in moderately turbulent or complex flows, when we have nothing else. If you want to do a very accurate description of a well-documented laminar flow, I know nothing better than floating-point approaches.

Gary Doolen:

There's a quantitative answer to that, and that's the error of one over the square root of N, for the time average for the flow past a boundary. Average over as many timesteps as you want for the desired accuracy.

Parviz Moin:

I'm curious, because in the quantitative examples you showed, the accuracy was 8 percent.

Uriel Frisch:

There are cases where the accuracy was a factor of 5-10 better than we would have expected by these kinds of arguments. In the problem of an impulsively-started plate, if you look at the stagnation point and how it moves away: there are well-documented experiments. And the accuracy is rather amazing-almost an order of magnitude better than you would have predicted. We don't understand why? It may be purely accidental. But I would really not recommend this technique for high accuracy. You can get these models to run very fast on all kinds of machines. That's the great advantage. For me, its an experimental tool. If we are able to make in three dimensions a machine comparable to what we now have for two dimensions, we'll actually be able to sit in front of the machine, and in a matter of seconds, watch the flow going on. It will be a rough version of the flow. It will have an accuracy of maybe 10 percent. But we'll have an idea of what's going on. Already on the RAP machine, the 2-dimensional machine that has slightly more than $1000 in components in it-you can look at a Karman street, stop the calculation, and in a matter of seconds, change the form of the boundary, and continue. That's why it's a good idea to make such a machine.

Parviz Moin:

Something seems paradoxical. Comments regarding difficulties with boundary conditions were made. How do you fit this with the suggestion that complex geometries are the areas of best application?

Uriel Frisch:

Fixed boundaries are not much of a problem. Moving boundaries can be done, but there is a price you have to pay. Another problem is the traditional problem of an in- and out-flow. With lattice gases you can rather cheaply get semi-decent in- and out-flow conditions. But if you're interested in doing it completely right, then I think it's essentially the same inflow and outflow problem that the floating-point people are fighting with. You also have to evaluate the computational overhead. Take your favorite floating point method and compare situations with and without boundaries. That's the computational overhead. With lattice-gas methods the computational overhead of introducing arbitrary, complex, non-moving boundaries is typically a few percent.

Session Six

Phenomenological Modelling: Present and Future

Discussion Leader: M. Coantic, IMST
Marseille

Phenomenological Modelling : Present .... and Future?

Brian E. Launder

Department of Mechanical Engineering, UMIST,
P.O. Box 88, Manchester M60 1QD, UK

Summary

The paper summarizes the present position of second-moment closure and outlines possible directions for future development. It is first argued that a simple form of second-moment treatment that has been widely used for computing industrial flows gives demonstrably superior predictive accuracy than any eddy-viscosity model. More complex schemes, now in the final phases of development, that exactly satisfy various limiting constraints, give a further marked improvement in our ability to mimic the response of turbulence to external inputs. The inclusion of such models into commercial software over the next few years is quite feasible.

The desirability of introducing a further second-rank tensor into the closure is considered; the conclusion reached is that, for most applications, the likely benefits would not justify the additional effort. The split-spectrum approach may, however, be attractive for certain flows with unusual spectral distributions of energy. The introduction of a second scale-related equation is arguably a more sensible approach since the extra computational cost is small while the added flexibility could bring significant benefits in modelling turbulence far from equilibrium.

Principal Nomenclature

a_{ij}	anisotropic Reynolds stress $(\overline{u_i u_j} - \frac{1}{3} \delta_{ij} \overline{u_k u_k})/k$
A	stress invariant quantifying the proximity of the stresses to the two-component limit, $1 - \frac{9}{8} (A_2 - A_3)$
A_2	second invariant $a_{ij} a_{ji}$
A_3	third invariant $a_{ij} a_{jk} a_{ki}$
c's	denote coefficients in turbulence model
$C_{ij}, C_{i\theta}$	convective transport of $\overline{u_i u_j}, \overline{u_i \theta}$
d	small pipe diameter
d_{ijk}	diffusion of Reynolds stress

D	large pipe diameter or distance between walls of plane channel
D_{ij}	$-\{\overline{u_i u_k}\ \partial U_k/\partial x_j + \overline{u_j u_k}\ \partial U_k/\partial x_i\}$
f	fluctuating body force per unit mass
F_{ij}	generation rate of Reynolds stresses
$F_{i\theta}$	generation rate of scalar flux by body forces
g_i	gravitational acceleration vector
ℓ	turbulent length scale $k^{3/2}/\epsilon$
k	turbulent kinetic energy, $\overline{u_i u_i}/2$
p	fluctuating pressure
P_{ij}	shear generation rate of $\overline{u_i u_j}$
$P_{i\theta}$	scalar flux generation rate of $\overline{u_i \theta}$ due to scalar ($P_{i\theta_1}$) and velocity ($P_{i\theta_2}$) gradients
Nu	Nusselt number
Nu_∞	fully developed value of Nusselt number in pipe flow
R	ratio of thermal:dynamic turbulent time scales, $k\epsilon_\theta/\overline{\tfrac12\theta^2}\epsilon$
R_t	turbulent Reynolds number $k^2/\nu\epsilon$
Re	mean Reynolds number UD/ν
Ro	rotation number (reciprocal of Rossby number) $\Omega D/\bar{U}$
u	streamwise fluctuating velocity
u_i	x_i component of fluctuating velocity
U	streamwise velocity
$\bar{U}$	bulk average streamwise velocity
U_i	x_i component of mean velocity
U_τ	friction velocity, $\sqrt{\tau_w/\rho}$
v	fluctuating velocity in y direction (normal to wall or down velocity gradient)
w	fluctuating velocity in z direction
W	streamwise velocity component (Fig. 4 only)
	swirl velocity component (Fig. 3)
x	streamwise direction
x_i	Cartesian direction coordinate
y	distance from wall normal to flow direction
z	direction of mean flow vorticity (in a simple shear flow)
ϵ	dissipation rate of k
$\epsilon_{ij}, \epsilon_{i\theta}$	dissipation rate of $\overline{u_i u_j}$, $\overline{u_i \theta}$
ϵ_θ	dissipation rate of $\tfrac12\overline{\theta^2}$
θ, Θ	fluctuating, mean value of scalar
ρ	density
σ_t	turbulent Prandtl number
τ_w	wall shear stress
$\varphi_{ij}, \varphi_{i\theta}$	non-dispersive contributions of fluctuating pressure due to $\overline{u_i u_j}$ and $\overline{u_i \theta}$ budgets
$\varphi_{ij1}, \varphi_{i\theta1}$	turbulence-turbulence part of φ_{ij}, $\varphi_{i\theta}$
$\varphi_{ij2}, \varphi_{i\theta2}$	mean-strain or "rapid" part of φ_{ij}, $\varphi_{i\theta}$
$\varphi_{ij3}, \varphi_{i\theta3}$	body force part of φ_{ij}, $\varphi_{i\theta}$
Ω	angular rotation rate

1. Introduction

The paper provides a personal view on second-moment closure: where we stand in terms of established and prototype models; what predictive capabilities and shortcomings have been demonstrated; and what steps might foreseeably be taken in the next few years to remove or reduce the acknowledged weaknesses. This sample is admittedly a very thin slice from the pie of phenomenological modelling, but a comprehensive treatment of the subject at large would require not a paper but a book - and a pretty thick one at that. Only at the level of second-moment closure can one begin to see the interconnections with other approaches to representing turbulence. It is also a level at which one can discuss what is *left out* of the model. Finally, despite the fact that it has been around for nearly forty years, it provides an approach to modelling turbulence that seems likely to become increasingly employed between now and the turn of the century. While it might be termed a "mature" topic in that a more-or-less standard version has been (or is in the course of being) incorporated into commercial CFD codes for use by clients in a black-box mode, it is also the subject of intensive research which is leading rapidly to the replacement of such hitherto "standard" versions even in software for complex shear flows.

While the paper's title signals our preoccupation here with the present and future of second-moment closure, a few words of historical perspective might provide a useful backcloth against which to set these discussions. Rotta's [1,2] pioneering 1951 papers provided both a powerful insight into the treatment of pressure fluctuations in the Reynolds stress equations and a transport equation for the evolution of a turbulent length scale. Perhaps the next landmark, a decade later, was Davidov's [3] proposal of a transport equation for the dissipation rate of turbulence energy as a replacement for the scale equation. However, his proposal, like Rotta's, could be subjected to only the most rudimentary checks because digital computers were in their infancy and numerical solvers for the Reynolds equations unheard of. It was thus not until the late sixties and early seventies, when appropriate hardware and software had been assembled, that computations of inhomogeneous shear flows began to appear [4-9]. These applications concerned two-dimensional thin shear flows without significant streamline curvature or body forces. While agreement with experiment was reasonably satisfactory, it must be said that, apart from one or two special features, the quality level was not, in fact, better than obtained with two-equation eddy-viscosity models, results from which were simultaneously appearing [7,8,10].

In other than thin shear flows, successful applications of second-moment closures were reported to predict turbulence-driven secondary flows in square ducts [11,12]; however, the computation of recirculating flow behind a disc [13] showed poor agreement with experiment - equally as poor as with the two-equation k-ϵ model. This inconsistent pattern persisted throughout the remainder of the seventies: for example, the successful capturing of diverse effects of buoyancy [14-16] and curvature [17] on turbulent shear flows were offset by an abysmal failure to predict the effects of swirl on the development of an axisymmetric jet [18]. At the landmark Stanford Conference in 1981 [19] the Evaluation Committee declined to acknowledge a demonstrated superiority of second-moment-type closures over eddy-viscosity schemes.

Some of the misleading signals from the Stanford '81 Conference, and, indeed, from the published literature of the '70's, arose from severe discretization error associated with the use of strongly diffusive numerical schemes. Since the entry to the '80's, these problems of numerical resolution have been declining rapidly, partly through the development of high-order (yet tolerably stable) schemes for discretizing convective transport, e.g. [20], and partly through the continual growth in execution speed and core available from mainframe computers. From numerous comparative computations over the last four or five years, it has now been demonstrated that, even with fairly rudimentary modelling of the processes, second-moment closure offers a far more reliable approach to predicting complex flows than any eddy-viscosity-based model. While the present workshop is not concerned with the problem of *how* to solve the complex system of equations arising from second-moment closure for flow in arbitrary geometrical configurations, the *results* of such computations are certainly of interest; several such examples are included in Section 2.

This effort to embed second-moment closures into large-scale, general-purpose solvers is one of the two principal research emphases of the 1980's. The other has been the renewed effort at improving the physical ingredients of closure schemes, a development that can perhaps be said to originate from Lumley's [21] extensive article and given further impetus by the direct simulations that have been emerging over the past five years from the Stanford/NASA Ames research conglomerate. The remainder of Section 2 is thus devoted to current modelling efforts, in various stages of development, that hold promise for improving the reliability with which we can predict industrially interesting flows.

Section 3, concerned with future developments, is intentionally brief, for experience teaches one how hard it is to try and look far into the future. Perhaps the message of the last twenty years is that progress is likely to be steady rather than spectacular. It is unlikely that second-moment closure will be widely seen as one of the "hot" topics of turbulence research for the 1990's; nevertheless, in the decade when CFD seems destined to reach genuine maturity, the extent to which computer modelling can replace experimental testing will depend crucially on how far one gets with second-moment closure in the next three or four years.

2. Present Status

2.1 Scope of the Enquiry

The community concerned with second-moment closure is heterogeneous in composition, ranging from those working purely on homogeneous flows to others needing to compute three-dimensional recirculating flows, perhaps with chemical reaction thrown in for good measure. The "present status" for these different ends of the spectrum is not the same so far as the models adopted are concerned. Accordingly, after a statement of the exact equations in §2.2, consideration is given first to the simplest form of second-moment closure. The reason for giving space to a model whose ingredients are all more than ten years old is that it has now been applied sufficiently widely for an unambiguous impression to be formed of its capabilities relative to that of an eddy-viscosity model. The remainder of Section 2 is devoted to more recent developments beginning with homogeneous flows and proceeding via free shear flows to near-wall and viscous effects.

2.2 The Second Moment Equations and the Basic Closure

For an incompressible turbulent flow whose large scales are unaffected by viscosity, the transport of the Reynolds stress tensor $\overline{u_i u_j}$ is conveniently expressed in the symbolic form:

$$\frac{\partial \overline{u_i u_j}}{\partial t} + C_{ij} = P_{ij} + F_{ij} + \frac{\partial}{\partial x_k} d_{ijk} + \varphi_{ij} - \epsilon_{ij} \qquad (1)$$

where $\quad C_{ij} \equiv U_k \frac{\partial \overline{u_i u_j}}{\partial x_k} \quad ; \quad \text{convective transport}$

a) Reynolds-Stress Equations

Process	Model	Originator (where appropriate)
d_{ijk}	$c_s \overline{u_k u_\ell} \dfrac{k}{\epsilon} \dfrac{\partial \overline{u_i u_j}}{\partial x_\ell}$	Daly & Harlow [5]
ϵ_{ij}	$\frac{2}{3} \delta_{ij} \epsilon$	Local isotropy
φ_{ij}	$\varphi_{ij_1} + \varphi_{ij_2} + \varphi_{ij_3} + \boxed{\varphi_{ijw}}$ wall flows only	
φ_{ij_1}	$-c_1 \epsilon a_{ij}$	Rotta [1]
φ_{ij_2}	$-c_2[(P_{ij} - \frac{1}{3} \delta_{ij} P_{kk}) - (C_{ij} - \frac{1}{3} \delta_{ij} C_{kk})]$	Naot et al [22], Fu et al [41] (The IP Model)
φ_{ij_3}	$-c_3(F_{ij} - \frac{1}{3} \delta_{ij} F_{kk})$	Launder [23]
φ_{ijw}	$\{c_1' \dfrac{\epsilon}{k}(\overline{u_k u_m} n_k n_m \delta_{ij} - \frac{3}{2} \overline{u_k u_i} n_k n_j - \frac{3}{2} \overline{u_k u_j} n_k n_i)$ $+ c_2'(\varphi_{km_2} n_k n_m \delta_{ij} - \frac{3}{2} \varphi_{ik_2} n_k n_j - \frac{3}{2} \varphi_{kj_2} n_k n_i)$ $+ c_3'(\varphi_{km_3} n_k n_m \delta_{ij} - \frac{3}{2} \varphi_{ik_3} n_k n_j - \frac{3}{2} \varphi_{kj_3} n_k n_i)\} \dfrac{k^{3/2}}{c_\ell \epsilon x_n}$ where n_k = unit vector normal to wall	Shir [24] Gibson & Launder [25]
ϵ	$\dfrac{D\epsilon}{Dt} = c_\epsilon \dfrac{\partial}{\partial x_\ell}\left[\overline{u_k u_\ell} \dfrac{k}{\epsilon} \dfrac{\partial \epsilon}{\partial x_\ell}\right] + \frac{1}{2} c_{\epsilon 1}(P_{kk} + F_{kk})\dfrac{\epsilon}{k} - c_{\epsilon 2} \dfrac{\epsilon^2}{k}$	Davidov [3] Hanjalić & Launder [6] [5]

Coefficients:

c_s	c_1	c_2	c_3	c_1'	c_2'	c_3'	c_ℓ	c_ϵ	$c_{\epsilon 1}$	$c_{\epsilon 2}$
0.22	1.8	0.6	c_2	0.5	0.3	0	2.5	0.15	1.44	1.92

b) <u>Scalar-Flux Equations</u>

Process	Model	Originator (where appropriate)
$d_{i\theta k}$	$c_\theta \; \overline{u_k u_\ell} \; \dfrac{k}{\epsilon} \; \dfrac{\partial \overline{u_i \theta}}{\partial x_\ell}$	[5]
$\epsilon_{i\theta}$	0	Local isotropy
$\varphi_{i\theta}$	$\varphi_{i\theta 1} + \varphi_{i\theta 2} + \varphi_{i\theta 3} + \boxed{\varphi_{i\theta w}}$	
$\varphi_{i\theta 1}$	$-c_{\theta 1} \; \dfrac{\epsilon}{k} \; \overline{u_i \theta}$	Monin [26]
$\varphi_{i\theta 2}$	$-c_{\theta 2}(P_{i\theta 2} - C_{i\theta})$	Owen [27] [23]
$\varphi_{i\theta 3}$	$-c_{\theta 3} \; F_{i\theta}$	[23]
$\varphi_{i\theta w}$	$\left\{ c'_{\theta 1} \; \dfrac{\epsilon}{k} \; \overline{u_k \theta} \; n_k n_i + c'_{\theta 2} \; \varphi_{k\theta 2} \; n_k n_i \right.$ $\left. + \; c'_{\theta 3} \; \varphi_{k\theta 3} \; n_k n_i \right\} \dfrac{k^{3/2}}{c_\ell \epsilon x_n}$	[25]

Coefficients:

c_θ	$c_{\theta 1}$	$c_{\theta 2}$	$c_{\theta 3}$	$c'_{\theta 1}$	$c'_{\theta 2}$	$c'_{\theta 3}$
0.15	2.9	0.4	0.4	0.25	0	0

Table 1: The basic second-moment closure

$$P_{ij} \equiv - \left\{ \overline{u_k u_i} \frac{\partial U_j}{\partial x_k} + \overline{u_k u_j} \frac{\partial U_i}{\partial x_k} \right\} \quad ; \quad \text{shear generation}$$

$$F_{ij} \equiv \left\{ \overline{f_i u_j} + \overline{f_j u_i} \right\} \quad ; \quad \text{body-force generation}$$

$$d_{ijk} \equiv - \left\{ \overline{u_i u_j u_k} + \frac{\overline{u_j p}}{\rho} \delta_{ik} + \frac{\overline{u_i p}}{\rho} \delta_{jk} \right\} \quad ; \quad \text{diffusion}$$

$$\varphi_{ij} \equiv \frac{\overline{p}}{\rho} \left\{ \frac{\partial u_i}{\partial x_j} + \frac{\partial u_j}{\partial x_i} \right\} \quad ; \quad \text{pressure-strain}$$

$$\epsilon_{ij} \equiv 2\nu \overline{\frac{\partial u_i}{\partial x_k} \frac{\partial u_j}{\partial x_k}} \quad ; \quad \text{dissipation}$$

The corresponding equation for the transport of the scalar flux $\overline{u_i \theta}$ takes the form

$$\frac{\partial \overline{u_i \theta}}{\partial t} + C_{i\theta} = P_{i\theta_1} + P_{i\theta_2} + F_{i\theta} + \frac{\partial}{\partial x_k} d_{ik\theta} + \varphi_{i\theta} - \epsilon_{i\theta} \qquad (2)$$

where $C_{i\theta} \equiv U_k \frac{\partial \overline{u_i \theta}}{\partial x_k}$; $P_{i\theta_1} \equiv -\overline{u_i u_k} \frac{\partial \theta}{\partial x_k}$; $P_{i\theta_2} \equiv -\overline{u_k \theta} \frac{\partial U_i}{\partial x_k}$

$$d_{ik\theta} = -\left\{ \overline{u_i u_k \theta} + \frac{\overline{p\theta}}{\rho} \delta_{ik} \right\} \quad ; \quad \varphi_{i\theta} \equiv \frac{\overline{p}}{\rho} \frac{\partial \theta}{\partial x_i} \quad ; \quad \epsilon_{i\theta} \equiv (\lambda + \nu) \overline{\frac{\partial u_i}{\partial x_k} \frac{\partial \theta}{\partial x_k}}$$

To close these equations, approximations must be provided for the processes d_{ijk}, φ_{ij} and ϵ_{ij} in equation (1) and for the corresponding terms in equation (2). During the 1980's a large number of turbulent-flow computations, including topographically complex flows, have been made with (virtually) a single set of approximations; these we here refer to as the Basic Model given in Table 1. Among the types of flow considered with this model have been boundary layers and wall jets on curved surfaces [28,29]; the curved wake and mixing layer [29-31]; the atmospheric surface layer under changing stratification [25], buoyancy-modified free surface flows [32-34]; natural convection on a flat plate [35,36]; the development of buoyant plumes [33]; the flow through an annular faired diffuser [37]; recirculating flow behind a variety of abrupt-expansion configurations [38-40]; free [41] and confined [42] swirling jets, including cases

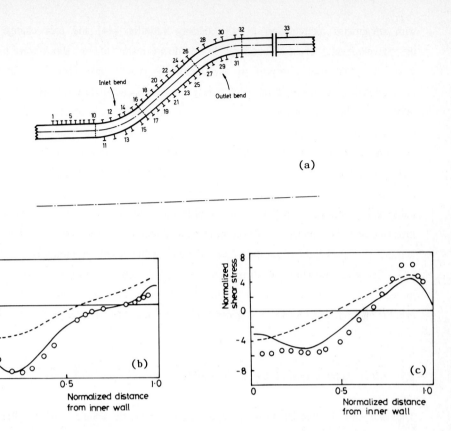

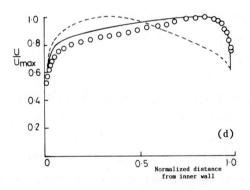

Fig. 1   Boundary layer development in a faired diffuser,
Jones and Manners [37]

(a) Flow configuration (b) Shear stress profiles at end of inlet bend
O O Experiment [50]; ———— Basic Model; - - - k-ε EVM (c) Shear
stress profiles at end of straight diffusing section (key as in (b))
(d) Mean velocity profiles at exit from diffuser (key as in (b))

with streamwise recirculation [41,43], density variation [44] and area change [43]; the development of a boundary layer downstream of a shock-wave-induced separation [45]; heat transport in asymmetrically heated pipe flow [46]; the flow in rectangular-sectioned ducts dominated by turbulence-induced secondary flows [47]; the three-dimensional flow around U-bends of square and circular cross-section [48,49]; and many, many others. If agreement with experiment has not always been complete, there is nevertheless a sufficient level of accord to make even the most sceptical of observers acknowledge something more powerful at work than just coincidence or elaborate interpolation. Of the proposals for the individual processes, none of them would stand up to the searching examination that can now be applied by direct numerical simulation. The model is thus to be viewed as a package; its utility to be inferred by the results it produces.

To give substance to these remarks, a few representative examples will be presented for cases where the predicted behaviour can be compared both with experiment and with predictions generated by the k-ϵ eddy-viscosity model.

2.3 Some Applications of the Basic Model

The first case is that of flow through the axisymmetric annular diffuser shown in Fig. 1a which has been the subject of a detailed computational study by Jones and Manners [37]; the experiments are those of Stevens and Fry [50]. The flow is unseparated, but the successive imposition of the inlet bend, a straight diverging section and an outlet bend create a searching test case. The k-ϵ model gives results that are arguably worse than no results at all. Because this scheme does not show selective sensitivity to streamline curvature, the shear-stress profile at the end of the inlet bend, Fig. 1b, shows far too little asymmetry, a feature that remains by the end of the diverging section, Fig. 1c. As a result, a short way downstream of the outlet bend the mean velocity distribution departs from symmetry in a precisely opposite way to the experiment, Fig. 1d. The computed behaviour with the second-moment closure, by contrast, follows with reasonable fidelity the measured evolution of the shear stress and consequently the downstream velocity profile broadly accords with the measured behaviour.

The second example is that of flow through a large aspect-ratio plane channel rotating in orthogonal mode, Fig. 2a, a case for which Johnston et al [51] have provided the most detailed experimental data (supplemented by Kim's [52] large-eddy simulation). Modifications to the flow structure that occur as the

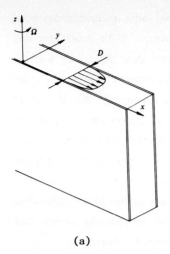

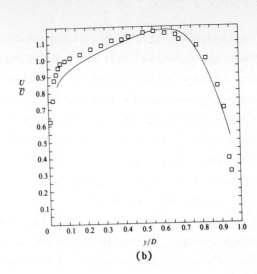

(a)   (b)

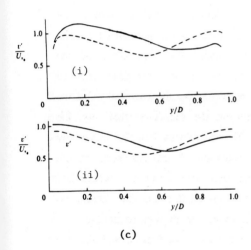

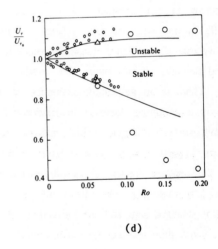

(c)   (d)

Fig. 2   Flow in a plane channel rotating in orthogonal mode

       (a)   Flow configuration
       (b)   Mean velocity profile for Ro = 0.21
                □ □ Experiments [51];   ———   Basic Model [53]
       (c)   Rms velocity fluctuations normal to wall
                (i) LES, Kim [52];   (ii) Basic Model [53];
                      ———   Ro = 0.07;   - - -   Ro = 0
       (d)   Dependence of friction velocity on Ro
                Symbols : Experiments from [51];   ———   Basic Model [53]

rotation number Ro $\equiv \Omega D/\bar{U}$ is progressively increased arise predominantly from the action of Coriolis forces in the Reynolds-stress budget. While results for this case were not generated with the k-ϵ model, we know that *that* model predicts rotation to have *zero* effect on the flow pattern. The reason is that there are no Coriolis terms in the kinetic energy equation (the contributions in the $\overline{u^2}$ and $\overline{v^2}$ budgets being $+4\overline{uv}\Omega$ and $-4\overline{uv}\Omega$ respectively which cancel on summation). Experiment shows that quite modest rotation rates in fact produce large asymmetries on the mean velocity profile, Fig. 2b. With the chosen coordinates $\overline{uv}$ is negative near the pressure surface, so the Coriolis source will tend to raise $\overline{v^2}$ and depress $\overline{u^2}$ while, in the shear-stress budget, the Coriolis source $-2(\overline{u^2}-\overline{v^2})\Omega$ will act to raise the (negative) magnitude of $\overline{uv}$ since $\overline{u^2}$ is ordinarily greater than $\overline{v^2}$. Near the suction surface the above effects are reversed. Figure 2c compares for a rotation number of 0.07 the predicted variation of v'($\equiv \sqrt{\overline{v^2}}$) obtained from the second-moment closure [53] with the large-eddy simulation of Kim [52]; for comparison, the non-rotating results (broken line) are also included. While agreement is not complete, the predicted modifications to the v' profile are broadly in line with the LES results. Figure 2d compares the normalized friction velocity on the pressure (unstable) and suction (stable) faces predicted by the second-moment closure with the available experimental and computer-generated data. There is an interesting difference in the behaviour on the two walls: while there is a continual decrease in friction velocity on the suction surface[†], on the pressure surface, after an initial increase, the friction velocity remains essentially uniform for Ro > 0.08. The reason for this "saturation" is that, as Ro is raised, near the pressure surface $\overline{u^2}$ is progressively depressed and $\overline{v^2}$ raised. Eventually $\overline{u^2}$ is less than $\overline{v^2}$. At this point the Coriolis contribution to the shear-stress equation changes sign and thereafter there is no further rise in wall friction.

Swirl provides another manifestation of an imposed rotation, though in this case the axis is usually aligned with the mean flow direction. In the experiment considered [54], a pair of isothermal, confined coaxial jets, the outer one of which is swirling, undergo mixing at a rate enhanced by the thick lip of the inner pipe. However, the fact that the core jet is non-swirling at discharge tends to inhibit mixing between streams and this is why, Fig. 3a, despite the thick lip, the

---

[†] At high rotation rates the experiments by Johnston et al [51] found a reversion to laminar flow on the suction surface and thus exhibit a precipitate decrease in wall friction. The computations cannot mimic this behaviour as the flow is forced to satisfy turbulent local-equilibrium boundary conditions at the edge of the viscosity-affected sublayer.

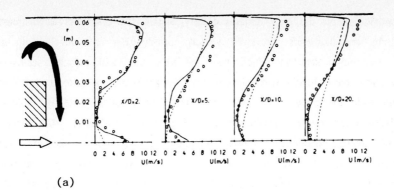

(a)

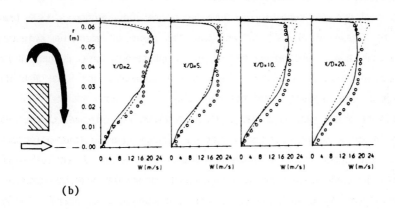

(b)

Fig. 3   Development of confined swirling coaxial jets

     (a) Axial velocity profiles
        O O   Experiments [54]
        ―――   Basic model ⎫
        - - - k-ϵ EVM ⎬ [42]
     (b) Swirl velocity profiles (key as in (a))

velocity maximum on the axis remains clearly visible ten diameters downstream. Moreover, by ten diameters the swirl velocity profile, Fig. 3b, develops to an interesting two-region state in which, near the axis, the swirl velocity seems to increase roughly parabolically with distance (*not* linearly) and thereafter (apart from a thin near-wall zone where the swirl is reduced to zero) is virtually uniform. The second-moment closure predictions of Hogg and Leschziner [42] broadly capture all these features. Their companion predictions with the k-ϵ eddy-viscosity model, however, reflect far too strong levels of mixing; consequently the jet has disappeared by x/D = 5, while by x/D = 20 virtually the whole pipe contents are in solid-body rotation.

The final example concerns the three-dimensional flow around a 180° bend measured by Chang et al [55]. Even with the plane of geometric symmetry present in this case, the magnitude of the computational task is here much greater than in the earlier examples. To reduce it somewhat, the usual algebraic truncation has been applied to the transport terms in the stress budget, Table 2. In the resultant model (known as an *algebraic* second-moment (ASM) closure) the turbulence energy k is the only second-moment quantity obtained from a transport equation. The main point of interest in this flow is the extent to which computations succeed in capturing the troughs in streamwise velocity that have developed by half-way around the bend. The first computational attempts [56,57] entirely failed to show any dip in velocity due to the application of the local-equilibrium near-wall boundary condition. The computations of Choi et al [49] shown in Fig. 4 have adopted instead a version of the mixing-length hypothesis to span the thin near-wall sublayer. Two sets of computations were made, one with the k-ϵ eddy-viscosity model (EVM) and the other with the ASM closure over the remainder of the duct. Figures 4a and 4b show profiles of the streamwise velocity along lines parallel to the flow symmetry plane. In this case

Original	Replacement
C_{ij}	$\dfrac{\overline{u_i u_j}}{\overline{u_p u_p}}\, C_{kk}$
$\dfrac{\partial}{\partial x_k}\, d_{ijk}$	$\dfrac{\overline{u_i u_j}}{\overline{u_p u_p}}\, \dfrac{\partial}{\partial x_k}\, d_{mmk}$

Table 2: Basic ASM truncation [7]

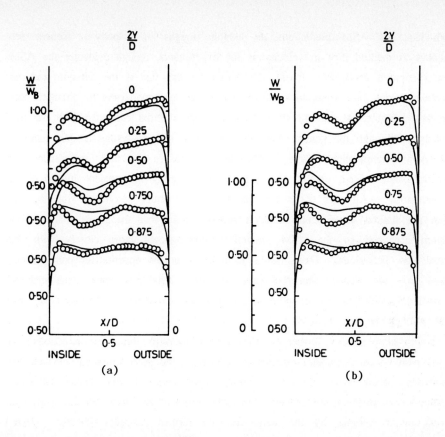

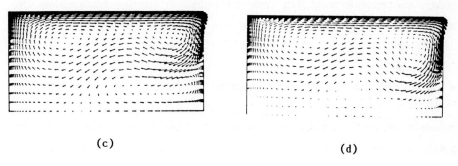

Fig. 4   Flow around square-sectioned U-bend

    (a)   Streamwise velocity profiles at 130°
             Symbols : Experiment [55]; —— k-ϵ EVM [49]
    (b)   Streamwise velocity profiles at 130°
             Symbols : Experiment [55]; —— ASM version of Basic
                                         Model [49]
    (c)   Computed secondary flow vectors : k-ϵ EVM
    (d)   Computed secondary flow vectors : ASM version of Basic Model

both k-ε and ASM predictions *do* exhibit troughs in velocity. Neither set displays as marked dips in velocity as the experiments, though evidently the ASM results come a good deal closer. The origin of the dips is the curvature-induced secondary flow that brings low-velocity fluid from the end walls to circulate near the axis. Figures 4c and 4d show secondary flow vectors computed with the two models. Although the general pattern for the two sets of results is quite similar, the ASM computations show much steeper changes in both magnitude and direction and, indeed, exhibit a further secondary eddy not present in the EVM computations. Although it is not possible to construct secondary flow vectors from the limited experimental data, it seems certain that the more chaotic flow pattern generated by the ASM predictions does accord more closely with the actual flow behaviour. These results naturally whet the appetite for computations based on a *full* second-moment closure. It is not certain, of course, that such an elaboration would further improve agreement with experiment - though the odds must strongly be that it would.

For those seeking further evidence of the markedly improved predictions (vis à vis eddy-viscosity models) obtainable with this basic second-moment closure, the numerous papers cited earlier describing applications should make satisfying reading. Alternatively, for readers with little time but an unsatiated curiosity, two recent review articles by the writer provide further examples [58,59]. Here, however, our attention must shift from applications in complex configurations to current fundamental modelling research generally involving far simpler flow fields.

2.4 Modelling Fluctuating Pressure Interactions

The role of pressure fluctuations in modifying the level of the second moments is arguably the liveliest debating point in second-moment closure. The discussion is facilitated by limiting attention to homogeneous flows and, following Chou [60], replacing the pressure fluctuations themselves, via the Poisson equation, by integrals of two-point velocity, temperature and fluctuating force-field products. Thus:

$$\varphi_{ij} = -\frac{1}{4\pi}\int\left[\frac{\partial^3 \overline{u_\ell' u_m' u_j}}{\partial r_m \partial r_\ell \partial r_i} + \frac{\partial^3 \overline{u_\ell' u_m' \partial u_i}}{\partial r_m \partial r_\ell \partial r_j}\right]\frac{dVol}{|\underline{r}|} - \frac{1}{2\pi}\frac{\partial U_\ell}{\partial x_m}\int\left[\frac{\partial^2 \overline{u_m' u_i}}{\partial r_\ell \partial r_j} + \frac{\partial^2 \overline{u_m' u_j}}{\partial r_\ell \partial r_i}\right]\frac{dVol}{|\underline{r}|}$$

$$\underbrace{\qquad\qquad\qquad\qquad}_{\varphi_{ij1}}\qquad\qquad\qquad\underbrace{\qquad\qquad\qquad\qquad}_{\varphi_{ij2}}$$

$$+ \frac{1}{4\pi} \int \left[\frac{\partial^2 \overline{f'_\varrho u_i}}{\partial r_\varrho \partial r_j} + \frac{\partial^2 \overline{f'_\varrho u_j}}{\partial r_\varrho \partial r_i} \right] \frac{dVol}{|\underline{r}|} + \text{ surface integrals}$$

$$\underset{\varphi_{ij3}}{}$$

$$\tag{3}$$

$$\varphi_{i\theta} = -\frac{1}{4\pi} \int \frac{\partial^3 \overline{u_\varrho u_m \theta}}{\partial r_\varrho \partial r_m \partial r_i} \frac{dVol}{|\underline{r}|} - \frac{1}{2\pi} \frac{\partial U_\varrho}{\partial x_m} \int \frac{\partial^2 \overline{u'_m \theta}}{\partial r_\varrho \partial r_i} \frac{dVol}{|\underline{r}|} + \frac{1}{4\pi} \int \frac{\partial^2 \overline{f'_\varrho \theta}}{\partial r_\varrho \partial r_i} \frac{dVol}{|\underline{r}|}$$

$$\underset{\varphi_{i\theta1}}{} \qquad\qquad\qquad \underset{\varphi_{i\theta2}}{} \qquad\qquad\qquad \underset{\varphi_{i\theta3}}{}$$

$$\tag{4}$$

$$+ \text{ surface integrals}$$

where primes denote quantities evaluated at a point $\underline{r}$ removed from that of interest, r_ϱ etc. are components of $\underline{r}$ and the volume integrations are over all r-space.

This representation leads naturally to the adoption of separate approaches to modelling the component integrals of φ_{ij} and $\varphi_{i\theta}$. While, with results generated by direct simulation, the various volumetric integrals can be evaluated separately, only very recently have workers begun to process their results in this way. The usual approach to approximating the *turbulence-interaction* parts (φ_{ij1}, $\varphi_{i\theta1}$) is thus to consider flows (whether directly measured or computer-simulated) where mean strain and force fields are absent and the turbulence field is relaxing from various initial states of anisotropy. Here and in later sections our attention is principally directed at modelling the dynamic processes (i.e. those concerned with closure of the $\overline{u_i u_j}$ equation). The general form of approximation usually adopted for φ_{ij1} is [21]:

$$\varphi_{ij1} = -c_1 \epsilon [a_{ij} + c_1^* (a_{ik} a_{kj} - \tfrac{1}{3} \delta_{ij} A_2)] \tag{5}$$

The leading term of eq.(5) is just Rotta's [1] proposal, though the coefficients are commonly taken to be functions of the stress invariants A_2 ($\equiv a_{ij} a_{ij}$) and A_3 ($\equiv a_{ij} a_{jk} a_{ki}$) and perhaps of a local turbulent Reynolds number R_t ($\equiv k^2/\nu\epsilon$) too, e.g. Lumley and Newman [61]. Most workers nowadays follow Lumley and Newman's practice of requiring $\varphi_{\alpha\alpha1}$ to vanish if the $\overline{u_\alpha u_\alpha}$ vanishes (no summation on α). This is normally achieved by arranging that c_1 be of the form

$$c_1 = A^n f(A_2, A_3, R_t) \tag{6}$$

455

where $A \equiv 1 - \frac{9}{8}(A_2 - A_3)$ vanishes in the limit of two-component turbulence. An alternative way of satisfying this limit is by taking eq.(5) in the particular form

$$\varphi_{ij1} = -\tilde{c}_1 \epsilon ((1 + \alpha A_2) a_{ij} + \beta(a_{ik} a_{kj} - \tfrac{1}{3} \delta_{ij} A_2)) \tag{7}$$

Then, provided the coefficients α and β take the limiting values $-\frac{3}{4}$ and $\frac{3}{2}$ in two-component turbulence, $\varphi_{\alpha\alpha1}$ will also vanish, Reynolds [62]. The two approaches are intrinsically different in that the former makes *all* components of φ_{ij1} vanish in the aforesaid limit, whereas the latter only applies to those elements containing the vanishing component of velocity, u_α; thus, at a free surface, there may remain some transfer between components lying in the plane of the surface itself. No usage of eq.(7) with the indicated values of α and β is known to the writer.

It is fair to say that, while there are many proposals for φ_{ij1} on the table based on eq.(5), none is particularly persuasive. Some advocate a different *basis* for approximating this process. Lee and Reynolds [63,64] conclude, from their direct numerical simulations of turbulence decay, that φ_{ij1} is not in fact well correlated by the anisotropy of the stress field. Their results show a closer linkage between the pressure-strain process and the anisotropy of the *dissipation* tensor, ϵ_{ij} There is, however, some disagreement among leading workers as to the relevance of these results to cases likely to be encountered in practice as the turbulence Reynolds number of the simulations was so low: roughly the same as that in a turbulent boundary layer at $y^+ \simeq 20$. In a somewhat different context the work of Weinstock and Burk [65,66] has also underlined the linkage between dissipation anisotropy and the magnitude of the return-to-isotropy coefficient c_1. Their theoretical results, while presented in the form of Rotta's model, Table 1, display variations in c_1, not only with the anisotropy of the stress field but also between different components of φ_{ij}. It appears to the present writer that, however valid these results may prove to be, a different organizational basis is needed before they can be absorbed into second-moment closures.

Given the uncertainties surrounding the approximation of φ_{ij1}, it does not seem fruitful to embark on an extensive review of approaches to modelling $\varphi_{i\theta1}$ about which several additional questions of strategy arise. One item seems worth flagging, however, if only because it seems to have been ignored by workers in the USA. In 1983 Jones and Musonge [67] argued that, in addition to heat fluxes, the *mean temperature gradient* ought to appear in the model of $\varphi_{i\theta}$, partly on physical grounds but more pressingly because they could find no other way of correctly mimicking the homogeneous thermal shear layer experiment of Tavoularis and Corrsin [68]. They proposed the inclusion of an additional term proportional to $-ka_{ij}\partial\Theta/\partial x_j$ with a coefficient of 0.23. Their position has been confirmed in subsequent analytical work by Dakos and Gibson [69] (though with a coefficient of about 0.08) and retained in computational studies by Gibson et al [70] and Craft et al [71].

While there may be widespread disagreement on how the non-linear "turbulence-turbulence" part of the pressure-strain process should be represented, there is a greater measure of accord on handling the *mean-strain* or *"rapid"* contribution, φ_{ij2}. All proposals begin by representing the process as

$$\varphi_{ij2} = \frac{\partial U_\ell}{\partial x_m} (a_{\ell j}^{mi} + a_{\ell i}^{mj}) \tag{8}$$

where the fourth rank tensor $a_{\ell j}^{mi}$ has the dimensions of Reynolds stress and, indeed, is composed of Reynolds-stress elements assembled in such a way that the tensor satisfies various kinematic and symmetry properties of the original integral, eq.(3).

The linear "quasi-isotropic" model [8,9,72] contains a single empirical coefficient which, when chosen by reference to a simple shear flow in local equilibrium, enables a range of non-equilibrium shear flows to be successfully predicted [9,17]. However, the model is less successful overall than the even simpler IP model given in Table 1 and gives spectacularly wrong predictions of the effects of swirl on the spreading rate of an axisymmetric jet [18]. There are other grounds for discarding the model; for example, it is impossible to make it consistent with the two-component limit [21] noted above in connection with φ_{ij1}. Several workers have therefore added non-linear terms in representing $a_{\ell j}^{mi}$ on the grounds that the shape and spacing of the iso-correlation surfaces of the two-point products appearing in the exact integral form of φ_{ij2} will themselves depend on the anisotropy of the stress field. Le Penven and Gence [73], Reynolds [62] and Lecointe et al [74] have preferred to determine the (sometimes substantial number of) extra unknown coefficients by reference to irrotational plane-strain distortions, while Shih et al [75,76], Fu et al [77,78] (see also Craft et al [71]), Ristorcelli [79] and Reynolds [80] have required compliance with the two-component limit.

Limiting attention to terms strictly quadratic in a_{ij} gives (Shih and Lumley [75], Fu et al [77]):

$$\varphi_{ij2} = -0.6(P_{ij} - \tfrac{1}{3}\delta_{ij}P_{kk}) + 0.3\,\epsilon\,a_{ij}(P_{kk}/\epsilon)$$

$$- 0.2\left\{\frac{\overline{u_k u_j}\;\overline{u_\varrho u_i}}{k}\left[\frac{\partial U_k}{\partial x_\varrho} + \frac{\partial U_\varrho}{\partial x_k}\right] - \frac{\overline{u_\varrho u_k}}{k}\left[\overline{u_i u_k}\frac{\partial U_j}{\partial x_\varrho} + \overline{u_j u_k}\frac{\partial U_i}{\partial x_\varrho}\right]\right\}$$

(9)

This form, which has no disposable coefficients, gives, as its leading term, the simple isotropization-of-production model [22]. However, in a simple shear flow in local equilibrium, $U_1(x_2)$, it gives values of $\overline{u_2^2}$ greater than $\overline{u_3^2}$, the reverse of what experiment indicates. This weakness was removed by Shih et al [76] by the addition of a corrective term, φ_{ij2}^c:

$$\varphi_{ij2}^c = -0.8A^{\frac{1}{2}}\left\{\tfrac{7}{15}(P_{ij} - \tfrac{1}{3}\delta_{ij}P_{kk}) + \tfrac{8}{15}(D_{ij} - \tfrac{1}{3}\delta_{ij}D_{kk})\right.$$

(10)

$$\left. + \tfrac{2}{5}k\left[\frac{\partial U_i}{\partial x_j} + \frac{\partial U_j}{\partial x_i}\right]\right\}$$

$$D_{ij} \equiv -\left(\overline{u_i u_k}\frac{\partial U_k}{\partial x_j} + \overline{u_j u_k}\frac{\partial U_k}{\partial x_i}\right)$$

a form that retains compliance with all the kinematic constraints. Craft et al [71] (see also Fu [78]) add cubic terms and thus obtain a form with two freely assignable coefficients. In most of the work by the UMIST group, one of these is set to zero to reduce the resultant expression to the relatively simple form, [77]:

$$\varphi_{ij2} = -0.6[P_{ij} - \tfrac{1}{3}\delta_{ij}P_{kk}] + 0.3\,\epsilon\,a_{ij}(P_{kk}/\epsilon)$$

$$- 0.2\left\{\frac{\overline{u_k u_j}\;\overline{u_\varrho u_i}}{k}\left[\frac{\partial U_k}{\partial x_\varrho} + \frac{\partial U_\varrho}{\partial x_k}\right] - \frac{\overline{u_\varrho u_k}}{k}\left[\overline{u_i u_k}\frac{\partial U_j}{\partial x_\varrho} + \overline{u_j u_k}\frac{\partial U_i}{\partial x_\varrho}\right]\right\}$$

$$- \lambda[A_2(P_{ij} - D_{ij}) + 3a_{mi}\,a_{nj}(P_{mn} - D_{mn})]$$

(11)

where the freely assignable parameter λ is assigned the constant value 0.6. Since $(D_{ij} - P_{ij})$ is proportional to the vorticity tensor, the extra term in (11) compared with (9) may be explicitly regarded as a correction for mean flow rotation. A feature of eq.(11) is that, compared with the IP model, it brings a good deal more subtlety and variety to the effects of even a simple shear on the stress field. This is well brought out by the computations of Craft et al [71] of the homogeneous shear flows of Champagne et al [81] and Harris et al [82]. Figure 5 compares the development of the anisotropic stress components for the case of approximate local equilibrium [81] where $\frac{1}{2}P_{kk} \simeq \epsilon$ and of the "high-strain-rate" case [82] where, according to Leslie's re-evaluation [83], $\frac{1}{2}P_{kk} \simeq 1.55\epsilon$. The experiments indicate that, while increasing the ratio of turbulence energy generation:dissipation rate increases the anisotropy of the normal stresses in the plane of the shear, it *decreases* that of the shear stress. In other words, the higher shearing causes a rotation of the planes of principal stress.[†] This feature is broadly captured by the new form of φ_{ij2} but is entirely missed by the IP model. The performance of this model in various *in*homogeneous flows will be considered later in §2.6.

Analogous models for mean strain effects on the pressure-temperature gradient correlation, $\varphi_{i\theta 2}$, have been devised by Shih and Lumley [75] and Craft et al [71] by applying different forms of realizability constraint. The former may be organized as:

$$\varphi_{i\theta 2} = 0.8 \, \overline{\theta u_k} \, \frac{\partial U_i}{\partial x_k} - 0.2 \, \overline{\theta u_k} \, \frac{\partial U_k}{\partial x_i} + 0.15 \, \frac{\epsilon}{k} \, \overline{\theta u_i} \, (P_{kk}/\epsilon)$$

$$+ 0.1 \, \overline{\theta u_k} \, a_{i\ell} \left[\frac{\partial U_\ell}{\partial x_k} - 4 \frac{\partial U_k}{\partial x_\ell} \right] + 0.2 \, \overline{\theta u_k} \, a_{k\ell} \, \frac{\partial U_i}{\partial x_\ell} \qquad (12)$$

while the latter results in the somewhat bulkier expression:

$$\varphi_{i\theta 2} = 0.8 \, \overline{\theta u_k} \, \frac{\partial U_i}{\partial x_k} - 0.2 \, \overline{\theta u_k} \, \frac{\partial U_k}{\partial x_i} + \frac{1}{6} \, \frac{\epsilon}{k} \, \overline{\theta u_i} \left[\frac{P_{kk}}{\epsilon} \right]$$

$$- 0.4 \, \overline{\theta u_k} \, a_{i\ell} \left[\frac{\partial U_k}{\partial x_\ell} + \frac{\partial U_\ell}{\partial x_k} \right]$$

$$+ 0.1 \, \overline{\theta u_k} \, a_{ik} \, a_{m\ell} \left[\frac{\partial U_m}{\partial x_\ell} + \frac{\partial U_\ell}{\partial x_m} \right]$$

$$- 0.1 \, \overline{\theta u_k} (a_{im} \, P_{mk} + 2 a_{mk} \, P_{im})$$

$$+ 0.15 \, a_{mk} \left[\frac{\partial U_k}{\partial x_\ell} + \frac{\partial U_\ell}{\partial x_k} \right] (a_{mk} \, \overline{\theta u_i} - a_{mi} \, \overline{\theta u_k})$$

$$- 0.05 \, a_{m\ell} \left[7 a_{mk} \left[\overline{\theta u_i} \, \frac{\partial U_k}{\partial x_\ell} + \overline{\theta u_k} \, \frac{\partial U_i}{\partial x_\ell} \right] - \overline{\theta u_k} \left[a_{m\ell} \, \frac{\partial U_i}{\partial x_k} + a_{mk} \, \frac{\partial U_i}{\partial x_\ell} \right] \right]$$

$$(13)$$

[†]   Direct numerical simulations by Lee et al [84] have confirmed this behaviour over a wide range of strain rates.

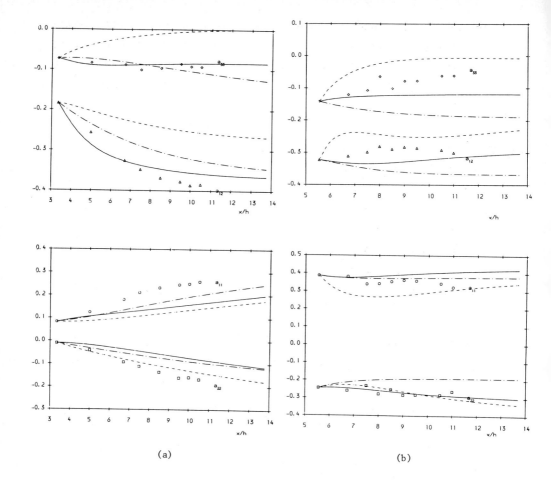

(a)   (b)

Fig. 5   Development of homogeneous shear layer

    (a) Turbulence near local equilibrium
        Symbols : Experiment [81]; ——— Craft et al [71] - eq.(11);
        —— · —— Basic Model (Table 1); - - - Shih et al [75,76]
    (b) High shear rate, $\frac{1}{2}P_{kk}/\epsilon \approx 1.55$
        Symbols : Experiments, Harris et al [82];
        ——— Craft et al [71]; —— · —— Basic Model - Table 1;
    - - - Shih and Lumley [75,76]

There are no adjustable coefficients in either model. The first two terms in each are the same, corresponding to Launder's [85] linear model (see also [86,87]). Craft et al [71] report predictions for heat transport in four homogeneous shear flows adopting, in conjunction with eq.(13), the following representation of $\varphi_{i\theta1}$:

$$\varphi_{i\theta1} = -1.6(1 + 0.5A_2^{\frac{1}{2}}) \frac{R\epsilon}{k} [\overline{u_i\theta} - 1.1a_{ik} \overline{u_k\theta} + 1.6a_{ik}a_{kj} u_j\theta]$$

$$(14)$$

$$- 0.16A^{\frac{1}{2}} R k a_{ij} \frac{\partial\theta}{\partial x_j}$$

where R is the thermal:dynamic time-scale-ratio supplied from the experiments in question.

Two examples from that study are considered in Figs. 6 and 7. The first is the high-strain-rate experiment of Tavoularis and Corrsin [68] $(\frac{1}{2}P_{kk} \simeq 1.8\epsilon)$ with co-aligned temperature and velocity gradients. In this case the computations of Craft et al [71] and Gibson et al [70] both give satisfactory agreement with the measured behaviour, while the Basic Model ("IP model" in the figure) and the Shih-Lumley model produce substantially too large down-gradient heat fluxes leading to a far too small turbulent Prandtl number, σ_t, and a too rapid rise in $\overline{\theta^2}$. In the later experiment by the same workers [88], Fig. 7, the mean temperature gradient occurs in the direction of the mean flow vorticity. Now all models except that of [71] give spuriously high down-gradient heat fluxes and correlation coefficients. While, as noted, the Shih-Lumley model [75] gives poor agreement with experiment, Craft et al [71] showed that the discrepancy arose purely from their model of $\varphi_{i\theta1}$. When their proposal for this process was replaced by eq.(14). Craft et al [71] report that nearly the same results are obtained as with their own model.

Ristorcelli [79] and Reynolds [80] have pointed out that none of the above models of φ_{ij2} (or $\varphi_{i\theta2}$) ensures that turbulence is materially indifferent to rotation in the two-component limit, Speziale [89]. They therefore employed the requirement that it *should* be to determine unknown coefficients in representations of $a_{\ell j}^{mi}$ based on complete expansions containing fourth-rank products in the Reynolds stresses. The writer has not yet seen any predictions from these models of actual shear flows. They should clearly have the potential for representing more faithfully the reactions of the stress field to the application of complex strain.

Corresponding expressions for the *body-force contributions*, to the pressure interactions φ_{ij3} and $\varphi_{i\theta3}$, can be obtained in an analogous way. In the case of buoyancy, the two-point integral in φ_{ij3} is essentially the same as that in $\varphi_{i\theta2}$, so results for the latter may be transformed to the former. The resultant expressions are bulky and are not reproduced here. The writer is not aware of any applications of these models so far. Indeed, it might be said that the experimental data of horizontal, buoyancy-modified shear flows are not accurate enough to enable any very searching tests to be applied (both the Basic Model, Table 1, and the linear Quasi-Isotropic Model [85,86,119] fit the widely spread data adequately). The recent direct numerical simulations of Gerz et al [120] are thus particularly valuable; their extension to the case of a *vertically* directed flow would be doubly welcome.

2.5 Diffusive Transport, d_{ij}

A great deal *could* be said about modelling transport processes in the second-moment equations. Several fairly elaborate truncations of the third-moment equations have been proposed [90,91], while Lumley's proper orthogonal decomposition [21] offers an alternative route, producing formulae of comparable complexity. Here, however, the writer will show. his prejudices by contributing almost nothing on the topic. His position is arrived at from an interplay of the following considerations.

- Experience over two decades in computing thin shear flows suggests that few of the important anomalies one finds in predictions of inhomogeneous flows can be traced to weaknesses in the diffusion model (e.g. [18]), at least if one excludes the near-wall viscosity-affected sublayer.

- In complex shear flows, especially those that require analysis in curvilinear or non-orthogonal frames, to adopt other than a simple model would consume resources out of proportion to any conceivable benefits.

- There are, in any event, few sets of reliable experimental data of the triple moments with which to refine more elaborate models.

- In cases where failures *do* seem to point to inadequacies in the diffusion model (as, for example, in the exceptionally high rate of spread of the "fountain" arising from the collision of two opposed wall jets, [92]) it seems

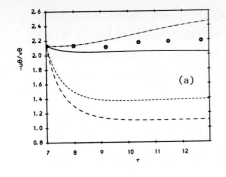

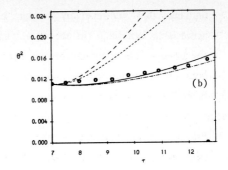

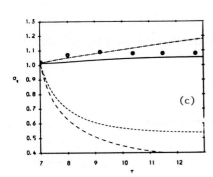

Fig. 6   Heat transport in strongly sheared
          homogeneous flows with co-aligned
          temperature and velocity gradients

o o o   Experiment, Tavoularis and Corrsin [68]
           Computations (from [71])
———   Eq.(13) and (14), Craft et al [71]
— · — · — Gibson et al [70]
- - - -   Basic Model (Table 1)

(a) Ratio of streamwise to cross-stream heat
flux; (b) Effective turbulent Prandtl number
for cross-stream transport; (c) Development
of $\overline{\theta^2}$

U(y), θ(z) U_c= 12.4   DUDY= 46.8   DTDX= 0.0   DTDY= 0.0   DTDZ= 8.2

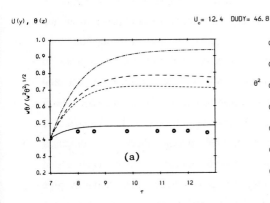

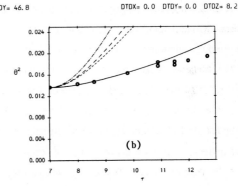

Fig. 7   Heat transport in strongly sheared homogeneous flow : temperature
          gradient aligned with mean vorticity vector

Experiments, Tavoularis and Corrsin [88]
Computations (from [71]) : Key as Fig. 6

(a) Heat flux correlation coefficient;   (b) Development of $\overline{\theta^2}$

unlikely that a relatively minor elaboration of the model will bring appreciable benefits. In the case cited, a third-moment closure seems to be suggested - an approach that is feasible when there are inhomogeneities in only one direction as in, say, the atmospheric boundary layer (André et al [93]) but not in general.

* Finally, in continuation of the last point, diffusion-related paradoxes seem frequently to be linked explicitly to coherent structural features in the flow and/or intermittency; these may require acknowledgement in ways that are not presently attempted. The concept of quasi-homogeneity that underpins current closure efforts is at its most tenuous when devising models for flow *in*homogeneities.

2.6 Dissipation, ϵ_{ij}, and Applications in Inhomogeneous Flows

Lumley [21] has argued that any anisotropic contributions to the dissipation tensor should be regarded as being absorbed into the pressure-strain model, a practice which is normally followed[†], if for no other reason than that, practically speaking, it is impossible to distinguish one process from the other. Accordingly, as in the Basic Model, the task of determining ϵ_{ij} reduces to that of obtaining ϵ. Numerous amendments to the form of the dissipation transport equation, given in Table 1, have been proposed over the years. There is, for example, no essential reason that mean shear should affect the dissipation rate in the same way that it does the turbulence energy - i.e. through the term P_{kk}. Pope [94], Hanjalić and Launder [95], Bardina et al [96], Aupoix et al [97], among others, have proposed alternatives. Unfortunately none of these proposals appears to have been tested over a wide range of flows, or, where they have, they have been found to give worse predictions than the Basic Model in some other flow. In particular, the Pope correction for the round jet in stagnant surroundings, though appearing physically plausible, seems to be a strictly one-flow correction. Even for the round jet in a co-flowing stream [31] it leads to much inferior agreement than the original version.

[†]   As a wall is approached, however, it is readily inferred from the exact form of these terms that φ_{ij} and $(\epsilon_{ij} - \frac{2}{3}\delta_{ij}\epsilon)$ behave in quite different ways; then there *is* merit in distinguishing between them (c.f. Section 2.7).

It has long been argued [98] that mean shear terms have no proper place in a dissipation-rate equation - since the dissipation process is concerned with fine-grained turbulence. A rebuttal of this position is that one is mainly interested in *anisotropies* in the stress field, a_{ij}, and that the only important contribution of "ϵ" in affecting the anisotropy is through the time scale associated with φ_{ij}. Now *that* is generally seen as a time scale of the energy-containing eddies on which deformations by mean shear may be expected to exert some influence. Yet, just as mean shear deforms the large-scale eddies, so must anisotropies in the large-scale eddies be expected to stretch eddies of somewhat finer scale and thus contribute to the rate of energy cascade across the spectrum. This line of thought leads inexorably to the conclusion that one or both of the stress invariants ought to appear in the ϵ equation. Although Lumley's group originally introduced a term containing A_2 as a *replacement* for mean strain effects [98], they soon acknowledged a need for both types of term [16]. The writer's group had intermittently experimented with the inclusion of A_2 for at least a dozen years [99,31] but found that existing proposals did not work well in the flows of interest while specially devised versions did not seem to offer decisive advantages over the basic form, for all its weaknesses. However, with the recent introduction of new models for φ_{ij}, there do now seem to be benefits across an interestingly wide range of flows. The form that has been employed for a variety of free-shear-flow computations at UMIST is:

$$\frac{D\epsilon}{Dt} = c_\epsilon \frac{\partial}{\partial x_k} \left[\overline{u_k u_\ell} \frac{k}{\epsilon} \frac{\partial \epsilon}{\partial x_\ell} \right] + \frac{1}{2} P_{kk} + F_{kk}) \frac{\epsilon}{k} - \frac{1.92}{(1 + 0.6A^{\frac{1}{2}}A_2)} \frac{\epsilon^2}{k} \quad (15)$$

Thus, direct generation by shear is of diminished importance† compared with the basic version, but, in turn, through the appearance of the stress invariants in the sink term, that process is also reduced in magnitude. Clear-cut improvements have been noted when this model was applied to self-preserving shear flows [29,100], the computed and measured spreading rates for which are summarized in Table 3. We note that the anomalously low rate of spread of the plane wake has been entirely removed, the round-jet "problem" has been substantially diminished and there are also marked improvements for the plume flows. Indeed, for the first time we have been able to adopt the same coefficients for the buoyant (F_{kk}) and shear (P_{kk}) contributions to the ϵ source in both horizontal and vertical flows (see for comparison the discussion by Rodi [101]).

---

† The generation coefficient, at 1.0, is still however more than twice as large as that adopted by Zeman and Lumley [16].

Flow	Standard (IP) Model	Recommended Exptl. Values	New Model
Plane Plume	0.078	0.120	0.118
Round Plume	0.088	0.112	0.122
Plane Jet	0.100	0.110	0.110
Round Jet	0.105	0.093	0.101
Plane Wake	0.078*	0.098	0.100*

* Computations by Nemouchi [29]

Table 3

A further interesting behavioural difference between the "basic" form of ϵ equation and that shown in eq.(15) is the different dynamic responses that it causes in non-equilibrium flows. For example, careful measurements by Wygnanski et al [102] of the development of plane wakes suggest that the asymptotic rate of spread depends on the nature of the wake generator (or, effectively, on the ratio of P_{kk}/ϵ a short distance downstream of the wake generator). Computations with the Basic Model by Nemouchi [121], Fig. 8a, predict essentially a unique growth pattern of the wakes irrespective of initial conditions beyond $x/\theta > 200$ (θ here denoting the momentum thickness of the half wake). In contrast, the predicted development with the new ϵ equation, shown in Fig. 8b, indicates that significantly different slopes persist to values of x/θ beyond 2000, depending on the initial conditions. Unfortunately the experimenters did not report sufficient details of their near-wake behaviour to allow precise initial conditions to be assigned. However, both the range of spreading rates reported and the somewhat surprising observation that the case with the highest initial turbulence energy shows the slowest apparent rate of growth are in accord with the computed behaviour.

While, to keep the scope of the paper within bounds, attention has been generally limited to the behaviour of the dynamic field in uniform density flows, the notable study of the variable-density-plane mixing layer [122] by Shih et al [123] demands mention. Straightforward application of volume-weighted averaging (accompanied by a careful discard of the least important terms) produced very close accord between predicted and measured velocity and concentration profiles. Although their results were not thus organized, the implication must be that the strong decrease in turbulent Prandtl number that accompanies the severe density inhomogeneity is being correctly predicted. Since there have previously been several rather unsuccessful attacks on this problem, it appears that the new

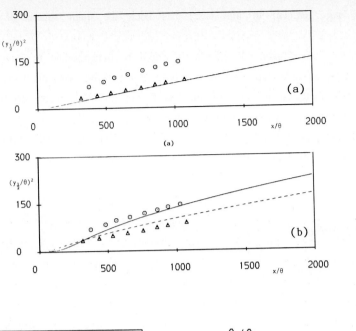

(a)

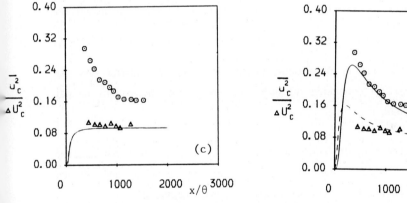

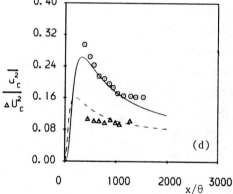

Fig. 8   Comparison of computed and measured development of plane wake
with different initial conditions

Symbols: Experiments, Wygnanski et al [102] for wakes generated by:
airfoil, ⊙ ;   solid strip, △
─────────   Computations with initial conditions obtained assuming local
equilibrium initial conditions. - - - - Computations with initial stress
levels at 4 times local equilibrium values; levels of ϵ unchanged

(a) Computed rate of growth with Basic Model; (b) Computed rate of growth
with new pressure-strain model and ϵ obtained from eq.(15); (c) Computed
development of streamwise normal stress on centre line with Basic Model;
(d) Computed development of streamwise normal stress on centre line with
new pressure-strain model and ϵ obtained from eq.(15)

modelling of $\varphi_{i\theta_2}$ (eq.(12) in this case) has been the main source of this improvement.

Despite the above favourable signals, it would be inappropriate to leave the impression that, at a stroke, all remaining turbulence modelling problems had been removed. It will doubtless be the case that, as further testing is made, anomalies of one kind or another become evident. Nevertheless, it seems clear that the generation of second-moment closures that is now emerging will bring a significantly greater reliability of prediction than currently offered by the Basic Model.

2.7 Near-Wall Effects

A rigid boundary exerts many different effects on turbulence, the most important of which are:

- it reduces the length scales of the fluctuation raising the dissipation rates;

- it reflects pressure fluctuations, thereby inhibiting the transfer (via the pressure-strain correlation) of turbulence energy into fluctuations normal to the wall;

- it enforces a no-slip condition, thus ensuring that within a wall-adjacent sublayer, however thin, turbulent stresses are negligible and viscous effects on transport processes become of vital importance.

Brief remarks will be made on these different aspects in the above order.

The constants in the "basic" form of the dissipation-rate equation were selected to give the correct variation of length scale in near-wall turbulence in local equilibrium. However, this equation returns progressively too high length scales as the near-wall level of $(\tfrac{1}{2}P_{kk}/\epsilon)$ falls towards zero.[†] Consequences of this are that boundary-layer separation will tend to be predicted too late [103]

[†] The evidence of this weakness has been acquired from computations based on the k-ϵ eddy-viscosity or ASM schemes. The anomaly seems certain to remain in a full second-moment closure, however.

and, in separated flows, too large heat transfer coefficients predicted. Yap [39] (see also [104]) proposed an additional source term, S_ϵ, that goes at least some way to removing this problem. The term takes the form:

$$S_\epsilon = \text{Max}\left[0.83 \frac{\epsilon^2}{k}\left(\frac{\ell}{c_\ell x_n} - 1\right)\left(\frac{\ell}{c_\ell x_n}\right)^2, 0\right]\qquad(16)$$

where $\ell = k^{3/2}/\epsilon$, c_ℓ equals 2.44 and x_n is the normal distance from the wall. The term vanishes in local-equilibrium wall turbulence because then $\ell = c_\ell x_n$; it also becomes small at large distances from the wall, since then $\ell/c_\ell x_n$ is much less than unity. If the near-wall length scale is larger than $c_\ell x_n$, however, the term is positive, leading to increased values of ϵ and reduced values of ℓ. An example of the dramatically beneficial effect of this term is provided in Fig.9 that compares the measured and computed variation of Nusselt number downstream of an abrupt pipe expansion [39].

A weakness of the Yap correction term is the need to prescribe a wall distance - a choice that is difficult to make when dealing with corrugated surfaces (such as are becoming popular in heat-exchanger tubing) or, indeed, any complex topography. An alternative version that does not exhibit this weakness is:

$$S_\epsilon \propto \frac{\epsilon}{k^{\frac{1}{2}}}\left(\frac{\partial k}{\partial x_n}\right)\left(\frac{\partial \ell}{\partial x_n}\right)\qquad,\qquad(17)$$

a form that Ince [105] has found works well in a number of flows. The term has negligible effect in at least the common free shear flows because then $\partial \ell/\partial x_n$ is very small. It is also uninfluential in local equilibrium because, in the region where $\partial \ell/\partial x_n$ is large, $\partial k/\partial x_n$ is small.[†] When near-wall generation rates are small, however, $\partial k/\partial x_n$ is positive where $\partial \ell/\partial x_n$ is large, thus raising ϵ and depressing the length scale.

The alternative form of ϵ equation, eq.(15), that gave such encouraging performance in free flows has only been tested in near-wall flows in conditions close to local equilibrium. It is thus not yet clear how large a correction, if any, would be needed for near-wall separated flows.

[†] The term must, however, be suppressed in the near-wall viscous sublayer as the gradients of k and ℓ are then both large and positive.

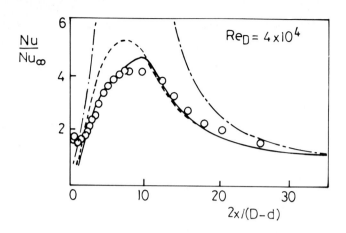

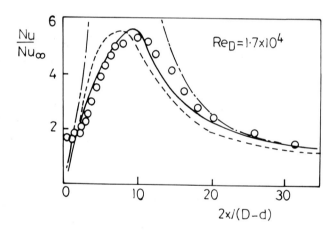

Fig. 9   Behaviour of Nusselt number downstream of abrupt pipe expansion
(D/d = 2.5)

O O	Experiment, Baughn et al
——	ASM : Basic Model including near-wall ϵ source
- - - -	k-ϵ EVM including near-wall ϵ source
— · —	k-ϵ EVM without near-wall ϵ source

A side effect of predicting incorrect levels of near-wall length scale is that the wall-reflection part of φ_{ij} will also be wrongly estimated. This is because the strength of the near-wall length scale is normally taken proportional to (ℓ/x_n).[†] The models proposed for the wall-reflection process itself, φ_{ijw} [9,25,106], should only be used in conjunction with the associated proposals for φ_{ij1} and φ_{ij2}. It seems, in fact, that the Basic Model shown in Table 1 requires an artificially strong wall-reflection process to help counteract the fact that the IP model of φ_{ij_2} gives equal levels of $\overline{u_2^2}$ and $\overline{u_3^2}$ under simple shear, $U_1(x_2)$ while φ_{ij1}, given by Rotta's [1] model, does not vanish at the wall. Tselepidakis [107] has shown that by using a properly vanishing form of φ_{ij}, and the correct asymptotic limits for ϵ_{ij} (see below), one can get fairly satisfactory predictions of the stress field in a plane channel flow *without including wall-reflection processes* in the model. It ought to be admitted that he does get *better* agreement by adding a traditional "wall-echo" component, though the strength of this term is much weaker than in earlier studies of flow near walls. This relative unimportance of the near-wall correction term must be seen as good news for the development of a general model for arbitrary surface topographies still looks some way off.

For values of the turbulent Reynolds number R_t ($\equiv k^2/\nu\epsilon$) below about 150 direct viscous effects become significant. The region is so thin and strongly inhomogeneous that accurate experimental data are hard to come by. Yet, over the last four years our knowledge of this region has progressed by leaps and bounds thanks to the direct numerical simulations of channel flow [108,109] and boundary layers [110]. There have simultaneously been renewed attempts at computing, via second-moment closure, the Reynolds stress field right across this semi-viscous sublayer to the wall itself. We mention *inter alia* the contributions of Prud'homme and Elghobashi [111], Shima [112], Launder and Shima [113], Lai and So [114] and Launder and Tselepidakis [115]. None of these schemes has been tested over an especially wide range of flows, the most extensive explorations being those in [113] for five boundary-layer cases. Figure 10 shows the computed behaviour obtained from this model for a strongly accelerated boundary layer undergoing laminarization. The gradual decay of the streamwise velocity fluctuations and the reversion of the velocity profile to that appropriate to a laminar boundary layer match the data of Simpson and Wallace [116] very closely.

[†]   Demuren and Rodi [47] have preferred $(\ell/x_n)^2$ in their computation of flow in rectangular-sectioned ducts.

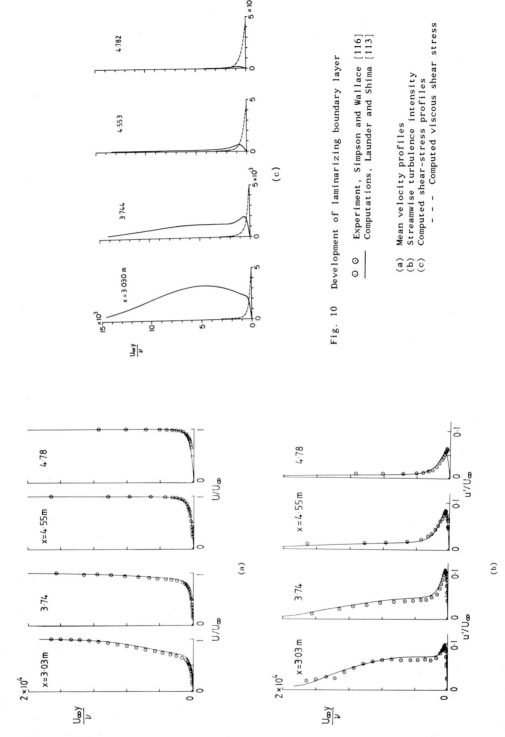

Fig. 10 Development of laminarizing boundary layer

⊙ ⊙ Experiment, Simpson and Wallace [116]
─── Computations, Launder and Shima [113]

(a) Mean velocity profiles
(b) Streamwise turbulence intensity
(c) Computed shear-stress profiles
─ ─ ─ Computed viscous shear stress

In ref.[115], and in ongoing work by its authors, an attempt is made to extend the UMIST closure for free flows discussed in §2.4 and §2.6 to the near-wall sublayer. There have been two principal problems encountered, neither of which is entirely resolved: how to make $\overline{u_2^2}/k$ decay sufficiently rapidly as the wall is approached and how to make ϵ approach a maximum value at the wall. An initially surprising discovery was that, while all the modelled processes were consistent with $(\overline{u_2^2}/k)$ being zero at the wall, they did not *enforce* it. Only by independently arranging for φ_{ij1} to vanish, by replacing ϵ appearing in eq.(5) by $\epsilon - 2\nu(\partial k^{1/2}/\partial x_k)^2$ was the desired decrease to zero achieved. The correct representation of dissipation rate was also an important factor. While many studies have assumed that, as the wall is approached, the componental dissipation rate is

$$\epsilon_{ij} = \frac{\overline{u_i u_j}}{k} \epsilon \quad , \tag{18}$$

this gives the wrong relative values in components where i or j denotes the direction normal to the wall. Instead, the following exact limiting form [117,118] was adopted:

$$\epsilon_{ij} = \frac{\epsilon}{k} (\overline{u_i u_j} + \overline{u_i u_k} \, n_k n_j + \overline{u_j u_k} \, n_k n_i$$
$$+ \delta_{ij} \overline{u_k u_\ell} \, n_k n_\ell)/[1 + \frac{5}{2} \frac{\overline{u_p u_q} \, n_p n_q}{k}] \tag{19}$$

The direct simulations [124] suggest that diffusive transport is of greater importance in this region than the simple models advocated in §2.5 indicate. Moreover, the role of *pressure* diffusion in both the $\overline{u_i u_j}$ and ϵ equations appears to be crucial as, unlike velocity fluctuations, the fluctuating pressure does not vanish at the wall. Lumley's [86] isotropic model for the pressure-diffusion

$$\frac{\overline{pu_j}}{\rho} = -0.2 \, \overline{u_j u_k^2} \tag{20}$$

at least introduces a net up-gradient turbulent transport of $\overline{u_2^2}$ in the near-wall sublayer which helps to depress $\overline{u_2^2}/k$. The applicability of this model to the sublayer region is, however, highly questionable.

The detailed near-wall budget for the processes in the ϵ equation deduced from direct numerical simulation will, in due course, transform the reliability of modelled forms of the equation. An accurate budget for ϵ is, however, significantly harder to evaluate than that for $\overline{u_i u_j}$ because the component terms in the former equation involve higher-order derivatives. While the processed ϵ-budget reported in [124] shows good internal consistency, it would be reassuring to have these results confirmed by a further channel-flow simulation with a considerably finer near-wall mesh.

3. Possible Future Developments

In trying to look ahead to what developments may be anticipated from now to the end of the century, it is perhaps helpful to distinguish three types of activity:

- The *application* and *assessment* of existing turbulence models outside of the sheltered homogeneous environment within which most have been nurtured until now.

- The *extension* and *refinement* of turbulence models within the framework of existing methodology in order to remove known weaknesses and to render the models applicable to new physical phenomena.

- The *development* of models with a more radically new complexion in order to achieve a greater range of applicability and reliability.

Brief remarks will be made under each of the above heads.

3.1 Application and Assessment

There is clearly a need to subject the new generation of models to extensive and searching examination against a wide range of well-documented test flows representative of those of industrial interest. Testing in complex strain fields is especially important. To bring the Basic Model from its formation to a point where its capabilities in complex flows are reasonably well established has taken fifteen years. There is no reason why the comparable time interval for today's models should not be shortened to four or five years - but it will require the active participation of many groups in the "user" community.

Some may see the new schemes (perhaps even the Basic Model) as unworkably complex for engineering flows. For whose who *do*, recent trends in methodology for RANS[†] solvers offer some encouragement. It appears that *non*-orthogonal (as opposed to curvilinear orthogonal) grids are gaining ascendancy [125]. With such schemes, one normally solves momentum equations for the Cartesian components. This means in turn that Cartesian elements of the stress tensor appear in those momentum equations - which provides the most transparent system within which to shift from tensor algebra to software. Moreover, our experience at UMIST in the past two years suggests that, in computing complex flows, greater algebraic complexity does not necessarily mean longer computing times. If one's model cannot give unrealizable values of $\overline{u_i u_j}$ even in the course of iteration to the final numerical solution, it is quite likely that the number of iterations needed to reach that solution will be reduced. So, even if the number of arithmetic operations per cycle is larger, the CPU time may be less.

There is, of course, no guarantee that the resultant computed behaviour will be in better accord with experiments than that generated by the Basic Model. But even if improvements in predictions are not always found, the diagnosis of the causes of any failure and the feedback to the modelling fraternity is vital to the advancement of second-moment closures.

3.2 Extension and Refinement

Every active worker in the field will have his own shopping list of weaknesses he would like to rectify; what follows is no more than such a personal statement.

An important area where current schemes are *ad hoc* and strongly dependent on experimental data is in turbulent impinging flows - especially where the main focus of interest is the heat- or mass-transport rates at the surface. Current models are almost exclusively based on information pertaining to flows *parallel* to walls. If attention shifts to impinging flows, there are thus numerous ways in which the model can go wrong. Mention has already been made in §2.6 of the fact that the basic ϵ equation tends to cause too large near-wall length scales in these circumstances. There must also be doubt about the applicability of Reynolds-number damping functions tuned by reference to parallel flows. What is very much needed is a full turbulence simulation of the region in the vicinity of a plate on which is impinging a flow with a well-defined turbulence field.

[†] Reynolds Averaged Navier Stokes

The whole area of low-Reynolds-number near-wall turbulence is one where the scalar flux modelling lags well behind that of the Reynolds stresses, even for flow parallel to the wall. The provision of direct numerical simulation data [126] gives a much needed stimulus for the necessary model development, even though at present a comprehensive processing of the direct-simulation results is still awaited. High-Schmidt-number mass transport is an especially challenging area for, as the Schmidt number increases, the region of the flow that is most influential in determining surface transport rates is confined to a thinner and thinner region adjacent to the wall, eventually being confined within what - so far as the velocity field is concerned - is the viscous sublayer. The further extension of sublayer modelling to include non-Newtonian fluids and flows with large variations in transport properties will become increasingly important over the next decade as CFD becomes extensively applied in the chemical and process industries.

The other main area of ignorance in relation to wall turbulence concerns the surface-reflection contribution to the pressure fluctuations. As has been noted earlier, it seems that earlier models have attached too much importance to this process. Nevertheless, it will seldom suffice to neglect it entirely. Current approaches are formulated only for plane surfaces and their extension to tightly curved boundaries or corners would be highly desirable, albeit not straightforward. A full turbulence simulation of flow in a *square*-sectioned channel would greatly assist modelling, especially if the surface-reflection contributions to φ_{ij} could be separately evaluated.

Shifting to free flows, the case of the axisymmetric wake seems to pose difficulties. This is a flow where the ratio of turbulence energy generation:dissipation rates is only about 0.2 and where, in consequence, the stress invariants are appreciably different from those in jets or even the plane wake. Our preliminary conclusions [127] are that to bring predictions for this flow into accord with experiment will require a careful re-tuning of the invariant-dependent functions both in φ_{ij1} and in the ϵ transport equation.

Another well known problem flow is the supersonic free shear layer. The progressive reduction in spreading rate as the Mach number is raised is arguably the one case where an unequivocal dependence on compressibility is demonstrated. The phenomenon has hitherto been correlated by making the leading coefficient in φ_{ij1} dependent upon the fluctuating Mach number [128]. While this *does* seem the likely candidate process, it would be helpful to have this confirmed by direct numerical simulation[†] prior to attempting a more fundamental analysis.

3.3 New Modelling Developments

What is presented here is not so much a collection of new proposals as ideas that have been on the table for some years wondering when their day would come.

A basic question is whether there is merit in introducing a second second-rank tensor (in addition to $\overline{u_i u_j}$) that one would determine by way of transport equations. The direct numerical simulations [63] suggest considerably greater anisotropy of the dissipation tensor than is conventionally supposed and, moreover, imply a strong connection with the pressure-strain process. These indications naturally raise the question of whether transport equations should be solved for the components of ϵ_{ij}. Proposals for closed forms of an equation for ϵ_{ij} have been made by Morse [99] and Lin and Wolfshtein [129]. In the writer's view, such an elaboration of the closure would be premature. Lee and Reynolds' [63] finding that φ_{ij1} correlated well with ϵ_{ij} does not necessarily mean that information on the anisotropy of the dissipation tensor was feeding back up-spectrum to the larger eddies. A far more plausible scenario is that, at the low Reynolds numbers of their simulations, the dissipation tensor looked very much like the pressure-strain tensor simply *because* spectral transfer took place over such a limited range of wave numbers. In order words, the DNS results seem entirely consistent with the conventional wisdom that the dog wags its tail rather than the reverse.

Another possible approach to refinement is what might be termed a "split-spectrum" model, first proposed by Hanjalić et al [130] and subsequently elaborated by Cler [131] and Schiestel [132]. Basically one divides the energy-containing part of the stress spectrum into two parts - a large-scale and medium-scale slice - and provides independent transport equations for both parts. Such a model opens up interesting possibilities: Schiestel [133] shows, for example, computations in which a decaying, non-isotropic turbulence field is brought rapidly to isotropy by the application of a suitable strain over a short

---

[†]   A study which, I gather, is in progress.

distance and then *departs from isotropy again* further downstream. This feature, which has been observed experimentally, cannot be captured by any "full-spectrum" treatment. It occurs because the strain field modifies the large-scale eddies more than the medium scale, so the apparent isotropy achieved by applying the strain was really a matter of producing *cancelling anisotropies* in the two spectral slices. Downstream the fine-scale fluctuations decay faster than the large-scale, causing a departure of the complete spectrum from isotropy. A number of important industrial flows where unusual spectral distributions are present (the wakes from boundary-layer manipulators, for example) seem suitable for attack with this type of model. The most problematical part of such a closure is the representation of the rate of transfer from large to medium scales (including, perhaps, the possibility of reverse transfer). To keep the closure within bounds, one should probably aim at representing the anisotropy in the transfer tensor in terms of the anisotropy of the stress field in the two spectral slices and then provide a transport equation for the spectral transfer of turbulence energy from large to medium scales - an equation that would look somewhat similar to that for ϵ in full-spectrum approaches. Schiestel [132] makes outline proposals for how the rates of stress transfer between the two slices can be explicitly linked to existing spectral transport theories.

An attractive feature of split-spectrum closure is its obvious linkage with large-eddy simulation, the "medium-scale" slice being roughly analogous to the part of the spectrum being covered by the "sub-grid-scale" model in a LES approach. There is thus the possibility of developing in parallel the common aspects of these two modelling levels. Yet, while the split-spectrum approach does allow greater scope for physical realism (for example, its explicit admission of variations in the shape of the stress spectra from one flow to another), it is difficult to see it being adopted for general use.

Finally, before leaving discussion of new tensor variables, mention should be made of the work at ARAP aimed at introducing a length-scale tensor, Λ_{ij}, into second-moment closure. It is not clear from the preliminary and possibly out-of-date reports [134,135] known to the present writer whether solution of transport equations for the components of Λ_{ij} is advocated or whether it is felt that the anisotropy of the length scale can be linked algebraically to that of the stress tensor. When the latter route is adopted, the transport of Λ_{ij} is simply expressed by way of the transport of the *scalar* length scale Λ_{kk}. In this case, the model is generically similar to those proposed by Schiestel [136] and Wu et al [137] in which, essentially, independent scale equations are developed for the large- and fine-scale motion. The addition of a further scale equation at least makes economic sense: if one is going to solve six transport equations for the stress tensor it little affects the computational burden of the model whether one solves one or two transport equations for scale-related quantities. A weakness of the proposals in [136] and [137], however, is that they were made within the framework of an eddy-viscosity model. As a result, one does not have available the stress invariants as possible model parameters since, for example, A_2, according to an eddy-viscosity stress-strain hypothesis, is just proportional to P_{kk}/ϵ, an invariant ratio which already appears as a parameter in the scale equations. The use of a second scale equation within the framework of second-moment closure could, however, become a standard feature of such models.

Virtually the whole of the paper has been concerned with approaches to determining the Reynolds stresses in a turbulent flow. Before closing, however, acknowledgement must be made that, increasingly, other characteristics of the flow, only indirectly linked to the second moments, may require estimation. If one is concerned with the mixing of a turbulent field with an irrotational external stream, one may wish to know the relative proportions of time that turbulent and non-turbulent fluid are present at any particular point in the flow and, perhaps, the average time interval between a switch-over from turbulent to non-turbulent fluid. This information is vital to tackle the "thermal-striping" problems associated with various types of reaction chamber. Equally one may be interested in determining the probability that a temperature or a species concentration is above some threshhold value. Problems of the above type will be increasingly tackled in the next decade. While computational studies of such problems will draw on second-moment closure, they will also have special features: an intermittency equation or an equation for the pdf of the fluctuating quantity of interest. Substantial progress in these directions has already been made by those whose main interest in turbulence is combustion, [138-140]. It is likely that substantial parts of their work can usefully feed back to the study of non-reacting shear flows.

4. Concluding Remark

The present article, like the others in this volume, was prepared for discussion at the workshop "Whither Turbulence? ... or Turbulence at the Cross Roads." While I suspect no-one intended the ominous ring to those words to be taken too seriously, it does nevertheless call for some kind of response.

Second-moment closure, so far as I am aware, is not at or near a crossroads. It is not suffering an identity crisis; its long convoy of vehicles is moving confidently forward. There were, of course, diversions and the occasional U-turn back down the road a way; and the wrong signposts on leaving the '81 Stanford Conference certainly confused some people. Some were slow to take advantage of the extra lane in the highway provided by the direct numerical simulation; others haven't yet noticed that the larger-capacity, realizable engine actually gives better mileage. Nevertheless, the route now to be followed is well marked. Wrong turnings are not anticipated.

To shift the metaphor a little, second-moment closure may not be the most exciting vehicle in the showroom; but, given the destination, the terrain in between and the climate, it's the only one with a chance of getting there. And that's likely to be as true in '99 as it is in '89!

Acknowlegements

My warm thanks are expressed to present and former members of UMIST's CFD group who, through the results they have generated, have helped shape the viewpoints expressed in this paper. Especial thanks go to Mrs. L.J. Ball who, over the last five years, has produced more camera-ready documents than I, and probably she, care to remember. This, her final paper as my secretary, has had to be produced from a fragmented and continually changing manuscript against the tightest of deadlines.

References

1. Rotta, J. Z. Phys. 129, 547-573, 1951.
2. Rotta, J. Z. Phys. 131, 51-77, 1951.
3. Davidov, B.I. Dokl. Akad. Nauk. SSSR 136, 47, 1961.
4. Donaldson, C. du P. "A computer study of an analytical model of boundary layer transition," AIAA J. 7, 271-278, 1969.
5. Daly, B.J. and Harlow, F.H. Phys. Fluids 13, 2634, 1970.
6. Hanjalić, K. and Launder, B.E. J. Fluid Mech. 52, 609, 1972.
7. Rodi, W. "The prediction of free turbulent shear flows by use of a 2-equation model of turbulence," PhD Thesis, Faculty of Engineering, University of London, 1972.
8. Launder, B.E., Morse, A.P., Rodi, W. and Spalding, D.B. Proc. 1972 NASA Conf. on Free Turbulent Shear Flows, Vol. 1, pp. 361-526, NASA SP-321, 1973.
9. Launder, B.E., Reece, G.J. and Rodi, W. J. Fluid Mech. 68, 537, 1975.
10. Launder, B.E. and Spalding, D.B. Comp. Meth. Appl. Mech. and Engrg. 3, 269, 1974.
11. Launder, B.E. and Ying, W.M. Proc. IMechE (London), 187, 455-461, 1973.

12. Naot, D., Shavit, A. and Wolfshtein, M.   Wärmeundstoffubertragung 7, 151, 1974.
13. Pope, S.B. and Whitelaw, J.H.   J. Fluid Mech. 73, 9-32, 1976.
14. Wyngaard, J. and Coté, O.   Boundary-Layer Met. 7, 289-308, 1974.
15. Launder, B.E.   J. Fluid Mech. 67, 569-581, 1975.
16. Zeman, O. and Lumley, J.L.   "Buoyancy effects in turbulent boundary layers: A second order closure study" in Turbulent Shear Flows-1 (ed. F. Durst et al), 295-306, Springer, Heidelberg, 1979.
17. Irwin, H.P.A. and Arnot-Smith, P.   Phys. Fluids 17, 624-630, 1975.
18. Launder, B.E. and Morse, A.P.   "Numerical prediction of axisymmetric free shear flows with a Reynolds stress closure" in Turbulent Shear Flows-1 (ed. F. Durst et al), 274-295, Springer, Heidelberg, 1979.
19. Kline, S.J., Cantwell, B. and Lilley, G.K.   (Editors), Proc. 1980-81 AFOSR-HTTM-Stanford Conf. on Complex Turbulent Flows, Stanford, 1981/82.
20. Leonard, B.P.   Comp. Meth. Appl. Eng. 19, 59, 1979.
21. Lumley, J.L.   Advances in Appl. Mech. 18, 123-175, 1978.
22. Naot, D., Shavit, A. and Wolfshtein, M.   Israel J. Tech. 8, 259-269, 1970.
23. Launder, B.E.   J. Fluid Mech. 67, 569-581, 1975.
24. Shir, C.C.   J. Atmos. Sci. 30, 1327-1339, 1973.
25. Gibson, M.M. and Launder, B.E.   J. Fluid Mech. 86, 491-511, 1978.
26. Monin, A.S.   Isv. Atmos. Ocean. Phys. 1, 85-94, 1965.
27. Owen, R.G.   "An analytical turbulent transport model applied to non-isothermal fully-developed duct flows," PhD Thesis, Dept. Mech. Engrg., The Pennsylvania State University, 1973.
28. Gibson, M.M., Jones, W.P. and Younis, B.A.   Phys. Fluids 24, 386-395, 1981.
29. Nemouchi, Z.   "The computation of turbulent thin shear flows associated with flow around multi-element aerofoils," PhD Thesis, Faculty of Technology, University of Manchester, 1988.
30. Gibson, M.M. and Rodi, W.   J. Fluid Mech. 103, 161-182, 1981.
31. Huang, G.P.G.   "The computation of elliptic turbulent flows with second-moment closure models," PhD Thesis, Faculty of Technology, University of Manchester, 1986.
32. Gibson, M.M. and Launder, B.E.   ASME J. Heat Transfer 98C, 81-87, 1976.
33. Hossain, M.S. and Rodi, W.   "Influence of buoyancy on the turbulent intensities in horizontal and vertical jets" in Heat Transfer and Turbulent Buoyant Convection (ed. D.B. Spalding and N. Afghan), Hemisphere, New York 1977.
34. McGuirk, J. and Papadimitriou, C.   "Buoyant surface layer under fully entraining and hydraulic jump conditions," Proc. 5th Symp. on Turbulent Shear Flows, 22.33-22.41, Cornell, 1985.
35. To, W.M. and Humphrey, J.A.C.   Int. J. Heat Mass Transfer 29, 593, 1986.
36. Humphrey, J.A.C. and To, W.M.   "Numerical prediction of turbulent free convection along a heated vertical flat plate," Proc. 5th Symp. on Turbulent Shear Flows, 22.19-22.26, Cornell University, 1985.
37. Jones, W.P. and Manners, A.   "The calculation of the flow through a two-dimensional faired diffuser" in Turbulent Shear Flows-6 (ed. J.C. André et al), 18-31, Springer, Heidelberg, 1989.
38. Launder, B.E., Leschziner, M.A. and Sindir, M.   "The UMIST-UCD computations: Comparison of computations with experiment," Proc. 1980-81 AFOSR-HTTM-Stanford Conf. on Complex Turbulent Flows (ed. S.J. Kline et al), Vol. III, 1390-1407, Stanford, 1982.
39. Yap, C.   "Turbulent heat and momentum transfer in recirculating and impinging flows," PhD Thesis, Faculty of Technology, University of

Manchester, 1987.

40. Leschziner, M.A., Kadja, M. and Lea, C.K. "A combined computational and experimental study of a separated flow in an expanding annular passage," Proc. 3rd IAHR Symp. on Refined Flow Modelling and Turbulence Measurements, 83-91, Tokyo, 1983.

41. Fu, S., Launder, B.E. and Leschziner, M.A. "Modelling strongly swirling recirculating jet flow with Reynolds-stress transport closures," Proc. 6th Symp. on Turbulent Shear Flows, Paper 17.6, Toulouse, 1987.

42. Hogg, S.I. and Leschziner, M.A. "Computation of highly swirling confined flow with a Reynolds stress closure." To appear in AIAA J., 1989.

43. Weber, R., Boysan, F., Swithenbank, J. and Roberts, P.A. Proc. 21st Symp. (International) on Combustion, The Combustion Institute, 1435-1443, 1986.

44. Hogg, S.I. and Leschziner, M.A. To appear in Int. J. Heat and Fluid Flow, 1989.

45. Benay, R., Cöet, M-C and Delery, J. "A study of turbulence modelling in transonic shock-wave boundary layer interactions" in Turbulent Shear Flows-6 (ed. J.C. André et al), 194-214, Springer, Heidelberg, 1989.

46. Baughn, J.W., Hoffman, M., Launder, B.E. and Samaraweera, D.S.A. "Three-dimensional turbulent heat transport in pipe flow: experiment and model validation," ASME Paper 78-WA-HT-15, ASME Winter Annual Meeting, San Francisco, 1978.

47. Demuren, O. and Rodi, W. J. Fluid Mech. 140, 189, 1984.

48. Iacovides, H. and Launder, B.E. "ASM predictions of turbulent flow and heat transfer in coils and U-bends," Proc. 4th Int. Conf. on Numerical Methods in Laminar and Turbulent Flows, 1023-1044, Pineridge Press, Swansea, 1985.

49. Choi, Y.D., Iacovides, H. and Launder, B.E. ASME J. Fluids Eng. 111, 1989.

50. Stevens, S.J. and Fry, P. J. Aircraft 10, 73, 1973.

51. Johnston, J.P., Halleen, R.M. and Lezius, D.K. J. Fluid Mech. 56, 533, 1972.

52. Kim, J. Proc. 4th Symp. on Turbulent Shear Flows, Karlsruhe, p. 6.14, 1983.

53. Launder, B.E., Tselepidakis, D.P. and Younis, B.A. J. Fluid Mech. 183, 63, 1987.

54. So, R.M.C., Ahmed, S.A. and Mongia, H.C. Exp. in Fluids 3, 221-230, 1985.

55. Chang, S.M., Humphrey, J.A.C. and Modavi, A. Physico-Chemical Hydrodynamics 4, 243, 1983.

56. Chang, S.M., Humphrey, J.A.C., Johnson, R.W. and Launder, B.E. Proc. 4th Symp. on Turbulent Shear Flows, pp. 6.20-6.25, Karlsruhe, 1983.

57. Johnson, R.W. "Turbulent convecting flow in a square duct with a 180° bend: an experimental and numerical study," PhD Thesis, Faculty of Technology, University of Manchester, 1984.

58. Launder, B.E. "The prediction of force field effects on turbulent shear flows via second-moment closure," Invited Paper, 2nd European Turbulence Conf., Berlin, 1988.

59. Launder, B.E. "Turbulence modelling of three-dimensional shear flows," Invited Paper, Proc. AGARD Conf. on 3-Dimensional Turbulent Shear Flows, Cesme, Turkey, 1988.

60. Chou, P-Y. Quart. App. Math. 3, 38, 1945.

61. Lumley, J.L. and Newman, G.R. J. Fluid Mech. 82, 161, 1977.

62. Reynolds, W.C. in 'Turbulence Models and Their Applications,' Eyrolles, Paris, 1984.

63. Lee, M. and Reynolds, W.C. "On the structure of homogeneous turbulence,"

Proc. 5th Symp. on Turbulent Shear Flows, pp. 17.7-17.12, Cornell, 1985.

64. Lee, M. and Reynolds, W.C. "Structure and modeling of homogeneous turbulence in strain and relaxation processes." Submitted to J. Fluid Mech., 1987.

65. Weinstock, J. and Burk, S. "Theoretical pressure-strain term: resistance to large anisotropies of stress and dissipation," Proc. 5th Symp. on Turbulent Shear Flows, pp. 12.13-12.18, Cornell, 1985.

66. Weinstock, J. and Burk, S. J. Fluid Mech. 154, 429-443, 1985.

67. Jones, W.P. and Musonge, P. "Modelling of scalar transport in homogeneous turbulent flows," Proc. 4th Symp. on Turbulent Shear Flows, pp. 17.18-17.24, Karlsruhe, 1983.

68. Tavoularis, S. and Corrsin, S.C. J. Fluid Mech. 104, 311, 1981.

69. Dakos, T. and Gibson, M.M. "On modelling the pressure terms of the scalar flux equations," Proc. 5th Symp. on Turbulent Shear Flows, pp. 12.1-12.6, Cornell, 1985.

70. Gibson, M.M., Jones, W.P. and Kanellopoulos, V.E. "Turbulent temperature mixing layer: measurement and modelling," Proc. 6th Symp. on Turbulent Shear Flows, Paper 9-5, 1987.

71. Craft, T., Fu, S., Launder, B.E. and Tselepidakis, D.P. "Developments in modelling the turbulent second-moment pressure correlations," UMIST Mech. Eng. Dept. Report TFD/89/1, 1989. (Submitted for publication).

72. Naot, D., Shavit, A. and Wolfshtein, M. Phys. Fluids 16, 738, 1973.

73. Le Penven, L. and Gence, J-N. C.R. Acad. Sci. Paris 297, Ser. II, 309-383, 1983.

74. Lecointe, Y., Piquet, J. and Visonneau, M. "Rapid-term modelling of Reynolds stress closures with the help of rapid-distortion theory," proc. 5th Symp. on Turbulent Shear Flows, pp. 12.7-12.12, Cornell, 1985.

75. Shih, T-H and Lumley, J.L. "Modeling of pressure correlation terms in Reynolds stress and scalar flux equations," Report FDA-85-3, Sibley School of Mech. Aero. Eng., Cornell University, 1985.

76. Shih, T-H, Lumley, J.L. and Chen, J-Y. "Second-order modeling of a passive scalar in a turbulent shear flow," Report FDA-85-15, Sibley School of Mech. Aero. Eng., Cornell University, 1985.

77. Fu, S., Launder, B.E. and Tselepidakis, D.P. "Accommodating the effects of high strain rates in modelling the pressure-strain correlation," UMIST Mech. Eng. Dept. Report TFD/87/5, 1987.

78. Fu, S. "Computational modelling of turbulent swirling flows with second-moment closures," PhD Thesis, Faculty of Technology, University of Manchester, 1988.

79. Ristorcelli, J.R. "A realizable rapid pressure model satisfying two-dimensional frame indifference and valid for three-dimensional three-component turbulence," Report FDA-87-19, Sibley School of Mech. Aero. Eng., Cornell University, 1987.

80. Reynolds, W.C. "Fundamentals of turbulence for turbulence modeling and simulation," Notes for Class ME261B Turbulence, Stanford University, 1989.

81. Champagne, F., Harris, V.G. and Corrsin, S.C. J. Fluid Mech. 41, 81, 1970.

82. Harris, V.G., Graham, J.A. and Corrsin, S.C. J. Fluid Mech. 81, 657, 1977.

83. Leslie, D.C. J. Fluid Mech. 98, 435-448, 1980.

84. Lee, M.J., Kim, J. and Moin, P. "Turbulence structure at high shear rate," Paper 22-6, Proc. 6th Symp. on Turbulent Shear Flows, Toulouse, 1987.

85. Launder, B.E. "Turbulence models and their experimental verification; II - Scalar property transport by turbulence," Mech. Eng. Dept. Report HTS/73/26, Imperial College, London, 1973.

86. Lumley, J.L. Lecture Series No. 76, Von Karman Inst., Rhode-St-Genese, Belgium, 1975.

87. Launder, B.E.   Lecture Series No. 76, Von Karman Inst., Rhode-St-Genese, Belgium, 1975.
88. Tavoularis, S. and Corrsin, S.C.   Int. J. Heat Mass Trans. 28, 265, 1985.
89. Speziale, C.G.   Geophys. Astrophys. Fluid Dynamics 23, 69-84, 1983.
90. Ettestad, D. and Lumley, J.L.   "Parameterization of turbulent transport in swirling flows - I. Theoretical considerations" in Turbulent Shear Flows-4 (ed. L.J.S. Bradbury et al), pp. 87-101, Springer, Heidelberg, 1985.
91. Dekeyser, I. and Launder, B.E.   "A comparison of triple-moment temperature-velocity correlations in the asymmetric heated jet with alternative closure models" in Turbulent Shear Flows-4 (ed. L.J.S. Bradbury et al), pp. 102-117, Springer, Heidelberg, 1985.
92. Gilbert, B.L.   "Detailed turbulence measurements in a two-dimensional upwash," AIAA Paper 83-1678, 1983.
93. André, J.C., de Moor, G., Lacarrère, P., Therry, G. and du Vachat, R.   "The clipping approximation and inhomogeneous turbulence simulations" in Turbulent Shear Flows-1 (ed. F. Durst et al), pp. 307-318, Springer, Heidelberg, 1979.
94. Pope, S.B.   AIAA J. 16, 279-281, 1978.
95. Hanjalić, K. and Launder, B.E.   J. Fluids Eng. 102, 34-40, 1980.
96. Bardina, J., Ferziger, J. and Reynolds, W.C.   "Improved turbulence models based on large-eddy simulation of homogeneous, incompressible, turbulent flows," Mech. Eng. Dept. Report TF-19, Stanford University, 1983.
97. Aupoix, B., Cousteix, J. and Liandrat, J.   "Effects of rotation on isotropic turbulence," Proc. 4th Symp. on Turbulent Shear Flows, pp. 9.7-9.12, Karlsruhe, 1983.
98. Lumley, J.L. and Khajeh-Nouri, B.J.   "Computational modeling of turbulent transport" (Proc. 2nd IUGG-IUTAM Symp. on Atmospheric Diffusion in Environmental Pollution), Advances in Geophysics 18A, 169, 1974.
99. Morse, A.P.   "Axisymmetric free shear flows with and without swirl," PhD Thesis, Faculty of Engineering, University of London, 1980.
100 Ince, N.Z. and Launder, B.E.   "The prediction of buoyancy-modified turbulence with a new second-moment closure."   Paper accepted for 7th Symp. on Turbulent Shear Flows, Stanford, 1989.
101 Rodi, W.   Proc. 2nd Symp. on Turbulent Shear Flows, pp. 10.37-10.42, London, 1979.
102 Wygnanski, I., Champagne, F. and Marasli, B.   J. Fluid Mech. 168, 31, 1986.
103 Rodi, W. and Scheuerer, G.   Proc. 4th Symp. on Turbulent Shear Flows, pp. 2.8-2.14, Karlsruhe, 1983.
104 Launder, B.E.   ASME J. Heat Trans. 110, 1112-1128, 1988.
105 Ince, N.Z.   Personal communication, 1988.
106 Shih, T-H and Lumley, J.L.   Phys. Fluids 29, 971-975, 1986.
107 Tselepidakis, D.P.   Personal communication, 1988.
108 Kim, J., Moin, P. and Moser, R.D.   J. Fluid Mech. 177, 133-166, 1987.
109 Moser, R.D. and Moin, P.   "Direct numerical simulation of curved channel flow," NASA Report TM 85974, 1984.
110 Spalart, P.   J. Fluid Mech. 172, 307, 1986.
111 Prud'homme, M. and Elghobashi, S.   "Prediction of wall-bounded turbulent flows with an improved version of a Reynolds stress model," Proc. 4th Symp. on Turbulent Shear Flows, pp. 1.7-1.12, Karlsruhe, 1983.
112 Shima, N.   J. Fluids Eng. 10, 38-44, 1988.
113 Launder, B.E. and Shima, N.   "A second-moment-closure study for the near-wall sublayer: development and application."   To appear in AIAA J., 1989.

114 Lai, Y.G. and So, R.M.C. "On near-wall turbulent flow modelling." Submitted for publication.

115 Launder, B.E. and Tselepidakis, D.P. "Contribution to the modelling of sublayer turbulent transport," Proc. Zoran Zarić Memorial Conf. on Wall Turbulence, Dubrovnik, 1988 (to be published by Hemisphere).

116 Simpson, R.L. and Wallace, D.B. "Laminarescent turbulent boundary layers: experiments on sink flows," Project SQUID, Tech. Report. SMU-1-PU, 1985.

117 Launder, B.E. and Reynolds, W.C. Phys. Fluids 26, 1157-1158, 1983.

118 Kebede, W., Launder, B.E. and Younis, B.A. "Large-amplitude period pipe flow: a second-moment-closure study," Proc. 5th Symp. on Turbulent Shear Flows, pp. 16.23-16.28, Cornell, 1985.

119 Gerz, T., Schumann, U. and Elghobashi, S.E. To appear in J. Fluid Mech., 1989.

120 Launder, B.E. Chapter 6: "Heat and Mass Transport" in Turbulence (ed. P. Bradshaw), Springer, Heidelberg, 1976.

121 Nemouchi, Z. Personal communication, 1988.

122 Rebollo, R.M. "Analytical and experimental investigation of a turbulent mixing layer of different gases in a pressure gradient," PhD Thesis, Cal. Inst. Tech., 1973.

123 Shih, T-H, Lumley, J.L. and Janicka, J. J. Fluid Mech. 180, 93-116, 1987.

124 Mansour, N.N., Kim, J. and Moin, P. J. Fluid Mech. 194, 15-44, 1988.

125 Leschziner, M.A. "Modelling turbulent recirculating flows by finite-volume methods," Invited Paper, 3rd Int. Symp. on Refined Flow Modelling and Turbulence Measurement (ed. Y. Iwasa et al), IAHR, 1988.

126 Kim, J. and Moin, P. "Transport of passive scalars in a turbulent channel flow," Proc. 6th Symp. on Turbulent Shear Flows, Paper 5.2, Toulouse, 1987.

127 ElBaz, A. Personal communication, 1989.

128 Bonnet, J-P. Summary Report of Computor Group 174, Proc. 1980/81 AFOSR-HTTM-Stanford Conference on Complex Turbulent Flows (ed. S.J. Kline et al), Vol. III, pp. 1407-1410, Stanford, 1982.

129 Lin, A. and Wolfshtein, M. "Theoretical study of the Reynolds stress equations" in Turbulent Shear Flows-1 (ed. F. Durst et al), 327-343, Springer, Heidelberg, 1979.

130 Hanjalić, K., Launder, B.E. and Schiestel, R. "Multiple-time-scale concepts in turbulent transport modelling in Turbulent Shear Flows-2 (ed. L.J.S. Bradbury et al), 36-49, Springer, Heidelberg, 1980.

131 Cler, A. Thèse Docteur Ingénieur, Ecole Nat. Sup. Aero. Espace, Toulouse, 1982.

132 Schiestel, R. Jo. de Méc. Th. et Appl. 2, 417-449, 1983.

133 Schiestel, R. Phys. Fluids 30, 722-731, 1987.

134 Donaldson, C. du P. and Sandri, G. "On the inclusion of eddy structure in second order closure models of turbulent flow," AGARD Report 1982.

135 Sandri, G. and Cerasoli, C. "Fundamental research in turbulent modelling," Aero. Res. Assoc. of Princeton, Inc., Report 438, 1981.

136 Schiestel, R. "Sur un nouveau modèle de turbulence appliqué au transfert de quantité, de mouvement et de chaleur," Thèse Docteur és Sciences, Université de Nancy, 1974.

137 Wu, C-T, Ferziger, J. and Chapman, D.R. Proc. 5th Symp. on Turbulent Shear Flows, pp. 17.13-17.19, Cornell, 1985.

138 Libby, P.A. J. Fluid Mech. 68, 273, 1975.

139 Janicka, J. and Kollmann, W. Proc. 4th Symp. on Turbulent Shear Flows, pp. 14.13-14.17, Karlsruhe, 1983.

140 Pope, S.B. Prog. Energy Combustion Sci. 11, 119, 1985.

Phenomenological Modeling: Present and Future

Comment 1

A. Roshko

Graduate Aeronautical Laboratories
California Institute of Technology
Pasadena, CA 91125, USA

Professor Launder's position paper on phenomenological modelling is an impressive survey and valuable account of the status of second-moment closure, principally as applied to Reynolds Stresses. In this respect, it supplements and updates the monograph of Professor Rodi (1980), in which the emphasis is on the status of first-order closures, in particular the $\kappa - \varepsilon$ model, as of 10 years ago. Between them the two works provide an excellent reference source containing the equations; the rationale for modelling decisions that are made; tables of the constants that have been selected; displays of flow computations and their comparison with experimental measurements for a varied number of flows; and extensive reference lists.

More than simply reference sources, the two works are critical accounts of difficulties, limitations and continuing efforts to improve performance. One is impressed by the effort, sophistication and ingenuity that is exercised in developing and in working over the equations for the turbulent correlations in order to discover the most effective and most economical ways to introduce the empiricism (the selection and evaluation of the constants) while satisfying various constraints; appropriate formulation of the latter is no small part of the overall exercise.

The ten examples chosen by Launder for assessing the performance of second-order closure with respect to experiment, and for comparison with the $\kappa - \varepsilon$ model, argue persuasively for the superior performance of second-order modelling, when one includes in the assessment the distributions of all quantities of interest, not just a single one, say a thickness distribution. It would be of interest and significance to know whether the contrary occurs, i.e. whether, for some flows, the overall performance of a first order model is better. It would be significant because one supposes that in second-order closure it is not just the larger number of constants that permits better "embodiment of experience" (one could have just as many constants at first order) but, rather, that accuracies and uncertainties in modelling the higher order correlations will be smoothed out and diminished in the integration to the lower order correlations (here, the shear stress) and thus a broader class of flows can be modelled. By this logic, third-order closures would perform even better, and so on. However, as implied at the beginning of this statement, it is not really clear whether such "convergence" always occurs. Are there counterexamples? Professor Cantwell, in the Introduction to his position paper, cites an observation from the 1980-81 Conference (Kline, Cantwell and Lilly, 1981) that "model effectiveness did not necessarily increase with increasing model complexity". In any case, there is probably little stomach, incentive or need to proceed to higher order closures.

It was probably expected that I would make some remarks, in the spirit of the Workshop, about traditional "turbulent modelling"[1] from the viewpoint of an "eddy chaser". However there is little for me to add to the views on this which are already so well expressed in some of the papers that have preceded mine, in particular the position papers of Narasimha and of Cantwell, and the discussion paper by Bridges, Husain and Hussain. In the following I will enlarge on only one aspect, call it "sensitivity" or "controllability" of turbulent shear flows, on which all these authors make some comment.

It is useful to preface the discussion by asking what accuracy we should expect from turbulence models, i.e. what degree of agreement between model predictions and measured flows. There can be little argument that, at present, agreement is at best at the 10% level, possibly worse for free turbulent shear flows. Can one reasonably expect better performance? The question is somewhat academic because the flows themselves are not *defined* to better precision. This is apparent, for example, for the mixing layer, for which the spreading rate and the level of Reynolds stress are uncertain to this degree (Brown & Roshko, 1974; Weisbrot et al, 1982). It is not a question of measurement accuracy but rather of flow *definition*, and it implies an incomplete understanding of factors affecting the flow. This is relevant to both experiment and modelling.

The point can be illustrated indirectly by commenting on recent progress in defining the relation between the Reynolds number and the Stouhal number for vortex shedding from a cylinder of circular cross section $S = S \ (Re)$, a problem that is over 100 years old. Judging from published values during just the past 20 years or so,

---

[1]The term "turbulence modelling" seems to have been pre-empted for modelling of the time-averaged Navier-Stokes equations. In fact, after the basic, unsteady N-S equations, all other approaches to the turbulence problem are turbulence modelling.

definition of this relation was at the 10% level even though the quantities involved (U, d, v and f) can rather easily be measured to better than 1% accuracy. The improvement came about from a series of events[1], that could not have been anticipated in any workshop, beginning with a paper of Sreenivasan (1985), leading to one by Van Atta and Gharib (1987), and then to the paper by Williamson (1988) in which $S(Re)$ at the 1% level was finally obtained, for the so-called "laminar" vortex shedding range, $50 < Re < 180$. The resolution came about from new understanding of relevant flow phenomena which made possible a correct flow definition. The impact of this on modelling the flow by direct simulation was that, whereas previous Direct Simulations were all pretty much in the 10% experimental band, now only some of them are at the 1% level. In DS "modelling" there is no limitation from model empiricism, only from computational adequacy. There may however also be problems of *definition*, known to computational researchers as problems of "inflow" and "outflow" conditions.

The point I want to make, concerning one role of coherent structures, is that they have led us to the understanding that free turbulent shear flows, are sensitive to influences that have so far been ignored, or not even anticipated, in traditional modelling of them,[2] derived from a long held view that these flows are robust, self generating, unique. But experiments like those of Crow and Champagne (1971) and of Oster and Wygnanski (1982), which demonstrate the *controllability* of these flows by small, spectrally sharp excitation, suggest also that under ordinary flow conditions they may be sensitive to and controlled by the broad spectral noise which always exists in experimental installations. The broad spectral content does not produce the anomalous effects of the spectrally sharp excitation but might be sufficiently different in different experiments to account for the 10% lack of definition.

Relevant to this are fundamental questions as to whether the turbulent, convectively unstable free shear layer is self generating, i.e. self sustaining via Biot-Savart feedback, or whether it is mainly responding to broad spectral excitation. It is probably the former when external forcing is very low, but in that case feedback from downstream geometry may have an effect (cf the observations of Dimotakis and Brown, 1976). A fuller discussion of these questions (Roshko, 1989) will appear in the Proceedings of the recent Grenoble Conference on Organized Structures and Turbulence in Fluid Mechanics. An earlier, very useful discussion appears in the work of Kaul (1988).

---

[1]These are described in Williamson, 1988.
[2]Bridges et al. in their discussion paper give a number of examples in which CS has led to new understanding which is important and useful even if not yet quantitatively incorporated into models.

The implications of these questions are far from being clear and settled. Their relevance in this discussion is to the expectations and limitations of modelling. At the 10% level, the present state of the art looks quite good, as exemplified in this position paper by Launder. In fact, at this stage it may be academic to argue about the physics which is or is not represented. At the 1% level it might be quite anther story. Science is replete with examples where improvement in precision of definition as well as accuracy of measurement have advanced science.

References

Crow, S.C. and Champagne, F.H. (1971). Orderly structure in jet turbulence. J. Fluid Mech. **48**, 549-591.

Dimotakis, P.E. and Brown, G.L. (1976). The mixing layer at high Reynolds number: large-structure dynamics and entrainment. J. Fluid Mech. **78**, 535-560.

Kaul, U.K. (1988). Do large structures control their own growth in a mixing layer? An assessment. J. Fluid Mech. **190**, 427-450.

Oster, D. and Wygnanski, I. (1982). The forced mixing layer between parallel streams. J. Fluid Mech. **123**, 91-130.

Rodi, W. (1980). *Turbulence Models and their Application in Hydraulics*. International Association for Hydraulic Research.

Roshko, A. (1989). Problems in defining turbulent mixing layers. *In Organized Structures and Turbulence in Fluid Mechanics*. International Conference, Grenoble, France, 18-21 September 1989.

Sreenivasan, K.R. (1985). Transition and turbulence in fluid flows and low-dimensional chaos. In *Frontiers in Fluid Mechanics*, (ed. S.H. Davis and J.L. Lumley) Springer 41-67.

Van Atta, C.W. and Gharib, M. (1987). Ordered and chaotic vortex streets behind circular cylinders at low Reynolds number. J. Fluid Mech. **174**, 113-133.

Weisbrot, I., Einav, S. and Wygnanski, I. (1982). The non unique spread of the two-dimensional mixing layer. Phys. Fluids **25**, 1691-1693.

Williamson, C.H.K. (1988). Defining a universal and continuous Strouhal-Reynolds number relationship for the laminar vortex shedding of a circular cylinder. Phys. Fluids **31**, 2742-2744.

Turbulence Modeling: Present and Future

Comment 2.

Charles G. Speziale

Institute for Computer Applications in Science and Engineering
NASA Langley Research Center
Hampton, VA 23665

1. Introductory Remarks

The thrust of the position paper by Launder is that second-order closure models represent the best hope for the reliable prediction of the complex turbulent flows of technological interest both now and in the foreseeable future. By building on the pioneering research of Rotta [1] and by introducing some fundamental new ideas, the work of Launder, Lumley, and others has without doubt made significant contributions to the advancement of second-order closures. In the position paper by Launder, a strong case is made for the superior predictive capabilities of second-order closures in comparison to two-equation models or eddy viscosity models. Most notably, turbulent flows involving rotations and streamline curvature have been shown by Launder and others [2,3] to be better described by second-order closure models. The same is true for turbulent flows with stratification and relaxation effects. Launder very aptly cites four active areas of research for the improvement of second-order closures: (i) models for the rapid pressure strain correlation, (ii) models for the turbulent diffusion terms, (iii) adjustments for near wall turbulence effects, and (iv) modeled transport equations for the turbulent dissipation rate or length scale.

In the sections to follow, by making use of some simple examples from homogeneous turbulence, the primary point made by Launder concerning the superior predictive capabilities of second-order closures will be amplified. Most notably, it will be shown that second-order closures are capable of describing the stabilizing or destabilizing effect of rotations on shear flow – a problem which cannot be even remotely analyzed by the simpler models. However, some lingering problems concerning the development of adequate models for the rapid pressure-strain correlation and the turbulent length scale will be emphasized (see Speziale [4]). In regard to the latter issue, the strengths and weaknesses of the commonly used modeled dissipation rate transport equation will be discussed and a definitive argument will be put forth as to why previous attempts at the development of an improved dissipation rate transport equation have failed. Alternative approaches based on a tensor length scale will also be discussed along with the author's views concerning the prospects for future research.

2. The Case for Second-Order Closure Models

The commonly used eddy viscosity models and two-equation models have three major deficiencies [5,6]:

(i)   the inability to properly account for rotational strains,
(ii) the inaccurate prediction of normal Reynolds stress anisotropies,
(iii) the inability to account for component Reynolds stress relaxation and amplification effects.

In so far as point (i) is concerned, it should be noted that the K-ε model is oblivious to the presence of rotational strains (e.g., it fails to distinguish between the physically distinct cases of plane strain, plane shear, and rotating plane shear). Other commonly used algebraic eddy viscosity models such as the Baldwin-Lomax model are also fundamentally incapable of describing the effect of rotations on sheared or strained turbulent flows [2,6]. As alluded to in point (ii), all eddy viscosity models, including the K-ε model, yield highly inaccurate predictions for the normal Reynolds stress anisotropies in simple turbulent shear flows. This makes it impossible to describe a variety of secondary flow phenomena (e.g., the K-ε model erroneously predicts that there are *unidirectional* mean turbulent flows in non-circular ducts in contradiction to experiments which indicate the presence of an additional secondary flow [5]; see Figure 1). These problems can be partially overcome by the use of two-equation turbulence models with a nonlinear algebraic Reynolds stress model (see Launder and Ying [7], Rodi [8], and Speziale [5]), but only for turbulent flows that are nearly in equilibrium. Non-equilibrium turbulent flows that have a spatially or temporally evolving structure (e.g., the flows with relaxation or amplification effects mentioned in point (iii)) cannot, in general, be described properly by two-equation models. For example, in an initially anisotropic turbulence, where at some time $t = t_0$ the mean velocity gradients are set to zero, the K-ε model erroneously predicts an instantaneous return to isotropy wherein

$$\tau_{ij} = -\frac{2}{3}K\delta_{ij}, \quad t \geq t_0 \tag{1}$$

or equivalently,

$$b_{ij} = 0, \quad t \geq t_0 \tag{2}$$

given that K is the turbulent kinetic energy, $\tau_{ij} \equiv -\overline{u_i u_j}$ is the Reynolds stress tensor, and $b_{ij} \equiv -(\tau_{ij} + \frac{2}{3}K\delta_{ij})/2K$ is the anisotropy tensor. In considerable contradiction to (1), experiments indicate that there is a very gradual return to isotropy – an effect that can be characterized much better by second-order closure models. In Figure 2, the temporal evolution of the second invariant of the anisotropy tensor II is shown corresponding to a relaxation from the plane strain experiment of Choi and Lumley [9]. From this graph, it is clear that the second-order closure model (which is based on the Rotta model for the slow pressure-strain correlation) does a reasonably good job in reproducing the experimental trends unlike the K-ε model which erroneously predicts that $II = 0$ for dimensionless time $\tau \geq 0$.

Now, the greater predictive capabilities of second-order closure models will be demonstrated by a simple, but non-trivial, example which is not often discussed in the turbulence modeling literature. The problem to be considered is homogeneous turbulent shear flow in a rotating frame (see Figure 3). This problem constitutes a non-trivial test of turbulence models since it involves arbitrary combinations of shear and rotation which can have either a stabilizing or destabilizing effect.

For any homogeneous turbulent flow, the standard K-ε model takes the general form [10]

$$\tau_{ij} = -\frac{2}{3}K\delta_{ij} + C_\mu \frac{K^2}{\varepsilon}\left(\frac{\partial \bar{v}_i}{\partial x_j} + \frac{\partial \bar{v}_j}{\partial x_i}\right) \tag{3}$$

$$\dot{K} = \tau_{ij}\frac{\partial \bar{v}_i}{\partial x_j} - \varepsilon \tag{4}$$

$$\dot{\varepsilon} = C_{\varepsilon 1}\frac{\varepsilon}{K}\tau_{ij}\frac{\partial \bar{v}_i}{\partial x_j} - C_{\varepsilon 2}\frac{\varepsilon^2}{K} \tag{5}$$

in all frames of references independent of whether or not they are inertial. In (3)-(5), $\bar{v}_i$ is the mean velocity field, ε is the turbulent dissipation rate, and $C_\mu, C_{\varepsilon 1}$ and $C_{\varepsilon 2}$ are constants which assume the values of 0.09, 1.44, and 1.92, respectively. For homogeneous turbulent shear flow in a rotating frame (as shown in Figure 3), the mean velocity gradient tensor is given by

$$\frac{\partial \bar{v}_i}{\partial x_j} = \begin{pmatrix} 0 & S & 0 \\ 0 & 0 & 0 \\ 0 & 0 & 0 \end{pmatrix} \tag{6}$$

and $\Omega_i = (0,0,\Omega)$ is the rotation rate of the reference frame relative to an inertial framing. Since (3)-(5) are independent of Ω, the standard K-ε model predicts the *same results for all rotation rates* and, hence, does not distinguish between turbulent shear flow in an inertial frame and rotating turbulent shear flow. Speziale and Mac Giolla Mhuiris [11] recently showed that the K-ε model has the following equilibrium solution for rotating shear flow:

$$(b_{11})_\infty = 0, \quad (b_{12})_\infty = -\frac{1}{2}(C_\mu \alpha)^{\frac{1}{2}}, \quad (b_{13})_\infty = 0 \tag{7}$$

$$(b_{22})_\infty = 0, \quad (b_{23})_\infty = 0, \quad (b_{33})_\infty = 0 \tag{8}$$

$$\left(\frac{SK}{\varepsilon}\right)_\infty = \left(\frac{\alpha}{C_\mu}\right)^{\frac{1}{2}} \tag{9}$$

where $\alpha = (C_{e2} - 1)/(C_{e1} - 1)$ and $(\cdot)_\infty$ denotes the equilibrium value obtained in the limit as $t \to \infty$. These equilibrium values are universal, i.e., are completely independent of the initial conditions, the shear rate, and the rotation rate. It was also shown in [11] that the long time solutions (corresponding to $t^* \equiv St \gg 1$) for the kinetic energy and dissipation rate in the K-ε model grow exponentially:

$$K \sim \exp\left[\sqrt{\frac{C_\mu}{\alpha}}(\alpha - 1)t^*\right] \tag{10}$$

$$\varepsilon \sim \exp\left[\sqrt{\frac{C_\mu}{\alpha}}(\alpha - 1)t^*\right]. \tag{11}$$

Hence, the K-ε model predicts the following physical picture for rotating shear flow: the turbulent kinetic energy and dissipation rate grow exponentially in time at a comparable rate; the anisotropy tensor b_{ij} and shear parameter SK/ε approach a universal equilibrium. While this characterization is qualitatively correct for pure shear flow (see Tavoularis and Corrsin [12]), it is quite incorrect for most values of Ω/S in rotating shear flow. Linear stability analyses and numerical simulations of the Navier-Stokes equations indicate that for values of the rotation rate that are discernibly outside of the range $0 \leq \Omega/S \leq 0.5$, the flow undergoes a restabilization wherein K and $\varepsilon \to 0$ as $t \to \infty$ (see Bardina, Ferziger, and Reynolds [13] and Bertoglio [14]).

It will now be demonstrated that, unlike the commonly used two-equation models, second-order closures are able to describe the stabilizing or destabilizing effect of rotations on turbulent shear flow. Speziale and Mac Giolla Mhuiris [11] recently considered a fairly general class of second-order closure models of the form

$$\dot{\tau}_{ij} = -\tau_{ik}\frac{\partial \bar{v}_j}{\partial x_k} - \tau_{jk}\frac{\partial \bar{v}_i}{\partial x_k} - \Pi_{ij} + \frac{2}{3}\varepsilon\delta_{ij} - 2(\tau_{ik}\varepsilon_{mkj}\Omega_m + \tau_{jk}\varepsilon_{mki}\Omega_m) \tag{12}$$

$$\dot{\varepsilon} = C_{e1}\frac{\varepsilon}{K}\tau_{ij}\frac{\partial \bar{v}_i}{\partial x_j} - C_{e2}\frac{\varepsilon^2}{K} \tag{13}$$

for any rotating homogeneous turbulence. In (12)-(13), Π_{ij} denotes the pressure-strain correlation which is assumed to be of the general form

$$\Pi_{ij} = \Pi_{ij}(\tau_{ij}, \frac{\partial \bar{v}_i}{\partial x_j}, \varepsilon) \tag{14}$$

493

and $C_{\varepsilon 1}$ and $C_{\varepsilon 2}$ are either constants or functions of the invariants of b_{ij}. This class of second-order closures encompasses a wide variety of models including the simplified form of the Launder, Reece, and Rodi model for which

$$\Pi_{ij} = C_1 \frac{\varepsilon}{K}(\tau_{ij} + \frac{2}{3}K\delta_{ij}) - C_2 \left[\tau_{ik}\left(\frac{\partial \bar{v}_j}{\partial x_k} + \varepsilon_{mkj}\Omega_m\right)\right.$$

$$\left. +\tau_{jk}\left(\frac{\partial \bar{v}_i}{\partial x_k} + \varepsilon_{mki}\Omega_m\right)\right] + \frac{2}{3}C_2\tau_{mn}\frac{\partial \bar{v}_m}{\partial x_n}\delta_{ij} \tag{15}$$

and $C_1 = 1.8$, $C_2 = 0.6$, $C_{\varepsilon 1} = 1.44$ and $C_{\varepsilon 2} = 1.92$. It was shown by Speziale and Mac Giolla Mhuiris [11] that this class of second-order closure models has two-equilibrium solutions for rotating shear flow: one where

$$\left(\frac{\varepsilon}{SK}\right)_\infty = 0 \tag{16}$$

which exists for *all* Ω/S and one where

$$\left(\frac{\varepsilon}{SK}\right)_\infty = \gamma_0 \left[\gamma_1 + \gamma_2\left(\frac{\Omega}{S}\right) - \left(\frac{\Omega}{S}\right)^2\right]^{\frac{1}{2}} \tag{17}$$

which exists for a small intermediate band of Ω/S which can range from $-0.1 \leq \Omega/S \leq 0.6$ (here, γ_0, γ_1, and γ_2 are directly related to the constants of the model). The former equilibrium solution for which $(\varepsilon/SK)_\infty = 0$, predominantly is connected with solutions wherein K and ε undergo a power law decay in time; the latter equilibrium solution (17), where $(\varepsilon/SK)_\infty$ is nonzero, is connected with solutions where K and ε grow exponentially in time at the same rate. In this intermediate band of Ω/S, these two solutions exchange stabilities in a fashion that qualitatively mimics the shear instability with its exponential time growth of disturbance kinetic energy.

In Figure 4(a), a bifurcation diagram is shown for the Launder, Reece, and Rodi model. This bifurcation structure qualitatively mimics the stabilizing or destabilizing effects of rotations on turbulent shear flow as discussed above. In stark contrast to the bifurcation that is properly predicted by the second-order closure, the equilibrium diagram for the K-ε model shows the erroneous prediction of a universal value for $(\varepsilon/SK)_\infty$ which is completely independent of Ω/S (see Figure 4(b)). As mentioned earlier, this universal equilibrium solution for the K-ε model corresponds to an unstable flow wherein K and ε grow exponentially in time.

In addition to yielding a superior qualitative description of the equilibrium structure of rotating turbulent shear flows, the quantitative values of the equilibrium states for pure shear flow predicted by the second-order closures are also substantially better than those obtained from the commonly used two-equation models. To illustrate this superiority of the second-order closures, the equilibrium values for b_{ij} and SK/ε obtained from the Launder, Reece, and Rodi model and the K-ε model are compared in Table 1 with the experimental results of Tavoularis and Corrsin [12] for homogeneous turbulent shear flow.

In Figures 5(a)-(c), the time evolution of the turbulent kinetic energy predicted by the Launder, Reece, and Rodi model and the K-ε model are compared with results from the large-eddy simulations of Bardina, Ferziger, and Reynolds [13] for three rotation rates: $\Omega/S = 0$, $\Omega/S = 0.25$, and $\Omega/S = -0.5$. A direct comparison of Figure 5(a) and Figure 5(b) with the large-eddy simulations shown in Figure 5(c), graphically demonstrates the superior capability of second-order closure models in predicting the stabilizing or destabilizing effect of rotations on shear flow.

3. Needed Modeling Improvements in Second-Order Closures

While the author is in full agreement with the main points of the Launder position paper concerning the superior capabilities of second-order closure models, it must be cautioned that these models have not yet matured to the point where reliable predictions can be made for a *variety* of complex turbulent flows. Several areas where improvements are needed (some of which were pointed out by Professor Launder), will be discussed in more detail in this section.

Second-order closure models are based on the Reynolds stress transport equation which takes the exact form

$$\frac{\partial \tau_{ij}}{\partial t} + \bar{v}_k \frac{\partial \tau_{ij}}{\partial x_k} = -\tau_{ik} \frac{\partial \bar{v}_j}{\partial x_k} - \tau_{jk} \frac{\partial \bar{v}_i}{\partial x_k} + \frac{\partial C_{ijk}}{\partial x_k} - \Pi_{ij} + \varepsilon_{ij} + \nu \nabla^2 \tau_{ij} \tag{18}$$

where

$$C_{ijk} \equiv \overline{u_i u_j u_k} + \overline{p u_i} \delta_{jk} + \overline{p u_j} \delta_{ik} \tag{19}$$

$$\Pi_{ij} \equiv \overline{p \left(\frac{\partial u_i}{\partial x_j} + \frac{\partial u_j}{\partial x_i} \right)} \tag{20}$$

$$\varepsilon_{ij} \equiv 2\nu \overline{\frac{\partial u_i}{\partial x_k} \frac{\partial u_j}{\partial x_k}} \tag{21}$$

are the third-order diffusion correlation, the pressure-strain correlation, and the dissipation rate correlation, respectively (ν is the kinematic viscosity of the fluid). In order for closure to be achieved at this "second moment" level (which forms the raison d'etre of second-order modeling), models must be developed wherein the higher-order correlations C_{ijk}, Π_{ij} and ε_{ij} are taken to be functionals of the Reynolds stress τ_{ij}, mean velocity gradients $\partial \bar{v}_i/\partial x_j$ and some length scale of turbulence Λ. The Reynolds stress is decomposed into isotropic and deviatoric parts as follows

$$\tau_{ij} = -\frac{2}{3}K\delta_{ij} - 2Kb_{ij} \tag{22}$$

and the turbulence length scale Λ is usually assumed to be of the form

$$\Lambda = C^* \frac{K^{\frac{3}{2}}}{\varepsilon} \tag{23}$$

where C^* is a dimensionless constant and $\varepsilon \equiv \frac{1}{2}\varepsilon_{ii}$ is the turbulent dissipation rate. Hence, consistent with the use of (23), the higher-order correlations can be taken to be functionals of $b_{ij}, \partial \bar{v}_i/\partial x_j, K$, and ε instead.

Typically, the third-order diffusion correlation is modeled by a gradient transport hypothesis wherein it is assumed that C_{ijk} is of the general form

$$C_{ijk} = C_{ijklmn}(\mathbf{b}, K, \varepsilon)\frac{\partial \tau_{lm}}{\partial x_n}. \tag{24}$$

Motivated by analyses based on homogeneous turbulence [15], virtually all of the commonly used models for the pressure-strain correlation are assumed to be of the form

$$\overline{p\left(\frac{\partial u_i}{\partial x_j} + \frac{\partial u_j}{\partial x_i}\right)} = \varepsilon A_{ij}(\mathbf{b}) + KM_{ijkl}(\mathbf{b})\frac{\partial \bar{v}_k}{\partial x_l} \tag{25}$$

where the first term on the right-hand-side of (25) represents the slow pressure strain while the second term represents the rapid pressure strain. Typically, the turbulence dissipation correlation ε_{ij} is assumed to be of the general form

$$\varepsilon_{ij} = \frac{2}{3}\varepsilon\delta_{ij} + f_s\varepsilon b_{ij} \tag{26}$$

where f_s is taken to be dimensionless constant or a function of the invariants of b_{ij}. For high Reynolds number turbulent flows that are sufficiently far from solid boundaries, f_s is taken to be zero. In order to achieve closure of the Reynolds stress transport equation, (24)-(26) must be supplemented with a modeled transport equation for the turbulent dissipation rate which is of the general form

$$\frac{\partial \varepsilon}{\partial t} + \bar{v}_k\frac{\partial \varepsilon}{\partial x_k} = \nu\nabla^2\varepsilon + \mathcal{P}_\varepsilon - \Phi_\varepsilon + \mathcal{D}_\varepsilon \tag{27}$$

where $\mathcal{P}_\varepsilon, \Phi_\varepsilon$ and $\mathcal{D}_\varepsilon$ represent the production, dissipation, and turbulent diffusion of ε.

Most of the existing second-order closure models can be constructed by expanding the unknown tensor coefficients on the right-hand-sides of (24)-(25) in a Taylor series in

b_{ij} (subject to the symmetry properties of C_{ijk} and Π_{ij}). The older models are actually first-order Taylor expansions in b_{ij}; for example in the Launder, Reece, and Rodi model [16]:

$$C_{ijk} = -C_s \frac{K}{\varepsilon}(\tau_{i\ell}\frac{\partial \tau_{jk}}{\partial x_\ell} + \tau_{j\ell}\frac{\partial \tau_{ik}}{\partial x_\ell} + \tau_{k\ell}\frac{\partial \tau_{ij}}{\partial x_\ell}) \qquad (28)$$

$$\Pi_{ij} = -C_1\varepsilon b_{ij} + C_2 K\overline{S}_{ij} + C_3 K(b_{ik}\overline{S}_{jk}$$

$$+ b_{jk}\overline{S}_{ik} - \tfrac{2}{3}b_{mn}\overline{S}_{mn}\delta_{ij}) + C_4 K(b_{ik}\overline{W}_{jk} + b_{jk}\overline{W}_{ik}) \qquad (29)$$

where $\overline{S}_{ij} \equiv \frac{1}{2}(\partial \overline{v}_i/\partial x_j + \partial \overline{v}_j/\partial x_i)$ and $\overline{W}_{ij} = \frac{1}{2}(\partial \overline{v}_i/\partial x_j - \partial \overline{v}_j/\partial x_i)$ are the mean rate of strain tensor and vorticity tensor, respectively. In the simplified version of the Launder, Reece, and Rodi model (which is now referred to as the "Basic Model" by Launder and his co-workers), $C_1 = 1.8$, $C_2 = 0.8$, $C_3 = 1.2$, $C_4 = 1.2$, and $C_s = 0.11$.

The modeled terms in the dissipation rate transport equation (27) are typically based on the assumption that the production (or dissipation) of the turbulent dissipation is proportional to the production (or dissipation) of the turbulent kinetic energy. A gradient transport hypothesis is typically invoked for the turbulent diffusion term on the right-hand-side of (27). These assumptions give rise to a modeled transport equation for the turbulent dissipation rate that takes the general form

$$\frac{\partial \varepsilon}{\partial t} + \overline{v}_i \frac{\partial \varepsilon}{\partial x_i} = C_{\varepsilon 1}\frac{\varepsilon}{K}\tau_{ij}\frac{\partial \overline{v}_i}{\partial x_j} - C_{\varepsilon 2}\frac{\varepsilon^2}{K}$$

$$-\frac{\partial}{\partial x_i}(C_\varepsilon \frac{K}{\varepsilon}\tau_{ij}\frac{\partial \varepsilon}{\partial x_j}) + \nu\nabla^2\varepsilon. \qquad (30)$$

In the Launder, Reece, and Rodi model, $C_{\varepsilon 1}$, $C_{\varepsilon 2}$, and C_ε are taken to be constants which assume the values of 1.44, 1.92, and 0.15, respectively. Some more recent models have taken $C_{\varepsilon 1}$ and $C_{\varepsilon 2}$ to be functions of some subset of the invariants of b_{ij} and $\partial \overline{v}_i/\partial x_j$; these newer models will be discussed in more depth in the next section.

Now, with the aid of this background material, the four active areas of research for the development of improved second-order closure models that were mentioned in the introductory remarks can be elaborated on. These areas are as follows:

(i) The development of improved models for the pressure-strain correlation of turbulence which account for nonlinear anisotropic effects. Since for most flows of engineering interest, $\|b\| \sim 0.2$, the use of a first-order Taylor expansion in **b** for A_{ij} and $M_{ijk\ell}$ is highly questionable. In fact, nonlinear terms in the model for A_{ij} are needed to predict the curved trajectories that occur in the phase space of the return to isotropy problem as shown in Figure 6 (see Choi and Lumley [9] and Sarkar and Speziale [17]). At least a quadratic nonlinearity in the model for M_{ijkl} is needed in order to satisfy the constraint of Material Frame Indifference (MFI) in the limit of two-dimensional turbulence (Speziale [18] and Haworth and Pope [19]). This MFI constraint is the mathematical embodiment of the well-known result that two-dimensional disturbances evolve *identically* in both a rotating frame and an inertial frame.* In addition, recent work on rotating turbulent shear flows (Speziale, Sarkar, and Gatski [20]) and Rapid Distortion Theory (Reynolds [21]) have suggested the possible need for new terms in (25) that are nonlinear in the mean velocity gradients $\partial \bar{v}_i / \partial x_j$.

(ii) The development of non-gradient transport models for the turbulent diffusion terms such as C_{ijk} need to be considered seriously. It is well-known that turbulent flows do not have a clear cut separation of scales; the largest eddies (which contain a significant portion of the turbulent kinetic energy) are of a comparable size to the geometrical scale of the flow. Consequently, one would expect a gradient transport hypothesis (which is rigorously derived as a first-order expansion in the ratio of fluctuating to mean length scales) to only constitute a crude approximation. Many difficulties in the prediction of turbulent mixing layers could be tied to this deficiency in the modeling of turbulent diffusion by means of gradient transport.

(iii) The development of asymptotically consistent near wall corrections to the turbulence models for Π_{ij}, ε_{ij} and C_{ijk} that are geometry-independent and do not have any ad hoc damping functions. Most of the commonly used corrections are either asymptotically inconsistent, geometry-dependent through an artificially imposed dependence on the unit normal to the wall, or otherwise ad hoc through the use of wall damping functions based on the turbulence Reynolds number or the distance from the wall (see Hanjalic and Launder [22]). Such empiricisms do not allow for the reliable prediction of wall transport properties (e.g., skin friction and heat transfer coefficients) that are extremely important in aerodynamic applications.

(iv) The development of improved modeled transport equations for the turbulence length scale is an issue of utmost importance. Most of the commonly used models have a scalar turbulence length scale Λ based on the dissipation ($\Lambda \propto K^{\frac{3}{2}}/\varepsilon$). There is the obvious objection that the construction of a turbulence macro-scale based on small-scale (one-point) information is conceptually wrong. Furthermore, this rather simplified definition of length scale contains *no directional information*. Although attempts have been made to develop a length scale equation based on an integral of the two-point velocity correlation tensor (Wolfshtein [23] and Donaldson and Sandri [24]), it can be shown that

---

*Consequently, the models that violate this constraint cannot be used in the analysis of geostrophic turbulence.

these specific models are equivalent to the standardly used dissipation rate transport model in the limit of homogeneous turbulence – a simplified case for which the modeled dissipation rate transport equation is already deficient. Fundamentally new research is needed on the development of modeled transport equations for some appropriate choice of integral length scales that contain the required directional information.

Finally, in regard to these four points, it should be mentioned that some significant progress has been made in the development of improved models for the rapid pressure-strain correlation by means of realizability (Lumley [25], and Shih and Lumley [26]), and the invariance considerations discussed in point (i). However, in the opinion of the author, very little progress, if any, has been made since the early 1970's in the development of better models for the turbulence length scale and diffusion effects. An examination of the modeled dissipation rate transport equation as a basis for the turbulence length scale will be discussed in the next section.

4. The Modeled Dissipation Rate Transport Equation

Now, the strengths and weaknesses of the commonly used modeled dissipation rate transport equation will be discussed. Furthermore, an attempt will be made to demonstrate at what level of approximation this model is derivable from the two-point correlation tensor which more properly contains information about the turbulent macroscale. Finally, an argument will be put forth as to why attempts at the development of improved modeled dissipation rate transport equations have failed during the past decade.

In order not to cloud the issue with the added difficulties that are associated with the integration of turbulence models to a solid boundary, the more simplified case of homogeneous turbulence will be considered. For any homogeneous turbulence, the standardly used version of the modeled dissipation rate transport equation (30) reduces to

$$\dot{\varepsilon} = C_{\varepsilon 1} \frac{\varepsilon}{K} \tau_{ij} \frac{\partial \bar{v}_i}{\partial x_i} - C_{\varepsilon 2} \frac{\varepsilon^2}{K}. \tag{31}$$

The transport equation for the turbulent kinetic energy has the exact form

$$\dot{K} = \tau_{ij} \frac{\partial \bar{v}_i}{\partial x_j} - \varepsilon \tag{32}$$

and is obtained from a contraction of (18). Equations (31)-(32) can be combined to yield a transport equation for the reciprocal turbulent time scale ε/K which is given by

$$\frac{d}{dt} \left(\frac{\varepsilon}{K} \right) = 2(1 - C_{\varepsilon 1}) \frac{\varepsilon}{K} b_{ij} \frac{\partial \bar{v}_i}{\partial x_j} + (1 - C_{\varepsilon 2}) \left(\frac{\varepsilon}{K} \right)^2. \tag{33}$$

Provided that $C_{e1} > 1$, this equation has two equilibrium solutions in the limit as $t \to \infty$:

$$\left(\frac{\varepsilon}{K}\right)_\infty = 0 \tag{34}$$

$$\left(\frac{\varepsilon}{K}\right)_\infty = -2\left(\frac{C_{e1} - 1}{C_{e2} - 1}\right)(b_{ij})_\infty \frac{\partial \overline{v}_i}{\partial x_j} \tag{35}$$

which are obtained by setting the time derivative on the left-hand-side of (33) to zero. $C_{e1} = 1$ constitutes a bifurcation point of equation (33); for $C_{e1} \leq 1$, the only realizable fixed point of (33) is $(\varepsilon/K)_\infty = 0$. The zero fixed point (34) predominantly corresponds to solutions for K and ε that undergo a power law decay in time [11] (i.e., ε and $K \to 0$ as $t \to \infty$). This branch of solutions allows for the prediction of isotropic decay as well as the flow restabilization that occurs at certain rotation rates in rotating plane shear and plane strain turbulence. Furthermore, since from (33) we have

$$\frac{d}{dt}\left(\frac{\varepsilon}{K}\right) = 0 \tag{36}$$

when $\varepsilon/K = 0$, and from (31) we have

$$\frac{d\varepsilon}{dt} = 0 \tag{37}$$

when $\varepsilon = 0$, we conclude that if $\varepsilon(0) > 0$ and $K(0) > 0$ then $\varepsilon(t) \geq 0$ and $K(t) \geq 0$ for all later times t. *Hence, the standardly used modeled ε-transport equation guarantees realizability with respect to K and ε by virtue of the fact that $\varepsilon/K = 0$ is, in dynamical systems terms, an "invariant plane."*

The non-zero fixed point (35) is associated with solutions for K and ε which grow exponentially in time [11]. More precisely, for the non-zero fixed point (35), it can be shown that

$$K \sim \exp(\lambda t), \quad \varepsilon \sim \exp(\lambda t) \tag{38}$$

for $\lambda t \gg 1$, where

$$\lambda = |2(b_{ij})_\infty \frac{\partial \overline{v}_i}{\partial x_j} + \left(\frac{\varepsilon}{K}\right)_\infty|. \tag{39}$$

This allows for the prediction of a structural equilibrium in homogeneous turbulent shear flow wherein SK/ε achieves an equilibrium value that is independent of both the initial conditions and the shear rate. Such an equilibrium for turbulent shear flows has been observed experimentally by Tavoularis and Corrsin [12] for weak to moderately strong shear rates.

It is thus clear that the two fixed points (34)-(35) of the commonly used ε-transport equation have certain properties that are crucial to the proper description of homogeneous turbulent flows. Now, it will be shown that recently proposed alterations to this modeled ε-transport equation destroyed one or the other of these key fixed points and, hence, were doomed to failure.

For example, based on the desire to account for rotational strains, Pope [27] proposed an alteration to the ε-transport equation wherein a term of the form

$$C_{\varepsilon 3} \frac{K^2}{\varepsilon} \overline{S}_{ij} \overline{W}_{jk} \overline{W}_{ki}$$

(where $C_{\varepsilon 3}$ is a constant) was added to the right-hand-side of (30). This eliminates the $(\varepsilon/K)_\infty = 0$ fixed point, for nonzero $\overline{S}_{ij}\overline{W}_{jk}\overline{W}_{ki}$, which can cause problems with realizability and can eliminate the ability to predict flow restabilizations in three-dimensional rotating shear flows. Hanjalic and Launder [28] proposed a modification wherein the term

$$-C_{\varepsilon 3} K \overline{W}_{ij}\overline{W}_{ij}$$

was added to the right-hand-side of (30) (where $C_{\varepsilon 3}$ is a different constant). It is a simple matter to show that, for any nonzero $\overline{W}_{ij}$, this alteration also eliminates the fixed point $(\varepsilon/K)_\infty = 0$. Consequently, this model can have problems with realizability and predicts that the turbulent kinetic energy and dissipation rate oscillates (for all times $t > 0$) in rotating isotropic turbulence – a prediction that is in substantial contradiction to the results of physical and numerical experiments which suggest a monatonic power law decay. Most recently, Launder and his co-workers proposed a modified version of the ε-transport equation for which

$$C_{\varepsilon 1} = 1, \ C_{\varepsilon 2} = C_{\varepsilon 2}(II, III). \tag{40}$$

In this case, the choice of $C_{\varepsilon 1} = 1$ eliminates the nonzero fixed point (35) (the only fixed point is $(\varepsilon/K)_\infty = 0$). Consequently, this model will not predict an exponential time growth of K and ε in homogeneous shear flow (and SK/ε will not approach a universal equilibrium) in apparent contradiction to physical and numerical experiments. Similar problems occur in rotating turbulent shear flows with a modification proposed recently by Bardina, Ferziger, and Rogallo [29].

It has been demonstrated that the standard form of the modeled dissipation rate transport equation has several crucial properties that have been destroyed by virtually every attempt to modify it. To understand this point more deeply, it would be helpful to see what approximations are necessary to derive the modeled ε-transport equation (31) from an analysis of the dynamics of the two-point correlation tensor

$$R_{ij} \equiv \overline{u_i(\mathbf{x})u_j(\mathbf{x} + \mathbf{r})}. \tag{41}$$

A study conducted not too long ago by Donaldson and Sandri [24] can shed some light on this issue. They introduced a tensor length scale Λ_{ij} defined by the solid angle integration

$$\frac{2}{3}K\Lambda_{ij} = \int_{-\infty}^{\infty} \frac{R_{ij}}{4\pi r^2} d^3r. \tag{42}$$

A modeled transport equation for Λ_{ij} was obtained as follows: (a) the exact transport equation for R_{ij} was closed by assuming that the higher-order unknown correlations could be expanded in a Taylor series in R_{ij} (subject to the appropriate symmetry properties) where only first-order terms were kept, and (b) the resulting modeled transport equation for R_{ij} was multiplied by $1/4\pi r^2$ and integrated over all of $\mathbf{r}$ space. The derived transport equation for Λ_{ij} obtained by this method is as follows for homogeneous turbulence [24]:

$$\begin{aligned}
\frac{d}{dt}(K\Lambda_{ij}) &= -K\Lambda_{ik}\frac{\partial \bar{v}_j}{\partial x_k} - K\Lambda_{jk}\frac{\partial \bar{v}_i}{\partial x_k} \\
&\quad + \sqrt{2}v_{c_2}\frac{K^{\frac{3}{2}}}{\Lambda}\Lambda_{ij} - \sqrt{2}\frac{K^{\frac{3}{2}}}{\Lambda}(\Lambda_{ij} - \Lambda\delta_{ij}) \\
&\quad - 2\sqrt{2}\, bK^{\frac{3}{2}}\delta_{ij}
\end{aligned} \tag{43}$$

where $\Lambda \equiv \frac{1}{3}\Lambda_{ii}$; v_{c_2} and b are constants. The modeled Reynolds stress transport equation that is solved in conjunction with (43) is obtained from a simple contraction of the modeled transport equation for R_{ij}; this equation takes the form [24]

$$\begin{aligned}
\frac{d\tau_{ij}}{dt} &= -\tau_{ik}\frac{\partial \bar{v}_j}{\partial x_k} - \tau_{jk}\frac{\partial \bar{v}_i}{\partial x_k} - \sqrt{2}\frac{K}{\Lambda}(\tau_{ij} + \frac{2}{3}K\delta_{ij}) \\
&\quad + \frac{4}{3}\sqrt{2}\, b\frac{K^{\frac{3}{2}}}{\Lambda}\delta_{ij}.
\end{aligned} \tag{44}$$

From a direct comparison of the contraction of (44) with (32), it follows that

$$\varepsilon = 2\sqrt{2}\, b\frac{K^{\frac{3}{2}}}{\Lambda} \tag{45}$$

for this tensor length scale model in homogeneous turbulence. By taking the time derivative of (45), the transport equation

$$\begin{aligned}
\dot{\varepsilon} &= 2\sqrt{2}\, b(\frac{3}{2}\frac{K^{\frac{1}{2}}}{\Lambda}\dot{K} - \frac{K^{\frac{3}{2}}}{\Lambda^2}\dot{\Lambda}) \\
&= \frac{1}{2}\frac{\varepsilon}{K}\tau_{ij}\frac{\partial \bar{v}_i}{\partial x_j} + \frac{\sqrt{2}}{6b}\frac{\varepsilon^2}{K^{\frac{3}{2}}}\Lambda_{ij}\frac{\partial \bar{v}_i}{\partial x_j} \\
&\quad - (\frac{3}{2} + \frac{1}{2}\frac{v_{c_2}}{b})\frac{\varepsilon^2}{K}
\end{aligned} \tag{46}$$

is obtained. If an anisotropy tensor is defined for the tensor length scale in the following manner

$$b_{ij}^{(\Lambda)} \equiv \frac{\Lambda_{ij} - \frac{1}{3}\Lambda_{kk}\delta_{ij}}{\Lambda_{kk}} \tag{47}$$

it can be shown that $b_{ij}^{(\Lambda)}$ is a solution of a transport equation that is of *exactly the same form* as that for b_{ij}. Hence, if we consider a homogeneous turbulence that evolves from an initially isotropic state, then

$$b_{ij}^{(\Lambda)} = b_{ij} \tag{48}$$

and

$$\Lambda_{ij}\frac{\partial \bar{v}_i}{\partial x_j} = 3\sqrt{2}\, b\frac{K^{\frac{1}{2}}}{\varepsilon}\tau_{ij}\frac{\partial \bar{v}_i}{\partial x_j}. \tag{49}$$

Then the dissipation rate transport equation (46) corresponding to this tensor length scale model simplifies to

$$\dot{\varepsilon} = 1.5\frac{\varepsilon}{K}\tau_{ij}\frac{\partial \bar{v}_i}{\partial x_j} - \frac{3}{2}(1 + \frac{v_{c_2}}{3b})\frac{\varepsilon^2}{K} \tag{50}$$

which is of the same form as the standardly used equation (31) with $C_{\varepsilon 1} = 1.5$ and $C_{\varepsilon 2} = (\frac{3}{2} + v_{c_2}/2b) \approx 1.80$.

It has thus been shown that the standard form of the modeled ε-transport equation can be derived from the two-point correlation tensor by making approximations that are comparable to those made in the derivation of the modeled Reynolds stress transport equation by Launder, Reece, and Rodi [16]. Consequently, it seems to be somewhat questionable to argue (as many have done) that the standard form of the ε-transport equation is the weak link in the commonly used second-order closures. There is no doubt that improvements in the specification of the turbulence length scale are needed, but these are more likely to come from general analyses of some set of modeled transport equations for the integral length scales which incorporate directional information. The kind of ad hoc adjustments in the modeled dissipation rate transport equation that have been considered during the past decade appear to be counterproductive.

5. Concluding Remarks

Projected advances in computer capacity make it highly unlikely, in the opinion of the author, that the complex turbulent flows of engineering interest will be solved on a routine basis by direct or large-eddy simulations for at least several decades to come. Furthermore, substantial theoretical difficulties with two-point closures for complex turbulent flows that are strongly inhomogeneous make their application to problems of engineering interest equally unlikely for the foreseeable future. It appears that Reynolds stress models are likely to remain the method of choice for the solution of the turbulence problems of technological interest for at least the next few decades. Direct numerical simulations of the Navier-Stokes equations will continue to be restricted to less geometrically complex turbulent flows, at lower Reynolds numbers, where they will be used to gain a better insight into the basic physics of turbulence.

Among the variety of existing Reynolds stress closures, it is the opinion of the author that second-order closure models constitute, by far, the most promising approach. The predictive capabilities of second-order closures have been enhanced somewhat over the past decade due to significant modeling improvements in the pressure-strain correlation. However, new prescriptions for the turbulent length scale (which incorporate some directional and two-point information) as well as new methods for integration to a solid boundary are direly needed before more reliable models can be obtained. In fact, these issues are of such overriding importance for wall-bounded turbulent flows that Reynolds stress model predictions can be degraded to the point where they are no better than those of the K-ε model – a state of affairs that has contributed to some of the misleading critical evaluations of second-order closures that have been published in the recent past.

It must be remembered that second-order closure models are *one-point closures* and thus can never yield accurate quantitative predictions for a wide variety of turbulent flows where the energy spectrum can change drastically. Nonetheless, with the implementation of the improvements discussed in this paper, it is quite possible that a new generation of second-order closures can be developed that will provide acceptable engineering answers for a significant range of turbulent flows that are of technological interest.

Acknowledgement

This research was supported by the National Aeronautics and Space Administration under NASA Contract No. NAS1-18605 while the author was in residence at ICASE, NASA Langley Research Center, Hampton, VA 23665.

References

[1] J. C. Rotta: Z. Phys. **129**, 547 (1951).

[2] B. E. Launder, D. P. Tselepidakis, and B. A. Younis: J. Fluid Mech. **183**, 63 (1987).

[3] M. M. Gibson, W. P. Jones, and B. A. Younis: Phys. Fluids **24**, 386 (1981).

[4] C. G. Speziale: Geophys. and Astrophys. Fluid Dyn. **23**, 69 (1983).

[5] C. G. Speziale: J. Fluid Mech. **178**, 459 (1987).

[6] C. G. Speziale: Theoret. Comput. Fluid Dynamics **1**, 3 (1989).

[7] B. E. Launder and Y. M. Ying: J. Fluid Mech. **54**, 289 (1972).

[8] W. Rodi: ZAMM **56**, T219 (1976).

[9] K. S. Choi and J. L. Lumley: in Turbulence and Chaotic Phenomena in Fluids (edited by T. Tatsumi), p. 267, North Holland (1984).

[10] K. Hanjalic and B. E. Launder: J. Fluid Mech. **52**, 609 (1972).

[11] C. G. Speziale and N. Mac Giolla Mhuiris: J. Fluid Mech. (in press).

[12] S. Tavoularis and S. Corrsin: J. Fluid Mech. **104**, 311 (1981).

[13] J. Bardina, J. H. Ferziger, and W. C. Reynolds: Stanford University Technical Report TF-19 (1983).

[14] J. P. Bertoglio: AIAA J. **20**, 1175 (1982).

[15] W. C. Reynolds: Lecture Notes for Von Karman Institute, AGARD Lecture Series No. 86, NATO (1987).

[16] B. E. Launder, G. Reece, and W. Rodi: J. Fluid Mech. **68**, 537 (1975).

[17] S. Sarkar and C. G. Speziale: ICASE Report No. 89-10, NASA Langley Research Center (1989).

[18] C. G. Speziale: Quart. Appl. Math. **45**, 721 (1987).

[19] D. C. Haworth and S. B. Pope: Phys. Fluids **29**, 387 (1986).

[20] C. G. Speziale, S. Sarkar, and T. B. Gatski: ICASE Report (in preparation).

[21] W. C. Reynolds: Private Communication (1988).

[22] K. Hanjalic and B. E. Launder: J. Fluid Mech. **74**, 593 (1976).

[23] M. Wolfshtein: Israel J. Tech. **8**, 87 (1970).

[24] C. DuP. Donaldson and G. Sandri: AGARD Report No. CP-308 (1982).

[25] J. L. Lumley, Adv. Appl. Mech. **18**, 123 (1978).

[26] T.-H. Shih and J. L. Lumley: Cornell University Technical Report FDA-85-3 (1985).

[27] S. B. Pope: AIAA J. **16**, 279 (1978).

[28] K. Hanjalic and B. E. Launder: ASME J. Fluids Engng. **102**, 34 (1980).

[29] J. Bardina, J. H. Ferziger, and R. S. Rogallo: J. Fluid Mech. **154**, 321 (1985).

Equilibrium Values	Standard $K-\varepsilon$ Model	Launder, Reece & Rodi Model	Experiments
$(b_{11})_\infty$	0	0.193	0.201
$(b_{22})_\infty$	0	-0.096	-0.147
$(b_{12})_\infty$	-0.217	-0.185	-0.150
$(SK/\varepsilon)_\infty$	4.82	5.65	6.08

Table 1.   Comparison of the predictions of the standard $K-\varepsilon$ model and the Launder, Reece, and Rodi model with the experiments of Tavoularis and Corrsin [12] on homogeneous shear flow.

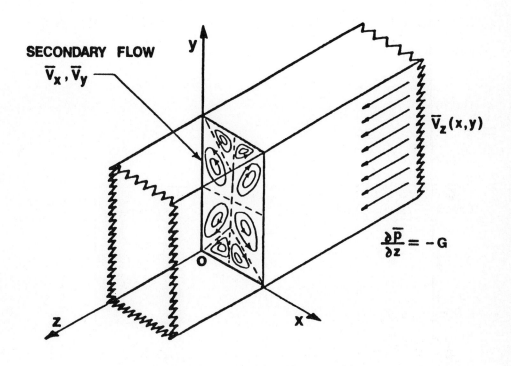

Figure 1.   Turbulent secondary flow in a rectangular duct.

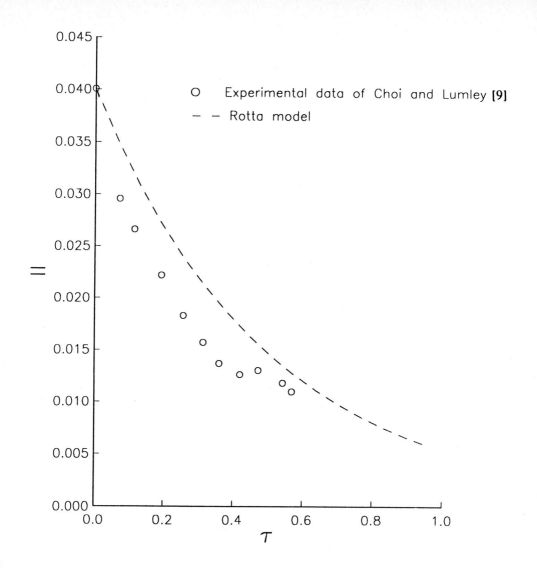

Figure 2. Temporal decay of the second invariant of the anisotropy tensor: Comparison of the Rotta model with the relaxation from plane strain experiment of Choi and Lumley [9].

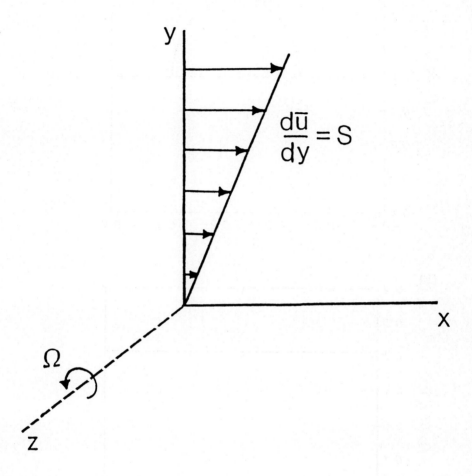

Figure 3.   Homogeneous turbulent shear flow in a rotating frame.

(a)

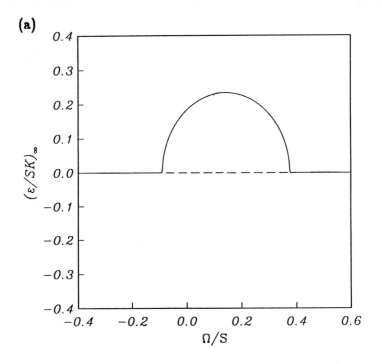

(b)

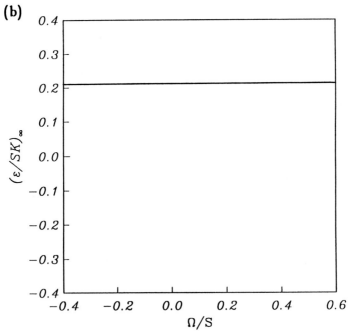

Figure 4.   Bifurcation diagrams for homogeneous turbulent shear flow in a rotating frame: (a) Launder, Reece and Rodi model, (b) K-ε model.

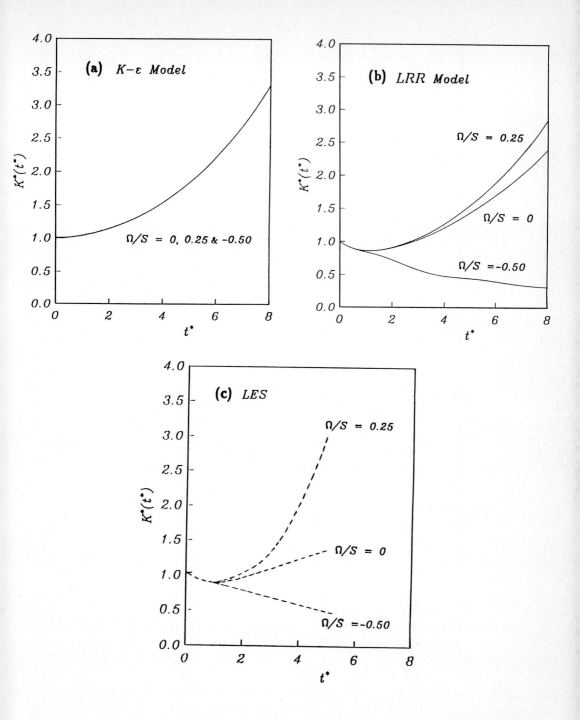

Figure 5. Time evolution of the turbulent kinetic energy for homogeneous turbulent shear flow in a rotating frame: (a) K-ε model, (b) Launder, Reece and Rodi model, (c) Large-eddy simulations of Bardina, Ferziger and Reynolds [13].

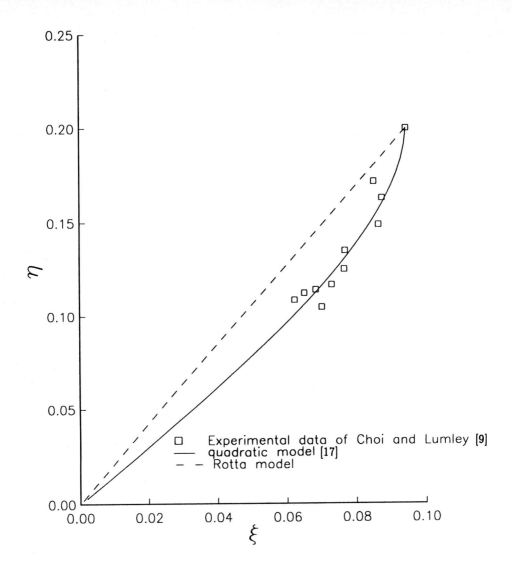

Figure 6.   Phase space diagram of the relaxation from plane strain experiment of Choi and Lumley [9] ($\xi = III^{1/3}, \eta = II^{1/2}$).

Simplex Discussion Regarding "Modeling: Present and Future" With Emphasis Upon High Speeds

Comment 3.

D.M. Bushnell

NASA Langley Research Center
Hampton, VA 23665-5225, USA

1. Introduction

The conclusions of the Launder position paper regarding turbulence modeling can be summarized as follows: (a) computer size and numerical sophistication have now advanced sufficiently to allow reasonable attempts at full $(\tau_{i,j})$ closures, as opposed to 2-Eq. or ASM methods; (b) we have very little idea, at the present time, of how accurate $(\tau_{i,j})$ closures are or can be, as most previous application attempts have been hampered by inadequate resolution/numerics; (c) for those cases where $(\tau_{i,j})$ closure results are quasi-believable, predictions are better, especially in the presence of force fields such as buoyancy or rotation; (d) a possible major difficulty with the $(\tau_{i,j})$ (or other Reynolds-averaged) approaches is the neglect of "intermittency" influences. The present discusser is in agreement with these conclusions. High-speed computational experience gleaned from comparisons of various models directly supports these views of Launder, i.e., "within the framework of single-point closure full Reynolds stress models are probably the only way to improve turbulence modeling (for shock/boundary-layer interactions) as the ASM approach is shown to be unsuited," Escande, 1987, ONERA TP 1987-96 and "the RSE model holds a distinct advantage over the 2-Eq. model for turbulent flows which experience sudden changes in the strain rate and/or shear," Wilcox/Rubesin, 1980, NASA TO-1517. Having agreed that full $(\tau_{i,j})$ closures are the current "best bet" for conventional "Reynolds-averaged" approaches, it is reasonable to briefly examine several obvious and some perhaps not so obvious problems and opportunities regarding $(\tau_{i,j})$ closures, particularly as related to high-speed applications.

2. The Spectral Issue

An implicit assumption of Reynolds-averaged closures is that the turbulence exhibits similar spectra or, somewhat equivalently, that the dynamic events comprising the turbulence are, in general, similar in nature. Research since the '60's on "coherent structures" has obviated this assumption; i.e., "Contrary to earlier ideas, turbulence is not a single or even a simple set of states, it is a very complex and variable set of states that react in sometimes unanticipated ways to a great variety of circumstances," "The quasi-coherent parts of the turbulent flows, what we call the large or medium eddy structures, are not the same in different kinds of turbulent flows," Kline, 1982, and, "It is not meaningful to talk of the properties of a turbulent flow independently of the physical situation in which is arises. (There are) a large number of turbulent flows and our problem is the possibly impossible task of fitting many phenomena into the procrustean bed of a universal turbulence theory," Saffman, 1980. As a superb illustration of these remarks, there exists a very interesting set of data taken by flying a hot-wire equipped aircraft in the proximity of the ocean surface for a wide variety of boundary conditions in the atmospheric boundary layer, e.g., unstable, neutral and stable atmosphere, large and small waves, waves going with, against, and skewed to the wind, etc. The spectral data indicate, for the first two-and-a-half decades in wave number, large variability between these various boundary conditions, whereas the high wave-number data (next one-and-a-half decades) are in good agreement, supporting quite well the concept of an LES closure with "universal" SGS modeling and not supporting well at all one of the basic tenants of the Reynolds stress closure. Therefore, the $(\tau_{i,j})$ closures as discussed by Launder cannot constitute the be-all and end-all of turbulence modeling, merely the current (and vitally important) best bet for "making numbers" and bringing to fruition some of the promise of CFD.

3. High-Speed Issues

Another implicit assumption in $(\tau_{i,j})$ closures is the existence of fields which are continuously differentiable. The occurrence of shock waves at high speeds tends to somewhat obviate this. In general, shock-turbulence interaction produces direct amplification of incident dynamic fields and direct production of dynamic motions via shock oscillations, all in addition to the usual effects of strong pressure gradients such as streamline curvature(s) and bulk dilatation. Turbulence amplification of factors of 2 and 3 due to shock interaction are not uncommon. These can differentially influence various portions of the spectrum and such influences are not included in the "conventional" ($\tau_{i,j}$) models. Shock oscillations are typically produced by shock interaction with, reflection from, or production by, turbulence fields or innately unstable/"bifurcated" flow fields.

Additional compressibility-specific problems which must either be "ad-hoc'ed" or addressed (in a more complete, non-Reynolds-averaged theory/closure?) include: (a) "eddy shocklets," alteration of eddy-phase interrelationships, (b) bulk dilatation effects, (c) influence of flow chemistry/combustion, (d) the persistence at high Mach number (with consequent thick boundary layers) of transitional flow artifacts, and (e) (possibly) bulk viscosity effects.

4. Coherent Vortical Structures

A further, very serious problem with conventional ($\tau_{i,j}$) closures is an inability, due to Reynolds averaging, to deal with, or even identify the occurrence of, coherent vortical structures. The most obvious instances of such entitles include <u>streamline</u> (not necessarily wall) curvature T-G instability induced quasi-stationary longitudinal vortices, and (unsteady) Karman shedding. Other instabilities embedded within turbulent flows are also possible, and the only obvious method of including these first-order effects are either ad-hoc or via LES. In fact, many if not most of the issues in turbulence modeling are resolvable via LES with SGS modeling, given computer enough and time. But these are not yet available and, therefore, we must ad-hoc for yet awhile.

5. Suggestions

Pursue ($\tau_{i,j}$) are perhaps $R_{i,j}$ closures to determine limits of applicability now that computers are becoming large enough to begin such an evaluation, i.e., Launder is correct. This probably necessitates a redo of the numerics to accommodate stiffer systems and use of dynamic adaptive grids. The calculations should resolve organized embedded large-scale vorticity (e.g., mega eddy simulation?) and include transitional artifacts and flow curvature influence(s). Many tools are available to aid in this endeavor including: (a) experimental data, the zeroth order use of even mean flow information will indicate when more physics must be included (via consistent disagreements), (b) numerical simulation results, (c) third order and 2-pt. correlation equations, (d) rapid distortion theory, and (e) restrictions such as invariance, realizability, the second law of thermodynamics, and behaviors at end/extreme conditions.

Discussion on "Phenomenological Modeling: Present and Future"

Reporter Alexander J. Smits

Princeton University
Dept. of Mechanical and
  Aerospace Engineering
The Engineering Quadrangle
Princeton, NJ 08544

Roland Stull:

This afternoon all of the models that we have heard fall in the same class; namely, local closures. First-order local closure (K-theory or eddy diffusivity) models the momentum fluxes as down-gradient of the mean momentum. The second-order local closure models the third moments as down-gradient of the local second moments, or local mean variables.

There is another completely different class of modeling or class of closure, and that is non-local turbulence closure. I mentioned before about the transilient matrix that describes the mixing between different points separated a finite distance in space. One can parameterize this matrix in terms of mean flow state or mean flow instability. When you do that, you can then make forecasts of the mean field in a turbulent flow that takes into account this non-local mixing.

That has been done. For the ocean, we found results as good as third-order local closure. For the atmosphere, results were as good as second-order local closure. We've used it in three-dimensional weather forecast models covering the whole United States. This is a new concept of non-local closure, which is different from all the other local closures.

When would you want to consider using a non-local kind of closure? Well, if any of you are dealing with turbulent flow that has a spectrum of eddy sizes where your greatest energy is in the largest wavelengths, or if you are dealing with turbulent flow that has large structures in it that are causing non-local mixing, then you might want to consider a non-local turbulence closure.

Michel Coantic:

O.K. Thank you. I know that your view and your work is becoming popular among the geophysical turbulence community, so maybe there is a problem of the journals in which those works have been published, in the Journal of Atmospheric Sciences and JPO, so maybe most people in this community are not familiar with that literature. You should maybe publish that for the industrial fluid dynamics community as well.

Stephen Traugott:

It may not be just a matter of publication, I mean have you tried to tackle problems that this community is concerned with? Hard ones, like separation over a backward facing step, or something? I was just wondering, have you tried to model flows like that?

Roland Stull:

No, I haven't tried those kinds of problems. Basically I've compared this with second-order over third-order closure over homogeneous and non-homogeneous boundary layers, and not just boundary layers but also free shear flows higher in the atmosphere. Those kinds of situations are buoyancy-driven, and/or shear-driven, and where body forces play a role.

Michel Coantic:

O.K. So, I think there are two points which are rather important, and to which I think that two people in the audience have started to give a few comments. First, the problem of non-equilibrium spectra, the way in which the output of spectral calculation could be included in one-point models, and second, the problem of the diffusion terms.

Jean Mathieu:

I will just make some comments about closure methods. First of all, I think that it will be interesting to bridge the gap between the two methods because it seems that the two processes are somewhat different from each other. I think that it is important to bridge this gap because it is possible to extract information from one method and to apply it to the other one. It is possible to extract many parameters which are used in one-point closure methods by dealing with two-point closure methods. For instance it was possible to discuss with Prof. John Lumley concerning some linear terms which may be detected with great accuracy by using a two-point closure method. Generally, I think that failures occur in one-point closure methods when the spectrum is very complex. That it is our own experience. If we have some difficulty for modeling in a one-point closure method that means that the spectrum, especially the spectrum of the velocity pressure correlations, is complex. In that case we can expect trouble for very simple modeling carried out in physical space. I think that spectrum methods are capable of taking into account the problem of a peak in the spectrum. It is possible to introduce a peak by using a grid-generated turbulence where the grid is equipped with small propellers. The eddy-damped quasi-Markovian modelling which may be used in the case of a two-point closure method is well-suited to predict complex situations which are not very easy to describe by a one-point closure method. One advantage of the method carried out in physical space is that it is well-suited for describing problems similar to those encountered in technical situations.

I think also that is interesting to emphasize the role of the two-point closure method to predict compressibility effects. In that case, we have two modes, a turbulent mode, which is in the plane normal to the wave number vector, and another mode, the compressible mode, which is colinear with the wave-number vector. We have to consider two kinds of interactions, intractions which are generally called the interaction of the turbulence type (in the normal plane) but we have also to model interactions between compressible and incompressible modes. In that case the splitting of turbulence in compressible motion and in travelling waves is rather easy. Thank you very much for your attention.

Michel Coantic:

Thank you. Now if Dr. Shih wants to say a few words concerning the diffusion terms.

Tsan-Hsing Shih:

I would like to say a few words about the model of the turbulent diffusion term in the Reynolds stress equation. First, however, I would like to mention that we have not heard anything about the pdf method in this workshop. The pdf model plays an important role in the study of turbulent combustion. I do not believe there will be a single universal model which will be able to solve all the turbulence problems. Instead, I think we should try to put different models together to attack the more complex turbulence problems.

In a second-order closure scheme, or full Reynolds stress model, we must model the turbulent diffusion term: the triple velocity correlation term. Lumley [Adv. Appl. Mech., Vol. 18, 1978, ed. C. S. Yih] proposed an algebraic model for this term. This model is based on first principles and derived from the equation for cumulants by keeping only zeroth-order terms. It does not introduce any extra model constants. In my experience, this model performs very well in the modeling of turbulent jets, wakes, mixing layers and boundary layers.

In constructing a turbulence model, realizability first introduced by Schumann and Lumley [Phys. Fluids, Vol. 8, 1965] is not only important, but also provides a useful tool for developing a more general model for more complex turbulent flows.

Roddam Narasimha:

Brian Launder has given us a thorough survey of second-moment closure methods. I have three questions for him. First, how do these methods handle situations where transport is definitely counter-gradient and non-local, such as for example in the wall-jet that was mentioned earlier (in my talk). The qualitative failure of the gradient-transport concept in such flows is clearly due to the presence of two vastly disparate scales in the flow: the larger one characteristic of the outer jet flow and the smaller one of the boundary-layer like flow near the wall. Some non-local transport model of the kind mentioned by Roland Stull (and proposed many years ago by Townsend) would seem more appropriate here.

The second question concerns your calculations of wake development, showing that the predictions are different depending on the initial conditions. However, to get a significantly different wake development it appears that the initial value of ε had to be eight times and that of Reynolds stress 100 times the respective standard values. These seem unphysically high.

Finally you mentioned the growing maturity of such models and their imminent incorporation into commercial software. Since many simpler models are being so widely offered commercially these days, don't you think that during the coming decade this kind of modeling activity can increasingly be left to industry for further development?

Brian Launder:

Perhaps I could respond to the second point first. I'm glad you raised it because if others study the figures in the detail that you have they will also conclude that absurdly high levels of initial turbulence energy were assigned. Indeed, on discovering it myself just as the pre-workshop version of the paper was being mailed to Cornell, I asked the student why a more realistic initial level had not been assigned. The response was that he had made calculations with <u>several</u> initial k levels: the factor of 100 was the highest level tested but essentially the same behaviour was found with augmentation ratios of around five. That is to say, the spreading rate reaches an asympototic (lower) level for relatively small energy augmentation ratios. Needless to say, in the published version of this meeting, I will be replacing the existing figure 8 with computational results where the initial conditions have been chosen to match those obtained in the experiment immediately downstream of the wake generator, as closely as we are able to.

I go next to the first point, this question about the wall jet, because if we had had more time I'd like to have commented in the presentation itself and I'm glad of the opportunity to do so now. The dislocation between the position of maximum velocity in a wall jet and the zero shear stress position is an interesting thing to note. I think Jean Mathieu was perhaps the first person to discover this in his own Ph.D. work a few years ago. It's interesting, but it's not important. How do I know that? Well, Wolfgang Rodi and others have made <u>algebraic</u> second-moment (ASM) calculations where you make a very crude approximation for the diffusion of the second moments, and where in fact you enforce the coincidence of the position of zero shear stress and maximum velocity. Overall one predicts essentially the same behavior with an algebraic second-moment closure as a full second-moment closure (which does allow the zero-shear-stress and maximum velocity locations to be non-coincident). The only difference is that you actually modify the shape of the mean velocity: you displace the mean velocity maximum a little bit to the point where the zero shear stress occurs. I see Wolfgang nodding, I think he's in agreement.

The third point. Well, as I noted there <u>are</u> codes based on second-moment turbulence models. The code FLUENT by CREARE does have a crude second-moment closure in; so, in a sense all of that development was done commercially, academics were called in as consultants. The people that <u>use</u> those types of codes are, on the whole, not in what I think of as industry: for example in the research laboratories of General Motors and GE. Maybe it's not <u>my</u> opinion of what industry can do but rather their opinion of how much they can do that's important. I think they would want to take over from academic circles models and methods that have been reasonably well tested up to and including three-dimensional flows. They would not expect to undertake the basic research. But I look around, and there are many people from industry here that perhaps might be able to comment. I see Sherif El Tahry, Gino Sovran and others.

Sherif El Tahry:

So far, the use of multi-dimensional CFD codes in the auto-industry has been restricted to research work. But there are now signs of interest, in these codes, from the operational staffs (that is, the people involved in the day-to-day designs).

At the research level at GM, we do a lot of code and model development, with some cooperation from academia. Ultimately, we hope to minimize this basic development work, and concentrate on adapting these techniques to our specific applications, and finding the optimum way of integrating them into the design, testing and diagnostic phases of the engineering work.

Wolfgang Rodi:

I have two comments on Brian's paper. First, I fully agree that second-moment closure is the way to go in the future, but I'd like to give also some consolation to people who may have a k- ε model code and who may have been disappointed or even shocked by some of the poor agreement that Brian has shown in some of the examples. I think that at least for cases of uni-directional shear-layer-type flows there is a simple fix to the k- ε model, because in these cases the algebraic stress model simplifies to a mere replacement of one of the constants in the k- ε model to a function of streamline curvature, or rotation, or buoyancy. So I believe the first two examples Brian has shown, that is, the faired diffuser and the rotating channel flow could also be fairly well predicted with a k- ε model by including such a correction. But, I say again, that this is limited to shear layer and uni-directional flows. Now, on the second point, I believe that in the calculations you have shown using the second-moment closure you did not resolve the near wall region but you used a wall function, except for the third example. Is that correct?

Brian Launder:

Well, the influential region in the third example was the jet mixing region away from the wall. As in the first two examples wall functions were adopted.

Wolfgang Rodi:

O.K. Now I think that's somewhat problematic. It may be alright when you have flow parallel to the wall, unseparated, but what do you do in cases when you have separated flow. You mentioned impinging flows, and the impression I got is that there is no second-moment closure ready at the moment which allows a realistic simulation of the viscous sublayer. The question is, what do you do then, especially when you want to have a code for practical applications where you often do have such situations, like flows with separation? One possibility would be to actually match a second-moment closure to a simpler model near the wall. Do you see that as a viable possibility?

Brian Launder:

I'm sure, Wolfgang, that is what we'll do for a long time. The 180^0 bend flow calculations I've shown were actually matched to the mixing length solution over a near-wall skin. We have made calculations of that flow with wall functions, and you get the whole flow pattern quite wrong there. Perhaps you recall that that was the flow where a k-ε model gave you <u>three</u> secondary flow eddies, and the algebraic second-moment closure gave you <u>four</u> when each are matched to the mixing-length model in the sublayer. You only get <u>one</u> eddy with wall functions, which is just nothing like the measured flow pattern. I think we are going to be stuck using eddy viscosity models across the sublayer for quite a while. Even if one isn't stuck, one may still prefer to use an eddy viscosity there, because at least the mean velocity vector in many cases, even in those bend flows, is essentially parallel to the surface.

Michel Coantic:

O.K. I think that Jean Mathieu wanted to say something.

Jean Mathieu:

I think that it was a bit surprising concerning the wall jet. That the point where the velocity is maximum and the point where the shear stress is null do not coincide. It was a bit surprising in 1955, but it is not surprising now, and I agree with Lumley that it is just an effect of the advected terms concerning the correlation $\overline{u_1 u_2}$. Memory effects are well known if turbulence is treated in a statistical way. It is possible to obtain approximately the same phenomenon by using a channel with a rough wall and a smooth wall, as Launder did 20 years ago. You have exactly the same problem but in this case turbulent diffusion terms are responsible.

Michel Coantic:

It's no longer considered a violation of the second law of thermodynamics, as it was felt at the time.

Bill Reynolds:

I want to come back to the use of wall functions in the second-order closure for the Reynolds stress, and just ask, what do you use for the Reynolds stresses at the patch point? Do you use the Reynolds stresses from the isotropic eddy viscosity, or do you use something else?

Brian Launder:

For the Reynolds stresses at the patch point, we normally presume local equilibrium, so that you get a certain stress ratio that will come out from an algebraic form of the closure, depending on the strain field.

Bill George:

I'd just like to point out that there might be a hidden connection, not obvious to everyone here, between the comments of Anatol Roshko and the improvements we saw in the models' ability to predict flows. In the slide Brian showed, a number of those improved predictions came about because of improved measurements which, in fact, brought the measurements more in line with what the models have been predicting all along. I happened to have been involved with a couple of those cases, and I know people that were involved in the others, and in not a single instance was the improvement in the measurements because of the measuring technique. In every case, it was an improvement because of a recognition of the effect of the boundary conditions on the measurements. I think that is a lesson to all of us. We really have to be careful that we know what we are measuring.

Anatol Roshko:

I just wanted to comment on Dr. Shih's remark or observation that it would be desirable to develop models that could handle combustion and reactions more realistically. In fact there are such models being developed. Earlier in this workshop Dr. Stull mentioned his "transilient" model which, I would guess from his description, models more accurately the mixing processs which are involved in turbulent combustion or reaction. Similarly, at Caltech, Gene Broadwell has been leading an effort to develop reaction/combustion models that incorporate the observed physics more accutately. Still another approach is being taken by Alan Kerstein at the Sandia National Laboratories.

Bob Bilger:

I would disagree with that. There seems to be rather a lot of confusion about what is lacking in second-order closures as far as treating combustion is concerned. There's a very large literature, which not a lot of people seem to have read, about handling combustion/turbulence interactions from the second-order closure point of view. The pdf transport equation is, if you like, a different way of doing the same sort of closure. Here at Cornell, Steve Pope, who's on sabbatical at Cambridge at the moment, is a leader in that regard. The big thing that's missing, in most people's view of things, is the same mistake that's in the paper by Broadwell and Breidenthal [J. Fluid Mech., Vol. 125, 1982]; that is, they say you can't use an eddy diffusivity model where in fact they haven't put in the fluctuations in the chemistry, and that's where they're going wrong. You can handle the fluctuations in the chemistry with appropriate modeling, by either doing a full pdf transport model as Pope does, or by just getting a few moments like you do in second-order closure. It needs to be recognized, that the combustion people are well up alongside, following in the steps of the isothermal people, in this regard. For further discussion of these matters see Bilger, R. W., Ann. Rev. of Fluid Mech., Vol. 21, 1989.

Brian Launder:

I'm very glad Bob commented on the advanced techniques and concepts being used by combustion people. I'd intended to remark on this in my presentation; I <u>do</u> in the paper. It isn't the case that the combustion people are following close behind. I think in many areas they are way ahead and I think we have a lot to learn from them.

Jack Herring:

A very quick question, to Charles (Speziale) if he would respond. In the derivation of these single point closures, you make an approximation that seems to me that the two-point covariances are very sharp functions compared to the mean gradients. How well is this approximation borne out by experience?

Charles Speziale:

Actually, if you look at the fundamental assumption that is made in deriving the basic second-order closures, you will see that it is assumed that the third-order moments are just small perturbations of their isotropic forms. That is the essential assumption, and I would say that for flows which are dominated by turbulence production, this approximation seems to do extremely well. To give an example where there can be a tremendous pitfall, I cite one of the cases that Bill Reynolds talked about yesterday - the case of rotating isotropic turbulence. Essentially, no existing second-order closure can handle that problem correctly (that is, to predict that the Reynolds stresses decay isotropically but with a reduced energy cascade so that the dissipation is lower). Thus, I would say that in production-dominated flows, the assumption that third-order moments are a small perturbation of their isotropic form seems to do quite well. But for flows where you essentially have two-point phenomena exhibited (such as a reduction in the transfer, etc.), they don't do so well.

Bill Reynolds:

I'd like to comment on that. One thinks about the shear flows as being driven by the mean strain and the mean deformation through the rapid terms. In inviscid rapid distortion theory where the initial state is isotropic, the spectrum is unimportant, and the Reynolds stress history is independent of the form of the spectrum. I think this may explain why one-point closures work so well in flows dominated by rapid terms.

Hassan Nagib:

I'd like to ask Brian a question and that is, as we delve into harder and harder problems like the internal combustion engine, how do you expect us to be able to deal with the wall function as a recipe? How would you derive the wall functions appropriate for cases where there are chemical reactions, and so forth? Can it be done?

Brian Launder:

I wouldn't have the courage, myself. I think integrating to the wall in most practical situations, in time, will be the only way forward.

Wolfgang Rodi:

I suppose with increase in computer power you can do that, integrating to the wall.

Michel Coantic:

If there are no more contributions from the floor, I have prepared a set of possibly provocative questions to try and create a little bit of excitement. We have been told that second-order modeling is the only hope for the next few years, and I entirely agree with that as far as practical applications are concerned, so I would like to ask, what happens to the coherent structures that we have discussed here? Is there any progress in including them in the picture of second-order closures or, if not, what is their use? I am trying to be provocative. Would somebody say something? My feeling in reading the plans for the meeting was that we should try to take into account, in the last session, what could be included from the previous ones, so what about the coherent structures? Another question is the following one: What can we hope to take out from dynamical systems to improve closures? Would someone say something?

Bill George:

Let me take a stab at at least one aspect of that problem. A couple of observations. Point 1: the people who showed us results of the multi-point closures, particularly two-point closures, could see in their simulations, results that resemble very closely the kind of coherent structures we see in the laboratory. So obviously those multi-point closures seem to be able to reproduce coherent features. In particular the Large Eddy Simulations, were showing us things that looked very much like coherent structures. Point 2 is that I don't think any single point closure is going to be able to reproduce those features. In fact, I think the calculation that Brian Launder did, where all of the different wakes went to the same asymptotic value, was very consistent with some work that Dale Taulbee and I did about two years ago where we showed that no single-point closure could reproduce different self-preserving states in the far wake (in George, W. K. and Arndt, R.E.A., Advances in Turbulence, Hemisphere, N.Y., 1978). So I think the point is, if we intend to include structures in any way, it's going to come about through the Large Eddy Simulations, or some kind of multi-point closure models.

Michel Coantic:

Yes, this was exactly what we were hoping somebody would say. As mentioned by several speakers, there are means for manipulating turbulent structures (for instance, by controlling conditions at the wall through riblets) that cannot be treated conveniently in the framework of one-point closures. I don't see any way to take this into account other than to try to make a marriage between the 40 year old lady with some Large Eddy Simulation. Is that the only way? Or does somebody have another idea?

Brian Cantwell:

Could I ask perhaps a related question and, that would be, how could second-order closures be used in sub-grid scale models?

Michel Coantic:

John, do you want to comment on that?

John Wyngaard:

Yes, but I want to say first that in my comments on Bill Reynolds' paper I showed an example of a diffusion process in a flow - the convective atmospheric boundary layer - that is filled with skewed, coherent structures, and I said the flow has an unusal transport asymmetry. I should have added that we looked at those results, those asymmetric diffusion properties, in the context of second-order closure and found two things:

1)   Some of the asymmetry is due to the transport (third moment divergence terms). This can be dealt with in large part by going to third-order closure for those terms, as John Lumley has advocated. That resolves explicitly the buoyancy effects on these third moments and give much improved predictions of them.

2) The rest of the asymmetry stems from the pressure covariance terms in the second-moment budgets. It seems to be a "geometric imprint" of the diffusion source; for the same turbulent velocity field, the pressure covariances behave differently depending on the source location.

It seems, then, that the diffusion geometry is important and I don't know how to incorporate that into second-order closure. One can achieve some gains by going to third-moment closure, but one still has to deal with the geometrical influence on the pressure covariances.

Now to respond directly to your question: one can apply second-order closure to the equations for the subgrid-scale fluxes that appear in the resolvable-scale equations. Jim Deardorff tried this 15 years ago in simulations of the atmospheric boundary layer but found it had little effect (beyond running up his computer bill!). Perhaps the second-moment equations he used did not introduce appreciably better physics than the standard (Smagorinsky) closure. It might be a different story with more modern closures.

Michel Coantic:

Thank you. Does somebody want to comment on the possibility of a link between dynamical systems and one-point closure?

Charles Speziale:

I would just like to make one comment, as trivial as it may seem, and that is namely when one applies second-order closures to homogeneous turbulence, you have a natural low-order dynamical systems problem. Just the notion of examining bifurcation structures, and the stability of fixed points (and understanding this better by making comparisons with results from direct simulations) could be very fruitful. In particular, we have seen the degenerate type of transcritical bifurcation that the second-order closures give rise to in rotating shear flow. It would be interesting to look at the bifurcation structure in more complicated flows. It's a natural low-order dynamical systems problem since there are no more than seven coupled non-linear ODE's. I think it should be looked at much more in the future.

Michel Coantic:

Gentlemen, I think the session is coming to an end. Peter Bradshaw wants to make an interesting communication before John Lumley takes the chair.

Peter Bradshaw:

This is supposed to be a comment on Brian Launder's papers. Perhaps I might say that he's introduced even more confusing analogies. We've had green flyers, we've had Achilles's, tortoises and hares, we've had red pigs, and very enjoyable they were. Brian has referred to second-order closure as being a long line of vehicles moving competently forward without saying whether it's a funeral procession or a line of garbage trucks.

To come to the point, I tested a number of you with a perfectly ordinary black and white flyer on the possibility of another meeting, a sort of Stanford III meeting on turbulence modeling, in 1992. After discussions with representatives from the funding agencies, Dennis Bushnell had the very sensible idea that this is something that ought to be attacked by some kind of collaborative research program, rather than simply by another meeting. So the 1992 meeting is off, but we hope that we can get some kind of collaboration between the funding agencies and between workers all over the world on turbulence modeling - either second-order or higher or lower order, whatever seems to be likely to give good results - to see if we can build on the results of this meeting and of previous work to get these more advanced models actually into use by industry and by the aerospace people. So stay tuned for some further specific effort, which I think we can really say is being triggered by the Whither Turbulence effort.

Michel Coantic:

Well, I think that John is expecting to talk. But since I still appear to have the microphone, I think that it is possibly my duty to express to him our thanks for organizing the meeting in such an interesting and brilliant and pleasant way for all of us.

John Lumley:

That's very kind. Thank you very much. I want to thank all of you - many of you came from great distances and worked very hard to make this work, and I think it <u>has</u> worked. I want to thank particularly the contract monitors - there aren't too many of them left, but perhaps those who are will convey my remarks to their colleagues. You've taken a lot of grief in the last few days, and I think you should bear in mind that people like me are a burden of the democratic process. At a certain age we start querulously criticizing things that we never liked much and we try to change them. Fortunately, you don't have to pay too much attention to us. It's been suggested that I might feel differently if I sat in Steve Traugott's seat for a year. That's true. I'm not sure it would be a good idea necessarily, because I would have a different point of view. I think my remarks, whether he agrees with them or not, are at least from my side of the table, so to speak, and I think we need people on my side as well as people on his side. After all, you don't require that I be able to hit high C in order to be an opera critic. I have actually great sympathy with the pressures under which the contract monitors work. I think it's important to say that, as far as I'm concerned, this meeting was not about money, certainly not about the amount of money that any particular person gets, and I don't think even about the total amount of money that the field gets, except very indirectly. I think it has more to do with the decision making process. Possibly we provided some ammunition these last three days that monitors could use to keep other monitors from raiding their budgets. I think the most important aspect of this, and the part that I had in mind when I designed the meeting, I wanted to pull us together a little bit to help to establish a sense of community in our diversity. We're certainly all very different, and I don't think in a healthy field you should expect any sort of agreement, but I think we can respect each other's paths and feel part of the same community, and I think if we do that, that other things will follow. So, thank you very much for coming.

I want to say just one more thing. I want to thank all of the support people who made this possible - my secretary, Gail Fish, my technician for more than a quarter century, Ed Jordan, who's running the TV camera, Panch Panchepakesan who's running the other one, Ray Ristorcella who organized all the graduate students into a support group, the graduate students themselves, so if you'll give them a big hand.

Lecture Notes in Mathematics

Lecture Notes in Physics